Current Topics in Membranes, Volume 48

Membrane Permeability

100 Years since Ernest Overton

Current Topics in Membranes, Volume 48

Series Editors

Arnost Kleinzeller
Department of Physiology
University of Pennsylvania
School of Medicine
Philadelphia, Pennsylvania

Douglas M. Fambrough
Department of Biology
The Johns Hopkins University
Baltimore, Maryland

Dale J. Benos
Department of Physiology and Biophysics
University of Alabama
Birmingham, Alabama

Current Topics in Membranes, Volume 48

Membrane Permeability

100 Years since Ernest Overton

Edited by

David W. Deamer
Department of Chemistry and Biochemistry
University of California at Santa Cruz
Santa Cruz, California

Arnost Kleinzeller
Department of Physiology
University of Pennsylvania
School of Medicine
Philadelphia, Pennsylvania

Douglas M. Fambrough
Department of Biology
The Johns Hopkins University
Baltimore, Maryland

ACADEMIC PRESS
San Diego London Boston New York Sydney Tokyo Toronto

This book is printed on acid-free paper. ♾

Academic Press
a division of Harcourt Brace & Company
525 B Street, Suite 1900, San Diego, California 92101-4495, USA
http://www.apnet.com

Academic Press
24-28 Oval Road, London NW1 7DX, UK
http://www.hbuk.co.uk/ap/

International Standard Book Number: 0-12-153348-4

PRINTED IN THE UNITED STATES OF AMERICA
99 00 01 02 03 04 EB 9 8 7 6 5 4 3 2 1

Contents

Contributors

Numbers in parentheses indicate the pages on which the authors' contributions begin

Clay M. Armstrong (283), Department of Physiology, University of Pennsylvania School of Medicine, Philadelphia, Pennsylvania 19104

David S. Cafiso (197), Department of Chemistry, University of Virginia, Charlottesville, Virginia 22901

David W. Deamer (77), Department of Chemistry and Biochemistry, University of California, Santa Cruz, Santa Cruz, California 95064

Brian W. Dubois (311), Department of Anesthesia and Critical Care, Massachusetts General Hospital, Boston, Massachusetts 02114-2636

Else K. Hoffmann (123), Department of Biochemistry, August Krogh Institute, University of Copenhagen, Copenhagen, 2100 Denmark

Morley D. Hollenberg (351), Department of Oncology, Pharmacology and Therapeutics, The University of Calgary, Calgary, Alberta T2N 4N1, Canada

Arnost Kleinzeller[1] (1), Department of Physiology, University of Pennsylvania School of Medicine, Philadelphia, Pennsylvania 19104

Thomas J. McIntosh (23), Department of Cell Biology, Duke University Medical Center, Durham, North Carolina 27710

Keith W. Miller (311), Department of Anesthesia and Critical Care, Massachusetts General Hospital, Boston, Massachusetts, 02114-2636

John W. Mills (123), Department of Biology, Clarkson University, Potsdam, New York 13699-5805

Michael H. New (49), NASA/Ames Research Center, Moffett Field, California 90435

Stefan Paula (77), Department of Membrane Biochemistry, Max-Planck-Institute, Martinsried, Germany

[1] Deceased.

Andrew Pohorille (49), NASA/Ames Research Center, Moffett Field, California 90435

Robert L. Post (397), Department of Physiology, School of Medicine, University of Pennsylvania, Philadelphia, Pennsylvania 19104-6085

Stephen M. Robbins (351), Departments of Oncology and Medical Biochemistry, University of Calgary, Calgary, Alberta T2N-4N1, Canada

Warren S. Sandberg (311), Department of Anesthesia and Critical Care, Massachusetts General Hospital, Boston, Massachusetts 02114-2636

Michael J. Saxton (229), Institute of Theoretical Dynamics, University of California, Davis, Davis, California 95616

Karl Schweighofer (49), Department of Pharmaceutical Chemistry, University of California, San Francisco, San Francisco, California 94143

A. S. Verkman (97), Cardiovascular Research Unit, University of California, San Francisco, San Francisco, California 94143-0521

Micheal A. Wilson (49), Department of Pharmaceutical Chemistry, University of California, San Francisco, San Francisco, California 94143

Tribute

Arnost Kleinzeller, 1914–1997

On February 1, 1997, Arnost Kleinzeller, M.D., Ph.D., D.Sc., and Professor Emeritus of Physiology at the University of Pennsylvania, died at the age of 82 after a long battle with cancer. He was an internationally known physiologist and biochemist whose groundbreaking scientific work demonstrated how kidney cells regulate the contents of water and salt and how they move sugars through tissue membranes.

A leader during the Cold War in building links between the scientific community in Czechoslovakia and the West, Professor Kleinzeller organized a major international scientific congress in Prague in 1960. The work of that congress helped establish the course of research over the next two decades for the emerging field of biological fluid and solute transfer.

The openings he developed with Western scientists brought him increasing political pressure from the orthodox Czech communist regime. In 1966 with the help of American colleagues, Professor Kleinzeller and his family escaped, settling first at the University of Rochester and then in 1967 moving to the University of Pennsylvania, where he was named Professor of Physiology.

At the University of Pennsylvania Professor Kleinzeller taught at the medical school and conducted research in the Department of Physiology. During summers he was a researcher at the Mount Desert Island Marine Biological Laboratory in Maine. He wrote over 150 scientific papers and monographs and was one of the editors of the scientific series *Current Topics in Membranes and Transport.* He continued to campaign for greater openness in international scientific exchange as a member of the Committee on Scientific Freedom and Responsibility of the American Association for the Advancement of Science. He was elected to Germany's prestigious Academia Leopoldina in 1966 and was elected fellow of the College of Physicians of Philadelphia.

Professor Kleinzeller was born December 6, 1914, in Ostrava, Czechoslovakia. In 1938 he graduated from the Masaryk University's Medical School in Brno and published his first scientific report. That same year, following the signing of the Munich Pact, he was forced to flee his native country to England, where continued his scientific studies and worked for the Czech government in exile. He completed his Ph.D. under the Nobel laureate Sir Hans Krebs and continued his postdoctoral work at Cambridge University.

Professor Kleinzeller is survived by his wife of 54 years, Lotte, two daughters, Anna Romancova of Prague and Jana Richman of Forest Hills, New York, and three grandsons.

Friends and Colleagues
Department of Physiology
University of Pennsylvania

Arnost Kleinzeller, to me, was synonymous with *Current Topics in Membranes.* He served as editor for many years and put the force of his intellect and passion for science into it. Working as co-editor with him for 7 years was an experience to be cherished. It was a troubling time because the sciences were in transition from an era of broad, scholarly, archival treatises to something else, involving new speeds of information of exchange and storage. How to guide *Current Topics in Membranes* through this changing world of science was a constant question. Arnost could be fiery in defense of the academic high ground against the inclination of publishers to peek at their bottom line or their occasional lapses of due respect for scholars. Arnost's standards and his sense of the dignity of intellectual endeavor were among his many awesomely admirable traits. What I enjoyed most, however, was the fun of discussing science with him. Whenever we got together to deal with the series, our discussions always turned to science questions: What is most important and exciting in our field? How might we capture this in a volume of the series? Who has the breadth of vision and the skills to be an outstanding guest editor? We always parted with the sentiment that the sheer fun of the science itself more than compensated for our burden of editorial responsibilities. Even during his final suffering, Arnost was intent upon seeing one more volume come into being: a volume celebrating 100 years since Overton's landmark publication in membrane biology. This idea of his, one of many that kept us in animated conversation, has now come to fruition. It is perfectly fitting that this volume be dedicated to the memory of another important figure in the history of membrane biology: Arnost Kleinzeller himself.

Douglas M. Fambrough, Ph.D.
Professor of Biology
The Johns Hopkins University

In his last publication to appear in his lifetime, Arnost Kleinzeller explored the history of the cell membrane, as exemplified by research on the erythrocyte membrane and the early contributions by William Hewson (Kleinzeller, 1996). Kleinzeller describes how the earlier insights into the existence of a cell membrane were disregarded for the greater part of the 19th century, as was the concept of a cell membrane as a distinct and functionally important cellular structure, a concept that was revived by Overton and which, in the 20th century, has played a signal role in cellular physiology. It is a concept to which Arnost Kleinzeller contributed throughout his professional life, as both experimenter and theorist. In a 1994 publication (Kleinzeller *et al.,* 1994), he and colleagues focused on the two choline transport systems in lung cell membrane vesicles; in 1992 he coauthored a study of the polarity of glucose transport in cultured renal epithelia (Miller *et al.,* 1992), glucose transport and kidney cells having been topics of life-long interest. Still in 1992, he published an experimental study on cellular volume regulation (Ziyadeh *et al.,* 1992), again a topic that challenged him throughout his professional career. These publications are witness not only to his undimmed intellectual drive and curiosity, but also to his tenacity, as he suffered from chronic and debilitating illness for several years after his formal retirement in 1985.

I first met Arnost Kleinzeller in the late 1960s, after he and his wife, Lotte, had emigrated to the United States. It was the second time he had had to leave his native Czechoslovakia because of a totalitarian regime. When we met, I was struck by his biochemical approach to physiological events and therefore asked him to join me as co-editor of the series that became *Current Topics in Membranes and Transport.* That the term membranes appeared in the series title may well have been due to Arnost's interests and expertise. As editor, Arnost Kleinzeller applied the same demands of rigor to others that he applied to himself, stimulating critical analysis and encouraging a well-formulated viewpoint, but mindful of contrary evidence.

In the Preface to the first volume of *Current Topics in Membrane and Transport* we wrote ". . . of necessity, recognition [the first step of biological transport of solutes] must occur at the system boundary, most frequently the membrane boundary of cells or organelles." Arnost Kleinzeller devoted his life to understanding how this boundary permitted communication, yet provided individuality. He shall be missed.

Kleinzeller, A. (1996). William Hewson's studies of red blood corpuscles and the evolving concept of a cell membrane. *Am. J. Physiol.* **271** (*Cell Physiol.* **40**), C1–C8.

Kleinzeller, A., Dodia, C., Chander, A., and Fisher, A. B. (1994). Na^+-dependent and Na^+-independent systems of choline transport by plasma membrane vesicles of A549 cell line. *Am. J. Physiol.* **267,** C1279–C1287.

Miller, J. H., Mullin, J. M., McAvoy, E., and Kleinzeller, A. (1992). Polarity of transport of 2-deoxy-D-glucose and D-glucose by cultured renal epithelia (LLC-PK1). *Biochim Biophys. Acta* **1110,** 209–217.

Ziyadeh, F., Mills, J. W., and Kleinzeller, A. (1992). Hypotonicity and cell volume regulation in shark rectal gland: Role of organic osmolytes and actin. *Am. J. Physiol.* **262,** F468–F479.

Felix Bronner, Ph.D.
Former editor of Current Topics in Membranes and Transport

It is difficult to know how to begin when writing about Arnost Kleinzeller, for he was a man who had a significant impact on so many different areas of my life.

I first came to know him while working in his laboratory at MDIBL in the mid-1970s, washing dishes and doing basic laboratory routine. I quickly came to find that nothing was routine to Arnost, that the smallest tasks were as important as the largest. He taught me the importance of discipline and perseverence, in life and in science, and these lessons have helped me throughout my career.

One trait that stands out is Arnost's ability to maintain a sense of humor while pushing forward with his ideas. His trademark rubbing of his hands and face in a moment of intense laboratory work was one of his many memorable characteristics. Arnost's unique personality made him a joy to work with and helped take the sting out of the "bad days" where everything that could go wrong did.

Although Arnost had high expectations, he had great compassion and treated me as family as well as student. I will remember Arnost as the person who started me on my way in many aspects of my life. He will surely be missed as a teacher and friend.

Jonathan M. Goldstein
Department of Neurology
Yale University School of Medicine

Arnost Kleinzeller was a wonderful colleague and friend. It was a pleasure to go to him with a particularly difficult scientific problem and reap the benefit of his encyclopedic knowledge and wide experience. He would often think about the problem and come back in a day or two with a suggestion that often turned out to be quite helpful and insightful.

Arnost had little tolerance for sloppy thinking or sloppy work, but that was balanced by his soft, compassionate side. He was the first to offer guidance to those who needed help and was particularly generous to students and junior colleagues working with him. This help often extended to guiding his "offspring" in both their careers and their personal lives.

I remember Arnost as a great friend who knew when to listen and when to offer advice. I miss our conversations, scientific and nonscientific. But most of all I miss the pleasure of sharing his zest for life.

Leon Goldstein
Division of Biology and Medicine
Brown University

In 1965 the Department of Physiology of the School of Medicine of the University of Pennsylvania needed to increase its strength in cell physiology and biophysics and began considering a number of potential faculty. We heard that Arnost Kleinzeller had managed to slip out of Czechoslovakia with his family and was now at the University of Rochester under the wing of Aser Rothstein. Arnost had a superb reputation as a junior colleague of Sir Hans Krebs in England, from whence at the end of World War II he returned loyally to Prague, taking posts at the Technical University and then Charles University and finally becoming Head of the Laboratory for Cell Metabolism of the Czechoslovakian Academy of Sciences. He established a great reputation abroad while there. He was also famous for having his laboratory operate in different languages on different days. He and Lotte visited us in Philadelphia and he eventually agreed to join our department in June of 1967, giving us instant recognition in cell physiology.

Arnost was a precise and demanding teacher and preceptor, no nonsense about science. He organized many activities, large and small, while at Penn. Probably the first was a wonderful course in cell physiology techniques, making available for the first time in our medical school bench-side instruction in many new methods, cell fractionation, membrane transport by radioactivity measurements, and many basic methods now commonplace, but then new. Students, and even some professors, came from all over the school. The faculty was drawn from many parts of the university, even from abroad.

He was an encyclopedia of information on physiology and biochemistry, having grown up with the latter field. His research concentrated on the kidney, which so neatly brought together physiology, biochemistry, and biophysics. He kept abreast of new techniques and was able to review with knowledge a wide area of physiology, which enabled him to edit this wonderful series with Felix Bronner. His great interest in history led him later to do research on Overton, whom he considered ahead of his time and deserving of more credit for developing our concept of the cell membrane than he received.

As a colleague he was tremendously loyal to our department and a great support to me as chairman, even when he thought my judgment was wrong, a not infrequent occurrence. I sought his advice all the time on scientific matters.

Arnost was the very model of a cultured European academic. He knew music and played the violin, though unfortunately I never heard him. He knew Greek and Latin, about which I often consulted him, and was familiar with the great classics of Western civilization.

As he grew older he suffered a considerable series of illnesses, including the ignominy and near devastation of a childhood infectious disease, all of which he bore with fortitude. He visited me frequently all these later years with recent humorous anecdotes and some pungent gossip. Except for a brief factual bulletin concerning the state of his health, he voiced no complaints.

It is impossible to say anything about Arnost's long career here without mentioning Lotte, who supported him in all things but never hesitated to be sure that he did not feel he was too far above the salt. When he was ill, she was an avenging angel.

Arnost Kleinzeller was a tremendous scientist, a wonderful friend, a delight to know, and I miss him.

Robert E. Forster II, M.D.
Isaac Ott Professor Emeritus
University of Pennsylvania

Previous Volumes in Series

Current Topics in Membranes and Transport

Volume 23 Genes and Membranes: Transport Proteins and Receptors* (1985)
Edited by Edward A. Adelberg and Carolyn W. Slayman

Volume 24 Membrane Protein Biosynthesis and Turnover (1985)
Edited by Philip A. Knauf and John S. Cook

Volume 25 Regulation of Calcium Transport across Muscle Membranes (1985)
Edited by Adil E. Shamoo

Volume 26 Na^+–H^+ Exchange, Intracellular pH, and Cell Function* (1986)
Edited by Peter S. Aronson and Walter F. Boron

Volume 27 The Role of Membranes in Cell Growth and Differentiation (1986)
Edited by Lazaro J. Mandel and Dale J. Benos

Volume 28 Potassium Transport: Physiology and Pathophysiology* (1987)
Edited by Gerhard Giebisch

Volume 29 Membrane Structure and Function (1987)
Edited by Richard D. Klausner, Christoph Kempf, and Jos van Renswoude

Volume 30 Cell Volume Control: Fundamental and Comparative Aspects in Animal Cells (1987)
Edited by R. Gilles, Arnost Kleinzeller, and L. Bolis

Volume 31 Molecular Neurobiology: Endocrine Approaches (1987)
Edited by Jerome F. Strauss, III, and Donald W. Pfaff

Volume 32 Membrane Fusion in Fertilization, Cellular Transport, and Viral Infection (1988)
Edited by Nejat Düzgünes and Felix Bronner

** Part of the series from the Yale Department of Cellular and Molecular Physiology*

Volume 33 Molecular Biology of Ionic Channels* (1988)
Edited by William S. Agnew, Toni Claudio, and Frederick J. Sigworth

Volume 34 Cellular and Molecular Biology of Sodium Transport* (1989)
Edited by Stanley G. Schultz

Volume 35 Mechanisms of Leukocyte Activation (1990)
Edited by Sergio Grinstein and Ori D. Rotstein

Volume 36 Protein–Membrane Interactions* (1990)
Edited by Toni Claudio

Volume 37 Channels and Noise in Epithelial Tissues (1990)
Edited by Sandy I. Helman and Willy Van Driessche

Current Topics in Membranes

Volume 38 Ordering the Membrane Cytoskeleton Tri-layer* (1991)
Edited by Mark S. Mooseker and Jon S. Morrow

Volume 39 Developmental Biology of Membrane Transport Systems (1991)
Edited by Dale J. Benos

Volume 40 Cell Lipids (1994)
Edited by Dick Hoekstra

Volume 41 Cell Biology and Membrane Transport Processes* (1994)
Edited by Michael Caplan

Volume 42 Chloride Channels (1994)
Edited by William B. Guggino

Volume 43 Membrane Protein–Cytoskeleton Interactions (1996)
Edited by W. James Nelson

Volume 44 Lipid Polymorphism and Membrane Properties (1997)
Edited by Richard Epand

Volume 45 The Eye's Aqueous Humor: From Secretion to Glaucoma (1998)
Edited by Mortimer M. Civan

Volume 46 Potassium Ion Channels: Molecular Structure, Function, and Diseases (1999)
Edited by Yoshihisa Kurachi, Lily Yeh Jan, and Michel Lazdunski

Volume 47 Amiloride-Sensitive Sodium Channels: Physiology and Functional Diversity (1999)
Edited by Dale J. Benos

CHAPTER 1

Charles Ernest Overton's Concept of a Cell Membrane

Arnost Kleinzeller[1]
Department of Physiology, University of Pennsylvania School of Medicine, Philadelphia, Pennsylvania 19104

I. A BRIEF BIOGRAPHY OF ERNEST OVERTON

A century has elapsed since (Charles) Ernest Overton (1865–1933) presented his views on the osmotic properties of cells in the period 1895–1902. He is justifiably recognized as a pioneer in the development of a comprehensive concept of the cell membrane (Höber, 1926; Collander, 1965; Smith, 1962; Kleinzeller, 1995). As noted in Collander's appraisal of Overton's achievements on the occasion of his 100th birthday, he established several milestones in the field:

1. A "lipid-impregnated boundary layer" is a determinant of the osmotic properties of living cells. Both diffusion- and metabolism-linked active processes function in solute exchange between cells and their immediate environment.
2. The lipids of the cell (membrane) are involved in the phenomenon of narcosis.

[1] Deceased.

1063-5823/99 $30.00

3. An exchange of sodium and potassium across the membranes of muscle and nerve cells is responsible for their excitability.

Most of Overton's papers were published in German. This chapter discusses the experimental basis for these crucial elements of a comprehensive concept of the structure and function of a cell membrane. Such an evaluation of necessity must take into account the framework of knowledge of the field at the time of Overton's creative life.

Ernest Overton (in his publications, Overton used only this first name) was born in 1865 in Stretton, Cheshire, England. On his mother's side he was distantly related to Charles Darwin. Because of his mother's chronic illness, the family moved in 1882 to Switzerland. In 1884, Overton entered the University of Zürich as a student of botany under the preceptorship of Professors A. Dodel and E. Strasburger (in Bonn, Germany), defended his Ph.D. thesis in 1889, and was appointed "Docent" (roughly equivalent to lecturer) in biology in 1890.

During his time in Zürich, Overton was scientifically most productive. In the period of 1888–1893, Overton published seven papers related to his botanical interests (1888, 1889, 1890a,b; 1891, 1893a,b). His observations (1893a,b) on the reduction of chromosomes (meiosis) in certain plant cells gained him the reputation of a skilled and innovative investigator. Overton returned to strictly botanical subjects only in three papers published in 1897, 1899a, and 1899b. He established his scientific status primarily by his papers on the osmotic properties of cells, research that was also begun in Zürich.

In 1901, Overton joined the renowned Department of Physiology at the University of Würzburg (Germany) as assistant to the electrophysiologist Professor M. von Frey. In 1907 he accepted a call to become chairman of the newly established Department of Pharmacology at the University of Lund (Sweden) where he served until his death in 1933. It was in Lund that Overton married Dr. Louise Petrén, a member of a well-known Swedish academic family, and raised his family of four children. His contributions to science from the Lund period were rather limited. Comments of his daughters (Dr. S. Thesleff of Lund, personal communication) would suggest that protracted health problems and difficulties in mastering the Swedish language may have been a contributing factor.

Although he had not received formal medical training, his contribution to the medical field was acknowledged by honorary medical degrees by the universities of Lund and Jena.

II. EXCHANGE OF SOLUTES ACROSS THE CELL BOUNDARY

Overton's interest in this field was prompted by his genetic experiments, which required a substance that would enter plant cells rapidly. The work of Nägeli and Cramer (1855), de Vries (1871, 1884), and Pfeffer (1877) had established the existence of an osmotic barrier between the protoplasm of plant cells and their environment. The evidence was based on observations of protoplasmic permeability: while healthy cell protoplasts were readily permeable to water, they were impermeable to many solutes, including the plant pigment anthocyan (Nägeli and Cramer, 1855) or sucrose from beet roots (de Vries, 1871). This permeability barrier was lost following cell death or major injury, as judged by changes in the appearance of the protoplasm and cessation of protoplasmic streaming.

The phenomenon of plasmolysis was firmly established, that is, a retraction of the protoplasm from the cellulose wall of plant cells exposed to impermeable external solutes, as shown in Fig. 1. Intracellular vacuoles also reduced their volume under plasmolytic conditions, and the phenomenon was reversible (deplasmolysis). The observations were interpreted as osmotically induced flows of water across a thin osmotic barrier, for which

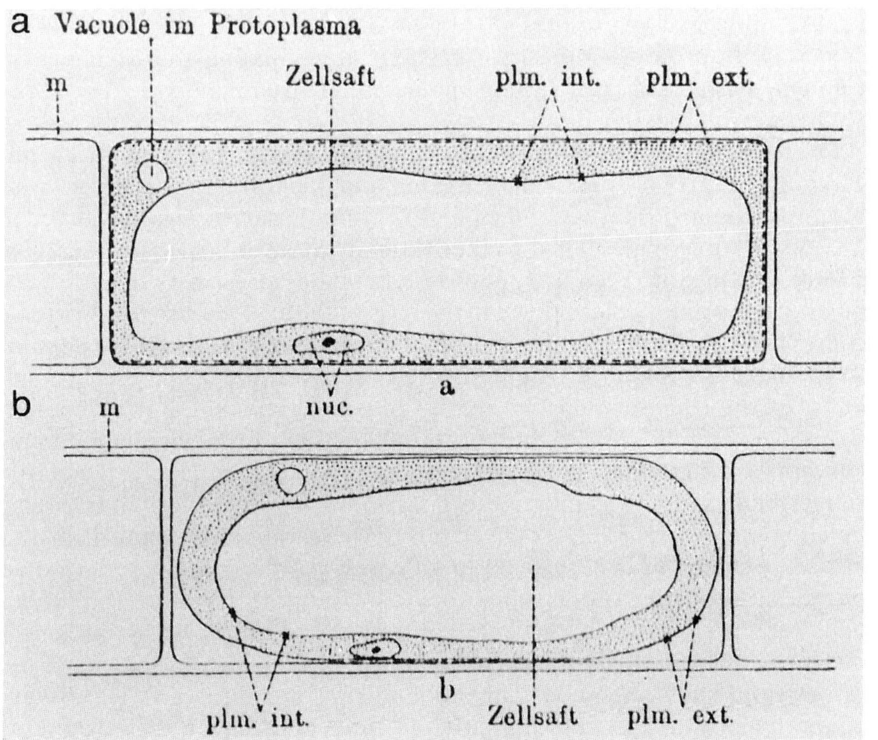

FIGURE 1 Scheme of a plant cell from the osmotic point of view. (a) Normal and (b) plasmolysed state.

Pfeffer used the equivalent terms plasma skin ("Plasmahaut", 1877, 1891), skin layer ("Hautschicht", 1886), or plasmatic membrane (1900). He suggested (1891) that the membranes facing the external and inner (vacuolar) spaces had identical properties and could actually fuse. In his view, the nature of the plasma membrane was proteinaceous, rejecting Quincke's proposal of an oil-like protoplasmic surface (1888). The idea that membranes were composed of proteins had many adherents, particularly when Ramsden (1904) demonstrated the spontaneous formation of solid films on the interface of protein solutions and air.

Investigators in the field employed solutions of sucross or salts as osmotic agents. There were some early observations that not all hypertonic solutes produced plasmolysis. For instance, Klebs (1887) found that glycerol was a poor plasmolytic agent, and this observation was confirmed and extended by de Vries (1888). At the time, no explanation of the phenomenon was offered.

Pfeffer's (1886) careful study of the cellular uptake of dyes allowed him to consider additional properties of the putative osmotic barrier. He proposed that in addition to the "static" properties of the plasma skin (reflected by simple osmotic phenomena), living cells also had the capacity to take up or exclude solutes. At the time, his suggestion that the plasma skin was a "protoplasmic organ capable of regulating the exchange of solutes" between the cell and its immediate environment was a cry in the scientific wilderness.

Overton's systematic exploration of the relationship between the chemical constitution of solutes, mainly organic, and their plasmolytic effects, with reference to the properties of the putative osmotic barrier, was presented in a series of three papers read to the Zürich Naturalist Society (1895, 1896, 1899c). Having observed that solutions of ethanol and other aliphatic alcohols did not produce plasmolysis in the cells of the alga *Spirogyra,* Overton's concept of the phenomenon was broad. Given van't Hoff's (1887) demonstration of basic similarities between the laws governing the physical behavior of gases (the relationship among pressure, volume, and number of molecules) and the osmotic properties of solutions, Overton illustrated his concept by comparing a plant cell to a soccer ball (Fig. 2). Here, the elastic rubber balloon (k) was inflated to 1.5 atmospheres pressure, in turn also inflating the external leather cover (l). Because rubber is impermeable to air whereas leather is permeable to all gases, any pressure increase would produce a shrinkage of the rubber balloon (to k_i). If, however, the rubber ball were permeable to the external gas, an equilibration of pressures outside and inside the rubber balloon would take place, and hence no retraction. Thus, the absence of balloon retraction indicates a permeability of the balloon to the gas. Similarly, a putative cell membrane (or skin) permeable to water would display plasmolysis when the cell was exposed to

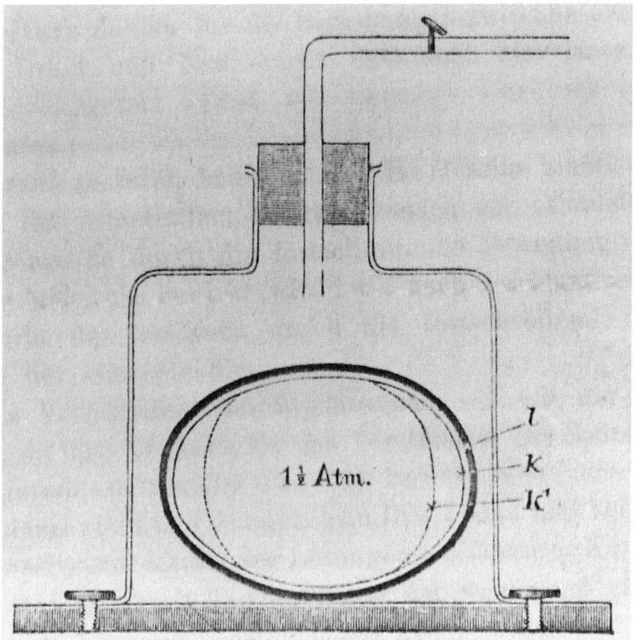

FIGURE 2 Demonstration of the relationship between osmotic and gas pressures as observed in plasmolytic experiments. A soccer ball with an outer leather covering that is permeable to air and an internal rubber bladder containing air are shown. When the pressure outside is increased (analogous to an increased osmolarity of the solution around a plant cell), the permeable leather cover does not change in volume, while the internal bladder shrinks.

an impermeable solute, If, however, the solute were permeable, plasmolysis would be slowed down or absent, with an ultimate equilibration of the solute concentration on both sides of the membrane. The rate of the observed volume changes of the balloon or protoplasm becomes a measure of the permeability of the structure separating the external and internal compartments. Hence, the use of plasmolysis as an experimental tool (the "osmometric method") focused on both the steady state of plasmolysis and its rate.

The osmometric technique employed by de Vries and Pfeffer consisted of observing changes of the cell volume of plants immersed in solutions of various solutes. This procedure presented certain problems. While in plant cells, plasmolysis and deplasmolysis could be followed visually, in animal cells the preparation had to be weighed to determine volume changes. Furthermore, the potential cellular toxicity of solutes had to be considered. Overton (1895, 1902a) refined the osmometric technique, taking into consideration the fact that the osmotic pressure of a solution equals the sum of partial pressures of the solutes present. The improved techniques permitted a considerable decrease in the actual concentrations of potentially toxic solutes, while allowing accurate measurements of changes of cell volume either by direct visual observation or by weighing (Overton, 1907).

An overview of Overton's results testing hundreds of different solutes led to his recognition of a relationship between the chemical structure of solutes and their osmotic properties. That is, lipid-soluble solutes (including physiologically and pharmacologically active substances such as anesthetics, narcotics, and certain alkaloids) penetrated the cell membrane fastest. In fact, some of these, such as ethanol, entered the cell at rates comparable to that of water. The study permitted statements about the relationship between the chemical nature of solutes and their permeability. The general rule appeared to hold widely that the lipid:water partition coefficient was a determinant of solute permeability, and Overton (1895) generalized that "the living protoplasm of all basic organisms (plant cells, protozoa, ciliary, and epithelial cells) is readily permeable for various organic solutes."

However, Overton found some clear exceptions to this rule, as physiological solutes such as glucose or amino acids did not appear to diffuse across the cellular osmotic barrier. The apparent absence of permeability to inorganic ions, particularly of Na^+ and K^+, was also puzzling. This was first noted by de Vries (1884) and endorsed by Overton (1895, 1899c). Earlier experiments using plasmolytic techniques had demonstrated that some plasma "membranes" were permeable to inorganic salts (de Vries, 1884). Overton sought an explanation for such anomalies by proposing permeability mechanisms different than those inolving osmosis, with a direct involvement of living cells. However, Osterhout (1911) later showed that Overton's experimental material, the alga *Spirogyra,* happened to be moderately permeable to a variety of inorganic salts.

Generalizing from his experimental data, Overton (1895) concluded that the ability (or lack thereof) of a solute to pass the osmotic barrier does not depend primarily on the size of the molecule (the concept of a molecular sieve, as visualized by Traube and Pfeffer). Furthermore, the osmotic barrier displays a permeability determined by the lipid-impregnated boundary layer of the protoplasm (Overton, 1899c). Overton thus adopted the concept that the determinant for a solute passing a membrane barrier is its solubility in the membrane, a view expressed previously by Pfeffer (1886). This idea was first suggested by Graham (1851) and by Lhermite (1855), who used the term "affinity between solute and membrane components." The concept was later resurrected by Nernst (1890).

The postulated lipid-impregnated boundary layer represented a general characteristic of plant and animals cells. This generalization was based on

Overton's (1895, 1899c) experiments on the osmotic properties of tadpole cells, as well as in part on Hamburger's (1887) work demonstrating that the osmotic properties of red blood cells were identical to those observed earlier in plant cells. Hamburger was not aware of the early experiments of Hewson (1773) on this topic (Kleinzeller, 1996). Overton (1902, 1904b) later extended this work to a study of the osmotic properties of muscle cells and to water flows in frogs exposed to salt solutions. Arguments in favor of the lipid-impregnated cell boundary layer might have been strengthened if reference had been made to the observations of Schmidt (1891) who demonstrated that the "plasma skin" of a variety of plants readily allowed the entry of neutral fat as well as fatty acids. Quincke (1888) first suggested that the protoplasm was enveloped by a very thin (less than 1 μm) layer of an essentially liquid, oil-like substance. His view was based on studies of cell movement, and the postulated lipid layer was not considered as a constituent of the cell osmotic barrier.

With remarkable prescience, Overton (1899c, 1900) suggested that the lipid material impregnating the cell membrane could have properties similar to cholesterol esters and/or phospholipids (lecithin), given their immiscibility with water and reflecting the fact that the cell surface was wettable by aqueous solute solutions. Overton (1900) sought to underpin this hypothesis by investigating the cellular uptake of organic dyes. Since 1887, Ehrlich (1902) had drawn attention to the correlation between the neurotropic action of basic vital dyes (aniline derivatives) and their affinity for cell lipids (lipotropic property); acidic (sulfonated) dyes were not taken up by the cells. Pfeffer (1886) has also described the vital staining of plant cells with basic aniline dyes. After referring to his own early experiments on vital cell staining, Overton (1890a,b) indeed found a distinct correlation between the rapid cellular uptake of basic aniline dyes and their solubility in lipids, particularly in solutions of lecithin and cholesterol esters, while acidic dyes were not taken up. This study was subsequently criticized (Ruhland, 1908, 1912; Küster, 1911) for not taking into account the colloidal nature of the dyes; Ruhland emphasized the role of molecular size in controlling the entry of vital stains into living cells.

Overton also noted that a certain amount of excess lipid had to be present in the cells in order to explain the "*de novo*" formation of a boundary layer of fragments of protoplasm formed on crushing cells. As observed by Nägeli and Cramer (1855) and Pfeffer (1877) these fragments (see Fig. 3) displayed osmotic properties.

While the permeability properties of many solutes were consistent with the established laws of diffusion, Overton (1899c) defined a set of phenomena where cellular activity was required to produce flux against prevailing osmotic gradients and related these active transfers of solutes to the secre-

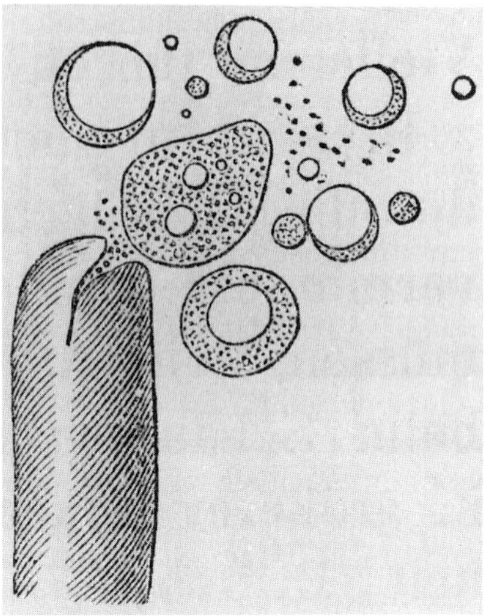

FIGURE 3 Formation of protoplasmic globular fragments from damaged cells. These membrane-bounded fragments have osmotic properties (Pfeffer, 1900).

tory or absorptive processes in animal cells. It followed that such processes were associated with the expenditures of metabolic energy. In particular, Overton (1902a) considered in detail the need for a special pathway for the entry of glucose into heart muscle in order to explain the discrepancy between the apparently low permeability of the muscle for this crucial metabolic substrate and the assessed rate of its utilization. Several hypothetical pathways were considered, all involving cell metabolism.

Overton shared the earlier view of Pfeffer (1886) that the permeability barrier needed to be regulated in order to meet the physiological requirements of the exchange of solutes between cells and their environment. Pfeffer's collaborator Nathansohn (1904) then pointed more specifically to the regulatory role of the protein membrane component for the selective permeability of lipid-insoluble substrates.

Overton's concept of the function and structure of a lipid-impregnated cell boundary layer differed in several respects from the views of earlier investigators. Schwann (1839) had put forward the early criteria of a membrane of plant and animal cells as a distinct cell structure that serves as a mechanical envelope of the fluid protoplasm and is the site of exchanges of solutes between the cell and its environment.

As opposed to Schwann (1848), Pfeffer (1877), or de Vries (1871), Overton did not consider such a boundary layer to be a distinct cell structure. In fact, Overton (1907) emphatically rejected Koeppe's (1897) deliberately neutral term of a limiting cell partition ("Wand") as the site of the exchange of anions in red blood cells. Presumably under the influence of Schultze's (1861) and Brücke's (1861) criticisms of the cell membrane theory (see also Kleinzeller, 1996), he was concerned about interpretating experimental results on red blood cells in the absence of their full "life activities." He also distanced himself from Pfeffer's (1877) concept of a very thin cell membrane (or skin) similar in its properties to Traube's (1867) precipitation membrane. However, Overton (1907) did cite his own experiments, demonstrating that the putative boundary layer was very thin. The idea of a structure distinguished by a physical cohesiveness different from that of the bulk protoplasm may not have been appealing to him. Höber (1926) later commented that although evidence for a plasma membrane in Pfeffer's sense did not exist because of visibility limitations (Pfeffer, 1886), such a term was justifiable on the basis of available observations in which nonoptical techniques could detect ~1-nm-thick layers (Freundlich, 1923).

The putative "boundary layer" was the site of the osmotic barrier of living cells. Previously, Nägeli and Cramer (1855) assumed the involvement of the whole protoplast in osmotic function, although they suggested that on injury of the cell, the protoplasm may form on its interface with water a membrane that then protects the cell content from damaging effects. Overton emphasized the unique role of this boundary layer by pointing out experiments of de Vries (1885) and Pfeffer (1886) that showed that the protoplasm could be destroyed effectively without affecting the osmotic properties of the system.

Pfeffer believed that many (but not all) osmotic phenomena could be explained in terms of the known laws of diffusion. Overton (1896) emphasized since 1895 that, as opposed to such Pfeffer's "static" view of the osmotic barrier, compelling physiological data on absorptive and secretary activities and the unequal distribution of cations between cells and their environment required a participation of cellular activity in "active" processes that proceeded in the opposite direction than predicted by the laws of diffusion. Examples such as absorption of salts from the intestine were given.

Previously, Hoppe-Seyler (1877–1881) suggested the participation of cell metabolism in secretary and absorptive processes. Overton then demonstrated his intuitive understanding of the role of the osmotic barrier by two statements: "Active participation of cells matters only for such solutes which are not at all, or only slowly, taken up by most cells by a strictly osmotic process (1899c, p. 102)." and "One might even say that most organisms try to avoid in their metabolism such compounds which can quickly pass through the plasma skin by the physical pathway; moreover, when such compounds transiently occur, the organisms try to convert these into other compounds which lack such capability" (1895, p. 200). This represented a new functional concept of the cell osmotic barrier. In particular, Overton did not perceive a single comprehensive concept of cell permeability, but envisaged several pathways (at least two) by which solutes passed across the osmotic barrier. Opponents of Overton's lipid membrane concept (Ruhland and Hoffmann, 1925) argued wrongly against his theory by pointing out that a variety of cell substrates did enter the cells despite their insolubility in lipids.

Overton's observations and concepts on the osmotic properties of cells were described during the period of 1895–1899, and a detailed account was given in an extensive review in 1907. Most of the actual experimental data were planned for full publication (1895, p. 159, 186) but Overton never devoted time for this chore.

Observations on the potassium-induced swelling of muscles (Overton, 1902b, 1904a) and the rapidly advancing information on the osmotic response of many cells, particularly red cells (Hamburger, 1887; Hedin, 1897), prompted Overton (1904a, 1907) to analyze cell volume changes in some detail. The question raised was whether cell "membranes" allow electrolytes to cross a nondissociated molecules, as individual ionic species or a combination of such pathways, and possible effects on the cell volume at equilibrium.

A simple model system where an osmotic barrier separated pure water from an electrolyte solution such as KCl demonstrated the various results when the partitioning "membrane" was assigned differing permeabilities for water, KCl, or K^+ and Cl^-. Furthermore, it was assumed that no external hydrostatic or other pressures were involved. For the simplest case, the membrane being permeable only to water, the net flux of water toward the electrolyte solution would occur. For a membrane permeable to water and KCl (but not the ionic species), there would be some next flux of water across the membrane while KCl would flow in the opposite direction and dissociate. At equilibrium, the concentrations of KCl, K^+, and Cl^- in both compartments would be identical.

For a second membrane permeable to water, K^+, and Cl^- but not to KCl, the equilibrium again would be characterized by equal concentrations of both ionic species. However, here the permeabilities of both ionic species must be considered: if these differed, a transient electrical potential gradient would occur. Finally, for a membrane permeable to water and only one of the ionic species, there would be a flow of water across the membrane, but the very small flux of the ionized species would generate an electrical potential, which would prevent further ionic fluxes.

This analysis of a model system introduced the novel concept that studies of the cell membrane permeability for an electrolyte should consider separately the fluxes, as well as the equilibrium distribution, of the nondissociated molecule and/or of the individual ionic species. Such studies then required actual measurements of the net water flux, the fluxes of electrolytes, and possible electrical potentials across the membrane. This imaginative approach became acceptable only decades later as ionic channels and cotransport systems became better understood. Overton also showed how to answer some of the questions by available experimental approaches. For instance, in Overton's (1902b, 1904b) study of the potassium-induced swelling of muscles, he demonstrated the role of membrane permeability for KCl by comparing effects of solutions of KCl with those employing an impermeant anion such as phosphate or sulfate.

Based on Overton's (1902b, 1904b) own experiments and a previous pioneering experimental study of Koeppe (1897), Overton also emphasized the electroneutral exchange of different ionic species of the same charges. Such exchange per se would not be associated with net volume changes.

Overton later returned to this topic. In his 1913 study of the hemolytic action of saponins on cells, he interpreted the hemolytic process as resulting from a specific interaction of the saponins with cholesterol in the cell surface. Overton (1918) also showed that treatment with saponins increased the permeability of tadpole skin and gill cells greatly for substances that did not otherwise cross the membrane. Strictly speaking, these findings were not internally consistent with a simple lipid membrane, but Overton did not offer any suggestion on this point. A later study on the uptake of organic acids and their esters in 1925 did not yield novel results.

Overton's concept of a "lipid-impregnated boundary layer of the protoplasm" (with the possible involvement of phospholipids and cholesterol), later known as the lipid membrane theory, is a pioneering contribution to the membrane field. Furthermore, the conceptual analysis of cell water (volume) changes as a function of the permeability properties of the membrane for solutes deserves recognition. The idea of an active cellular uptake (or extrusion) of solutes involving cell metabolism was less original, but his convincing presentation of this concept may have been crucial for the subsequent recognition of this phenomenon. Overton's (1895, 1896, 1901) studies on the relationships between the chemical structure of solutes and their osmotic properties also represented a milestone in general pharmacology.

III. MEYER–OVERTON THEORY OF NARCOSIS

Overton's principal contribution to this topic was published in a single scholarly monograph in 1901; the actual experimental work and underlying ideas logically developed from studies presented in his first lecture on the osmotic phenomena in 1895, and his general views on narcotic action were discussed in some detail in his 1899c lecture.

Texts of pharmacology available near the end of the 19th century (Hermann, 1874; Binz, 1891) mention several ideas on the possible mechanism by which narcotic effects are produced, that is, by a transient inhibition of oxidative processes or coagulation of protoplasm (Bernard, 1875). However, several early observations related narcotic effects to protoplasmic (particularly nervous tissue) lipids, in which the narcotics affected cell function indirectly. Bibra and Harless in 1847 (Overton 1901; Winterstein, 1919) claimed that brain lipids of ether-narcotized dogs were somehow extracted by the ether and suggested that the anesthetic effect is determined by the solubility of tissue lipids in narcotics. This line of investigation was taken up by Hermann (1866). Having demonstrated the presence of lecithin in the "stroma" of red blood cells, he described the hemolytic action of aliphatic narcotics and suggested that lecithin, cholesterol, and other fatty materials were the site of action of narcotics due to their affinity to lipids. Pohl (1890) reported an actual accumulation of the narcotic in the brain lipids of chloroform-narcotized dogs. Richet (1893) pointed out the apparent inverse relationship between the toxicity of narcotics and their solubility in water.

The lipid theory of narcosis was presented independently by Meyer (1899) and Overton (1899c, 1901). While Overton had generated ample experimental data that enabled him to make his general comments in his 1899 lecture, Meyer apparently was not aware of Overton's work when he published his results in 1899. Overton (1901) then acknowledged the availability of Meyer's manuscripts prior to their publication. Because the pharmacological properties of narcotics indicated their primary action on the central nervous system, both authors started with the established knowledge that most narcotics show a high solubility in lipids and suggested that their action is based on this property, keeping in mind the high lipid content of nervous tissue. Such views were opposed to earlier ideas that emphasized the solubility of lipids in narcotics (see earlier).

It followed that narcotics, being mostly chemically unreactive, could act by affecting the physical state of tissue lipids. An interaction between narcotics and lipids implied an important role of narcotic partitioning between a lipid phase and cellular water. Experimental data of Meyer (1899), Baum (1899, in Meyer's laboratory), and, most extensively, of Overton (1901) documented amply these points, particularly the close correlation between the effect of narcotics and their lipid:water partition coefficient. Overton's contribution was made possible by his improved methods for determining the lipid–water partition coefficient in tissues. His elegant "physiological" method of determining partition coefficient appears to be particularly useful. Here, small tadpoles or freshwater invertebrates were employed as indicators, and the minimal aqueous concentration of the narcotic required for complete narcosis was determined. Equilibration of the aqueous solution of known narcotic concentration with a measured volume of an oil-like solvent then permitted the actual partition coefficient to be determined. When interpreting his data, Overton (1895) emphasized his observation that narcotics passed rapidly into cells through the "lipid-impregnated protoplasmic boundary layer" and might then act either on the osmotic properties of this layer or on intracellular components. Overton's lipoid theory of narcosis embraced two essential components: the relationship of the narcotic effect to their lipid solubility and the lipoid theory of permeability. This last point left open the actual mechanism by which narcotics produce their effects (Overton, 1899). Alcock (1906), Höber (1907), and other investigators later began studies of the effect of narcotics on the ionic processes underlying the excitability of muscles as well as on cell permeability (Höber, 1926).

Overton's experience with narcotic effects contributed to two subsequent studies (with I. Bang) on the effects of snake venom (Bang and Overton, 1911a,b). A short abstract on the effect of narcotics (Overton, 1925b) demonstrated his continuous interest in the field. The lipid theory of narcosis stimulated new experimentation and also produced many comments (Winterstein, 1919), thus allowing the field to develop further.

IV. ROLE OF CATIONS IN THE EXCITABILITY PROCESS

This masterly study was presented in four papers (1902–1904) and was prompted by an effort to extend earlier research on plants and whole animals to isolated animal tissues. In the light of previous work on muscles, the focus was broadened to analyze the relationship between osmotic phenomena and the excitability of muscle and nerves.

Hales (1733) first noticed that injection of salt-free water into the blood circulation of dogs produced a massive swelling of all tissues, with spontaneous convulsions of muscles; these phenomena were absent when salt solutions were injected. Nasse (1869) extended such studies by exploring the concentrations of various salts necessary to preserve muscle excitability. Another indication of the possible role of ions in the excitation process was provided by the observation in 1880 of Biedermann (1895) who demonstrated spontaneous rhythmic contractions of frog muscles immersed in slightly hypertonic solutions of NaCl and postulated a specific role of sodium salts in the contraction process.

The osmotic properties of isolated frog muscles were studied later by Loeb (1898) and Cooke (1898), who analyzed their data in terms of diffusion of salts and water into muscles immersed in solutions of varying salt concentrations. They interpreted their results in terms of the muscle as a sac containing isotonic fluid, enveloped by the semipermeable perimysium.

Overton (1902a, 1904a,b) pointed out shortcomings in these previous studies, stemming primarily from a lack of appreciation of the muscle as a multiphase structure where the fibers are surrounded by perifibrillar fluid. In order to study the osmotic effects on muscles, it was necessary to measure actual weight changes during the course of an experiment. Overton (1902a) explored the effects of hypotonic NaCl solutions and demonstrated that the muscles swelled less than predicted on the basis the laws of diffusion. That is, at equilibrium the volume increase should have been proportional to the reciprocal of the change of osmotic pressure. Hence, the evaluation of osmotically induced fluid uptake by the tissue must take into account diffusion of salt solutions from the surrounding medium into the perifibrillar space as well as the semipermeable properties of the muscle fibers. This research in effect opened a new chapter of cell physiology: the study of cell volume and its regulation.

With his improved understanding of the underlying osmotic processes, Overton (1895, 1899c) established the similarity between the osmotic properties of frog muscles with those of plant cells, including the great permeability of the muscle fibers for short chain alcohols and narcotics. As mentioned earlier, Overton (1902b) observed a relatively small permeability of resting muscle fibers for physiologically important substrates such as electrolytes, amino acids, or glucose and concluded that cellular activity must be involved in the inward transport of these solutes.

Immersion of muscles in isotonic solutions of sucrose (or other neutral organic solutes) produced a rapid loss of perifibrillar NaCl as well as muscle excitability and electrical current conductance (Overton, 1902b). This phenomenon was fully reversible on reimmersion of the tissue in isotonic NaCl solutions. Given the available knowledge that most of the tissue NaCl was present in the perifibrillar fluid while the bulk of the fibrillar electrolyte was potassium chloride, it was reasonable to conclude that Na^+ in the extrafibrillar space played a role in muscle excitability. Experimental evidence bore out such conclusion. For instance, any sodium salt could reverse the sucrose block of excitability. Reversal of the sucrose block could also be produced by Li^+ and $NH_4{}^+$, whereas K^+, Rb^+, or Cs^+ produced a paralysis of the muscle in the presence of Cl^-, as already observed by Grandeau

(1864). Mg^{2+} was innocuous, whereas solutions of Ca^{2+}, Sr^{2+}, and particularly Ba^{2+} were toxic.

Overton (1902b) did not observe a loss of excitability of nerves upon immersion in sucrose solution; however, he pointed out that this observation did not necessarily imply differences in the actual excitatory process in both tissues and correctly considered the possibility that the perineural sheath may retain sodium around the fibers. Subsequently, Overton (1903) presented evidence for the essential role of Na^+ in the excitability of nerves and the central nervous system.

When interpreting the observed specific role of Na^+ in the excitatory process, Overton (1902b) put forward a novel concept of an exchange of Na^+ against K^+ at the muscle fiber–perifibrillar fluid interface during a particular short phase of the muscle contraction process. Overton was fully aware of the logical problems arising from this suggestion by pointing out that repeated contractions might lead to a gradual increase of fibrillar Na^+ unless a mechanism was available to reconstitute the previous ionic distribution, and he specifically noted that the secretary properties of glandular tissues could be a related phenomenon.

His increasing understanding of osmotic processes in muscle permitted Overton (1902b, 1904a) to explore the actual mechanism by which potassium salts produced a curare-like paralysis of muscle, first recognized in Bernard's laboratory (Grandeau, 1864). Previous data demonstrated an unusual swelling of frog muscles immersed in isotonic KCl (Loeb, 1898) associated with an initial brief spontaneous twitching. Overton found that at room temperature, the paralysis of muscle fibers immersed in 0.11 *M* KCl developed within minutes, was not associated with any volume change, and was fully reversible by reimmersion in isotonic NaCl. For the complete paralysis of most muscles, a threefold increase of the KCl in physiological solutions was sufficient. Prolonged exposure to 0.11 *M* KCl produced an irreversible swelling as well as loss of excitability and impulse conductance. Potassium salts with permeable anions (Cl^-, Br^-, I^-, NO_3^-) shared properties with KCl. However, potassium salts of impermeant anions such as phosphate or sulfate produced muscle paralysis that was reversible even after several days, and no fiber swelling was seen. This implied a role of K^+ permeability in the observed phenomenon. A series of experiments demonstrated that K^+ effects on muscle fibers and nerves were identical. The fact that K^+ did not affect the contractile system in sperm motility pointed to a direct effect of K^+ on the excitability process.

From these findings, Overton (1904a) drew a series of important conclusions. Because no apparent entry of KCl was observed during the initial paralytic phase (there was no weight change), the observed effects appeared to be due to changes of K^+ concentration in the perifibrillar fluid. Overton suggested that during some transient phase of the actual excitation process the permeability of the fiber "boundary layer" increased, allowing a decrease in the existing normal diffusion potential of K^+. Such a mechanism of the excitatory process may also involve "an exchange of sodium ions in the perifibrillar fluid with potassium ions in the muscle fiber" (Overton, 1902b). The major irreversible swelling phase would then reflect a net entry of KCl (plus water) into the fibers, a point that was elaborated only several decades later by Boyle and Conway (1941).

Overton was aware of Bernstein's views on the role of K^+ in the excitatory process, as Bernstein's (1902) paper had appeared in the same issue of *Pfluegers Archiv* as Overton's 1902b report, but he did not comment on the complementary hypotheses. Bernstein (1902) and Brünings (1903) both postulated that the selective permeability of the muscle fiber for K^+ was lost on activation, with the cell membrane now becoming highly permeable for all ionic species. Both concepts shared the assumption that the ionic permeability of the cell membrane was regulated in the course of the excitatory process.

It may appear that it took an inordinate length of time for the recognition of Overton's ideas about the role of cations in the excitability process. A cursory search indicates that in the years following the publication of Overton's views many investigators had recognized that excitability is indeed associated with an exchange of Na^+ for K^+ (Rothenberg, 1950; Keynes, 1951). However, only Hodgkin (1951) finally recognized Overton's contribution. The most plausible explanation for this delay appears to be the fact that Overton's evidence for ionic exchange was indirect. Furthermore, there was a need to sort out the ionic events involved in the resting and action potentials of excitable cells, as well as a desire for an overwhelming accumulation of experimental evidence by an increasing level of experimental sophistication to convince even the most conservative minds of the new concept.

Overton's imaginative new hypothesis of the ionic mechanism of the excitation in muscles and nerves found justification in the brilliant studies of Hodgkin and Huxley (1952). In a conversation with this author, Andrew Huxley commented that "If people had listened to what Overton had to say in 1902, the work of Alan (Hodgkin) and my own would have been obsolete." This tribute to Overton referred specifically to the discovery of the overshoot and its attribution to sodium entry. Andrew Huxley has also suggested (personal communication) that the lack of general acceptance of Overton's views may have arisen from the inadequate and indeed misleading reference to Overton's work in William Rayliss's influential "Principles of General Physiology."

V. OVERTON'S SCIENTIFIC PERSONALITY

Judging from his papers on osmotic phenomena, Overton focused his whole attention on scientific observations. It would appear that he was a compulsive experimenter, preferring to carry out all experiments alone rather than with students or collaborators. The only instance of cooperation with a colleague is reflected in the two joint papers with I. Bang (Bang and Overton, 1911a,b).

Overton's papers reveal an uncommon reluctance to underpin often brilliant generalizations and intuitive statements with convincing reports on actual concrete measurements. As also noted by Collander (1965), a portion of his published work, particularly that originating in Zürich (1895–1900), are transcripts of his lectures to the Zürich Naturalist Society and represent preliminary communications that summarize previous information and focus on crucial open questions. The promised full publication of the enormous amount of data on the osmotic properties of cells (1895, 1899c) never appears, thus exposing Overton to criticisms (Ruhland and Hoffmann, 1925; Collander, 1965). An exception is his 1901 monograph on narcosis, and the four papers on muscle and nerve in 1902a and 1904a, possibly reflecting good advice of Overton's preceptor, Professor Frey. However, Overton's shortcomings in this respect did not restrain him from criticizing other authors of supplying insufficient experimental details.

Despite the breadth and depth of Overton's erudition, ranging from botany, physiology and pharmacology to organic and physical chemistry, his papers have rather surprising occasional omissions of pertinent references. Thus, the exposition of his views on what he termed "active" permeability, involving a direct involvement of cellular function, did not mention similar views on the subject put forward by Pfeffer (1886). When emphasizing the role of cellular mechanisms involved in some absorptive and secretary processes (Overton, 1895, 1899c), the absence of a reference to E. W. Reid's (1890, 1892, 1900) ingenious experimental work and its scholarly interpretation is unexpected. This omission is still more glaring in Overton's (1907) otherwise scholarly review of absorptive and secretary processes. The same impression is gained when noticing that in the presentation of the lipid surface layer hypothesis, Overton did not mention Quincke's (1888) hypothesis of lipids as a thin boundary layer of the protoplasm. However, Overton (1902) criticized other investigators for similar minor sins of omission in referring to previous work.

Another of Overton's traits emerges when noticing that he did not care to respond to, or even mention, critical comments of his peers (Collander, 1965). Thus, given Overton's concern about the passage of lipid-insoluble physiological solutes (salts, sugars, amino acids) across cell membranes, one would have expected at least some comments on Nathanson's hypothesis when this topic was comprehensively discussed in the 1907 review.

The impact of major contributions to science depends to a considerable extent not only on the actual new observations and novel ideas, but also on the ability to communicate them persuasively to a typically conservative audience of one's peers. Overton was a pioneer in the experimental exploration of the structure and function of the cell membrane and in the creative formulation of new working hypotheses in the field. At the same time, he was a master in convincingly presenting his ideas to the broader scientific audience (but not necessarily to his peers), thus stimulating experimentation and new ideas.

Acknowledgments

The author is particularly grateful to Professor Stephen Thesleff, University of Lund, Sweden, for his valuable comments on Overton, his predecessor as chairman of the newly created Department of Pharmacology. Overton's move from Wurzburg to Lund certainly added the burden of mastering the Swedish language in addition to the formidable task of setting up a new department. By making available biographical comments written by Overton's daughters Harriet and Margaret, insight could be gained into their father's complex personality and work habits. The author also thanks Professor Andrew F. Huxley (Cambridge) for his scholarly comments and suggestions and for advice from esteemed colleagues; A. D. Bangham (Cambridge), M. M. Civan (Philadelphia), N. Hofer (Koln), E. E. Hoffman (Copenhagen), R. L. Post (Philadelphia), S. Silbernagl (Wurzburg), and H. H. Ussing (Copenhagen).

References

Alcock, N. H. (1906). The action of anaesthetics on living tissue. Part II. The frog skin. *Proc. R. Soc. London B* **78,** 159–169.

Bang, I., and Overton, E. (1911a). Studien über die Wirkungen des Kobragiftes. *Biochem. Z.* **31,** 243–293.

Bang, I., and Overton, E. (1911b). Studien über die Wirkungen des Crotalusgiftes. *Biochem. Z.* **34,** 428–461.

Baum, F. (1899). Zur Theorie der Alkoholnarkose. II. Mitt. *Arch. Exp. Pathol. Pharmakol.* **42,** 119–137.

Bernard, C. (1875). "Leçons sur les Anesthésiques et sur l'Asphyxie." J.-B. Baillière et fils, Paris.

Bernstein, J. (1902). Untersuchungen zur Themodynamik der bioelektrischen Ströme. *Pfluegers Arch. Geamte Physiol. Menschen Tiere* **92,** 521–562.

Biedermann, W. (1895). "Elektrophysiologie," pp. 89–92. Fischer, Jena.

Binz, C. (1891). "Vorlesungen über Pharmakologie," 2. Aufl. A. Hirschwald, Berlin.

Boyle, P. J., and Conway, E. J. (1941). Potassium accumulation in muscle and associated changes. *J. Physiol.* (*London*) **100,** 1–63.

Brücke, E. (1861). Die Elementarorganismen. *Sitzungsber. Akad. Wiss. Wien* **16,** 381–406.

Brünings, W. (1903). Beiträge zur Elektrophysiologie. II. Mitt. Ueber Ruhe strom und Reizung. *Pfluegers Arch. Gesamte. Physiol. Menschen Tiere* **100,** 367–427.

Collander, P. R. (1965). Ernest Overton (1855–1933): A pioneer to remember. *Leopold. Proc. German Acad. Sci. Leopold. Ser* 3, 242–254.

Cooke, E. (1898). Experiments upon the osmotic properties of the living frog's muscle. *J. Physiol.* (*London*) **23,** 137–149.

Ehrlich, P. (1902). Über die Beziehungen von chemischer Constitution, Vertheilung und pharmacologischer Wirkumg. *In* "The Collected Papers of Paul Ehrlich" (H. H. Dale. *et al.,* eds.), Vol. 1, pp. 570–595, 1960, Pergamon, London.

Freundlich, H. (1923). "Kapillarchemie," 3rd Ed., Akad. Verlagsges, Leipzig.

Graham, T. (1851). On osmotic force. *Philos. Mag.* **4** *Ser.* **8,** 151–155.

Grandeau, M. L. (1864). Expériences sur l'action des sels de potassium, de sodium et de rubidium injectés dans les veins. *J. Anat. Physiol. Norm. Pathol. Homme Anim* **1,** 378–385.

Hales, S. (1733). "*Statical Essays,*" Vol. 2, Exp. XIV, pp. 114–116; Exp. XXI, p. 154. W. Innys and R. Manby, London.

Hamburger, H. J. (1887). Ueber die durch Salz- und Rohrzucker-Lösungen bewirkten Veränderungen der Blutköperchen. *Arch. Anat. Physiol.,* 31–47.

Hedin, S. G. (1897). Ueber die Permeabilität der Blutkörperchen. *Pfluegers Arch. Gesamte. Physiol. Menschen Tiere* 68, 229–338.

Hermann, L. (1866). Ueber die Wirkungsweise einer Gruppe von Giften. *Arch. Anat., Physiol. Wiss. Med.,* 27–40.

Hermann, L. (1874). "Lehrbuch d. Experimentalen Toxicologie." A. Hirschwald, Berlin.

Hewson, W. (1773). On the figure and composition of the red particles of the blood, commonly called the red globules. *Philos. Trans. R. Soc. London* **63,** 306–324.

Höber, R. (1907). Beiträge zur physikalischen Chemie der Erregung und der Narkose. *Pfluegers Arch. Gesamte Physiol. Menschen Tiere* **120,** 492–516.

Höber, R. (1926). "Physikalische Chemie der Zelle und Gewebe," 6. Aufl. W. Engelmann, Leipzig.

Hodgkin, A. L. (1951). The ionic basis of electrical activity in nerve and muscle. *Biol. Rev.* **26,** 339–409.

Hodgkin, A. L., and Huxley, A. F. (1952). A quantitative description of membrane current and its application to conduction and excitation in nerve. *J. Physiol.* (*London*) **117,** 500–544.

Hoppe-Seyler, F. (1877–1881). "Physiologische Chemie," I, pp. 173–174; II, pp. 348–353. A. Hirschwald, Berlin.

Keynes, R. D. (1951). The ionic movements during nervous activity. *J. Physiol.* (*London*) **114,** 119–150.

Klebs, G. (1887). Beiträge zur Physiologie der Pflanzenzelle. *Unters. a.d. Bot. Inst. Tübingen* **2,** 489–568.

Kleinzeller, A. (1995). The postulate of a cell membrane. *In* "A History of Biochemistry: Exploring the Cell Membrane" (Comprehensive Biochemistry, Volume 39) (A. Kleinzeller, ed.), pp. 27–90. Elsevier, Amsterdam.

Kleinzeller, A. (1996). William Hewson's studies of red blood corpuscles and the evolving concept of a cell membrane. *Am. J. Physiol.* (*Cell Physiol.*) **271,** C1–C8.

Koeppe, H. (1897). Der osmotische Druck als Ursache des Stoffaustausches zwischen rothen Blutköperchen und Salzlösungen. *Pfluegers Arch. Gesamte Physiol. Menschen Tiere* **67,** 189–206.

Küster, E. (1911). Über die Aufnahme von Anilinfarben in lebende Pflanzenzell en. *Jahrb. Wiss. Bot.* **50,** 261–288.

Lhermite, M. (1855). Recherches sur l'endosmose. *Ann. Chim. Phys. (3)* **43,** 420–431.

Loeb, J. (1898). Physiologische Untersuchungen über Ionenwirkungen. 1. Mitt. *Pfluegers Arch. Gesamte Physiol. Menschen Tiere* **69,** 1–27.

Meyer, H. H. (1899). Zur Theorie der Alkoholnarkose. I. Mitt. *Arch. Exp. Pathol. Pharmakol.* **42,** 109–118.

Nathansohn, A. (1904). Ueber die Regulation der Afnahme anorganischer Salze durch die Knollen von *Dahlia. Jahrb. Wiss. Bot.* **39,** 607–644.

Nägeli, C., and Cramer, C. (1855). "Pflanzenphysiologische Untersuchungen," 1. Heft, Schulthess, Zürich.

Nasse, O. (1869). Beiträge zur Physiologie der contractilen Substanz. *Pfluegers Arch. Gesamte Physiol. Menschen Tiere* **2,** 97–121.

Nernst, W. (1890). Ein osmotischer Versuch. *Z. Phys. Chem.* **6,** 37–40.

Osterhout, W. J. V. (1911). The permeability of living cells to salts in pure and balanced solutions. *Science* **34,** 187–189.

Overton, E. (1888). Ueber den Konjugationsvorgang bei *Spirogyra. Ber. Dtsch. Bot. Ges.* **6,** 68–72.

Overton, E. (1889). Beitrag zur Kenntnis der Gattung *Volvox. Bot. Centralbl.* **39,** 65–72, 113–118, 145–150, 177–182, 209–214, 241–246, 273–277.

Overton, E. (1890a). Beitrag zur Histologie und Physiologie der Characeen. *Bot. Centralbl.* **44,** 1–10, 33–38.

Overton, E. (1890b). Mikrotechnische Mitteilungen aus dem botanischen Laboratorium der Universität Zürich. *Z. Wiss. Mikrosk.* **7,** 9–16.

Overton, E. (1891). Beitrag zur Kenntnis der Entwicklung und Vereinigung der Geschlechtsprodukte beim *Lilium Martagon. In* "Festschr. zur Feier des 50jährigen Doktorjubiläums der Herren Prof. v. Nägeli u. Prof. v. Kölliker," pp. 177–187. A. Muller, Zürich.

Overton, E. (1893a). On the reduction of the chromosomes in the nuclei of plants. *Ann. Bot.* **7,** 139–143.

Overton, E. (1893b). Ueber die Reduktion der Chromosomen in den Kernen der Pflanzen. *Vierteljahrsschr. Naturforsch. Ges. Zuerich* **38,** 169–186.

Overton, E. (1895). Ueber die osmotischen Eigenschaften der lebenden Pflanzen und Tierzelle. *Vierteljahrsschr. Naturforsch. Ges. Zuerich* **40,** 159–201.

Overton, E. (1896). Ueber die osmotischen Eigenschaften der Zelle in ihrer Bedeutung für die Toxikologie und Pharmakologie. *Vierteljahrsschr. Naturforsch. Ges. Zuerich* **41,** 383–406.

Overton, E. (1897). Notizen über die Grünalgen des Oberengadins. *Ber. Schweiz. Bot. Ges.* **7,** 50–68.

Overton, E. (1899a). Notizen über die Wassergewächse des Oberengadins. *Vierteljahrsschr. Naturforsch. Ges. Zuerich* **44,** 211–228.

Overton, E. (1899b). Beobachtungen und Versuche über das Auftreten von rothem Zellsaft bei Pflanzen. *Jahrb. Wiss. Bot.* **33,** 171–231.

Overton, E. (1899c). Ueber die allgemeinen osmotischen Eigenschaften der Zelle, ihre vermutliche Ursachen und ihre Bedeutung für die Physiologie. *Vierteljahrsschr. Naturforsch. Ges. Zuerich* **44,** 88–114.

Overton, E. (1900). Studien über die Aufnahme der Anilinfarben durch die lebende Zelle. *Jahrb. Wiss. Bot.* **34,** 669–701.

Overton, E. (1901). "Studien über die Narkose, zugleich ein Beitrag zur allgemeinen Pharmakologie." Gustav Fischer, Jena, Engl. Transl.: "Studies of Narcosis" R. L. Lipnick, Wood Library-Museum and Chapman & Hall, Washington, DC, 1990.

Overton, E. (1902a). Beiträge zur allgemeinen Muskel- und Nervenphysiologie. I. Abh. Ueber die osmotischen Eigenschaften der Muskeln. *Pfluegers Arch. Gesamte Physiol. Menschen Tiere* **92,** 115–280.

Overton, E. (1902b). Beiträge zur allgemeinen Muskel- und Nervenphysiologie. II. Mitt. Ueber die Unentbehrlichkeit von Natrium- (oder Lithium-) Ionen für den Contractionsact des Muskels. *Pfluegers Arch. Gesamte Physiol. Menschen Tiere* **92,** 346–386.

Overton, E. (1903). Ueber die Unentbehrlichkeit der Natriumionen für die Tätigkeit des zentralen und peripheren Nervensystems. *Verh. Ges. Dtsch. Naturforsch. Aerzte* **75,** 416–419.

Overton, E. (1904a). Beiträge zur allgemeinen Muskel- und Nervenphysiologie. III. Mitt. Studien über die Wirkung der Alkali- und Erdalkalisalze auf Skelettmuskeln und Nerven. *Pfluegers Arch. Gesamte Physiol. Menschen Tiere* **105,** 176–290.

Overton, E. (1904b). Neun und dreissig Thesen über die Wasserökonomie der Amphibien und die osmotischen Eigenschaften der Amphibienhaut. *Verh. Physik. Med. Ges. Wuerzburg N. F.* **36,** 277–295.

Overton, E. (1907). Über den Mechanismus der Resorption und der Secretion. *In* "Handbuch der Physiologie des Menschen," (W. Nagel, ed.), Vol. 2, pp. 744–898. Fr. Vieweg u. Sohn, Braunschweig, Germany.

Overton, E. (1913). Studien über einige Wirkungen der Saponine. *Lunds Univ. Arsskrift N.F., Avd. 2,* **9,** (Pt. 7), 1–27.

Overton, E. (1918). Untersuchungen über die Resorption und die relative Stärke einiger Herzgifte. *Lunds Univ. Arsskrift N.F., Avd. 2,* **14,** (Pt. 7), 1–48.

Overton, E. (1925a). Über den Mechanismus der Aufnahme und über das Verhalten der Ester im Organismus. *Skand. Arch. Physiol.* **46,** 333–334. (Abstract).

Overton, E. (1925b). Eine reversible, resorptive Lähmung der motorischen Nervenenden beim Frosche durch gewisse indifferente Narcotika. *Skand. Arch. Physiol.* **46,** 335 (abstract).

Pfeffer, W. F. P. (1877). "Osmotische Untersuchungen." W. Engelmann, Leipzig.

Pfeffer, W. F. P. (1886). "Über Aufnahme von Anilinfarben in lebende Zellen." *Unters. a. d. Bot. Inst. Tübingen* **2,** 179–331.

Pfeffer, W. (1891). Zur Kenntniss der Plasmahaut und der Vacuolen nebst Bemerkungen über den Aggregatzustand des Protoplasmas und über osmotische Vorgänge. *Abh. Math.-Phys. Kl. Köngl. Sächs. Ges. Wiss.* **16,** 185–344.

Pfeffer, W. (1900). "The Physiology of Plants." (from 2nd German Ed. of "Pflanzenphysiologie," (Transl. by A. J. Ewart). Oxford. Univ. Press, (Clarendon), Oxford.

Pohl, J. (1890). Ueber Aufnahme und Verteilung des Chloroforms im thierischen Organismus. *Arch. Exp. Pathol. Pharmakol.* **28,** 239–255.

Quincke, G. (1888). Ueber periodische Ausbreitung an Flüssigkeitsoberflächen und dadurch hervorgerufene Bewegungserscheinungen. *Ann. Phys.* (*Leipzig*) **35,** 580–642.

Ramsden, W. (1904). Abscheidung fester Körper in den Oberflächenschichten von Lösungen und "Suspensionen." (Beobachtungen über Oberflächenhäutchen Blasen, Emulsionen und mechanische Koagulation). *Z. Phys. Chem.* **47,** 336–346.

Reid, E. W. (1890). Osmosis experiments with living and dead membranes. *J. Physiol.* (*London*) **11,** 312–351.

Reid, E. W. (1892). Preliminary report on experiments upon intestinal absorption without osmosis. *Br. Med. J.* 1133–1134.

Reid, E. W. (1900). On intestinal absorption, especially on the absorption of serum, peptone and glucose. *Philos. Trans. R. Soc. London B* **192,** 211–297.

Richet, C. (1893). Sur le raport entre la toxicité et les propriétés physiques des corps. *C. R. Soc. Biol.* (*Paris*) **54,** 775–776.

Rothenberg, M. A. (1950). Studies on permeability in relation to nerve function. II. Ionic movements across axonal membranes. *Biochim. Biophys. Acta* **41,** 96–114.

Ruhland, W. (1908). Beiträge zur Kenntniss der Permeabilität der Plasmahaut. *Jahrb. Wiss. Bot.* **46,** 1–34.

Ruhland, W. (1912). Studien über die Aufnahme von Kolloiden durch die pflan zliche Plasmahaut. *Jahrb. Wiss. Bot.* **51,** 376–431.

Ruhland, W., and Hoffmann, C. (1925). Die Permeabilität von Beggiatoa mirabilis. Ein Beitrag zur Ultrafiltrationtheorie des Plasmas. *Planta* **1,** 1–83.

Schmidt, H. R. (1891). Ueber Aufnahme und Verarbeitung von fetten Oelen durch Pflanzen. *Flora* **74,** 300–370.

Schultze, M. (1861). Ueber Muskelkörperchen, und das was man als Zelle zu nennen habe. *Arch. Anat., Physiol. Wiss. Med.,* 1–27.

Schwann, T. (1848). "Microscopical Researchers into the Accordance in the Structure and Growth by Animals and Plants" (Engl. Transl. of the 1839 text by H. Smith). Sydenham Soc., London.

Smith, H. W. (1962). The plasma membrane, with notes on the history of botany. *Circulation* **26,** 987–1012.

Traube, M. (1867). Experimente zur Theorie der Zellbildung und Endosmose. *Arch. Anat., Physiol. Wiss. Med.,* 87–165.

Vries, H. de (1871). Sur la perméabilité due protoplasma des betteraves rouges. *Arch. Neerl. Sci. Exactes Nat.* **6,** 117–126.

Vries, H. de (1884). Eine Methode zur Analyse der Turgorkraft. *Jahrb. Wiss. Bot.* **14,** 429–601.

Vries, H. de (1885). Plasmolytische Studien über die Wand der Vacuolen. *Jahrb. Wiss. Bot.* **16,** 465–598.

Vries, H. de (1888). Ueber den isotonischen Coefficient des Glycerins. *Bot. Z.* **46,** 230–235, 245–253.

Winterstein, H. (1919). "Die Narkose," J. Springer, Berlin.

CHAPTER 2

Structure and Physical Properties of the Lipid Membrane

Thomas J. McIntosh
Department of Cell Biology, Duke University Medical Center, Durham, North Carolina 27710

I. BILAYERS IN BIOLOGICAL MEMBRANES

A. Introduction

A lipid bilayer consists of two apposed lipid monolayers with the hydrophilic lipid head groups facing the surrounding fluid spaces on each side of the bilayer and the hydrophobic lipid hydrocarbon chains localized in the bilayer center. Such an arrangement of lipid molecules represents an ideal functional and structural boundary for cells. The lipid bilayer in membranes is quite narrow (only 50 Å in total thickness) and flexible, allowing cells and cell organelles to respond and structurally accommodate to chang-

Current Topics in Membranes, Volume 48

1063-5823/99 $30.00

ing situations. Because it contains a thin (about 30 Å) layer of hydrocarbon in its interior, the bilayer provides an extraordinary semipermeable barrier for cells, allowing water to cross the cell membrane readily, yet preventing the transmembrane passage of ions and large hydrophilic molecules. The ion gradients across membranes, produced by membrane-embedded ion pumps, are critical to cell functions, including the production of ATP, nerve conduction, and muscle contraction. Moreover, the bilayer is a suitable environment for a variety of membrane-bound enzymes and recognition molecules.

B. Membrane Structure

Early evidence for the presence of a lipid bilayer at the boundary of cells came from the work of Gorter and Grendel (1925), who made the critical and fundamental observation that the amount of lipid that could be extracted from red blood cells was sufficient to cover the cell surface twice. They argued that the most probable explanation for this observation was that the cell surface was covered with a double layer of lipids. Although there were mistakes in their estimations of membrane surface area, their basic observation was valid and influenced subsequent work on membrane structure tremendously.

One of the earliest and most famous structural models of biological membranes was published by Danielli and Davson (1935). The permeability and surface tension properties of lipid films led these investigators to propose a model in which the membrane contained a lipid bilayer with protein molecules arranged on the surfaces of the bilayer (actually they argued that the lipid layer was between "unimolecular and trimolecular thickness," but their much reproduced diagram depicted a lipid bilayer).

In the 1950s and 1960s electron microscopy was first used to examine the surface of cells. The observation that many different membranes had the same appearance in thin sections, two darkly staining bands about 20 Å thick surrounding a pale staining zone about 35 Å thick, led J. D. Robertson (1957, 1965) to propose that all biological membranes had a lipid bilayer at their core (the "unit membrane" concept). The idea was that the hydrophobic interior of the bilayer should not pick up much stain, whereas the hydrophilic lipid head group region with associated proteins should stain intensely. At that time there was some debate at the molecular interpretation of thin section images (Sjostrand, 1963; Sjostrand and Barajas, 1970), and some authors in the 1960s (Lucy, 1964; Sjostrand, 1963) (see reviews by Hendler, 1971; Stoeckenius and Engelman, 1969) argued that the membrane might contain globular subunits or lipid micelles. However, electron microscopic images of freeze-fracture replicas (Branton, 1966) and extensive X-ray diffraction analysis of membranes of nerve myelin (Finean and Burge, 1963; Worthington and Blaurock, 1969; Caspar and Kirschner, 1971), retinal rods (Gras and Worthington, 1969; Corless, 1972), erythrocytes (Wilkins *et al.*, 1971), and sarcoplasmic reticulum (Dupont *et al.*, 1973; Worthington and Liu, 1973; Herbette *et al.*, 1977) provided strong evidence for the presence of lipid bilayers in plasma and organelle membranes.

Later models of membrane structure, such as the "fluid mosaic" model of Singer and Nicolson (1972) and refinements of that model (Capaldi and Green, 1972; Bretscher and Raff, 1975; Israelachvili, 1977; Bloom *et al.*, 1991), have all incorporated the lipid bilayer as the structural core of the membrane. In these models the bilayer provides a fluid matrix containing globular integral membrane proteins. There has been extensive and ongoing study concerning the organization and disposition of specific proteins in biological membranes. However, for many years it has been clear that mammalian plasma and organelle membranes contain lipid bilayers that serve as a structural framework and permeability barrier.

Because of the importance of the lipid bilayer as the structural core of biological membranes, there has been sustained effort to understand the structure and physical properties of phospholipid bilayers. This chapter summarizes some of the major findings of this work.

II. PHASE BEHAVIOR OF MEMBRANE PHOSPHOLIPIDS

A. Membrane Lipid Composition

Biological membranes of eukaryotic cells contain three main types of lipids: phospholipids, glycolipids, and cholesterol. The relative amounts of these three components vary from membrane to membrane, with phospholipids being the most abundant lipid in many mammalian membranes and glycolipids and cholesterol being found in appreciable amounts in plasma membranes. The zwitterionic phospholipids phosphatidylcholine (PC) and phosphatidylethanolamine (PE) are the most common membrane phospholipids, but the charged lipids phosphatidylglycerol (PG), phosphatidylserine (PS), and phosphatidylinositol (PI) are also present, and the charged phospholipid cardiolipin is a major phospholipid in the inner mitochondrial membrane.

B. Lipid Polymorphism: Individual Lipids

Because of their presence in most all biological membranes, the structure and self-aggregation properties of membrane phospholipids have been investigated thoroughly by a number of laboratories using a variety of techniques. Pioneering X-ray diffraction studies (Schmitt and Palmer, 1940; Bear *et al.,* 1941; Palmer and Schmitt, 1941) showed that hydrated PE, PC, sphingomyelin, and various membrane lipid extracts give diffraction patterns with spacings consistent with stacks of lipid bilayers. Thus, these X-ray results in the 1940s were supportive of the contemporary Danielli–Davson lipid bilayer model (Danielli and Davson, 1935) of membrane structure. However, in the late 1960s and early 1970s systematic X-ray studies (Luzzati, 1968; Luzzati *et al.,* 1968; Rand and Luzzati, 1968; Rand and Segupta, 1972; Tardieu *et al.,* 1973; Ranck *et al.,* 1974) showed that both synthetic phospholipids and lipids isolated from membranes exhibit polymorphism. That is, depending on the lipid head group, hydrocarbon composition, temperature, pH, and water content, phospholipids can form lamellar phases (bilayers) with their hydrocarbon chains in either a partly ordered (gel) conformation or a highly disordered (liquid–crystalline) conformation or else form nonlamellar (nonbilayer) phases.

1. Lamellar (Bilayer) Phases

The gel to liquid–crystalline phase transition temperature of lipid bilayers has been measured for a variety of phospholipids and glycolipids (Ladbrooke *et al.,* 1968; Hinz and Sturtevant, 1972; Chapman *et al.,* 1974; Shipley *et al.,* 1974; Jain, 1975; Mabrey and Sturtevant, 1976; Curatolo *et al.,* 1977; Keogh and Davis, 1979; Chen *et al.,* 1980; Mason *et al.,* 1981; Hauser *et al.,* 1982; Huang and Mason, 1986; Lewis *et al.,* 1987; Lohner *et al.,* 1987; Epand, 1990; Kariel *et al.,* 1991; Huang *et al.,* 1993). The phase transition temperature and enthalpy depend on a number of factors, including lipid head group, hydrocarbon chain length, number of double bonds, and amount of cholesterol present. Some of the important general conclusions that have emerged from this work include: (1) the longer the hydrocarbon chain, the higher the phase transition temperature (Ladbrooke and Chapman, 1969; Hinz and Sturtevant, 1972; Janiak *et al.,* 1976; Keogh and Davis, 1979; Stumpel *et al.,* 1983); (2) the presence of double bonds decreases the transition temperature (Calhoun and Shipley, 1979a; Reed and Shipley, 1989); (3) the type of lipid head group influences the phase transition temperature (Ladbrooke and Chapman, 1969; Chapman *et al.,* 1974; Calhoun and Shipley, 1979a; Hauser *et al.,* 1982; Maggio *et al.,* 1985; Epand, 1990), and (4) the presence of increasing concentration of cholesterol tends to decrease the enthalpy of the phase transition (Ladbrooke *et al.,* 1968; Calhoun and Shipley, 1979b). Because of their hydrocarbon chain compositions (chain length and number of double bonds), most lipids found in biological membranes have melting temperatures well below physiological temperature, meaning that the lipids are in the liquid–crystalline phase at

37°C. However, because some membrane lipids, such as sphingomyelin and uncharged glycolipids, melt at much higher temperatures, the isolated lipids form gel or ordered phases at 37°C (Shipley *et al.,* 1974; Curatolo *et al.,* 1977; Calhoun and Shipley, 1979a; Ruocco *et al.,* 1983; Ruocco and Shipley, 1986). Also, lipids that have the capacity to hydrogen bond with neighboring lipids (such as PE and glycolipids) tend to have higher phase transition temperatures than lipids that do not hydrogen bond with neighbors (such as PC).

2. Nonlamellar Phases

Nonlamellar phases, such as hexagonal phases (hexagonally close-packed arrays of lipid tubes) and cubic phases (three-dimensional networks of connected lipids rods or globules) have been observed and characterized for several phospholipids by X-ray diffraction (Reiss-Husson, 1967; Gruner *et al.,* 1988; Lindblom and Rilfors, 1989; Gawrisch *et al.,* 1992), freeze-fracture electron microscopy (Deamer *et al.,* 1970; Lindblom and Rilfors, 1989), and nuclear magnetic resonance (NMR) (Cullis and DeKruijff, 1978a; Cullis *et al.,* 1980). The polymorphism of lipid systems has been analyzed from both a theoretical and an experimental perspective (see reviews by Mariani *et al.,* 1988; Lindblom and Rilfors, 1989; Seddon, 1990). In terms of biological membranes, the most important features of this work deal with the lipid phases observed at physiological temperature, hydration, pH, and ionic strength. In general, under physiological conditions, cholesterol forms crystalline structures (Loomis *et al.,* 1979), various membrane glycolipids form micelles, hexagonal phases, gel or liquid–crystalline bilayers (Curatolo, 1987a; Curatolo and Jungalwala, 1985; Curatolo *et al.,* 1977), PCs form liquid–crystalline bilayers (Tardieu *et al.,* 1973), and membrane PEs form either liquid–crystalline bilayers or hexagonal phases (Cullis and DeKruijff, 1978b; Seddon *et al.,* 1983, 1984; Gruner *et al.,* 1988; Marsh and Seddon, 1982; Rand *et al.,* 1990), depending on the degree of unsaturation in their hydrocarbon chains.

Differences in the phase behavior of the most common membrane phospholipids, PC and PE, were initially related to the shapes of the phospholipid molecules. Cullis and DeKruijff (1979) and Israelachvili *et al.* (1980) noted that lipids that are approximately cylindrical in shape (the excluded area of the head group region is approximately the same as the excluded area of the hydrocarbon region) tend to form bilayers, whereas molecules that are cone shaped (the area of the head group is smaller than the area of the hydrocarbon chain region) tend to form hexagonal phases. Thus, PCs and PEs with saturated hydrocarbon chains (which are approximately cylindrical in shape) form bilayers, whereas PEs with unsaturated chains (which are cone shaped because of their relatively small head group and bulky hydrocarbon chain regions) tend to form hexagonal phases. Gruner (1985, 1989) has pointed out that in addition to molecular shape, other factors, such as electrostatic interactions and hydrogen bond formation, may be involved in determining the aggregation properties of phospholipids. He explained transitions between lipid phases in terms of a "competition between the elastic energy of bending the interfaces and energies resulting from the constraints of interfacial separation" (Gruner, 1989). Both Gruner (1985) and Seddon (1990) argue that the main factor that drives the transition from a bilayer to a nonbilayer phase is the tendency of one or both monolayers of the bilayer to curl, with the curling arising from an imbalance in lateral forces between the head group and hydrocarbon regions of each lipid monolayer.

C. Lipid Polymorphism: Mixtures of Lipids

Importantly, mixtures of phospholipids, glycolipids, and cholesterol approximating the mix found in biological membranes form liquid–crystalline lamellar phases under physiological conditions (Ladbrooke *et al.,* 1968; Ladbrooke and Chapman, 1969; Lecuyer and Dervichian, 1969; Levine and Wilkins, 1971; Curatolo, 1987a). Cholesterol is a key player in ensuring that such lipid mixtures form liquid–crystalline bilayers. Cholesterol is typically found in high concentrations (>25 mole percent total lipid) in plasma membranes. Thus, even though some plasma membranes, such as nerve myelin membranes, contain a high concentration of lipids (sphingomyelin and glycolipids) that form gel phase bilayers, the presence of cholesterol keeps these membranes in a fluid, liquid–crystalline bilayer phase.

Thus, a key conclusion of these extensive studies on lipid polymorphism is that lipid mixtures with the compositions found in biological membranes form lipid bilayers, although specific isolated membrane lipid components, such as unsaturated PEs, tend to form nonlamellar phases. This tendency of specific membrane components to form nonbilayer phases has been postulated to affect several critical membrane functions, such as membrane fusion and the activity of membrane-bond proteins (Gruner, 1985; Hui, 1987).

III. STRUCTURE OF PHOSPHOLIPID BILAYERS

The organization of bilayers composed of phospholipids, mixtures of phospholipids with cholesterol, and membrane glycolipids has been analyzed by spectroscopic techniques, including NMR and electron spin resonance (reviewed by Bloom *et al.,* 1991), and by X-ray and neutron diffraction. This chapter focuses on detailed structural data that have been accumulated over the years using diffraction methods.

Early diffraction work centered on determining the critical structural characteristics of the bilayers: the width of the bilayer and the area per lipid molecule. In notable early work, Levine *et al.* (1968) and Levine and Wilkins (1971) used X-ray diffraction data to calculate one-dimensional electron density profiles across bilayers of synthetic and naturally occurring PCs, and Rand and Luzzati (1968) determined profiles for lipids extracted from erythrocyte membranes. These relatively low-resolution profiles not only proved that these lipids formed bilayers, but also gave quite accurate values for the separation of the lipid phosphate groups from apposing monolayers of the bilayer and provided information on the localization and orientation of cholesterol in phospholipid bilayers. Higher resolution electron density profiles provided additional details on the organization of hydrated or partially hydrated PC bilayers (Lesslauer *et al.,* 1972; Franks, 1976; Torbet and Wilkins, 1976; Sakurai *et al.,* 1977; Franks *et al.,* 1978; McIntosh, 1978). Such studies showed that the lipid thickness and area per lipid varied systematically as a function of the number of carbons in the lipid acyl chains (McIntosh, 1978; Lewis and Engelman, 1983; Sun *et al.,* 1996) and that the area per molecule increased with the number of double bonds in the acyl chains (Lewis and Engelman, 1983; McIntosh *et al.,* 1995a).

Three-dimensional atomic resolution structures obtained for single crystals of PC (Pearson and Pascher, 1979), PE (Hitchcock *et al.,* 1974), and cerebrosides (Pascher and Sundell, 1977), show that these molecules crystallize in a bilayer configuration with solid phase hydrocarbon chains. Such single crystal studies provide information on the intra- and intermolecular interactions within the bilayer and, for example, show the hydrogen bond network in bilayers of cerebrosides (Pascher and Sundell, 1977). Electron diffraction studies of epitaxially crystallized phospholipid multilayers have shown details of phospholipid organization, including the orientation of the lipid head groups (Dorset, 1987a,b; Dorset and Zhang, 1990). Particularly relevant to lipids in biological membranes, X-ray diffraction studies have provided details on the organization of phospholipids in hydrated, liquid–crystalline bilayers such as found at the core of biological membranes. High-resolution studies of partially hydrated liquid–crystalline PC bilayers provided the packing, distributions, and thermal motions of specific portions of the lipid molecules in the bilayer, such as the terminal methyl groups, double bonds, and head group moieties (Wiener *et al.,* 1991; Wiener and White, 1992a,b).

Another method for analyzing X-ray diffraction data, often called the gravimetric method, was developed by Luzzati (1968). In this method, the repeat period for lipid–water mixtures is determined as a function of water content and, with the assumption that the lipid and water form separate layers, the area per lipid molecule and the partial lipid thickness are calculated from the volume fraction of water and the repeat period. This method has limitations in that the lipid area cannot be calculated in the presence of excess water and the precise determination of water content is difficult (Klose *et al.,* 1988; Koenig *et al.,* 1997). Nonetheless, this method has been valuable in providing estimates of lipid areas per molecule for a variety of lipid systems, including natural and synthetic PCs (Reiss-Husson, 1967; Small, 1967; Tardieu *et al.,* 1973; Janiak *et al.,* 1979), PC/cholesterol mixtures (Lecuyer and Dervichian, 1969), PE (Reiss-Husson, 1967), sphingomyelin (Reiss-Husson, 1967), and glycolipids (Reiss-Husson, 1967). One of the most important early discoveries was that the area per phospholipid molecule in the bilayer decreased with increasing cholesterol concentration (Lecuyer and Dervichian, 1969).

Formalisms that combine X-ray and dilatometry data have allowed very precise determinations of the areas per lipid molecule and the number of waters per lipid molecule (McIntosh and Simon, 1986a; Nagle and Wiener, 1988; Nagle *et al.,* 1996). An interesting finding has been that the area per molecule for fully hydrated PE bilayers is significantly less than that of fully hydrated PC bilayers of similar hydrocarbon chain composition (McIntosh and Simon, 1986a; Nagle and Wiener, 1988; Nagle, 1993; Nagle *et al.,* 1996). That is, presumably because of intermolecular hydrogen bonding, PE molecules pack more tightly in the plane of the bilayer than PC molecules.

Neutron diffraction results have provided valuable information on the distribution of water with respect to the bilayer and also on the conformation of specific regions of the lipid molecule. Because of the large difference in coherent scattering length of deuterium (D) and hydrogen (H), D_2O–H_2O exchange can be used to localize water in hydrated bilayer preparations, and selectively deuterated lipid can locate specific atoms in the bilayer. For PC bilayers, neutron density profiles show that water enters the lipid head group region of the bilayer and penetrates up to the hydrocarbon core (Zaccai *et al.,* 1975; Worcester and Franks, 1976). The incorporation of cholesterol into the bilayer decreases the depth of water penetration into the bilayer (Simon *et al.,* 1982; Simon and McIntosh, 1986). Elegant studies with specifically deuterated PC bilayers have localized labels in both head group and hydrocarbon chain regions of the bilayer (Buldt *et al.,* 1978, 1979; Zaccai *et al.,* 1979). These studies show that in both gel and liquid–crystalline PC bilayers the head group is oriented approximately parallel to the plane of the bilayer (Buldt *et al.,* 1979) and that the glycerol backbone is oriented such that two acyl chains are out of register by about 1.8 Å (Buldt *et al.,* 1978).

An important structural feature of bilayers in biological membranes is that they are asymmetric; the lipid composition of the outer monolayer (facing the extracellular space) differs from that of the inner monolayer (facing the cytoplasm) (Bergelson and Barsukov, 1977; Rothman and Lenard, 1977; Devaux and Zachowski, 1994). For example, in red blood cell membranes, PC and SM are enriched in the outer monolayer, whereas PE and PS are enriched in the inner monolayer (Bergelson and Barsukov, 1977; Rothman and Lenard, 1977) and glycolipids are found in the outer monolayer (Steck and Dawson, 1974). As outlined in Section VI, this lipid asymmetry has functional significance in membrane–membrane interactions and in the association of cytoplasmic proteins with the plasma membrane (Bazzi *et al.,* 1992).

IV. STABILITY AND MECHANICAL PROPERTIES OF BILAYERS

Experiments comparing the mechanical properties of cell membranes and lipid bilayers have provided critical information both on membrane structure and on the effects of specific lipids on bilayer toughness. Of particular note, micropipette aspiration experiments, developed originally by Mitchison and Swann (1954), have been used by a number of investigators (LaCelle, 1970; Leblond, 1973; Waugh and Evans, 1976; Evans and Hochmuth, 1977; Evans and Skalak, 1979a; Evans and Needham, 1987; Needham and Hochmuth, 1989) to study both plasma membranes and lipid bilayers.

As described in detail by Evans and Skalak (1979b), membranes organized as described in the Singer–Nicolson fluid mosaic model (Singer and Nicolson, 1972) would behave as a two-dimensional fluid layer. However, mechanical experiments show that plasma membranes have certain properties of solids, such as elasticity. For example, the elastic shear modulus of a biological membrane is one to two orders of magnitude greater than the surface viscosity of a liquid–crystalline lipid bilayer (Evans and Skalak, 1979b). Because the liquid–crystalline bilayer is a fluid structure, these solid characteristics must be due to connections between proteins associated with the bilayer (Evans and Hochmuth, 1977; Evans and Skalak, 1979b). Membrane-associated cytoplasmic proteins, such as spectrin in red blood cells (Marchesi *et al.,* 1970) and dystrophin in skeletal muscle (Mandel, 1989), likely perform such a scaffolding role and interconnect specific intrinsic membrane proteins. Thus, membranes are now thought to contain intrinsic proteins embedded in a sea of fluid lipids with underlying cytoplasmic proteins providing structural rigidity and support (Evans and Skalak, 1979b).

Micropipette aspiration experiments with lipid vesicles have provided details on how specific lipids affect bilayer mechanical properties, such as bending rigidity, surface shear rigidity, area expansion moduli, and tensile failure (see Needham and Zhelev, 1996). Researchers (Kwok and Evans, 1981; Evans and Needham, 1987; Needham and Nunn, 1990) have determined the thermal area expansivity, elastic area compressibility, and lysis tension for vesicles composed of lipids with similar hydrocarbon chain compositions to those typically found in biological membranes, namely phosphatidylcholine derived from egg yolks (EPC) or its synthetic analog stearoyloleoylphosphatidylcholine (SOPC). It has been found that bilayers composed of mixtures of PE and PC with similar hydrocarbon chains have similar thermal area expansivities and lysis tensions, indicating that such phospholipid mixtures have similar cohesive properties to single-component bilayers (Evans and Needham, 1987). However, vesicles composed of other specific membrane lipids have quite different mechanical properties. For example, vesicles composed of lipids with highly unsaturated acyl chains have a significantly smaller area compressibility modulus, a lower bending modulus, and a smaller lysis tension (Needham and Nunn, 1990; McIntosh *et al.,* 1995a) than SOPC vesicles. However, the incorporation of cholesterol into SOPC vesicles increases the compressibility modulus and the lysis tension significantly (Needham and Nunn, 1990). That is, cholesterol increases the bilayer toughness (Needham and Nunn, 1990) while maintaining the bilayer in a liquid–crystalline phase. This important observation should be noted in relation to Overton's prescient speculations in 1899 on the significance and possible roles of cholesterol in cell membranes.

V. INTERBILAYER INTERACTIONS

The adhesive and repulsive interactions between lipid bilayers are of interest for several reasons. For instance, the close approach and adhesion of apposing lipid bilayers, a final stage in the fusion of biological membranes, is affected by the nonspecific short-range pressures between the bilayer surfaces (Parsegian and Rand, 1983; Rand and Parsegian, 1989). Other pressures, such as the electrostatic and van der Waals pressures, can be felt at long distances and influence the initial approach of cell membranes (LeNeveu *et al.,* 1977), as well as the binding of proteins to the bilayer surface.

Sophisticated methods have been developed to measure both short- and long-range interbilayer interactions. Methods include a surface force apparatus that measures interactions between bilayers coated on curved solid surfaces (Israelachvili *et al.,* 1979; Marra and Israelachvili, 1985) and an osmotic stress–X-ray diffraction method that measures pressure–distance relations between apposing bilayers in multilamellar lipid vesicles (LeNeveu *et al.,* 1977; Parsegian *et al.,* 1979; McIntosh and Simon, 1986b). Although each method has its strengths and weaknesses (see reviews by Rand and Parsegian, 1989; Leikin *et al.,* 1993; McIntosh and Simon, 1994a; Israelachvili and Wennerstrom, 1996), the two methods are complimentary and similar pressure–distance data have been obtained with the two methods (Horn *et al.,* 1988).

Several critical features of interbilayer interactions have been elucidated by the mid-1990s. For bilayers containing charged phospholipids, such as phosphatidylglycerol or phosphatidylserine, there is a long-range repulsive pressure that depends on the amount of charged lipid in the bilayer and the salt concentration in the aqueous phase (Cowley *et al.,* 1978; Marra, 1986; Israelachvili and Wennerstrom, 1996). This pressure can be explained by electrostatic repulsion and double-layer theory.

For uncharged lipid bilayers there is strong short-range repulsion that is not dependent on ionic strength and cannot be explained by electrostatic theory (Cowley *et al.,* 1978; LeNeveu *et al.,* 1977; Parsegian *et al.,* 1979; Parsegian and Rand, 1983; Rand and Parsegian, 1989). Many theoretical (Marcelja and Radic, 1976; Schiby and Ruckenstein, 1983; Gruen and Marcelja, 1983; Dzhavakhidze *et al.,* 1986; Kornyshev and Leikin, 1989; Leikin and Kornyshev, 1990, 1991; Israelachvili and Wennerstrom, 1992; Kornyshev *et al.,* 1992; Lipowsky and Grotehans, 1993) and experimental studies (LeNeveu *et al.,* 1977; Parsegian *et al.,* 1979, 1991; Marra and Israelachvili, 1985; McIntosh and Simon, 1986b, 1993; Israelachvili and Marra, 1986; McIntosh *et al.,* 1987, 1989, 1995a; Rand *et al.,* 1988; Rand and Parsegian, 1989) have been undertaken to determine the physical origin of these interactions. Two general types of repulsive interactions are thought to exist between neutral bilayers: hydration pressure and entropic (steric) pressures. Hydration pressure is due to the partial polarization of water molecules by the hydrophilic bilayer surface (Marcelja and Radic, 1976; Schiby and Ruckenstein, 1983; Gruen and Marcelja, 1983; Kornyshev and Leikin, 1989; Leikin and Kornyshev, 1991). Entropic pressures have been shown or postulated to arise from several sources, including the direct interaction between head groups from apposing bilayers (McIntosh *et al.,* 1987, 1989), thermally driven undulations or fluctuations of the entire membrane surface (Helfrich, 1973; Helfrich and Servuss, 1984; Evans and Parsegian, 1986), and protrusions of individual lipid molecule into the fluid space between bilayers (Israelachvili and Wennerstrom, 1990, 1992).

Several reviews have been presented on these hydration and entropic interactions (Rand and Parsegian, 1989; Cevc, 1993; McIntosh and Simon,

1994a, 1996a; Parsegian and Rand, 1995; Israelachvili and Wennerstrom, 1996). Experimental studies (McIntosh and Simon, 1993; McIntosh *et al.,* 1995a) have indicated that the hydration pressure extends about 3 or 4 Å from the membrane surface. Other key experimental findings are that the range and magnitude of the total repulsive force between bilayers depend critically on both the type of lipid head group and the composition of the lipid acyl chains. Lipids with large, bulky head groups, such as glycolipids (Kuhl *et al.,* 1994; McIntosh and Simon, 1994b) or lipids with covalently attached hydrophilic polymer chains (Kenworthy *et al.,* 1995; Needham *et al.,* 1992), have strong interbilayer pressures due to the steric interactions between the bulky head groups. In terms of acyl chains, PCs with polyunsaturated chains have much longer-ranged repulsive pressures than PCs with saturated chains or chains containing one double bond (McIntosh *et al.,* 1995a). In the case of PCs with polyunsaturated chains, analysis of observed pressure–distance data (McIntosh *et al.,* 1995a) indicate that the range of the repulsive force can be explained by theoretical treatments of entropic interactions due to membrane undulations (Evans, 1991). Another interesting point is that lipids that can hydrogen bond to each other, such as PE (McIntosh and Simon, 1996b) and the glycolipid lactosyl ceramide (Yu *et al.,* 1998), have extremely large adhesion energies that prevent apposing bilayers from swelling apart or disjoining. This is thought to be due to hydrogen bond formation between apposing bilayers (McIntosh and Simon, 1996b; Yu *et al.,* 1998).

VI. ROLES OF SPECIFIC LIPIDS IN MEMBRANE BILAYERS

As noted earlier, two fundamental roles of lipid bilayers in biological membranes are that they form the basic structural matrix of membranes and serve as a permeability barrier for many water-soluble molecules. These basic roles of bilayers could be accomplished by bilayers formed from almost any membrane phospholipid. Therefore, a fundamental question of membrane biology has been the reason for the presence of such a vast variety of membrane lipids, including cholesterol and many different classes of phospholipids and glycolipids. The roles of specific lipids in biological membranes has been the focus of numerous studies, and a number of functional roles for specific classes of membranes have been demonstrated or postulated (see reviews by Curatolo, 1987b; Spector and Yorek, 1985; Merrill and Schroeder, 1993); Bienvenue and Marie, 1994; Dowhan, 1997). This section summarizes some of the key functional roles of specific membrane lipids.

Cholesterol, which is found in high concentrations (greater than 25 mol%) in many plasma membranes, but in low concentrations (less than 5%) in mitochondrial or endoplasmic reticulum membranes, is now known to have an important role in controlling the order (fluidity) of membrane lipids. As noted earlier, cholesterol has a condensing effect on liquid–crystalline bilayers, reducing the area per lipid molecule in the plane of the membrane. Also, the presence of cholesterol reduces the magnitude of the gel to liquid–crystalline phase transition temperature of phospholipids (Calhoun and Shipley, 1979b; Ladbrooke *et al.*, 1968). Thus, plasma membranes with appreciable quantities of lipids (such as glycolipids and sphingomyelin) that melt at temperatures above physiological temperature are kept in a fluid (liquid–crystalline phase) by the presence of cholesterol. In addition, cholesterol helps increase the mechanical toughness of lipid bilayers (Needham and Nunn, 1990).

Cardiolipins, phospholipids found in the inner mitochondrial membrane, are thought to have a role in proton conduction in the respiratory chain (Hoch, 1992). Conditions that alter membrane cardiolipins have been shown to slow mitochondrial respiration. Moreover, the integrity of the inner mitochondrial membrane depends specifically on cardiolipins (Hoch, 1992).

Phosphatidylinositol (PI), a phospholipid found in relatively low concentrations in membranes, has been shown to anchor specific proteins to the membrane bilayer (Fergusun *et al.*, 1985; Low and Kincaid, 1985; Low and Saltiel, 1988; Romero, 1988; Selvaraj *et al.*, 1988; Deckert *et al.*, 1996). Several functionally diverse proteins are thought to be bound to the membrane by covalent attachment to a glycosylated phosphatidylinositol (GPI) moiety. Although most of these proteins are bound to the cell surface, evidence also shows that cytoplasmic proteins may be bound to membranes through this mechanism (Low and Saltiel, 1988). One structural feature of cell surface GPI-anchored proteins is that they tend to cluster into domains or caveolae in the cell plasma membrane. GPI-anchored proteins have been shown to be involved in several important cellular functions, including signal transduction, adhesion, phagocytosis, and transport (Selvaraj *et al.*, 1988; Deckert *et al.*, 1996).

Membrane glycolipids play a number of important roles in membranes, such as cell adhesion, cell–cell recognition, interactions with the extracellular matrix, and modulation of transmembrane signaling (see reviews by Curatolo, 1987b; Hannun and Bell, 1989; Zeller and Marchase, 1992; Jungalwala, 1994; Maggio, 1994). For example, gangliosides, glycosphingolipids containing sialic acid found in the plasma membrane of all vertebrate tissues, are thought to be involved in cell adhesion and tissue development as well as influencing signal transduction processes through interactions with membrane proteins (Zeller and Marchase, 1992). Because most of these functions involve interactions of the cell with its surroundings, it is not surprising to find that glycolipids occur primarily in the outer monolayer of the cell plasma membrane.

Evidence has also accumulated concerning the role of lipids in the functioning of membrane proteins (see reviews by Spector and Yorek, 1985; Merrill and Schroeder, 1993; Bienvenue and Marie, 1994). Membrane lipids are thought to modulate the function of a variety of proteins, including receptors involved in cell signaling, enzymes, and extrinsic proteins. For example, an increase in the unsaturated fatty acid content of membrane phospholipids modifies the binding properties of the insulin receptor and other membrane receptors (Spector and Yorek, 1985). A number of membrane protein activities are modified by changes in membrane cholesterol content, phospholipid head group type, or degree of phospholipid hydrocarbon chain unsaturation (Moore *et al.,* 1981; Poon *et al.,* 1981; Baldwin and Hubbell, 1985; Spector and Yorek, 1985; Rudy *et al.,* 1989; Yeagle, 1989; Bienvenue and Marie, 1994; Brown, 1994; Litman and Mitchell, 1996). Specific lipids have also been shown to be critical to the interaction of cytoplasmic proteins with the plasma membrane. For example, the interaction of protein kinase C in signal transduction depends on the presence of phosphatidylserine and phosphatidylethanolamine in the inner monolayer of the cell membrane (Bazzi and Nelsestuen, 1991; Bazzi *et al.,* 1992). It has been argued that this is one reason that these lipids are located preferentially on the inner monolayer (Bazzi *et al.,* 1992).

Another potential role for PE and other lipids that tend to form nonbilayer phases (see earlier) is the promotion of membrane fusion. The addition of PE has been shown to enhance fusion between liposomes (Hui *et al.,* 1981; Bentz *et al.,* 1985; Ellens *et al.,* 1986; Conner and Huang, 1987; Allen *et al.,* 1990), presumably because of its propensity to form nonbilayer phases (Hui *et al.,* 1981; Ellens *et al.,* 1986; Conner and Huang, 1987; Allen *et al.,* 1990; Seddon, 1990), which could destabilize the bilayer at the point of fusion, and because of its large adhesion energy (Evans and Needham, 1987; Marra and Israelachvili, 1985; Rand and Parsegian, 1989; Seddon, 1990; McIntosh and Simon, 1996b), which would tend to hold PE-containing bilayer close together.

Finally, and quite importantly, several membrane phospholipids and glycolipids are sources of second messengers critical to cellular signaling (see reviews by Berridge and Irvine, 1984, 1989; Nishizuka, 1984, 1986; Majerus et al. 1986; Hannun and Bell, 1989; Merrill and Jones, 1990; Merrill, 1991; Haeffner, 1993; Ghosh et al., 1997). The first membrane lipid to be discovered as a source of second messengers was PI (Nishizuka, 1984, 1986). In this signal transduction process, stimulation of membrane-bound receptors initiates the hydrolysis of PI, leading to the production of two second messengers, diacylglycerol and inositol 1,4,5-trisphosphate (Berridge and Irvine, 1984, 1989). Diacylglycerol then stimulates protein kinase C, whereas inositol 1,4,5-trisphosphate acts by releasing calcium from internal stores (Nishizuka, 1984). Membrane sphingolipids, such as SM and various glycolipids, also are the source of second messengers, such as sphingosine and lysoshingolipids, that act by inhibiting protein kinase C (Hannun and Bell, 1989; Merrill, 1991). More recently it has been shown that the most common membrane lipid, phosphatidylcholine, can also be the source of second messengers (Mene *et al.,* 1989; Rogers and Hammerman, 1989; Asaoka *et al.,* 1992; Azzi *et al.,* 1992; Haeffner, 1993; Liscovitch *et al.,* 1993; Zeisel, 1993; Li *et al.,* 1995).

VII. SUMMARY

Since the prescient observations of Overton more than a century ago, an enormous amount of information has been obtained on the structure, properties, and functional roles of the lipid layer at the boundary of cells. The lipid bilayer forms a thin, flexible, semipermeable barrier that compartmentalizes all cells and many cell organelles. Moreover, it is now known that the bilayer is actively involved in a variety of cell functions and that its lipid components are important sources of second messengers used in cell signaling.

Acknowledgment

This work was supported by NIH Grant GM27278.

References

Allen, T. M., Hong, K., and Papahadjopoulos, D. (1990). Membrane contact, fusion, and hexagonal phase transitions in PE liposomes. *Biochemistry* **29,** 2976–2985.

Asaoka, Y., Oka, M., Yoshida, K., Sasaki, Y., and Nishizuka, Y. (1992). Role of lysophatidylcholine in T-lymphocyte activation: Involvement of phospholipase A2 in signal transduction through protein kinase C. *Proc. Natl. Acad. Sci. U.S.A.* **89,** 6447–6451.

Azzi, A., Boscoboinik, D., and Hensey, C. (1992). The protein kinase C family. *Eur. J. Biochem.* **208,** 547–557.

Baldwin, P. A., and Hubbell, W. L. (1985). Effects of lipid environment on the light-induced conformational changes of rhodopsin. *Biochemistry* **24,** 2633–2639.

Bazzi, M. D., and Nelsestuen, G. L. (1991). Extensive segregation of acidic phospholipids in membranes induced by protein Kinase C and related proteins. *Biochemistry* **30,** 7961–7969.

Bazzi, M. D., Youakim, M. A., and Nelsestuen, G. L. (1992). Importance of phosphatidylethanolamine for association of protein kinase C and other cytoplasmic proteins with membranes. *Biochemistry* **31,** 1125–1144.

Bear, R. S., Palmer, K. J., and Schmitt, F. O. (1941). X-ray diffraction studies of nerve lipides. *J. Cell. Comp. Physiol.* **17,** 355–368.

Bentz, J., Ellens, H., Lai, M.-Z., and Szoka, F. C. (1985). On the correlation between HII phase and the contact-induced destabilization of phosphatidylethanolamine-containing membranes. *Proc. Natl. Acad. Sci. U.S.A.* **82,** 5742–5745.

Bergelson, L. D., and Barsukov, L. I. (1977). Topological asymmetry of phospholipids in membranes. *Science* **197,** 224–230.

Berridge, M. J., and Irvine, R. F. (1984). *Nature (London)* **312,** 315–321.

Berridge, M. J., and Irvine, R. F. (1989). Inositol phosphates and cell signalling. *Nature (London)* **341,** 197–205.

Bienvenue, A., and Marie, J. S. (1994). Modulation of protein function by lipids. *Curr. Top. Membr.* **40,** 319–354.

Bloom, M., Evans, E., and Mouritsen, O. G. (1991). Physical properties of the fluid-bilayer component of cell membranes: A perspective. *Q. Rev. Biophys.* **24,** 293–397.

Branton, D. (1966). Fracture faces of frozen membranes. *Proc. Natl. Acad. Sci. U.S.A.* **55,** 1048.

Bretscher, M. S., and Raff, M. C. (1975). Mammalian plasma membranes. *Nature (London)* **258,** 43–49.

Brown, M. F. (1994). Modulation of rhodopsin function by properties of the membrane bilayer. *Chem. Phys. Lipids* **73,** 159–180.

Buldt, G., Gally, H. U., Seelig, A., Seelig, J., and Zaccai, G. (1978). Neutron diffraction studies on selectively deuterated phospholipid bilayers. *Nature (London)* **271,** 182–184.

Buldt, G., Gally, H. U., Seelig, J., and Zaccai, G. (1979). Neutron diffraction studies on phosphatidylcholine model membranes. I. Head group conformation. *J. Mol. Biol.* **134,** 673–691.

Calhoun, W. I., and Shipley, G. G. (1979a). Fatty acid composition and thermal behavior of natural sphingomyelins. *Biochim. Biophys. Acta* **555,** 436–441.

Calhoun, W. J., and Shipley, G. G. (1979b). Sphingomyelin-lecithin bilayers and interaction with cholesterol. *Biochemistry* **18,** 1717–1722.

Capaldi, R. A., and Green, D. E. (1972). Membrane proteins and membrane structure. *FEBS Lett.* **25,** 205–209.

Caspar, D. L. D., and Kirschner, D. A. (1971). Myelin membrane structure at 10 Å resolution. *Nature (London)* **231,** 46–52.

Cevc, G. (1993). Electrostatic characterization of liposomes. *Chem. Phys. Lipids* **64,** 163–186.

Chapman, D., Urbina, J., and Keogh, K. M. (1974). Biomembrane phase transitions. Studies of lipid-water systems using differential scanning calorimetry. *J. Biol. Chem.* **249,** 2512–2521.

Chen, S. C., Sturtevant, J. M., and Gaffney, B. J. (1980). Scanning calorimetric evidence for a third phase transition in PC bilayers. *Proc. Natl. Acad. Sci. U.S.A.* **77,** 5060–5063.

Conner, J., and Huang, L. (1987). Acid-induced fusion of liposomes. *In* "Cell Fusion." Plenum, New York.

Corless, J. M. (1972). Lamellar structure of bleached and unbleached rod photoreceptor membranes. *Nature (London)* **237,** 229–231.

Cowley, A. C., Fuller, N. L., Rand, R. P., and Parsegian, V. A. (1978). Measurement of repulsive forces between charged phospholipid bilayers. *Biochemistry* **17,** 3163–3168.

Cullis, P. R., and DeKruijff, B. (1978a). Polymorphic phase behaviour of lipid mixtures as detected by 31P-NMR. Evidence that cholesterol may destabilize bilayer structure in membrane systems containing PE. *Biochim. Biophys. Acta* **507,** 207–218.

Cullis, P. R., and DeKruijff, B. (1978b). The polymorphic phase behavior of PEs of natural and synthetic origin. *Biochim. Biophys. Acta* **513,** 31–42.

Cullis, P. R., and DeKruijff, B. (1979). Lipid polymorphism and the functional roles of lipids in biological membranes. *Biochim. Biophys. Acta* **559,** 399–420.

Cullis, P. R., DeKruijff, B., Hope, M. J., Nayar, R., Rietveld, A., and Verkleij, A. J. (1980). Structural properties of phospholipids in the rat liver inner mitochondrial membrane. A ^{31}P-NMR study. *Biochim. Biophys. Acta* **600,** 625–635.

Curatolo, W. (1987a). The physical properties of glycolipids. *Biochim. Biophys. Acta* **906,** 111–136.

Curatolo, W. (1987b). Glycolipid function. *Biochim. Biophys. Acta* **906,** 137–160.

Curatolo, W., and Jungalwala, F. B. (1985). Phase behavior of galactocerebrosides from bovine brain. *Biochemistry* **24,** 6608–6613.

Curatolo, W., Small, D. M., and Shipley, G. G. (1977). Phase behavior and structural characteristics of hydrated bovine brain gangliosides. *Biochim. Biophys. Acta* **468,** 11–20.

Danielli, J. F., and Davson, J. (1935). A contribution to the theory of permeability of thin films. *J. Cell. Comp. Physiol.* **5,** 495–508.

Deamer, D. W., Leonard, R., Tardieu, A., and Branton, D. (1970). Lamellar and hexagonal lipid phases visualized by freeze-etching. *Biochim. Biophys. Acta* **219,** 47–60.

Deckert, M., Ticchioni, M., and Bernard, A. (1996). Endocytosis of GPI-anchored proteins in human lymphocytes: Role of glycolipid-based domains, actin cytoskeleton, and protein kinases. *J. Cell Biol.* **133,** 791–799.

Devaux, P. F., and Zachowski, A. (1994). Maintenance and consequences of membrane phospholipid asymmetry. *Chem. Phys. Lipids* **73,** 107–120.

Dorset, D. L. (1987a). Electron diffraction structure analysis of phospholipids. *J. Electron Microsc. Tech.* **7,** 35–46.

Dorset, D. L. (1987b). Molecular packing of a crystalline ether-linked phosphatidylcholine an electron diffraction study. *Biochim. Biophys. Acta* **898,** 121–128.

Dorset, D. L., and Zhang, W. (1990). Lamellar packing of a chiral N,N-dimethylphosphatidylethanolamine: Electron diffraction. Evidence for a lecithin-type headgroup conformation. *Biochim. Biophys. Acta* **1028,** 299–303.

Dowhan, W. (1997). Molecular basis for membrane phospholipid diversity: Why are there so many lipids? *Annu. Rev. Biochem.* **66,** 199–232.

Dupont, Y., Harrison, S. C., and Hasselbach, W. (1973). Molecular organization in the sarcoplasmic reticulum membrane studied by X-ray diffraction. *Nature (London)* **244,** 555–558.

Dzhavakhidze, P. G., Kornyshev, A. A., and Levadny, V. G. (1986). The role of the interface in the nonlocal electrostatic theory of hydration force. *Phys. Lett. A* **118,** 203–208.

Ellens, H., Bentz, J., and Szoka, F. C. (1986). Fusion of PE-containing liposomes and mechanism of the L_{α}-H_{II} phase transition. *Biochemistry* **25,** 4141–4147.

Ellens, H., Siegel, D. P., Alford, D., Yeagle, P. L., Boni, L., Lis, L. J., Quinn, P. J. and Bentz, J. (1989). Membrane fusion and inverted phases. *Biochemistry* **28,** 3692–3703.

Epand, R. M. (1990). Hydrogen bonding and the thermotropic transitions of phosphatidylethanolamines. *Chem. Phys. Lipids* **52,** 227–230.

Evans, E. (1991). Entropy-driven tension in vesicle membranes and unbinding of adherent vesicles. *Langmuir* **7,** 1900–1908.

Evans, E. A., and Hochmuth, R. M. (1977). A solid–liquid composite model of the red cell membranes. *J. Membr. Biol.* **30,** 351–379.

Evans, E., and Needham, D. (1987). Physical properties of surfactant bilayer membranes: Thermal transitions, elasticity, rigidity, cohesion, and colloidal interactions. *J. Phys. Chem.* **91,** 4219–4228.

Evans, E. A., and Parsegian, V. A. (1986). Thermal–mechanical fluctuations enhance repulsion between bimolecular layers. *Proc. Natl. Acad. Sci. U.S.A.* **83,** 7132–7136.

Evans, E. A., and Skalak, R. (1979a). Mechanics and thermodynamics of biomembranes: Part 2. *In* "CRC Critical Reviews in Bioengineering." CRC Press, Boca Raton, Florida.

Evans, E. A., and Skalak, R. (1979b). Mechanics and Thermodynamics of Biomembranes: Part 1. *In* "CRC Critical Reviews in Bioengineering." CRC Press, Boca Raton, Florida.

Fergusun, M. A. J., Low, M. G., and Cross, G. A. M. (1985). Glycosyl-sn-1,2-dimyristoylphosphatidylinositol is covalently linked to Trypanosome brucei variant surface glycoprotein. *J. Biol. Chem.* **260,** 14547–14555.

Finean, J. B., and Burge, R. E. (1963). The determination of the Fourier transform of the myelin layer from a study of the swelling phenomena. *J. Mol. Biol.* **7,** 672.

Franks, N. P. (1976). Structural analysis of hydrated egg lecithin and cholesterol bilayers I. X-ray diffraction. *J. Mol. Biol.* **100,** 345–358.

Franks, N. P., Arunachalam, T., and Caspi, E. (1978). A direct method for determination of membrane electron density profiles on an absolute scale. *Nature (London)* **276,** 530–532.

Gawrisch, K., Parsegian, V. A., Hajduk, D. A., Tate, M. W., Gruner, S. M., Fuller, N. L., and Rand, R. P. (1992). Energetics of a hexagonal-lamellar-hexagonal-phase transition sequence in DOPE membranes. *Biochemistry* **31,** 2856–2864.

Ghosh, S., Strum, J. C., and Bell, R. M. (1997). Lipid biochemistry: Functions of glycerolipids and sphingolipids in cellular signaling. *FASEB J.* **11,** 45–50.

Gorter, E., and Grendel, F. (1925). *J. Exp. Med.* **41,** 439.

Gras, W. J., and Worthington, C. R. (1969). X-ray analysis of retinal photoreceptors. *Proc. Natl. Acad. Sci. U.S.A.* **63,** 233.

Gruen, D. W. R., and Marcelja, S. (1983). Spatially varying polarization in water. *J. Chem. Soc. Faraday Trans. 2* **79,** 225–242.

Gruner, S. M. (1985). Intrinsic curvature hypothesis for biomembrane lipid composition: A role for non-bilayer lipids. *Proc. Natl. Acad. Sci. U.S.A.* **82,** 3665–3669.

Gruner, S. M. (1989). Stability of lyotropic phases with curved interfaces. *J. Phys. Chem.* **93,** 7562–7570.

Gruner, S. M., Tate, M. W., Kirk, G. L., So, P. T. C., Turner, D. C., Keane, D. T., Tilcock, C. P. S., and Cullis, P. R. (1988). X-ray diffraction study of the polymorphic behavior of N-methylated DOPE. *Biochemistry* **27,** 2853–2866.

Haeffner, E. W. (1993). Diacylglycerol: Formation and function in phospholipid-mediated signal transduction. *Comp. Biochem. Physiol. C* **105,** 337–345.

Hannun, Y. A., and Bell, R. M. (1989). Functions of sphingolipids and sphingolipid breakdown products in cellular regulation. *Science* **243,** 500–507.

Hauser, H., Paltauf, F., and Shipley, G. G. (1982). Structure and thermotropic behavior of PS bilayer membranes. *Biochemistry* **21,** 1061–1067.

Helfrich, W. (1973). Elastic properties of lipid bilayers: Theory and possible experiments. *Z. Naturforsch* **28C,** 693–703.

Helfrich, W., and Servuss, R.-M. (1984). Undulations, steric interactions and cohesion of fluid membranes. *Nuovo Cimento Soc. Ital. Fis. A* **3,** 137–151.

Hendler, R. W. (1971). Biological membrane ultrastructure. *Physiol. Rev.* **51,** 66–97.

Herbette, L., Marquardt, J., Scarpa, A., and Blasie, J. K. (1977). A direct analysis of lamellar x-ray diffraction from hydrated oriented multilayers of fully functional sarcoplasmic reticulum. *Biophys. J.* **20,** 245–272.

Hinz, H.-J., and Sturtevant, J. M. (1972). Calorimetric studies of dilute aqueous suspensions of bilayers formed from synthetic L-α-lecithin. *J. Biol. Chem.* **247,** 6071.

Hitchcock, P. B., Mason, R., Thomas, K. M., and Shipley, G. G. (1974). Structural chemistry of 1,2-dilauroyl-DL-phosphatidylethanolamine: molecular conformation and intermolecular packing of phospholipids. *Proc. Natl. Acad. Sci. U.S.A.* **71,** 3036–3040.

Hoch, F. L. (1992). Cardiolipins and biomembrane function. *Biochim. Biophys. Acta* **1113,** 71–133.

Horn, R. G., Israelachvili, J. N., Marra, J., Parsegian, V. A., and Rand, R. P. (1988). Comparison of forces measured between phosphatidylcholine bilayers. *Biophys. J.* **54,** 1185–1186.

Huang, C.-H., and Mason, J. T. (1986). Structure and properties of mixed-chain phospholipid assemblies. *Biochim. Biophys. Acta* **864,** 423–470.

Huang, C., Wang, Z.-Q., Lin, H.-N., and Brumbaugh, E. E. (1993). Calorimetric studies of fully hydrated phosphatidylcholines with highly asymmetric acyl chains. *Biochim. Biophys. Acta* **1145,** 298–310.

Hui, S. W. (1987). Non-bilayer-forming lipids: Why are they necessary in biomembranes. *Comments Mol. Cell. Biophys.* **4,** 233–248.

Hui, S. W., Stewart, T. P., Boni, L. T., and Yeagle, P. L. (1981). Membrane fusion through point defects in bilayers. *Science* **212,** 921–922.

Israelachvili, J. N. (1977). Refinement of the fluid-mosaic model of membrane structure. *Biochim. Biophys. Acta* **469,** 221–225.

Israelachvili, J., and Marra, J. (1986). Direct methods for measuring conformational forces (hydration forces) between membrane and other surfaces. *In* "Methods in Enzymology" (L. Packer, ed.), Vol. 127, pp. 353–361. Academic Press, San Diego.

Israelachvili, J. N., and Wennerstrom, H. (1990). Hydration or steric forces between amphiphilic surfaces? *Langmuir* **6,** 873–876.

Israelachvili, J. N., and Wennerstrom, H. (1992). Entropic forces between amphiphilic surfaces in liquids. *J. Phys. Chem.* **96,** 520–531.

Israelachvili, J., and Wennerstrom, H. (1996). Role of hydration and water structure in biological and colloidal interactions. *Nature (London)* **379,** 219–225.

Israelachvili, J. N., Tandon, R. K., and White, L. R. (1979). Measurement of forces between two surfaces in aqueous poly(ethylene oxide) solutions. *Nature (London)* **277,** 120–121.

Israelachvili, J. N., Marcelja, S., and Horn, R. G. (1980). Physical principles of membrane organization. *Q. Rev. Biophys.* **13,** 121–200.

Jain, M. K. (1975). Role of cholesterol in biomembranes and related systems. *In* "Current Topics in Membranes and Ion Transport." Academic Press, New York.

Janiak, M. J., Small, D. M., and Shipley, G. G. (1976). Nature of the thermal pretransition of synthetic phospholipids: Dimyristoyl- and dipalmitoyllecithin. *Biochemistry* **15,** 4575–4580.

Janiak, M. J., Small, D. M., and Shipley, G. G. (1979). Temperature and compositional dependence of the structure of dimyristoyl lecithin. *J. Biol. Chem.* **254,** 6068–6078.

Jungalwala, F. B. (1994). Expression and biological functions of sulfoglucuronyl glycolipids (SGGLs) in the nervous system—a review. *Neurochem. Res.* **19,** 945–957.

Kariel, N., Davidson, E., and Keough, K. M. W. (1991). Cholesterol does not remove the gel-liquid crystalline phase transition of phosphatidylcholines containing two polyenoic acyl chains. *Biochim. Biophys. Acta* **1062,** 70–76.

Kenworthy, A. K., Hristova, K., Needham, D., and McIntosh, T. J. (1995). Range and magnitude of the steric pressure between bilayers containing phospholipids with covalently attached poly(ethyleneglycol). *Biophys. J.* **68,** 1921–1936.

Keogh, K. M. W., and Davis, P. J. (1979). Gel to liquid-crystalline phase transitions in water dispersions of saturated mixed-acid phosphatidylcholines. *Biochemistry* **18,** 1453–1458.

Klose, G., Konig, B., Meyer, H. W., Schulze, G., and Degovics, G. (1988). Small-angle x-ray scattering and electron microscopy of crude dispersions of swelling lipids and the influence of the morphology on the repeat distance. *Chem. Phys. Lipids* **47,** 225–234.

Koenig, B. W., Strey, H. H., and Gawrisch, K. (1997). Membrane lateral compressibility determined by NMR and X-ray diffraction: Effect of acyl chain polyunsaturation. *Biophys. J.* **73,** 1954–1966.

Kornyshev, A. A., and Leikin, S. (1989). Fluctuation theory of hydration forces: The dramatic effect of inhomogeneous boundary conditions. *Phys. Rev. A* **40,** 6431–6437.

Kornyshev, A. A., Kossakowski, D. A., and Leikin, S. (1992). Surface phase transitions and hydration forces. *J. Chem. Phys.* **97,** 6809–6820.

Kuhl, T. L., Leckband, D. E., Lasic, D. D., and Israelachvili, J. N. (1994). Modulation of interaction forces between bilayers exposing short-chained ethylene oxide headgroups. *Biophys. J.* **66,** 1479–1488.

Kwok, R., and Evans, E. (1981). Thermoelasticity of large lecithin vesicles. *Biophys. J.* **35,** 637–652.

LaCelle, P. L. (1970). Alteration of membrane deformability in hemolytic anemias. *Semin. Hematol.* **7,** 355–373.

Ladbrooke, B. D., and Chapman, D. (1969). Thermal analysis of lipids, proteins, and biological membranes. A review and summary of some recent results. *Chem. Phys. Lipids* **3,** 304–367.

Ladbrooke, B. D., Williams, R. M., and Chapman, D. (1968). Studies on lecithin-cholesterol-water interactions by differential scanning calorimetry and x-ray diffraction. *Biochim. Biophys. Acta* **150,** 333–340.

Leblond, P. (1973). The discocyte–echinocyte transformation of the human red cell: Deformability characteristics. *In* "Red Cell Shape" (N. Bressis, R. I. Weed, and P. Leblond, eds.), pp. 95–129. Springer-Verlag, Berlin.

Lecuyer, H., and Dervichian, D. G. (1969). Structure of aqueous mixtures of lecithin and cholesterol. *J. Mol. Biol.* **45,** 39–57.

Leikin, S., and Kornyshev, A. A. (1990). Theory of hydration forces. Nonlocal electrostatic interaction of neutral surfaces. *J. Chem. Phys.* **92,** 6890–6898.

Leikin, S., and Kornyshev, A. A. (1991). Mean-field theory of dehydration transitions. *Phys. Rev. A* **44,** 1156–1168.

Leikin, S., Parsegian, V. A., and Rau, D. C. (1993). Hydration forces. *Annu. Rev. Phys. Chem.* **44,** 369–395.

LeNeveu, D. M., Rand, R. P., Parsegian, V. A., and Gingell, D. (1977). Measurement and modification of forces between lecithin bilayers. *Biophys. J.* **18,** 209–230.

Lesslauer, W., Cain, J. E., and Blasie, J. K. (1972). X-ray diffraction studies of lecithin bimolecular leaflets with incorporated fluorescent probes. *Proc. Natl. Acad. Sci. U.S.A.* **69,** 1499–1503.

Levine, Y. K., and Wilkins, M. H. F. (1971). Structure of oriented lipid bilayers. *Nature (London) New Biol.* **230,** 69–72.

Levine, Y. K., Bailey, A. I., and Wilkins, M. H. F. (1968). Multilayers of phospholipid bimolecular leaflets. *Nature (London)* **220,** 577–578.

Lewis, B. A., and Engelman, D. M. (1983). Lipid bilayer thickness varies linearly with acyl chain length in fluid phosphatidylcholine vesicles. *J. Mol. Biol.* **166,** 211–217.

Lewis, R. N. A., Mak, N., and McElhaney, R. N. (1987). A differential scanning calorimetry study of the thermotropic phase behavior of model membranes composed of phosphatidylcholine containing linear saturated fatty acyl chains. *Biochemistry* **26,** 6118–6126.

Li, W., Liu, X., and Yanoff, M. (1995). Phosphatidylcholine hydrolysis and DNA synthesis in cultured retinal capillary pericytes. *Microvasc. Res.* **49,** 350–363.

Lindblom, G., and Rilfors, L. (1989). Cubic phases and isotropic structures formed by membrane lipids—possible biological relevance. *Biochim. Biophys. Acta* **988,** 221–256.

Lipowsky, R., and Grotehans, S. (1993). Hydration vs. protrusion forces between lipid bilayers. *Europhys. Lett.* **23,** 599–604.

Liscovitch, M., Ben-Av, P., Danin, M., Faiman, G., Eldar, H., and Livneh, E. (1993). Phospholipase D-mediated hydrolysis of phosphatidylcholine role in cell signalling. *J. Lipid Mediators* **8,** 177–182.

Litman, B. J., and Mitchell, D. C. (1996). A role for phospholipid polyunsaturation in modulating membrane protein function. *Lipids* **31,** S193–S197.

Lohner, K., Schuster, A., Degovics, G., Muller, K., and Laggner, P. (1987). Thermal phase behavior and structure of hydrated mixtures between DPPC and DHPC. *Chem. Phys. Lipids* **44,** 61–70.

Loomis, C. R., Shipley, G. G., and Small, D. M. (1979). The phase behavior of hydrated cholesterol. *J. Lipid Res.* **20,** 525–535.

Low, M. G., and Kincaid, P. W. (1985). Phosphatidylinositol is the membrane-anchoring domain of the Thy-1 glycoprotein. *Nature* (*London*) **318,** 62–64.

Low, M. G., and Saltiel, A. R. (1988). Structural and functional roles of glycosyl-phosphatidylinositol in membranes. *Science* **239,** 268–275.

Lucy, J. A. (1964). Globular lipid micelles and cell membranes. *J. Theor. Biol.* **7,** 360–373.

Luzzati, V. (1968). X-ray diffraction studies of lipid-water systems. *In* "Biological Membranes." Academic Press, New York.

Luzzati, V., Gulik-Krzywicki, T., and Tardieu, A. (1968). Polymorphism of lecithins. *Nature* (*London*) **218,** 1031–1034.

Mabrey, S., and Sturtevant, J. M. (1976). Investigation of phase transitions of lipids and lipid mixtures by sensitivity differential scanning calorimetry. *Proc. Natl. Acad. Sci. U.S.A.* **73,** 3862–3866.

McIntosh, T. J. (1978). The effect of cholesterol on the structure of phosphatidylcholine bilayers. *Biochim. Biophys. Acta* **513,** 43–58.

McIntosh, T. J., Magid, A. D., and Simon, S. A. (1989). Cholesterol modifies the short-range repulsive interactions between phosphatidylcholine membranes. *Biochemistry* **28,** 17–25.

McIntosh, T. J., and Simon, S. A. (1986a). Area per molecule and distribution of water in fully hydrated dilauroylphosphatidylethanolamine bilayers. *Biochemistry* **25,** 4948–4952.

McIntosh, T. J., and Simon, S. A. (1986b). The hydration force and bilayer deformation: A reevaluation. *Biochemistry* **25,** 4058–4066.

McIntosh, T. J., and Simon, S. A. (1993). Contribution of hydration and steric (entropic) pressures to the interaction between phosphatidylcholine bilayers: Experiments with the subgel phase. *Biochemistry* **32,** 8374–8384.

McIntosh, T. J., and Simon, S. A. (1994a). Hydration and steric pressures between phospholipid bilayers. *Annu. Rev. Biophys. Biomol. Struct.* **23,** 27–51.

McIntosh, T. J., and Simon, S. A. (1994b). Long- and short-range interactions between phospholipid/ganglioside GM1 bilayers. *Biochemistry* **33,** 10477–10486.

McIntosh, T. J., and Simon, S. A. (1996a). Short-range pressures between lipid bilayer membranes. *Colloids Surf. A: Physicochem. Eng. Aspects* **116,** 251–268.

McIntosh, T. J., and Simon, S. A. (1996b). Adhesion between phosphatidylethanolamine bilayers. *Langmuir* **12,** 1622–1630.

McIntosh, T. J., Magid, A. D., and Simon, S. A. (1987). Steric repulsion between phosphatidylcholine bilayers. *Biochemistry* **26,** 7325–7332.

McIntosh, T. J., Advani, S., Burton, R. E., Zhelev, D. V., Needham, D., and Simon, S. A. (1995a). Experimental tests for protrusion and undulation pressures in phospholipid bilayers. *Biochemistry* **34,** 8520–8532.

McIntosh, T. J., Kenworthy, A. K., and Needham, D. (1995b). Measurement of the range and magnitude of the repulsive pressure between PEG-coated liposomes. *In* "Stealth Liposomes." CRC Press, Boca Raton, Florida.

Maggio, B. (1994). The surface behavior of glycosphingolipids in biomembranes: A new frontier of molecular biology. *Prog. Biophys. Mol. Biol.* **62,** 55–117.

Maggio, B., Ariga, T., Sturtevant, J. M., and Yu, R. K. (1985). Thermotropic behavior of glycosphingolipids in aqueous dispersions. *Biochemistry* **24,** 1084–1092.

Majerus, P. W., Connolly, T. M., Deckmyn, H., Ross, T. S., Bross, T. E., Ishii, H., Bansal, V. S., and Wilson, D. B. (1986). The metabolism of phosphoinositide-derived messenger molecules. *Science* **234,** 1519–1526.

Mandel, J. L. (1989). Dystrophin—the gene and its product. *Nature* (*London*) **339,** 584–586.

Marcelja, S., and Radic, N. (1976). Repulsion of interfaces due to boundary water. *Chem. Phys. Lett.* **42,** 129–130.

Marchesi, S. L., Steers, E., Marchesi, V. T., and Tillack, T. W. (1970). Physical and chemical properties of a protein isolated from red cell membranes. *Biochemistry* **9,** 50–63.

Mariani, P., Luzzati, V., and Delacroix, H. (1988). Cubic phases of lipid-containing systems. Structure analysis and biological implications. *J. Mol. Biol.* **204,** 165–189.

Marra, J. (1986). Direct measurement of the interaction between phosphatidylglycerol bilayers in aqueous electrolyte solutions. *Biophys. J.* **50,** 815–825.

Marra, J., and Israelachvili, J. (1985). Direct measurements of forces between phosphatidylcholine and phosphatidylethanolamine bilayers in aqueous electrolyte solutions. *Biochemistry* **24,** 4608–4618.

Marsh, D., and Seddon, J. M. (1982). Gel-to-inverted hexagonal phase transitions in PEs and fatty acid-PC mixtures demonstrated by ^{31}P-NMR spectroscopy and x-ray diffraction. *Biochim. Biophys. Acta* **690,** 117–123.

Mason, J. T., Huang, C.-H., and Biltonen, R. L. (1981). Calorimetric investigations of saturated mixed-chain phosphatidylcholine bilayer dispersions. *Biochemistry* **20,** 6086–6092.

Mene, P., Simonson, M. S., and Dunn, M. J. (1989). Phospholipids in signal transduction of mesangial cells. *Am. J. Physiol.* **256,** F375–F386.

Merrill, A. H. (1991). Cell regulation by sphingosine and more complex sphingolipids. *J. Bioenerg. Biomembr.* **23,** 83–104.

Merrill, A. H., and Jones, D. D. (1990). An update of the enzymology and regulation of sphingomyelin metabolism. *Biochim. Biophys. Acta* **1044,** 1–12.

Merrill, A. H., and Schroeder, J. J. (1993). Lipid modulation of cell function. *Annu. Rev. Nutr.* **13,** 539–559.

Mitchison, J. M., and Swann, M. M. (1954). The mechanical properties of the cell surface. I. The cell elastimeter. *J. Exp. Biol.* **31,** 443–460.

Moore, B. M., Lentz, B. R., Hoechli, M., and Meissner, G. (1981). Effect of lipid membrane structure on the adenosine 5′-triphosphate hydrolyzing activity of the calcium-stimulate adenosine triphosphatase of sacroplasmic reticulum. *Biochemistry* **20,** 6810–6817.

Nagle, J. F. (1993). Area/lipid of bilayers from NMR. *Biophys. J.* **64,** 1476–1481.

Nagle, J. F., and Wiener, M. C. (1988). Structure of fully hydrated bilayer dispersions. *Biochim. Biophys. Acta* **942,** 1–10.

Nagle, J. F., Zhang, R., Tristram-Nagle, S., Sun, W., Petrache, H. I., and Suter, R. M. (1996). X-ray structure determination of fully hydrated Lα phase dipalmitoylphosphatidylcholine bilayers. *Biophys. J.* **70,** 1419–1431.

Needham, D., and Hochmuth, R. M. (1989). Electro-mechanical permeabilization of lipid vesicles. Role of membrane tension and compressibility. *Biophys. J.* **55,** 1001–1009.

Needham, D., and Nunn, R. S. (1990). Elastic deformation and failure of lipid bilayer membranes containing cholesterol. *Biophys. J.* **58,** 997–1009.

Needham, D., and Zhelev, D. V. (1996). The mechanochemistry of lipid vesicles examined by micropipet manipulation techniques. *In* "Vesicles" (M. Rosoff, ed.), pp. 373–439. Dekker, New York.

Needham, D., McIntosh, T. J. and Lasic, D. D. (1992). Repulsive interactions and mechanical stability of polymer-grafted lipid membranes. *Biochim. Biophys. Acta* **1108,** 40–48.

Nishizuka, Y. (1984). The role of protein kinase C in cell surface signal transduction and tumour production. *Nature* (*London*) **308,** 693–697.

Nishizuka, Y. (1986). Structure and properties of protein kinase C. *Science* **233,** 305–333.

Palmer, K. J., and Schmitt, F. O. (1941). X-ray diffraction studies of lipide emulsions. *J. Cell. Comp. Physiol.* **17,** 385–394.

Parsegian, V. A., and Rand, R. P. (1983). Membrane interaction and deformability. *Ann. N.Y. Acad. Sci.* **416,** 1–12.

Parsegian, V. A., and Rand, R. P. (1995). Interaction in membrane assemblies. *In* "Structure of and Dynamics of Membranes." Elsevier Science, Amsterdam.

Parsegian, V. A., Fuller, N., and Rand, R. P. (1979). Measured work of deformation and repulsion of lecithin bilayers. *Proc. Natl. Acad. Sci. U.S.A.* **76,** 2750–2754.

Parsegian, V. A., Rand, R. P., and Fuller, N. L. (1991). Direct osmotic stress measurements of hydration and electrostatic double layer forces between bilayers of double-chained ammonium acetate surfactants. *J. Phys. Chem.* **95,** 4777–4782.

Pascher, I., and Sundell, S. (1977). Molecular arrangements in sphingolipids. The crystal structure of cerebroside. *Chem. Phys. Lipids* **20,** 175–191.

Pearson, R. H., and Pascher, I. (1979). The molecular structure of lecithin dihydrate. *Nature (London)* **281,** 499–501.

Poon, R., Richards, J. M., and Clark, W. R. (1981). The relationship between plasma membrane lipid composition and physical properties. II. Effect of phospholipid fatty acid modulation on plasma membrane physical properties and enzymatic activities. *Biochim. Biophys. Acta* **649,** 58–66.

Ranck, J. L., Mateu, L., Sadler, D. M., Tardieu, A., Gulik-Krzywicki, T., and Luzzati, V. (1974). Order–disorder conformational transitions of the hydrocarbon chains of lipids. *J. Mol. Biol.* **85,** 249–277.

Rand, R. P. and Luzzati, V. (1968). X-ray diffraction study in water of lipids extracted from human erythrocytes. The position of cholesterol in the lipid lamellae. *Biophys. J.* **8,** 125–137.

Rand, R. P., and Parsegian, V. A. (1989). Hydration forces between phospholipid bilayers. *Biochim. Biophys. Acta* **988,** 351–376.

Rand, R. P. and Segupta, S. (1972). Cardiolipin forms hexagonal structures with divalent cations. *Biochim. Biophys. Acta* **255,** 484–492.

Rand, R. P., Fuller, N., Parsegian, V. A., and Rau, D. C. (1988). Variation in hydration forces between neutral phospholipid bilayers: Evidence for hydration attraction. *Biochemistry* **27,** 7711–7722.

Rand, R. P., Fuller, N. L., Gruner, S. M., and Parsegian, V. A. (1990). Membrane curvature, lipid segregation, and structural transitions for phospholipids under dual-solvent stress. *Biochemistry* **29,** 76–87.

Reed, R. A., and Shipley, G. G. (1989). Effect of chain unsaturation on the structure and thermotropic properties of galactocerebrosides. *Biophys. J.* **55,** 281–292.

Reiss-Husson, F. (1967). Structure des phases liquide-cristallines de differents phospholipides, monoglycerides, sphingolipides, anydres ou en presence d'eau. *J. Mol. Biol.* **25,** 363–382.

Robertson, J. D. (1957). The cell membrane concept. *J. Physiol. (London)* **140,** 58–59.

Robertson, J. D. (1965). Current problems of unit membrane structure and substructure. *In* "Intracellular Membranous Structure." Japan Society for Cell Biology, Okayama, Japan.

Rogers, S. A., and Hammerman, M. R. (1989). Growth hormone activates phospholipase C in proximal tubular basolateral membranes from canine kidney. *Proc. Natl. Acad. Sci. U.S.A.* **86,** 6363–6369.

Romero, G. (1988). Phosphatidylinositol-glycan anchors of membrane proteins: Potential precursors of insulin mediators. *Science* **240,** 509–511.

Rothman, J. E., and Lenard, J. (1977). Membrane asymmetry. *Science* **195,** 743–753.

Rudy, B., Dubois, H., Mink, R., Trommer, W. E., McIntyre, J. O., and Fleischer, S. (1989). Coenzyme binding by 3-hydroxybutyrate dehydogenase, a lipid-requiring enzyme: Lecithin acts as an allosteric modulator to enhance the affinity for coenzyme. *Biochemistry* **28,** 5354–5366.

Ruocco, M. J. and Shipley, G. G. (1986). Thermal and structural behavior of natural cerebroside 3-sulfate in bilayer membranes. *Biochim. Biophys. Acta* **859,** 246–256.

Ruocco, M. J., Shipley, G. G., and Oldfield, E. (1983). Galactocerebroside–phospholipid interaction in bilayer membranes. *Biophys. J.* **43,** 91–101.

Sakurai, I., Iwayanagi, S., Sakurai, T., and Seto, T. (1977). X-ray study on egg-yolk lecithin: Unit cell data and electron density profile. *J. Mol. Biol.* **117,** 285–291.

Schiby, D., and Ruckenstein, E. (1983). The role of the polarization layers in hydration forces. *Chem. Phys. Lett.* **95,** 435–438.

Schmitt, F. O., and Palmer, K. J. (1940). X-ray diffraction studies of lipide and lipide-protein systems. *Cold Spring Harbor Symp. Quant. Biol.* **8,** 94–101.

Seddon, J. M. (1990). Structure of the inverted hexagonal phase, and non-lamellar phase transitions of lipids. *Biochim. Biophys. Acta* **1031,** 1–69.

Seddon, J. M., Cevc, G., and Marsh, D. (1983). Calorimetric studies of the the gel-fluid and lamellar-inverted hexagonal phase transition in dialkyl- and diacylPEs. *Biochemistry* **22,** 1280–1289.

Seddon, J. M., Cevc, G., Kaye, R. D., and Marsh, D. (1984). X-ray diffraction study of the polymorphism of hydrated diacyl- and dialkylphosphatidylethanolamines. *Biochemistry* **23,** 2634–2644.

Selvaraj, P., Rosse, W. F., Silber, R., and Springer, T. A. (1988). The major Fc receptor in blood has a phosphatidylinositoyl anchor and is deficient in paroxysmal nocturnal haemoglobinuria. *Nature* (*London*) **333,** 565–567.

Shipley, G. G., Avecilla, L. S., and Small, D. M. (1974). Phase behavior and structure of aqueous dispersions of sphingomyelin. *J. Lipid Res.* **15,** 124–131.

Simon, S. A., and McIntosh, T. J. (1986). The depth of water penetration into lipid bilayers. *In* "Methods in Enzymology" (L. Packer, ed.), Vol. 127, pp. 511–521 Academic Press, San Diego.

Simon, S. A., McIntosh, T. J., and Latorre, R. (1982). Influence of cholesterol on water penetration into bilayers. *Science* **216,** 65–67.

Singer, S. J., and Nicolson, G. L. (1972). The fluid mosaic model of the structure of cell membranes. *Science* **175,** 720–731.

Sjostrand, F. S. (1963). A new repeat structural element of mitochondrial and certain cytoplasmic membranes. *Nature* (*London*) **199,** 1262–1271.

Sjostrand, F. S., and Barajas, L. (1970). A new model for mitochondrial membranes based on structural and on biochemical evidence. *J. Ultrastruct. Res.* **32,** 293–306.

Small, D. M. (1967). Phase equilibria and structure of dry and hydrated egg lecithin. *J. Lipid Res.* **8,** 551–557.

Spector, A. A., and Yorek, M. A. (1985). Membrane lipid composition and cellular function. *J. Lipid Res.* **26,** 1015–1035.

Steck, T. L., and Dawson, G. (1974). Topographical distribution of complex carbohydrates in the erythrocyte membrane. *J. Biol. Chem.* **249,** 2135–2146.

Stoeckenius, W., and Engelman, D. M. (1969). Current models for the structure of biological membranes. *J. Cell Biol.* **42,** 613–646.

Stumpel, J., Eibl, H., and Nicksch, A. (1983). X-ray analysis and calorimetry on phosphatidylcholine model membranes. The influence of length and position of acyl chains upon structure and phase behavior. *Biochim. Biophys. Acta* **727,** 246–254.

Sun, W.-J., Tristram-Nagle, S., Suter, R. M., and Nagle, J. F. (1996). Structure of gel phase saturated lecithin bilayers: Temperature and chain length dependence. *Biophys. J.* **71,** 885–891.

Tardieu, A., Luzzati, V., and Reman, F. C. (1973). Structure and polymorphism of the hydrocarbon chains of lipids: A study of lecithin-water phases. *J. Mol. Biol.* **75,** 711–733.

Torbet, J., and Wilkins, M. H. F. (1976). X-ray diffraction studies of lecithin bilayers. *J. Mol. Biol.* **62,** 447–458.

Waugh, R., and Evans, E. (1976). Viscoelastic properties of erythrocyte membranes of different vertebrate animals. *Microvasc. Res.* **12,** 291–321.

Wiener, M. C., and White, S. H. (1992a). Structure of a fluid dioleoylphosphatidylcholine bilayer determined by joint refinement of x-ray and neutron diffraction data. II. Distribution and packing of terminal methyl groups. *Biophys. J.* **61,** 428–433.

Wiener, M. C., and White, S. H. (1992b). Structure of a fluid dioleoylphosphatidylcholine bilayer determined by joint refinement of x-ray and neutron diffraction data. III. Complete structure. *Biophys. J.* **61,** 434–447.

Wiener, M. C., King, G. I., and White, S. H. (1991). Structure of a fluid dioleoylphosphatidylcholine bilayer determined by joint refinement of x-ray and neutron diffraction data. I. Scaling of neutron data and the distributions of double bonds and water. *Biophys. J.* **60,** 568–576.

Wilkins, M. H. F., Blaurock, A. E., and Engelman, D. M. (1971). Bilayer structure in membranes. *Nat. New Biol.* **230,** 72–76.

Worcester, D. L., and Franks, N. P. (1976). Structural analysis of hydrated egg lecithin and cholesterol bilayers II. Neutron diffraction. *J. Mol. Biol.* **100,** 359–378.

Worthington, C. R., and Blaurock, A. E. (1969). A structural analysis of nerve myelin. *Biophys. J.* **9,** 970.

Worthington, C. R., and Liu, S. C. (1973). Structure of sarcoplasmic reticulum membranes at low resolution (17 Å). *Arch. Biochem. Biophys.* **157,** 573–579.

Yeagle, P. L. (1989). Lipid regulation of cell membrane structure and function. *FASEB J.* **3,** 1833–1842.

Yu, Z. W., Calvert, T. L., and Leckband, D. (1998). Molecular forces between membranes displaying neutral glycosphingolipids: Evidence for carbohydrate attraction. *Biochemistry* **37,** 1540–1550.

Zaccai, G., Blasie, J. K., and Schoenborn, B. P. (1975). Neutron diffraction studies on the location of water in lecithin bilayer model membranes. *Proc. Natl. Acad. Sci. U.S.A.* **72,** 376–380.

Zaccai, G., Buldt, G., Seelig, A., and Seelig, J. (1979). Neutron diffraction studies on phosphatidylcholine model membranes II. Chain conformation and segmental disorder. *J. Mol. Biol.* **134,** 693–706.

Zeisel, S. H. (1993). Choline phospholipids: Signal transduction and carcinogenesis. *FASEB J.* **7,** 551–557.

Zeller, C. B., and Marchase, R. B. (1992). Gangliosides as modulators of cell function. *Am. J. Physiol.* **262,** C1341–C1355.

CHAPTER 3

Insights from Computer Simulations into the Interaction of Small Molecules with Lipid Bilayers

Andrew Pohorille*, Michael H. New*, Karl Schweighofer†, and Michael A. Wilson†

*Exobiology Branch, NASA Ames Research Center, Moffett Field, California 94035;
†Department of Pharmaceutical Chemistry, University of California, San Francisco, San Francisco, California 94143

I. INTRODUCTION

It is quite unusual in biology that ideas put forward a century ago still remain subjects of active research. Ernest Overton's hypotheses about

1063-5823/99 $30.00

the distribution of small solute molecules in biological membranes and their permeation across these membranes are two notable exceptions. In 1899 Overton found that the permeabilities of membranes to different solutes correlate with the water/octanol partition coefficients of the solutes. At approximately the same time, Overton and Meyer proposed independently that the solubilities in a lipid phase of solutes that induce anesthethesia correlate with their anesthetic potencies (Overton, 1901; Meyer, 1899).

Both correlations had far-reaching implications arising from the biological importance of the processes to which they are related. Although transport of many nutrients, such as sugars and amino acids, across cell membranes is mediated actively by specialized proteins, passive permeation remains the most common mechanism by which small solutes cross cell membranes. Water, oxygen, carbon dioxide, nitrous oxide, and many toxins are examples of molecules transported by this simple mechanism. Furthermore, the unassisted permeation of drugs across cell membranes is the dominant mechanism of drug delivery. The Overton rule implies that the rates by which all these compounds cross membranes can be predicted readily from simple measurements of partition coefficients between water and a bulk, nonpolar liquid.

The Meyer–Overton relationship for anesthetics has probably been even more influential. The correlation between the potencies of anesthetic compounds and their solubilities in olive oil, considered a model of the membrane interior, was found to be remarkably accurate for all conventional anesthetics and for different animals (Koblin, 1990). This naturally led to the hypothesis that anesthetics act inside the lipid bilayer of neuronal tissue by disrupting its structure and making it permeable to the ions that mediate the transmission of nervous signals (Meyer, 1937). Several different mechanisms for this effect have been postulated (Janoff and Miller, 1982).

In recent years, a growing body of evidence has accumulated that cannot be reconciled with the original ideas of Overton. In particular, it was found that the relationship between permeation and partition coefficients embodied in the Overton rule does not hold well for small solutes (Lieb and Stein, 1969; Walter and Gutknecht, 1986). Instead, the permeability of the membrane to these solutes is inversely proportional to their size. In the field of anesthesia, an increasing number of experiments indicate that anesthetics exert their action by interacting with membrane receptors rather than by any nonspecific disruption of the lipid bilayer (Franks and Lieb, 1994). Perhaps the most spectacular evidence of the interaction of anesthetics with receptors has been provided by Mihic and Harris (1997), who showed that single point mutations in glycine and GABA receptors abolished their response to anesthetics, without affecting their gating capabilities. Further, several compounds have been discovered that exert no anesthetic action, even though they were predicted to be good anesthetics by the Meyer–Overton relationship (Koblin *et al.,* 1994; Liu *et al.,* 1994). Several other anesthetic compounds have been found to have potencies that differ markedly from that expected from their lipid-phase solubilities (Koblin *et al.,* 1994).

In view of these findings, the relationships established by Overton need to be reconsidered. It is natural to concentrate on the properties that were unknown to Overton and, consequently, were not accounted for in his hypothesis: the detailed, molecular structure of lipid bilayers, the surrounding water, and membrane proteins. Computer simulations are a particularly convenient approach to this task as they can reveal levels of molecular detail often unavailable from experiments. This chapter does not attempt to review all the studies on this subject. Several papers cover this territory very well (Pohorille and Wilson, 1993; Stouch, 1997; Stouch and Bossalino, 1996; Marrink and Berendsen, 1996). Instead, this chapter shows, through selected examples, how computer simulations enrich the picture of solute behavior in membranes originally drawn by the work of Overton.

From the theoretical point of view, the permeation of small, nonelectrolyte solutes across membranes and the distribution of anesthetics in the lipid bilayer are different, but closely related, phenomena. Because anesthetics equilibrate rapidly across different cellular compartments, their distributions are determined by the thermodynamic equilibrium properties of the system. If the free energy of the solute, $\Delta A(z)$, along the coordinate z perpendicular to the membrane is calculated, the concentration distribution, $c(z)$, relative to the concentration in a bulk phase (*e.g.,* water or the gas phase), c_b, can be obtained as

$$\frac{c(z)}{c_b} = e^{-\Delta A(z)/k_B T}, \tag{1}$$

where T is the temperature and k_B is the Boltzmann constant.

Although in most practical approaches to the calculation of membrane permeability to solute molecules a knowledge of $\Delta A(z)$ is also necessary, it alone is insufficient. A description of the local, diffusive behavior of the solute in different regions of the membrane must be available as well. Thus, both the equilibrium and the dynamic behavior of the system must be known.

The next section presents a brief description of the methods of molecular simulations used to study the interactions of solutes with membranes. This is followed by a discussion of the distribution of anesthetics in the lipid bilayer and the mechanisms by which nonelectrolytes permeate membranes. The chapter ends with a summary.

II. METHODS OF SIMULATION

A. Molecular Dynamics Method

The method of choice for computational studies of membrane systems at a molecular level has been the molecular dynamics method (Roux and Karplus, 1994; Allen and Tildesley, 1987). Computationally efficient, this method can be adapted easily to account for any experimentally determined constraints on molecular conformation and is at the heart of a number of more advanced simulation methodologies, some of which are described next. The main idea behind the molecular dynamics method is simply the solution of Newton's equations of motion for all of the atoms in the system under study. In this way, a time history of the system, a trajectory, can be computed. Many thermodynamic and dynamic properties of the system can be computed from the trajectory using the methods of classical statistical mechanics.

For simulations of small molecules in membranes, the system consists of the molecules of the solute, a bilayer and solvating water molecules—all represented at the atomic level. The system as a whole is placed in a box—the primary simulation cell—whose dimensions in the x,y plane parallel to the interface are of the order of 40×40 Å and which contains approximately 5×10^3 atoms. To obtain results for a macroscopic system from simulations of such a box, the contents of the primary cell are replicated periodically in space, forming an infinite lattice of identical simulation cells. In this way, as a molecule exits the primary cell from one side, its periodic image enters the primary cell from the opposite side. This is a standard approach to handling edge effects in molecular dynamics.

B. Potential Energy Functions

In order to solve Newton's equations of motion for an atom, the forces acting on that atom, F, must be known. These forces depend, in turn, on the interactions between all the atoms in the system and are computed from a potential energy function, U, which describes the stretching of chemical bonds, the bending of bond angles, the rotation of dihedral angles, and the electrostatic and van der Waals interactions between atoms:

$$U = \sum_{bonds} K_b(r - r_{eq})^2 + \sum_{angles} K_a\,(\phi - \phi_{eq})^2 + \sum_{torsions} \frac{V_n}{2}[1 + cos(n\theta - \theta_{eq})] + \sum_i \sum_{j>i} \left[\frac{A_{ij}}{R_{ij}^{12}} - \frac{B_{ij}}{R_{ij}^{6}} + \frac{q_i q_i}{R_{ij}}\right]. \quad (2)$$

Here r, ϕ, and θ are bond lengths, planar bond angles, and dihedral angles, respectively, r_{eq}, ϕ_{eq}, and θ_{eq} are their values at equilibrium, R_{ij} is the distance between atoms i and j, and q_i is the partial charge on atom i. K_b, K_a, V_n, A_{ij}, and B_{ij} are atom-type specific parameters. For polarizable, atomic solutes, such as Xe, an additional term depending on the atomic polarizabilities must be added (Guillot *et al.,* 1991; Berry *et al.,* 1980). By adopting this functional form it is assumed that existing chemical bonds cannot be broken or new ones made. Much work has been invested in the construction of potential energy functions of this type and they have been used successfully to reproduce a wide range of properties of pure membranes (Bassolino-Klimas *et al.,* 1995; Tu *et al.,* 1996; Tobias *et al.,* 1997; Merz, 1997; Tieleman *et al.,* 1997; Wilson and Pohorille, 1994) and their mixtures with small solutes and peptides (Bassolino-Klimas *et al.,* 1995; Marrink and Berendsen, 1996; Pohorille and Wilson, 1996; Damodaran *et al.,* 1995; Shen *et al.,* 1997).

In simulations of the anesthetic membrane–water system, water was represented by the TIP4P potential (Jorgensen *et al.,* 1983). To describe membranes made of glycerol 1-monooleate (GMO), potential energy functions developed by Wilson and Pohorille (1994) were employed while simulations of phospholipids were performed using potential energy functions developed by Berkowitz and co-workers (Essmann *et al.,* 1995a). The potential energy functions for the anesthetics were designed such that they reproduced the solvation energies of these molecules in water and a nonpolar phase (Pohorille and Wilson, 1996; Chipot *et al.,* 1997; Pohorille *et al.,* 1996, 1997, 1998). Other calculations discussed in this chapter utilized similar potential energy functions.

The accurate computation of the electrostatic interactions between groups carrying large charges, such as phospholipid head groups or the side chains of ionizable amino acids, poses a challenge to the simulator due to the long distances over which these interactions are nonnegligible. For the relatively small system sizes and periodic boundary conditions employed in computer simulations, this long interaction range means that charged or polar groups interact significantly with their periodic images. To compute correctly the electrostatic interaction energy, and hence the force, a method was borrowed from studies of the physics of ionic crystals: the Ewald sum. In this method, interactions between all the periodic images of the atoms are summed correctly. As implemented directly, however, this method is exceedingly computationally demanding. Advances, however, have reduced its computational cost through the use of a spatial grid. In this new method, called particle mesh Ewald (PME) (Essmann *et al.,* 1995b), the Ewald sum is evaluated through the solution of a differential equation on a grid, using the efficient fast Fourier transform method. In simulations of phospholipids, PME was employed to compute long-range, electrostatic interactions.

C. Molecular Dynamics for Membrane Systems

Once a potential energy function is specified, the equations of motion in the molecular dynamics method are solved iteratively for all the atoms in the primary cell. Using the current positions of the atoms, the forces are computed. The forces and the current velocities are then used to compute the new positions and velocities of the atoms after a small increment of time. This increment is determined by the fastest motions in the system—generally the stretching of chemical bonds between heavy atoms and hydrogen atoms—and is usually limited to a few femtoseconds. In order to compute thermodynamic averages with sufficient accuracy, trajectories of several nanoseconds must be computed. Thus the state of the system must be advanced on the order of 10^8 steps. Increases in computer performance and simulation methodology, however, have rendered simulations of this length, and longer, quite feasible.

In standard molecular dynamics, the energy of the system is approximately conserved while the temperature fluctuates. Experiments, however, are most often performed under conditions of constant temperature, not constant energy. Moreover, during simulations of molecules inserted into membranes, the size and shape of the membrane, and therefore the dimensions of the primary simulation cell, must be allowed to change. When a solute is placed into a membrane, lipids surrounding the solute are compressed relative to their state in the pure membrane. For the membrane to relax to its equilibrium state, the surface area, and perhaps the thickness, of the membrane changes. Even when the membrane has relaxed, its surface area continues to fluctuate around the equilibrium value. To account for these fluctuations, the standard method needs to be modified. In extended Lagrangian methods (Martyna *et al.,* 1994, 1996), equations of motion for the atoms are derived that couple the motions of the atoms to fluctuations in the total energy and shape and size of the primary cell. These equations of motion are no longer Newton's equations and special techniques are needed to solve them. An alternative is to combine standard molecular dynamics with a probabilistic sampling of total energies and cell geometries. This approach is called hybrid Monte Carlo and has been used in the simulation of polymers (Forrest and Suter, 1994) and small, water-soluble proteins (Brass *et al.,* 1993) under conditions of constant temperature. Hybrid Monte Carlo procedures have been developed for studying membrane systems under conditions of constant temperature, with constant pressure normal to the membrane and constant membrane surface tension (New and Pohorille, 1998). Studies of distributions of anesthetics and nonanesthetics in model bilayers at clinically relevant concentrations were performed using hybrid Monte Carlo and are discussed in Section III.

A wide range of properties may be computed directly from a molecular dynamics simulation trajectory (Allen and Tildesley, 1987). Thermodynamic properties, such as kinetic temperature, pressure, or membrane surface tension, may be computed as simple averages over the series of configurations composing the trajectory. Structural quantities, such as nuclear magnetic resonance (NMR) order parameters of the lipid chains, may also be computed directly. Even transport coefficients, such as the lateral diffusion constant of lipids, may be computed, either as the time integral of an equilibrium time correlation function or as the long-time limit of a mean-square variation. In particular, the local diffusion coefficient, $D(z)$, in the direction perpendicular to the plane of the membrane can be calculated from the fluctuations of the force, $F(z,t)$, acting in the z direction on a solute constrained to lie at depth z in the bilayer:

$$D(z) = (k_B T)^2 / \int_0^\infty < F(z,t)F(z,0) > dt, \tag{3}$$

where $<\cdot>$ denotes statistical averaging.

D. Free Energy Calculations

Often, the main goal of computer simulations is to determine the free energy of a solute molecule as a function of its position in the membrane. Because free energies cannot be expressed as statistical averages of mechanical properties, special techniques for their evaluation are required.

One common approach to the computation of free energies is called umbrella, or importance, sampling (Chipot *et al.*, 1997; Allen and Tildesley, 1987). In this method, a series of molecular dynamics simulations is performed, in which the system is constrained to lie in several, overlapping ranges (“windows”) along a predefined coordinate. For example, to compute the free energy change accompanying the movement of a small organic molecule across an interface, the position of the molecule in a direction perpendicular to the interface would be defined as such coordinate. For each window, the probability, P, of finding the solute at different values of the chosen coordinates will be obtained. This probability is connected uniquely with the change of the free energy, ΔA, by the relation:

$$\Delta A = -k_B T \log P. \quad (4)$$

ΔA over all windows is obtained from the requirement that it must be a continuous function of the chosen coordinate.

A limiting case of umbrella sampling is to reduce the size of the window to a single point. This leads to the average force method (Marrink and Berendsen, 1994) in which the derivative of the free energy is computed directly from

$$\frac{d\Delta A(z_0)}{dz} = -N_{Av} < F_z(z_0) >, \quad (5)$$

where $< F_z(z_0) >$ is the average force, in the z direction, exerted on the molecule when it is constrained to lie at z_0 and N_{Av} is the number of samples used in the averaging. By fixing z at a set of z_0's, the free energy change over the whole range may be reconstructed by the integration of Eq. (5). The force, F_z, can fluctuate greatly, however, thus requiring long simulations for each value z_0.

Another, very powerful, method for estimating the free energies of solute molecules is the particle insertion method (Widom, 1963, 1982; Pohorille and Wilson, 1996). In contrast with the previous approach, only the system *without* the solute needs to be simulated. Then, advantage is taken of the relationship between the free energy of a solute molecule added to the system, ΔA, and the corresponding change in the potential energy, δU, of this system:

$$\Delta A = -k_B T \log < \exp(-\delta U / k_B T) > . \quad (6)$$

Returning to the example of computing the free energy change accompanying the movement of a small organic molecule across an interface, the particle insertion method starts from a molecular dynamics trajectory of the interface without the solute. For each configuration appearing in the trajectory, a number of attempted insertions of the ligand into the configuration are made and the change in potential energy accompanying each insertion is recorded. For a sufficiently large set of configurations and attempted insertions, the free energy can be computed accurately from Eq. (6). The particle insertion method has been applied successfully to calculations of the free energies of methane, fluoromethanes, and noble gases in water–membrane systems (Pohorille and Wilson, 1996, 1998; Schweighofer and Pohorille, 1998).

A major advantage of the particle insertion method is that it may yield free energies of several different solutes at many different locations (insertion sites) from a single simulation of the system in the absence of these solutes. This method, however, also has a major limitation. Because insertions that result in the overlap between the added particle and another atom yield very large values of δU, they add little to the estimate of ΔA. Thus the particle insertion method is efficient only for solutes sufficiently small that they can fit into the voids that form transiently in the system.

III. DISTRIBUTION OF SOLUTES IN A BILAYER

A. *Interfacial Activity of Polar, Nonamphiphilic Solutes*

The distribution of solutes across a water–membrane system depends strongly on their chemical structure. Ions and very polar molecules reside mostly in water, and their concentrations in the lipid bilayer are negligible (Deamer and Volkov, 1995). In contrast, nonpolar solutes, such as hexane (Simon *et al.,* 1977), partition preferentially into the membrane interior while their solubilities in water are very low. Amphiphilic molecules tend to concentrate at the water–membrane interface in a manner such that their polar parts are exposed to water, while their nonpolar parts are buried in the lipid phase (Israelachvili, 1992).

Although nonpolar, short-chain alkanes and amphiphilic, unbranched alcohols are anesthetics—alcohols, in fact, are quite potent, even more so than predicted from the Meyer–Overton relationship (Fang *et al.,* 1996); most anesthetics, including those used in clinical practice, do not fall into any of the simple catagories described earlier. These molecules exhibit some degree of polarity or polarizability, but cannot be divided into spatially separated polar and nonpolar segments and, therefore, cannot be classified as amphiphiles. The distribution of such solutes in the lipid bilayer is not immediately obvious and requires separate studies. These studies were performed on nearly 30 molecules of anesthetic interest, which differed widely in chemical structure, anesthetic potency, and solubility in both water and nonpolar phases (Pohorille *et al.,* 1996, 1997, 1998; Chipot *et al.,* 1997; Pohorille and Wilson, 1996). The list of these molecules is given in Table I. The simulated systems were at the physiological temperature of 310 K.

For each molecule, the investigation consisted of several steps. First, the solubilities of the solutes in water and hexane, chosen as a model, nonpolar medium, were calculated and compared with the experimentally measured solubilities. These simulations tested the potential energy functions used to describe the solute and its interactions with the environment. Both calculated and measured solubilities are listed in Table I. In most cases,

TABLE I

Free Energies of Solvation of Anesthetic Solutes in Water and Hexane

	Solubilities (kcal/mol)				
	Water		Hexane		
Molecule	Calc.	Exp.	Calc.	Exp.	MAC (atm)
Cyclopropane[a]	0.6	0.95	−2.2	−1.95	0.17
Nitrous oxide[a]	0.4	0.51	−0.7	−0.76	1.48
(R)-desflurane[a]	0.7	0.92	−2.1	−1.75	0.057
(R)-isoflurane[a]	0.5	0.31	−3.2	−2.67	0.0146
Methanol[a]	−5.2	−4.94	−1.3	−1.35	0.0019
Ethanol[a]	−5.2	−4.96	−2.1	−2.23	0.00096
Butanol[a]	−4.9	−4.74	−3.8	−3.87	0.00014
Hexanol[a]	−3.9	−3.74	−5.4	−5.46	0.00002
Ar[b]	2.2	2.16	0.5	0.54	27.0
Kr[b]	1.8	1.85	−0.02	−0.14	7.31
Xe[b]	1.4	1.48	−0.6	−0.87	1.61
CH_4[c]	2.1	2.30	0.0	0.00	9.89
CH_3F[c]	−0.1	−0.04	−0.8	−0.67	1.04
CH_2F_2[c]	0.1	−0.05	−0.4	−0.55	0.718
CHF_3[c]	1.0	0.97	−0.1	−0.19	1.65
CF_4[c]	3.2	3.29	0.4	0.57	58.8
CF_2Cl_2[d]	1.7	1.91	−0.4	−0.47	1.11
CHF_2CH_2F[e]	−0.6	−0.56	−1.5	−1.34	0.115
CF_2ClCF_2Cl[d]	3.2	2.67	−1.8	−2.15	1.27
CF_3CF_3[e]	4.1	4.07	−0.7	−0.18	NA
n-Butane[d]	−0.6	−0.56	−1.5	−1.34	0.115
$CHF_2CF_2CF_2CHF_2$[d]	2.0	1.13	−2.8	−2.07	0.0587
$CF_2ClCFClCFClCF_2Cl$[d]	2.2	2.35	−3.8	−3.35	0.03
$CF_3CFClCFClCF_3$[d]	4.2	3.96	−3.1	−3.17	NA
1-Chloro-1,2,2-trifluorocyclobutane[d]	−0.3	−0.28	−3.8	−3.55	0.0151
Octafluorocyclobutane[d]	3.6	3.95	−1.5	−1.20	NA
1,2-Dichlorohexafluorocyclobutane[d]	2.7	2.73	−3.4	−3.25	NA

[a] From Pohorille *et al.* (1997).
[b] From Pohorille *et al.* (1998).
[c] From Pohorille and Wilson (1996).
[d] From Chipot *et al.* (1997).
[e] From Pohorille *et al.* (1996).

the agreement between these two quantities is quite good. The standard deviation for all the compounds is 0.4 kcal/mol and the largest discrepancies do not exceed 0.8 kcal/mol. Next, the free energies for transferring the solute across the water–hexane and water–membrane interfaces were determined. In most simulations, a glycolipid, glycerol 1-monooleate, was the membrane-forming material, although some simulations for membranes built of the phospholipid 1-palmitoyl 2-oleoyl sn-glycero 3-phosphocholine (POPC) in the liquid–crystalline state were also performed. The comparison between water–membrane and water–hexane interfaces illuminates the influence of the lipid head groups and the partial ordering of the lipid tails on the distribution of anesthetics. In the final step, the calculated profiles of $\Delta A(z)$ were converted to concentrations using Eq. (1) and analyzed for correlations with anesthetic potencies.

In Fig. 1, the profiles of $\Delta A(z)$ across the water–GMO interface are shown for three clinical anesthetics: isoflurane, desflurane, and nitrous oxide. The profiles of $\Delta A(z)$ were obtained by the "window" method described in the previous section. All three profiles exhibit a minimum near the interface. As shown in Fig. 2, a similar result was obtained for the same compounds at the water–hexane interface. From Eq. (1) it follows that these compounds

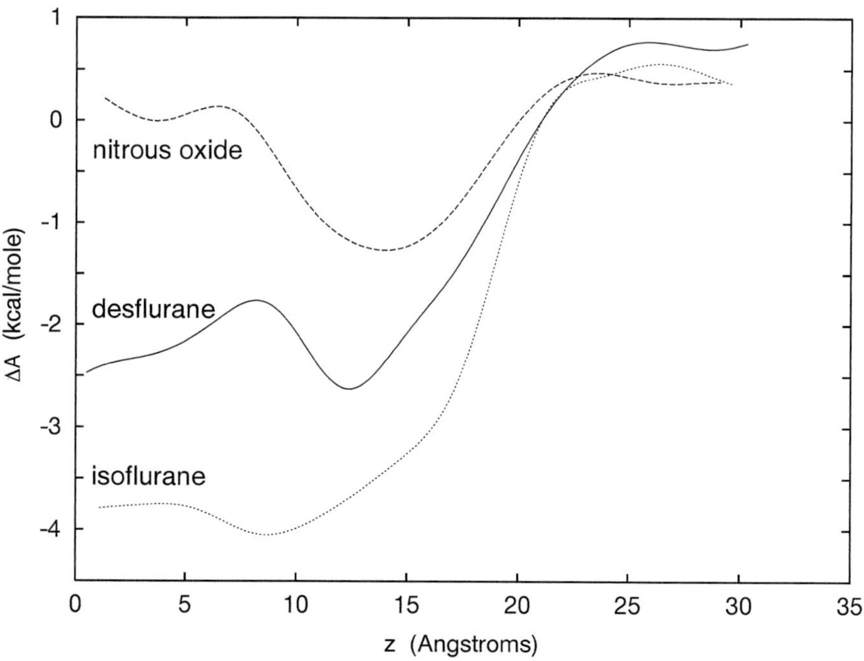

FIGURE 1 Free-energy profiles for the clinical anesthetics isoflurane (dotted line), desflurane (solid line), and nitrous oxide (dashed line) at the water–GMO interface. The bilayer center is at $z = 0$ and the head groups are between $z = 10$ Å and $z = 17$ Å.

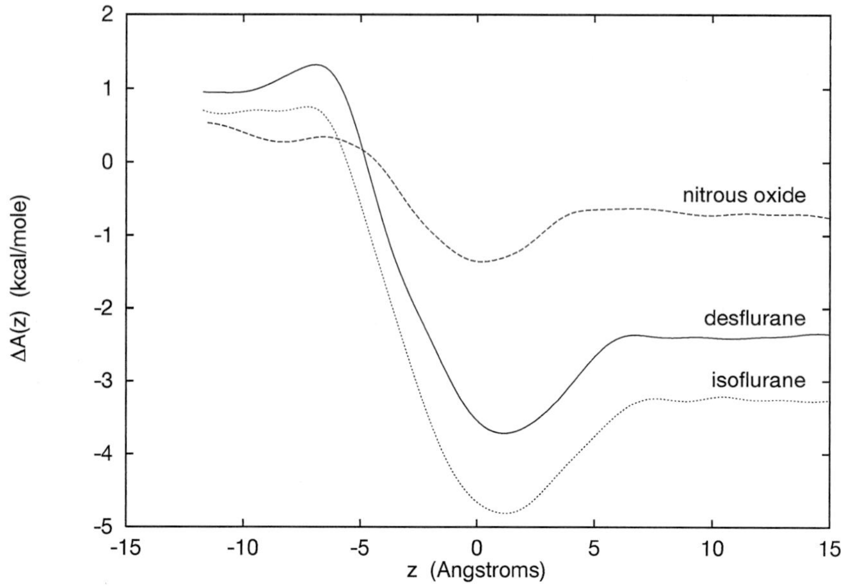

FIGURE 2 Free-energy profiles for the clinical anesthetics isoflurane (dotted line), desflurane (solid line), and nitrous oxide (dashed line) at the water–hexane interface. The interface is at $z = 0$.

are active interfacially, *i.e.,* tend to concentrate at the interface between water and a nonpolar medium. An increase in the interfacial activity of small, neutral solutes as their polarity increased was found in computer simulations of other anesthetics (Pohorille *et al.,* 1996, 1997, 1998; Chipot *et al.,* 1997; Pohorille and Wilson, 1996). The tendency to accumulate at the interface was also observed in NMR studies on isoflurane and other clinical anesthetics, halothane, enflurane, and xenon, in phospholipid bilayers (Baber *et al.,* 1995; Xu and Tang, 1997).

B. Origin of the Interfacial Activity of Polar Solutes

The origin of the interfacial activity of polar solutes can be analyzed conveniently through the consideration of five nearly spherical, isosteric molecules—methane and its four fluorinated derivatives: CH_3F, CH_2F_2, CHF_3, and CF_4 (Pohorille and Wilson, 1996). These molecules are sufficiently small that $\Delta A(z)$ for each of them can be calculated accurately using the particle insertion method outlined in the previous section. From the results, shown in Fig. 3, it is clear that the dipolar CH_2F_2 and CHF_3 (dipole moments of approximately 2 D) concentrate at the interface whereas the nonpolar CH_4 and CF_4 partition into the interior of the membrane. These results can be clarified by dividing the process of dissolving the solute into two hypothetical stages. In the first stage, a molecule with the same van der Waals parameters as the real solute, but without any atomic partial charges, is dissolved. The free energy in this stage is dominated by the reversible work of creating a cavity in the solvent large enough to accommodate the solute. In the second stage, the original values of the atomic partial charges are restored. The free energies associated with the two stages will be called nonelectrostatic and electrostatic to stress the nature of the dominant interactions, and the total free energy of dissolving the real solute is the sum of these two terms. It should be kept in mind, however, that each term contains both electrostatic and nonelectrostatic contributions from solvent reorganization.

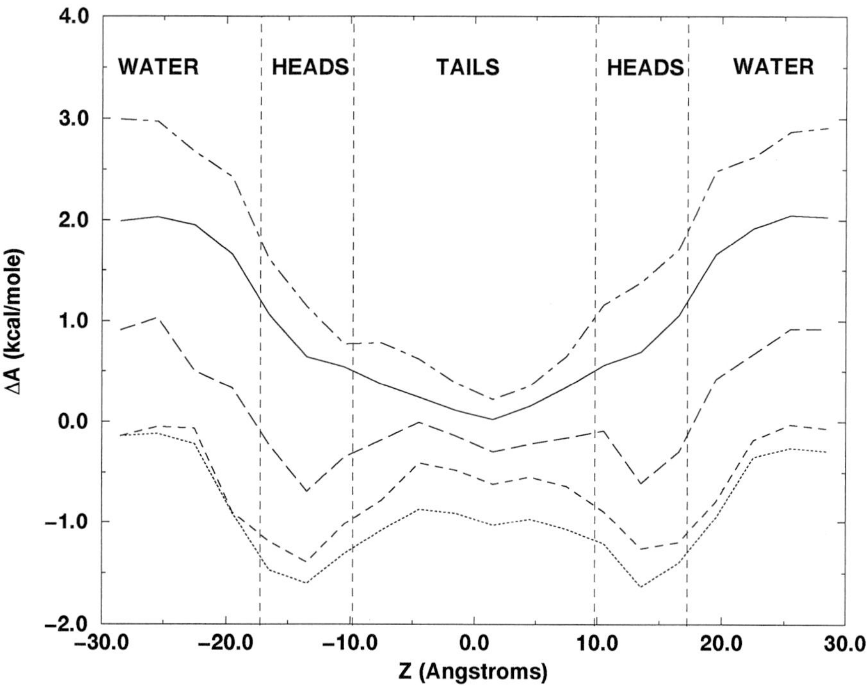

FIGURE 3 Free-energy profiles are methane, CH_4 (solid line), and its fluorinated derivatives: CH_3F (dotted line), CH_2F_2 (dashed line), CHF_3 (long-dashed line), and CF_4 (dotted–dashed line) at the water–GMO interface.

The calculated electrostatic and nonelectrostatic contributions to $\Delta A(z)$ for CH_4 and CH_2F_2 are shown in Fig. 4. For both molecules, the nonelectrostatic term *decreases* as the solutes are transferred from water to the membrane interior. This is a reflection of the hydrophobic effect: it is more difficult to disrupt solvent structure and create a cavity in water than in a nonpolar medium. The same argument explains the poor solubilities of nonpolar solutes in water, compared to other organic solvents (Pratt and Pohorille, 1992). The electrostatic contribution to $\Delta A(z)$ behaves oppositely. It *increases* (becomes less negative) during the transfer of the solute from the polar, aqueous solvent to the nonpolar membrane interior. Both contributions change mostly in the interfacial region. For the dipolar CH_3F, the magnitude of changes in both terms are similar. This yields a minimum in $\Delta A(z)$ located at an interfacial position in which the absolute values of the rates of change in these terms are equal. For nonpolar CH_4, the electro-

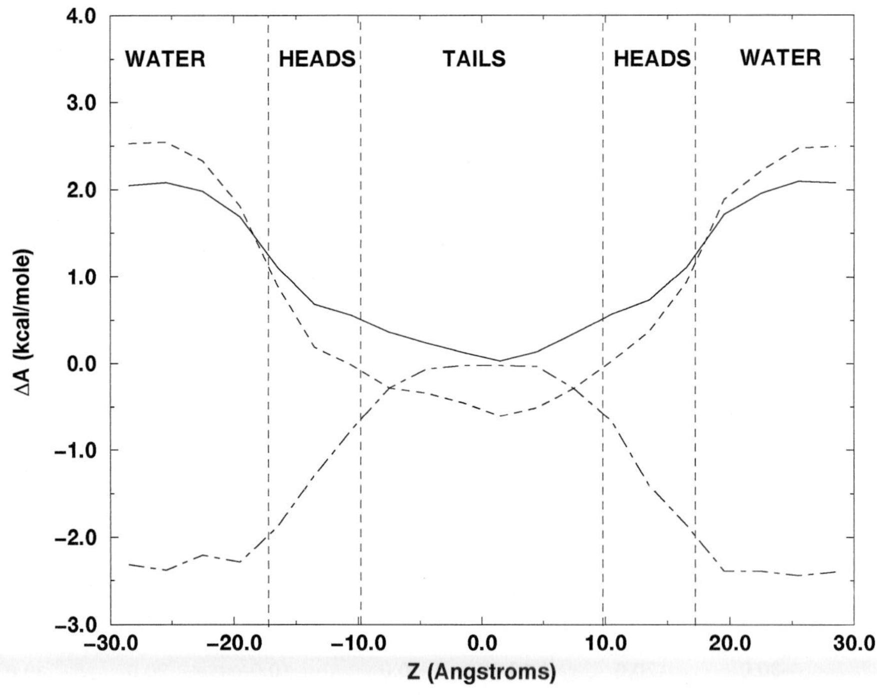

FIGURE 4 Components of the free-energy profiles of CH_4 and CH_2F_2 at the water–GMO interface. The nonelectrostatic component for CH_4 is shown with a solid line and that for CH_2F_2 with a dashed line. The electrostatic component for CH_2F_2 is shown with a dashed–dotted line. The electrostatic component for CH_4, essentially nil, is not shown.

static term is negligible and the behavior of $\Delta A(z)$ is determined by the cavity formation term, which exhibits a minimum in the membrane interior.

It can be noticed from Fig. 4 that the nonelectrostatic term is not constant in the hydrocarbon core of the membrane, but instead decreases toward the center of the bilayer. This decrease is due to the increasing disorder of the lipid chains, and the decreasing atomic density, as the depth in the membrane increases. The lower density, in turn, creates more voids between the tails of the lipid molecules and lowers the reversible work of cavity formation. A similar result was noted in other molecular dynamics studies of small solutes in membranes (Marrink and Berendsen, 1996) and was obtained from a mean-field lattice theory (Marqusee and Dill, 1985). This result indicates that packing of the membrane has an important effect on the distribution of solutes. This conclusion should apply not only to the interior of the bilayer, but also to the head group region. As can be seen in Fig. 5, this is indeed the case. Figure 5 shows $\Delta A(z)$ for transferring CH_4, CH_2F_2, CF_4, and another simple anesthetic, Xe, across the interfaces

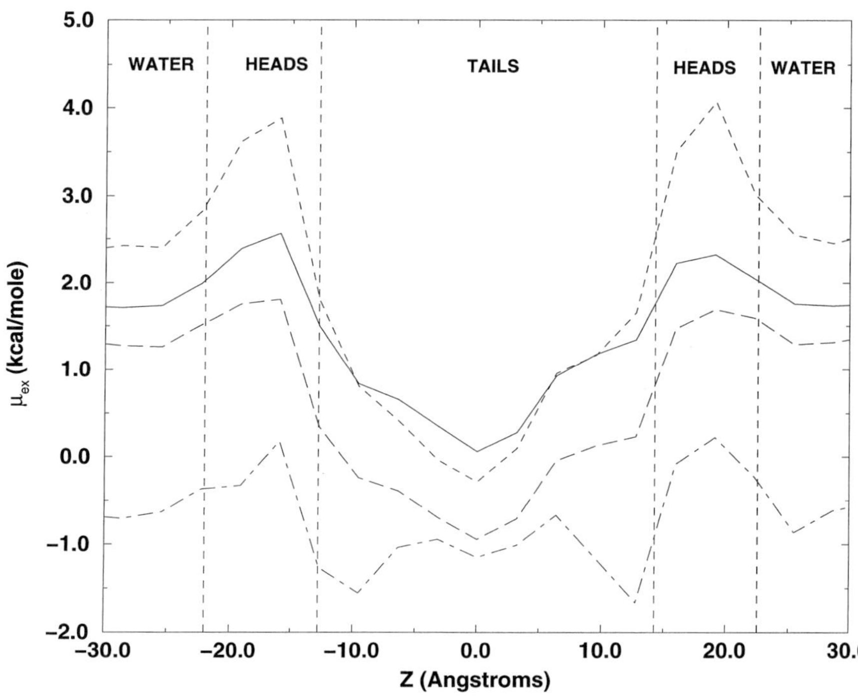

FIGURE 5 Free energies of small anesthetics across the water–POPC interface at 310 K: CH_4 (solid line), CH_2F_2 (dotted line), CF_4 (dashed line), and Xe (long-dashed line).

between water and the POPC bilayer. The profiles are similar to those in the water–GMO system, presented in Fig. 4. The main difference is the presence of a barrier on the water side of POPC bilayers. A very similar barrier was found for bilayer systems made of dipalmitoylphosphatidylcholine (DPPC) (Schweighofer and Pohorille, 1998). This barrier can be attributed to the dense packing of phospholipid head groups, which, in turn, increases the reversible work of cavity formation. This can be gauged from the probability distribution of finding cavities of different sizes in the membrane, which is connected with the reversible work of cavity formation by a relations analogous to Eq. (4) (Pohorille and Pratt, 1990). Other computer simulations (Marrink and Berendsen, 1996) and NMR studies of the distribution of different anesthetics in phospholipid bilayers (Baber *et al.,* 1995; North and Cafiso, 1997; Tang *et al.,* 1997; Xu and Tang, 1997) are consistent with this interpretation.

C. *Distribution of Anesthetics and Nonanesthetics at Clinically Relevant Concentrations*

The difference between the distributions of polar and nonpolar solutes can be seen very clearly by placing two structurally similar solutes, 1,2,3-trifluoroethane (TFE) and perfluoroethane (PFE), at anesthetically relevant concentrations in the water–GMO system. TFE is polar and is an anesthetic whereas PFE is nonpolar and, contrary to the predictions from the Meyer–Overton relationship, exerts no anesthetic action (has no immobilizing effect). The number of TFE molecules in the system was calculated from the profiles of $\Delta A(z)$, previously obtained for a single TFE solute, such that it corresponded to a partial pressure of 1 MAC. MAC (the maximum alveolar concentration) is the average applied partial pressure of an anesthetic needed to suppress the movement of animals in response to a painful stimulus. This quantity is used broadly as a measure of anesthetic potency; lower values of MAC indicate higher potencies. The values of MAC for TFE and other anesthetics are listed in Table I. The same procedure for identifying the number of molecules in the system cannot be applied to PFE as this compound is nonanesthetic and, therefore, does not have a MAC. Instead, the highest attainable partial pressure was used: the vapor pressure. In total, 6 TFE or 32 PFE molecules were placed in the simulation box, whose initial x,y dimensions, parallel to the plane of the membrane, were 36.9×36.9 Å. By applying hybrid Monte Carlo, the dimensions of the box were allowed to fluctuate during the course of the simulations such that the water–bilayer interfacial tension remained constant. This allowed the membrane to adjust to the presence of the anesthetics.

The distributions of both molecules, averaged over 10^4 hybrid Monte Carlo steps, are shown in Fig. 6. Although these distributions have not equilibrated fully, as manifested by their asymmetry with respect to the center of the bilayer, the difference is striking: TFE is located predominantly near the interface whereas PFE partitions into the membrane interior. Clearly, the distributions of these two compounds in the membrane are not in agreement with expectations based on the Meyer–Overton relationship.

D. Interfacial Revision to the Meyer–Overton Relationship

The results discussed so far not only indicate that the Meyer–Overton relationship needs to be revised, but also suggest an alternative hypothesis. If anesthetics tend to concentrate at the interface, then perhaps their anesthetic potencies correlate better with interfacial solubilities than with solubilities in the nonpolar membrane interior. This hypothesis can be tested

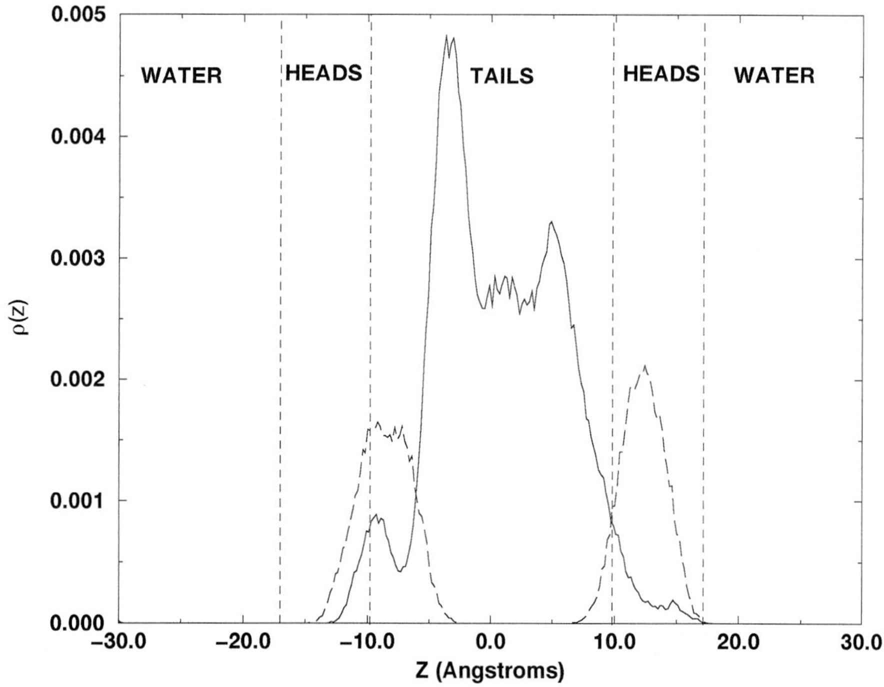

FIGURE 6 Density profiles for the nonanesthetic perfluoroethane (solid line) and the anesthetic, 1,1,2-trifluoroethane (dashed line) at the water–GMO interface at 310 K.

using the calculated profiles of $\Delta A(z)$ from which the interfacial solubilities, K_{int}, are obtained as

$$K_{\text{int}} = \frac{c}{z_2 - z_1} \int_{z_1}^{z_2} \exp[\beta \Delta A(z)] dz, \tag{7}$$

where c is a constant that defines the units of K_{int}. In general, the calculated solubilities depend on the choice of the interfacial region $[z_1, z_2]$ along the z direction. In practice, however, this choice has only a minor influence on the values of K_{int} for different compounds with respect to each other.

The correlation of anesthetic potencies with K_{int} at the water–GMO and water–hexane interfaces is shown in Figs. 7 and 8. For comparison, the correlation with the solubilities in olive oil (the Meyer–Overton relationship) for the same compounds is also shown in Fig. 9. As can be seen, all compounds that obey the Meyer–Overton relationship also obey the interfacial relationship. More importantly, the interfacial correlation also applies to the compounds for which the Meyer–Overton relationship fails. Among these compounds are 1,2-dichloro-perfluoroethane (CF_2Cl-CF_2Cl) and 1,2,3,4-tetrachloroperfluorobutane (CF_2Cl-CFCl-CFCl-CF_2Cl), which

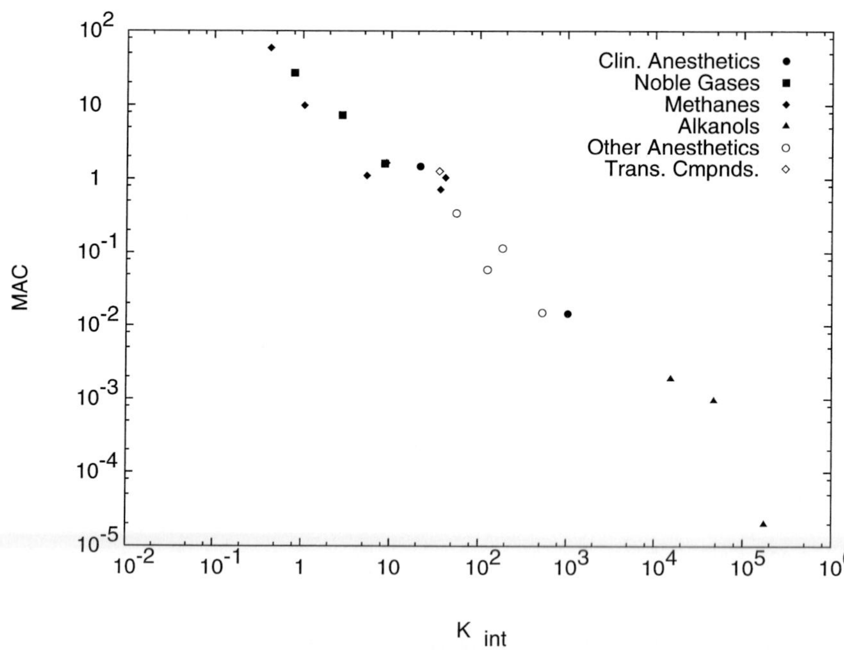

FIGURE 7 Correlation between solubilities at the water–GMO interface, K_{int}, and MAC.

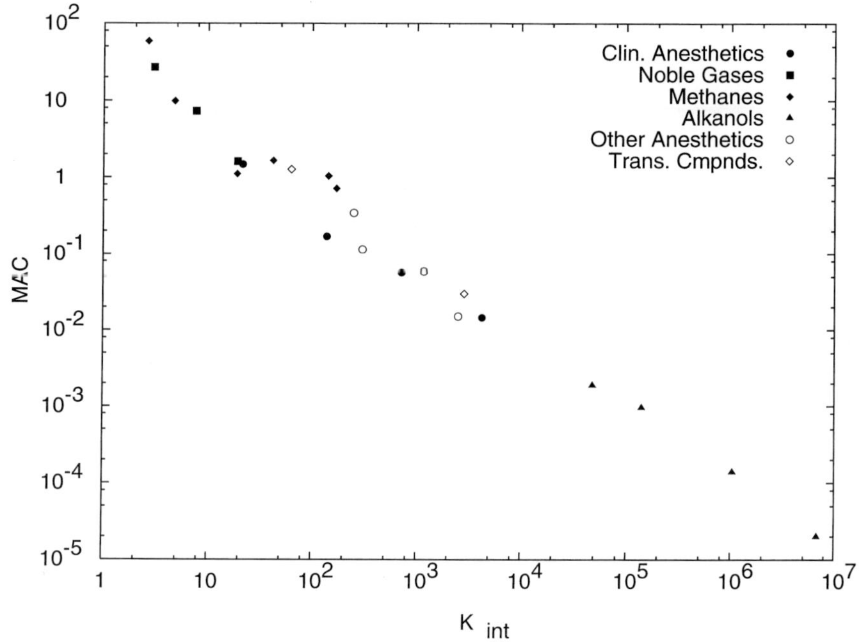

FIGURE 8 Correlation between solubilities at the water–hexane interface, K_{int}, and MAC.

are less potent than predicted by the Meyer–Overton relationship, and straight chain alcohols, which are more potent than predicted. Furthermore, the interfacial relationship can explain the lack of immobilizing activity of nonanesthetics: these compounds can never reach the anesthetizing concentration at the interface.

E. Implications of the Interfacial Activity of Anesthetics

In view of the mounting evidence that anesthetics act on membrane receptors, the results in Fig. 7 raise a question about the relationship between the interfacial correlation and the mechanism of anesthesia. It is, of course, possible that no such relationship exists and the observed correlation is not relevant to the mechanism of anesthesia. More likely, the interfacial correlation indicates that anesthetics act at the water–membrane interface either by interacting directly with membrane receptors or by influencing lipids, which, in turn, perturb the receptor. Because the interfacial relationship holds very well for different aqueous interfaces, such as water–

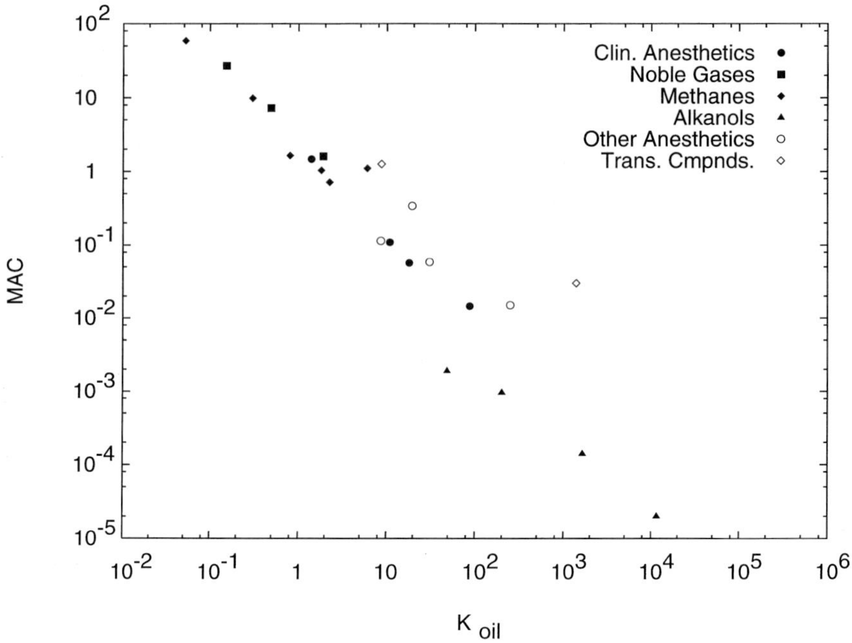

FIGURE 9 The Meyer–Overton correlation based on solubilities in oil, K_{oil}.

membrane and water–hexane interfaces, it is quite possible that it will be equally accurate for the interfacial system containing water, membrane, and the interfacial portion of a membrane protein. It should be further noted that molecular models of receptors (Ortells and Lunt, 1995, 1996; Ortells *et al.,* 1997; Sankararamakrishnan and Sansom, 1996; Tikhonov and Zhorov, 1998) place the sites of the single point mutations that abolish the response of GABA and glycine receptors to anesthetics (Mihic and Harris, 1997) near the water–membrane interface.

Although less popular, indirect mechanisms are also possible. It has been shown that the proper composition of the lipid bilayer is needed for functioning of the nicotinic acetylcholine receptor (Saffron *et al.,* 1997), which is related closely to the GABA and glycine receptors. This indicates that lipids can influence receptor action. Several mechanisms by which anesthetics may modulate this influence have been proposed. According to one view, by concentrating in the head group region, anesthetics can slightly modify the electrostatic surface potential and, by doing so, induce conformational transitions in receptors (Qin *et al.,* 1995). Another elegant hypothesis has been put forward by Cantor (1997), who proposed that interfacially located anesthetics change the lateral stress tensor profile across the membrane, which, in turn, changes the equilibrium between the open and the closed states of a channel receptor. At present, however, all hypotheses about the molecular mechanism of anesthesia remain speculative and a considerable body of experimental and theoretical work will have to be performed before we arrive at a well-founded molecular theory of anesthesia.

IV. MEMBRANE PERMEABILITY OF SMALL MOLECULES

A. Models of Permeation

The Overton rule has a direct conceptual connection with the traditional description of membrane permeability: the solubility–diffusion model (Hanai and Dayton, 1966; Finkelstein and Cass, 1967). In this model, it is assumed that the rate-limiting step in solute permeation through membranes is the diffusion of the solute in the bilayer. The bilayer is treated as a homogeneous, isotropic lamella of width w_b, sharply separated from the aqueous medium. The diffusion constant is assumed to be the same everywhere in the bilayer and is equal to D_b. If any interfacial barriers that may be encountered by the solute during its entrance into or exit from the membrane are neglected, the permeability, P, is given as

$$P = \frac{K_{\mathrm{bw}} D_b}{w_b}, \tag{8}$$

where K_{bw} is the bilayer/water partition coefficient. Because variations in the widths of different membranes and in diffusion constants for different solutes are typically smaller than variations in the partition coefficient, changes in K_{bw} have the largest influence on changes in P. If K_{bw} is further approximated as the partition coefficient between a bulk, nonpolar phase (*e.g.*, octanol or liquid alkane) and water, the solubility–diffusion model yields the Overton rule. The best correlations with membrane permeabilities were found for partition coefficients between hexadecane or 1,9-decadiene and water (Walter and Gutknecht, 1986; Xiang and Anderson, 1994a).

The assumptions of the solubility–diffusion model are quite restrictive and do not allow for the incorporation of information about the molecular structure of the membrane. A more general formulation that applies to inhomogeneous media and only requires that thermodynamic gradients in the system be small was given by Marrink and Berendsen (1994). The permeability was represented as:

$$P = 1/\int_{z_1}^{z_2} \frac{\exp[(\Delta A(z) - \Delta A_w)/k_B T]}{D(z)}\, dz, \tag{9}$$

where ΔA_w is the free energy of the solute in bulk water and $D(z)$ is the diffusion coefficient of the solute, which generally depends on z. The limits of integration, z_1 and z_2, correspond to positions in bulk water on both sides of the bilayer.

The most distinguishing feature of this formula is that it incorporates information about both the equilibrium and the dynamic behavior of the solute in the bilayer. Equilibrium properties of the solute are described by $\Delta A(z)$, which were discussed extensively in the previous section. The dynamic properties are characterized by the diffusion coefficient, which is not assumed to be uniform over the membrane, but, instead, is considered to be a function of the depth. This quantity can be calculated using Eq. (3).

B. *Distribution of Free Volume in Membranes and Its Influence on Solute Diffusion*

Simulation studies of the permeation of water, ammonia, oxygen and model, Lennard–Jones particles through a phospholipid bilayer revealed that the membrane should not be considered as an isotropic, homogeneous medium, but instead is described much better by a three-region model (Marrink and Berendsen, 1994, 1996). Due to its high density, the interfacial, head group region is characterized by a high viscosity. In this region the diffusion constant is low. The next region comprises the portion of the hydrocarbon chains adjacent to the head groups. This region is highly ordered and motions of the chains are strongly correlated. In this respect, its properties are similar to those of a soft polymer. This region is also characterized by the anomalous diffusion of solutes, which cannot be well described by the Stokes–Einstein relation connecting the diffusion rate with the volume of the penetrant. This will be discussed further later. The third region is formed by the hydrocarbon tails of lipid molecules and is located in the center of the bilayer. Because the tails are highly disordered, properties of this region resemble those of a liquid alkane. Because the atomic density in the tail region is lower than elsewhere in the membrane, diffusion in this region is the fastest. Diffusion constants in different regions can differ substantially, sometimes by as much as a factor of five. The idea of different, although not well separated, membrane compartments also emerges from experimental studies (Miller, 1986).

The rate of diffusion is not the only characteristic that distinguishes the three regions of the lipid bilayer. Another property that changes with the depth in the membrane is the distribution of the free volume. This property has already been discussed briefly in the context of the reversible work of cavity formation and its contribution to $\Delta A(z)$. Detailed studies on the size and shape of cavities in a lipid bilayer (Marrink *et al.,* 1996) revealed that large cavities are quite rare in the dense head group region. The probability of finding a cavity in this region decreases exponentially with the volume of the cavity. A different distribution of the free volume is observed in the chain region adjacent to the head groups. Here, the probability of finding a cavity scales as a power of the volume of the cavity. A relatively large number of cavities in this region are aspherical, with their long axis aligned preferentially with the lipid chains along the bilayer normal. In contrast, cavitities in the low-density central region of the membrane are distributed nearly isotropically and their average size is considerably larger.

Although the distribution of the free volume is an equilibrium property of pure membranes, it provides valuable clues to the mechanism of permeation of small solutes across lipid bilayers. This was noticed for the first time in molecular dynamics simulations of benzene in DPPC membranes (Bassolino-Klimas *et al.,* 1993, 1995). Occasionally, benzene molecules experienced large, rapid displacements (jumps) that cannot be explained by the conventional picture of Einsteinian diffusion. Instead, these jumps can be related to the presence of cavities in the membrane. The solute may become trapped in one such cavity, where it undergoes rattling motions until a pathway opens to another, nearby cavity. This usually requires a rearrangement of the surrounding lipid molecules, often through the *gauche-trans* isomerization of their segments. It could be predicted that jumps play the most important role in diffusion through the best ordered region of the hydrocarbon chains, near the lipid head groups. This, indeed, appears to be the case for very small solutes (Marrink and Berendsen, 1996), although for benzene a considerable number of jumps was also found in the center of the bilayer (Bassolino-Klimas *et al.,* 1995). This general picture of diffusion through membranes, emerging from computer simulations, is consistent with the proposal of Lieb and Stein (1969) that small solutes move in the bilayer by "hopping" between voids. This same mechanism is not available to large solutes that cannot encounter cavities of their size. This conclusion was confirmed in simulations of the large drug molecules adamantane and nifedipine (Alper and Stouch, 1995).

The anomalous, hopping diffusion mechanism may explain the unusually strong dependence of the permeation rate of small solutes on their size—a dependence that is in variance with the Overton rule. In this picture, the availability of cavities large enough to accommodate a solute decreases rapidly with the size of the solute, thus making large, hopping motions less likely for larger solutes. It should be noted that another explanation of the unusual size dependence of permeabilities, based entirely on the solubility part of the permeation process, has also been proposed (Xiang and Anderson, 1994b). The existence of solute jumping motions in the bilayer, however, appears, at present, to be well documented.

V. SUMMARY

Only recently, nearly 100 years after Overton formulated his remarkably insightful ideas about the behavior of solutes in membranes, have we started to develop a detailed, molecular understanding of the complex, multicomponent, inhomogeneous systems called biomembranes. Equipped with this new understanding, we are now ready to refine Overton's concepts and reconcile them with recent experimental data. Computer simulations play an important role in this process by providing accurate, molecularly detailed information about the equilibrium and dynamical properties of solutes in model lipid bilayers.

Computer simulations reveal that solutes are distributed nonuniformly in a membrane and that their permeation across the bilayer cannot be characterized satisfactorily by a single diffusion coefficient. Instead, it is more appropriate to consider the membrane as consisting of at least three different regions comprising, respectively, lipid head groups, rigid hydrocarbon chain segments adjacent to the head groups, and fluid hydrocarbon tails in the center of the bilayer. Nonpolar solutes accumulate preferentially in the tail region, whereas polar, but not necessarily amphiphilic, solutes concentrate close to the water–membrane interface, near the outer or inner boundaries of the head group region. This tendency explains why anesthetic potencies of some compounds deviate considerably from the Meyer–Overton relationship. In fact, potencies of anesthetic compounds correlate better with their interfacial solubilities than with solubilities in the membrane interior, as postulated by Meyer and Overton. As a consequence, it appears more likely that anesthetics act on their targets, presently believed to be membrane receptors, near the water–membrane interface rather than inside the bilayer.

The different degrees of packing and rigidity of different regions in the bilayer influence permeation of small solutes across membranes. These solutes often do not diffuse via an Einsteinian mechanism but, instead, translocate via hopping motions between voids in the membrane, coupled with translational and conformational motions of the surrounding lipids. This, in turn, leads to the observed deviations from the Overton rule for small solutes, which exhibit an unusually strong dependence of their permeation rate on their size.

Even though Overton's original ideas have been recently revised, we should bear in mind that they have formed the point of departure for most of the modern studies on the behavior of small solutes in membranes, thus facilitating progress in this field. We should also appreciate that what we know now is undoubtedly not the whole story and much work needs to be done before we gain a satisfactory understanding of interactions between small solutes and different membrane components. For example, despite extensive, century-old efforts, the molecular site and mechanism of anesthesia still remain elusive.

Acknowledgments

This work was supported by a National Institutes of Health grant (GM47818-01) and a grant from the National Aeronautics and Space Administration Exobiology Program. Computational resources from the National Cancer Institute's Frederick Biomedical Supercomputing Center and the Numerical Aerospace Simulation Facility at NASA Ames Research Center are gratefully acknowledged.

References

Allen, M. P., and Tildesley, D. J. (1987). "Computer Simulation of Liquids." Oxford Univ. Press (Clarendon), Oxford.

Alper, H. E., and Stouch, T. R. (1995). Orientation and diffusion of a drug analogue in biomembranes: Molecular dynamics simulations. *J. Phys. Chem.* **99,** 5724–5731.

Baber, J., Ellena, J. F., and Cafiso, D. S. (1995). Distribution of general anesthetics in phospholipid bilayers determined using ^{2}H NMR and ^{1}H-^{1}H NOE spectroscopy. *Biochemistry* **34,** 6533–6539.

Bassolino-Klimas, D., Alper, H. E., and Stouch, T. R. (1993). Solute diffusion in lipid bilayer membranes—an atomic level study by molecular dynamics simulation. *Biochemistry* **32,** 12624–12637.

Bassolino-Klimas, D., Alper, H. E., and Stouch, T. R. (1995). Mechanism of solute diffusion through lipid bilayer membranes by molecular dynamics simulation. *J. Am. Chem. Soc.* **117,** 4118–4129.

Berry, R. S., Rice, S. A., and Ross, J. (1980). "The Chemistry of Acid Rain: Sources and Atmospheric Processes." Wiley, New York.

Brass, A., Pendleton, B. J., Chen, Y., and Robson, B. (1993). Hybrid Monte Carlo simulations theory and initial comparison with molecular dynamics. *Biopolymers* **33,** 1307–1315.

Cantor, R. S. (1997). The lateral pressure profile in membranes: A physical mechanism of general anesthesia. *Biochemistry* **36,** 2339–2344.

Chipot, C., Wilson, M. A., and Pohorille, A. (1997). Interactions of anesthetics with the water–hexane interface. A molecular dynamics study. *J. Phys. Chem. B* **101,** 782–791.

Damodaran, K. V., Merz, K. M., and Gaber, B. P. (1995). Interaction of small peptides with lipid bilayers. *Biophys. J.* **69,** 1299–1308.

Deamer, D. W., and Volkov, A. G. (1995). Proton permeation of lipid bilayers. *In* "Permeability and Stability of Lipid Bilayers" (E. A. Disalvo and S. A. Simon, eds.), pp. 161–177. CRC Press, Boca Raton, Florida.

Essmann, U., Perera, L., and Berkowitz, M. L. (1995a). The origin of the hydration interaction of lipid bilayers from MD simulation of dipalmitoylphosphatidylcholine membranes in gel and liquid crystalline phases. *Langmuir* **11,** 4519–4531.

Essmann, U., Perera, L., Berkowitz, M. L., Darden, T., Lee, H., and Pedersen, L. G. (1995b). A smooth particle mesh Ewald method. *J. Chem. Phys.* **103,** 8577–8593.

Fang, Z., Ionescu, P., Chortkoff, B., Kandel, L., Sonner, J., Laster, M. J., and Eger, E. I., III. (1996). *Anesth. Analg.* **84,** 1042–1048.

Finkelstein, A., and Cass, A. (1967). Effect of cholesterol on the water permeability of thin lipid membranes. *Nature (London)* **216,** 717–718.

Forrest, B. M., and Suter, U. W. (1994). Hybrid Monte Carlo simulations of dense polymer systems. *J. Chem. Phys.* **101,** 2616–2629.

Franks, N. P., and Lieb, W. R. (1994). Molecular and cellular mechanisms of general anesthesia. *Nature (London)* **367,** 607–614.

Guillot, B., Guissani, Y., and Bratos, S. (1991). A computer-simulation study of hydrophobic hydration of rare gasses and of methane. 1. Thermodynamic and structural properties. *J. Chem. Phys.* **95,** 3643–3648.

Hani, T., and Dayton, D. A. (1966). The permeability to water of bimolecular lipid membranes. *J. Theor. Biol.* **11,** 1370.

Israelachvili, J. (1992). "Intermolecular and Surface Forces." Academic Press, New York.

Janoff, A. S., and Miller, K. W. (1982). A critical assessment of the lipid theories of general anesthetic action. *In* "Biological Membranes" (D. Chapman, ed.), Vol. 4, pp. 417–477. Academic Press, New York.

Jorgensen, W. L., Chandrasekhar, J., Madura, J. D., Impey, R. W., and Klein, M. L. (1983). Comparison of simple potential functions for simulating liquid water. *J. Chem. Phys.* **79,** 926–935.

Koblin, D. D. (1990). Mechanisms of action. *In* "Anesthesia" (R. D. Miller, ed.), pp. 51–83. Churchill-Livingston, New York.

Koblin, D. D., Chortkoff, B. S., Laster, M. J., Eger, E. I., Halsey, M. J., and Ionescu, P. (1994). Polyhaolgenated and perfluorinated compounds that disobey the Meyer-Overton hypothesis. *Anesth. Analg. (N.Y.)* **79,** 1043–1048.

Lieb, W. R., and Stein, W. D. (1969). Biological membranes behave as non-porous polymer sheets with respect to the diffusion of non-electrolytes. *Nature (London)* **224,** 240–243.

Liu, J., Laster, M. J., Koblin, D. D., Eger, E. I., Halsey, M. J., Taheri, A., and Chortkoff, B. S. (1994). A cutoff in potency exists in the perfluoroalkanes. *Anesth. Analg. (N.Y.)* **79,** 238–244.

Marqusee, J. A., and Dill, K. A. (1985). Solute partitioning into chain molecule interphases: Monolayers, bilayer membranes and micells. *J. Chem. Phys.* **85,** 434–444.

Marrink, S. J., and Berendsen, H. J. C. (1994). Simulation of water transport through a lipid membrane. *J. Phys. Chem.* **98,** 4155–4168.

Marrink, S. J., and Berendsen, H. J. C. (1996). Permeation process of small molecules across lipid membranes studied by molecular dynamics simulations. *J. Phys. Chem.* **100,** 16729–16738.

Marrink, S. J., Sok, R. M., and Berendsen, H. J. C. (1996). Free volume properties of a simulated lipid membrane. *J. Chem. Phys.* **104,** 9090–9099.

Martyna, G. J., Tobias, D. J., and Klein, M. L. (1994). Constant pressure molecular dynamics algorithms. *J. Chem. Phys.* **101,** 4177–4189.

Martyna, G. J., Tuckerman, M. E., Tobias, D. J., and Klein, M. L. (1996). Explicit reversible integrators for extended system dynamics. *Mol. Phys.* **87,** 1117–1157.

Merz, K. M. (1997). Molecular dynamics simulations of lipid bilayers. *Curr. Opin. Struct. Biol.* **7,** 511–517.

Meyer, H. H. (1899). Theorie der Alkoholnarkose. *Arch. Exp. Pathol. Pharmakol.* **42,** 109–118.

Meyer, K. H. (1937). Contributions to the Theory of Narcosis. *Trans. Faraday Soc.* **33,** 1062–1068.

Mihic, S. J., and Harris, R. A. (1997). Molecular mechanisms of anesthetic action on ligand-gated ion channels. *Neurotransmissions* **13,** 1–17.

Miller, D. M. (1986). The measurement of the rate of transport of solutes in both directions across the aqueous–nonaqueous liquid interface and its significance to membrane permeability. *Biochim. Biophys. Acta* **856,** 27–35.

New, M. H., and Pohorille, A. (1998). Hybrid Monte Carlo methods for interfacial systems. To be published.

North, C., and Cafiso, D. S. (1997). Contrasting membrane localization and behavior of halogenated cyclobutanes that follow or violate the Meyer-Overton hypothesis of general anesthetic potency. *Biophys. J.* **72,** 1754–1761.

Ortells, M. O., and Lunt, G. G. (1995). A mixed helix-β-sheet model of the transmembrane region of the nicotinic acetylcholine receptor. *Trends Neurosci.* **18,** 121–127.

Ortells, M. O., and Lunt, G. G. (1996). A mixed helix-β-sheet model of the transmembrane region of the nicotinic acetylcholine receptor. *Protein Eng.* **9,** 51–59.

Ortells, M. O., Barrantes, G. E., Lunt, G. G., and Barrantes, F. J. (1997). Molecular modelling of the nicotinic acetylcholine receptor transmembrane region in the open state. *Prot. Eng.* **10,** 511–517.

Overton, E. (1899). *Vierteljahrsschr. Naturforsch. Ges. Zuerich* **44,** 88–135.

Overton, E. (1901). "Studien über die Narkose zugleich ein Betrag zur Allgemeinen Pharmakologie." Gustav Fischer, Jena.

Pohorille, A., and Pratt, L. R. (1990). Cavities in molecular liquids and the theory of hydrophobic solubilities. *J. Am. Chem. Soc.* **112,** 5066–5074.

Pohorille, A., and Wilson, M. A. (1993). Viewpoint 9—Molecular structure of aqueous interfaces. *J. Mol. Struct.* (*Theochem.*) **284,** 271–298.

Pohorille, A., and Wilson, M. A. (1996). Excess free energy of small solutes across water–membrane and water–hexane interfaces. *J. Chem. Phys.* **104,** 3760–3773.

Pohorille, A., and Wilson, M. A. (1998). Free energies of the noble gasses Kr, Ar, and Xe at water–hexane and water–membrane interfaces. Unpublished results.

Pohorille, A., Cieplak, P., and Wilson, M. A. (1996). Interactions of anesthetics with the membrane–water interface. *Chem. Phys.* **204,** 337–345.

Pohorille, A., Wilson, M. A., and Chipot, C. (1997). Interaction of alcohols and anesthetics with the water–hexane interface: A molecular dynamics study. *Prog. Colloid Polym. Sci.* **103,** 29–40.

Pohorille, A., Wilson, M. A., New, M. H., and Chipot, C. (1998). Concentrations of anesthetics across the water–membrane interface; the Meyer–Overton hypothesis revisited. *Toxicology Lett.* **100–101,** 421–430.

Pratt, L. R., and Pohorille, A. (1992). Theory of hydrophobicity—transient cavities in molecular liquids. *Proc. Natl. Acad. Sci. U.S.A.* **89,** 2995–2999.

Qin, Z. H., Szabo, G., and Cafiso, D. S. (1995). Anesthetics reduce the magnitude of the membrane dipole potential—measurements in lipid vesicles using voltage-sensitive spin probes. *Biochemistry* **34,** 5536–5543.

Roux, B., and Karplus, M. (1994). Molecular dynamics simulations of the gramicidin channel. *Annu. Rev. Biophys. Biomol. Struct.* **23,** 731–761.

Saffron, E., Addona, G. H., Kloczewiak, M. A., Bugge, B., and Miller, K. W. (1997). The cholesterol dependence of activation and fast desensitization of the nicotinic acetylcholine receptor. *Biophys. J.* **73,** 2446–2455.

Sankararamakrishnan, R., and Sansom, M. S. P. (1996). The pore domain of the nicotinic acetylcholine receptor: Molecular modeling pore dimensions, and electrostatics. *Biophys. J.* **71,** 1659–1671.

Schweighofer, K. S., and Pohorille, A. (1998). Free energies of noble gasses and fluoromethanes at the water–POPC and water–DPPC membrane interfaces. Unpublished results.

Shen, L., Bassolino-Klimas, D., and Stouch, T. R. (1997). Transmembrane helix structure, dynamics, and interactions: Multi-nanosecond, molecular dynamics simulations. *Biophys. J.* **73,** 3–20.

Simon, S. A., Stone, W. L., and Busto-Latorre, P. (1977). A thermodynamic study of the partition of *n*-hexane into phosphatidylcholine and phosphatidylcholine-cholesterol bilayers. *Biochim. Biophys. Acta* **468,** 378–388.

Stouch, T. R. (1997). Solute transport and partitioning in lipid bilayers: Molecular dynamics simulations. *Prog. Colloid Polym. Sci.* **103,** 116–120.

Stouch, T. R., and Bossalino, D. (1996). Movement of small molecules in lipid bilayers: molecular dynamics simulation studies. *In* "Biological Membranes. A Molecular Perspective from Computation and Experiment" (K. M. Merz and B. Roux, eds.), pp. 255–278. Birkhaeuser, Boston.

Tang, P., Yan, B., and Xu, Y. (1997). Different distribution of fluorinated anesthetics and nonanesthetics in model membrane: A ^{19}F NMR study. *Biophys. J.* **72,** 1676–1682.

Tieleman, D. P., Marrink, S. J., and Berendsen, H. J. C. (1997). A computer perspective of membranes: Molecular dynamics studies of lipid bilayer systems. *Biochim. Biophys. Acta* **1331,** 235–270.

Tikhonov, D. B., and Zhorov, B. S. (1998). Kinked-helices model of the nicotinic acetylcholine receptor ion channel and its complexes with blockers: Simulation by the Monte Carlo minimization method. *Biophys. J.* **74,** 242–255.

Tobias, D. J., Tu, K., and Klein, M. L. (1997). Assessment of all-atom potentials for modeling membranes: Molecular dynamics simulations of solid and liquid alkanes and crystals of phospholipid fragments. *J. Chim. Phys. Phys.-Chim. Biol.* **94,** 1482–1502.

Tu, K., Tobias, D. J., Blasie, J. K., and Klein, M. L. (1996). Molecular dynamics investigation of the structure of a fully hydrated gel-phase dipalmitoylphosphatidylcholine bilayer. *Biophys. J.* **70,** 595–608.

Walter, A., and Gutknecht, J. (1986). Permeability of small nonelectrolytes through lipid bilayer membranes. *J. Membr. Biol.* **90,** 207–218.

Widom, B. (1963). Some topics in the theory of fluids. *J. Chem. Phys.* **39,** 2808–2812.

Widom, B. (1982). Potential distribution theory and the statistical mechanics of fluids. *J. Phys. Chem.* **86,** 869–872.

Wilson, M. A., and Pohorille, A. (1994). Molecular dynamics of a water–lipid bilayer interface. *J. Am. Chem. Soc.* **116.** 1490–1501.

Xiang, T.-X., and Anderson, B. D. (1994a). The relationship between permeant size and permeability in lipid bilayer membranes. *J. Membr. Biol.* **140,** 111–122.

Xiang, T.-X., and Anderson, B. D. (1994b). Molecular distributions in interphases—statistical mechanical theory combined with molecular dynamics simulation of a model lipid bilayer. *Biophys. J.* **66,** 561–572.

Xu, Y., and Tang, P. (1997). Amphiphilic sites for general anesthetic action? Evidence from ^{129}Xe-^{1}OH intermolecular nuclear Overhauser effects. *Biochim. Biophys. Acta* **1323,** 154–162.

CHAPTER 4

Membrane Permeability Barriers to Ionic and Polar Solutes

Stefan Paula and David W. Deamer
Department of Chemistry and Biochemistry, University of California at Santa Cruz
Santa Cruz, California 95064

I. INTRODUCTION

The permeability of membranes to solutes is a fundamental property of all cellular life. Living cells depend on membranes that provide a substantial barrier to ionic solutes such as protons, sodium, potassium, and calcium in order to maintain the ion gradients necessary for bioenergetic functions, nerve cell activity, and regulatory processes. Paradoxically, the same membranes must offer much lower barriers to the flux of water, oxygen, and carbon dioxide, which need to move rapidly into and out of the cell. The lipid bilayer itself can act as a semipermeable barrier membrane for the general functions just described, but most biological membranes also incorporate specialized transport systems in the lipid bilayer to facilitate the

1063-5823/99 $30.00

specific uptake of nutrient solutes such as glucose and amino acids. Channel proteins and ion pumps are present as well to produce ionic concentration gradients and membrane potentials characteristic of the life process.

Overton had no idea of the structural complexity of biological membranes, yet was still able to recognize that membranes somehow provide a barrier to the free diffusion of solutes. It is for that remarkable insight that we honor him with this centennial volume. This chapter focuses on the general barrier properties of membranes that are associated with the lipid bilayer, and other chapters discuss the contributions of embedded proteins. Our knowledge of bilayer permeability properties has advanced far beyond what Overton established for a series of neutral polar solutes. We now know that despite being approximately the same size, solutes such as sodium ions, hydrogen ions, water, and oxygen permeate the lipid bilayer at rates that vary by 12 orders of magnitude. How could such a vast difference in permeation rates be imposed by a seemingly simple bimolecular layer of amphiphilic molecules? Furthermore, while monovalent cations such as sodium and potassium ions permeate a lipid bilayer at approximately the same rate, a monovalent anion such as chloride permeates approximately 2 to 3 orders of magnitude faster, depending on the composition of the bilayer. Surprisingly, hydrogen ions (protons) penetrate the same bilayer at rates 5 orders of magnitude faster than other monovalent cations.

Given these remarkable facts, can the results be accounted for by a single unified theory or does the vast range in bilayer permeabilities arise from the interplay of multiple properties intrinsic to the bilayer itself as well as the interaction of solutes with both aqueous and nonpolar phases? The answer, of course, is that no single model of transbilayer transport can describe all solute permeation processes. This chapter first discusses the basic model proposed by Overton, which does a surprisingly good job of accounting for transport of small neutral solutes. It then describes exceptions to the Overton rule and shows how recent advances have expanded our understanding of this fundamental process.

II. OVERTON'S CONCEPT OF MEMBRANE PERMEABILITY

Overton's ideas about membrane permeability arose from experimental evidence gathered in numerous investigations on the osmotic properties of both plant and animal cells at the end of the last century (Overton, 1895, 1899). Many of his experiments were aimed at the assessment of the plasmolytic potency of nonionic solutes in cells. In a typical experiment, a cell was placed in a hypertonic solution and the resultant shrinkage of the cell (plasmolysis) was observed either visually or by weight measurements.

Solute molecules that were relatively membrane impermeable caused permanent plasmolysis. For instance, the volume of root hairs from *Hydrocharis morus* submerged in a 7.5% solution of sucrose decreased to a limiting value and remained so almost indefinitely (Overton, 1899). Many solutes, however, could diffuse across membranes at measurable rates, and the time it took for the cells to regain their original size was used to estimate the permeability of a given solute. Overton carried out numerous experiments and established a relationship between permeability and chemical structure for a large variety of solutes. His findings are known today under the term "Overton's rule." This rule simply states that the permeability coefficient of a solute is linearly related to its partition coefficient between oil and water. In other words, solutes that dissolve well in organic solvents such as hexane or octanol can penetrate membranes more rapidly than solutes that are less soluble in such solvents.

As illustrated in Fig. 1, with only a few exceptions, Overton's rule holds true for a surprisingly large number of compounds. In addition to providing a very practical method to predict the relative permeability of a given solute, Overton's observations also make a profound statement about the nature of the selective permeation barrier that a cell membrane imposes on solutes. The correlation between permeability and partition coefficient suggests that solute molecules need to dissolve into a hydrophobic layer present in all membranes as they enter or leave the cell.

At the time Overton put his ideas forward they were not always undisputed. Today, they are an important and fundamental part of our general understanding of membrane structure and function. In a slightly more sophisticated verison, they are reflected in the well-known "solubility–diffusion" mechanism for transmembrane solute transport. Many of the experimental techniques pioneered by Overton are still in use today. Typical applications include studies on the passive permeability of membranes or activity measurements of transmembrane channels. The same techniques can also be used in vesicular lipid bilayers such as liposomes (Bangham, *et al.,* 1967), which now permit studies of isolated, reconstituted transport proteins without the interference of other membrane proteins present in biological membranes.

III. PERMEATION BY THE SOLUBILITY–DIFFUSION MECHANISM

This section begins by reviewing how the permeability coefficient for a solute can be calculated according to the solubility–diffusion mechanism. This approach is based on simple laws of diffusion and has been used extensively to describe the permeation of many compounds, particularly

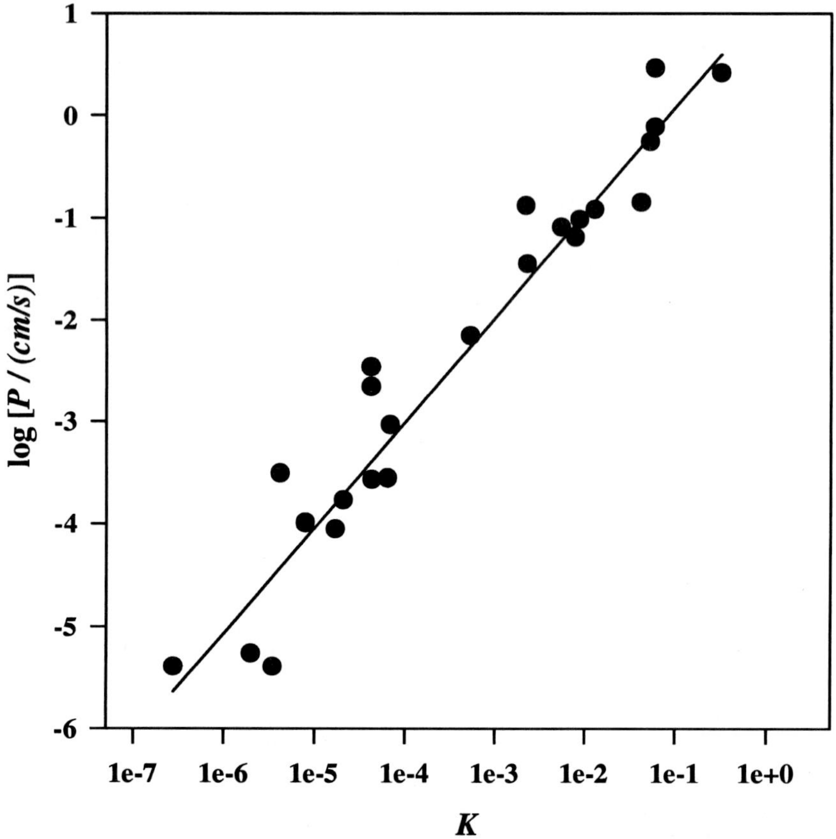

FIGURE 1 Relationship between permeability and partition coefficients for a broad selection of solutes (data from Orbach and Finkelstein, 1980; Walter and Gutknecht, 1986).

small polar molecules (Finkelstein, 1987). Predictions of the solubility–diffusion mechanism are then compared to experimental findings, both for ions and for nonelectrolytes. Finally, the solubility–diffusion mechanism is related to the observations made by Overton nearly a century ago.

A. Theory

The solubility–diffusion model postulates that the permeating species must first partition into the hydrophobic region of the membrane, diffuse across it, and then leave the membrane by dissolving into the aqueous phase on the other side of the membrane. Figure 2 describes this scenario schematically. To simplify matters, the properties of the hydrophobic region are approximated by those of an oil-like liquid such as octanol or a long-chain fluid hydrocarbon. These properties include the dielectric constant ε, which is assumed to be constant throughout the entire hydrophobic moiety of the bilayer, having a value similar to a liquid hydrocarbon. It is postulated furthermore that the structure of the bilayer is static and has well-defined boundaries at which the change from a hydrophilic to hydrophobic environment occurs rather abruptly. In addition, it is assumed that the diffuison through the hydrophobic region of the bilayer is the rate-limiting step of the overall process. Such simplifications might seem at first to call into question the reliability of this approach, but in fact the solubility–diffusion model turned out reasonably satisfactory in predicting the relative rates of solute transport across lipid bilayers.

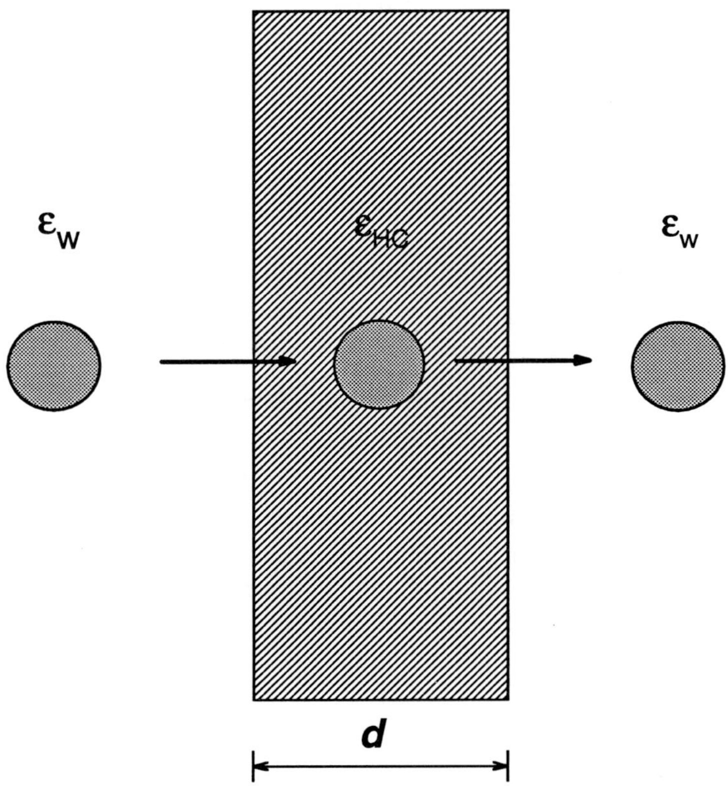

FIGURE 2 Permeation of a solute by the solubility–diffusion mechanism.

Transport phenomena can be described by Fick's first law of diffusion:

$$J = -D\left(\frac{dc}{dx}\right), \tag{1}$$

where J is the flux, expressed in moles of permeating species per second and bilayer area, and D is the diffusion coefficient of the permeating particle in the hydrophobic region. The driving force of the flux is the gradient of the chemical potential across the membrane. In the absence of electrical fields, the potential gradient is reduced to a concentration gradient. Assuming that the concentration of the permeant inside the bilayer changes linearly (Fig. 3), this gradient can be expressed as the concentration difference at the two boundaries divided by the thickness of the hydrophobic region

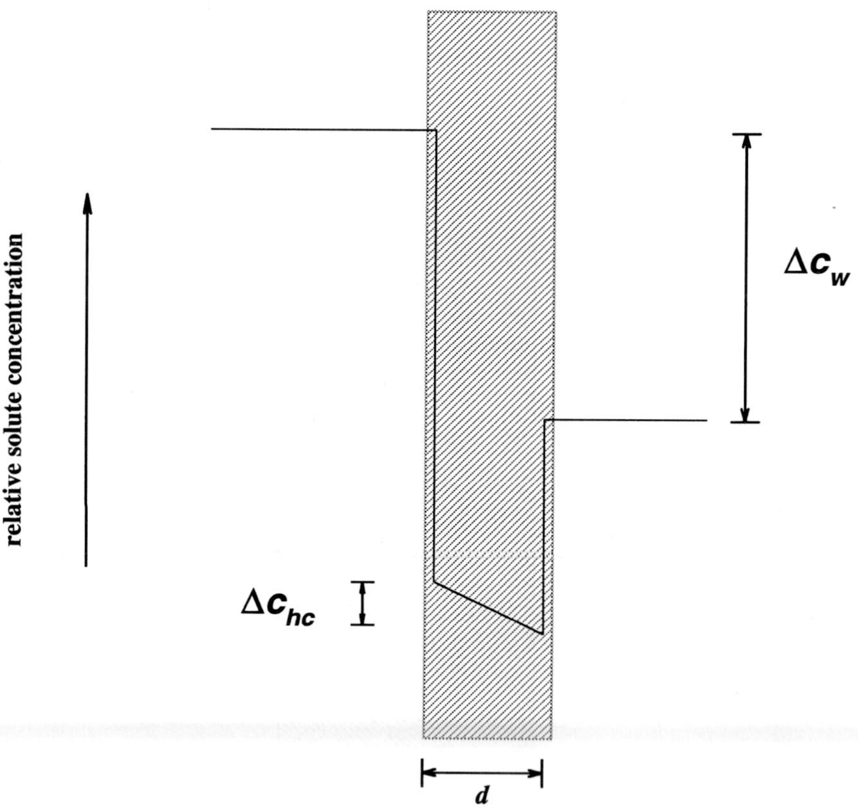

FIGURE 3 Concentration profile of a solute as it crosses a bilayer according to the solubility–diffusion mechanism.

d, as indicated in Eq. (3). To obtain the concentration difference inside the bilayer (Δc_{hc}), it is easiest to use the concentration difference of the permeant in the aqueous phase (Δc_w), which is known from the experimental conditions. The concentration gradient inside the hydrophobic part can then be calculated using the partition coefficient K of the permeant in a water/hydrocarbon system:

$$\Delta c_{\mathrm{hc}} = \Delta c_{\mathrm{w}} K. \tag{2}$$

Equation (1) can then be written as

$$J = DK \frac{\Delta c_{\mathrm{w}}}{d}. \tag{3}$$

Because it is customary to characterize the permeability of bilayers by their permeability coefficient P rather than by flux, J needs to be divided by the concentration difference in the aqueous phase to give the final expression:

$$P = \frac{J}{\Delta c_{\mathrm{w}}} = \frac{DK}{d}. \tag{4}$$

Due to the low solubility of many solutes in hydrocarbon-like liquids, experimentally determined values of D are often unavailable. Instead, values measured in water are often used as a substitute. It has been pointed out by Finkelstein (1987) that this approximation is reasonable in most cases. Water is one of the rare instances in which D has been measured in both aqueous and hydrophobic environments, and the numbers are different by less than an order of magnitude. Considering the difficulties involved in obtaining an accurate value for K, this difference is within the limits of experimental error.

The structure of a number of bilayers is known from X-ray studies (Lewis and Engelman, 1983), providing the basis for a reasonable estimate of the average thickness of the hydrophobic region, d. However, it should be kept in mind that Wiener and White (1992) have shown that the thickness is just that, an average, and that the bilayer undergoes substantial fluctuations over time, so that at any given instant there are regions both significantly thinner and thicker than the average value. These have the potential to give rise to transient defects in the bilayer barrier, which will become important later.

A related point is that it is not a simple matter to decide where to draw the boundary between the aqueous and the hydrophobic phase, as it is certainly a somewhat arbitrary choice. It is generally agreed that the polar head group region should not be included in the hydrophobic part. For instance, if the bilayer is composed of phosphatidylcholines, the second carbon atom in the fatty acid is often regarded as the location at which this transition from hydrophobic to hydrophilic occurs.

In many cases, particularly in those where the permeants have a very low solubility in hydrophobic solvents, the partition coefficient K is calculated rather than measured. This can be achieved by computing the Gibbs free energy, ΔG, that is necessary to transfer a solute from an aqueous into a hydrophobic phase:

$$K = \exp\left[\frac{-\Delta G}{RT}\right]. \tag{5}$$

The total free energy ΔG can be split into several contributions, including electrostatic, hydrophobic, and solute-specific terms:

$$\Delta G = \Delta G_{\mathrm{B}} + \Delta G_{\mathrm{I}} + \Delta G_{\mathrm{D}} + \Delta G_{\mathrm{H}} + \Delta G_{\mathrm{SP}}. \tag{6}$$

Table I summarizes how the individual energy contributions can be calculated explicitly. The first three terms in Eq. (6) are of electrostatic nature: ΔG_{B} represents the Born energy of transfer of an ion or dipole from water into a bulk hydrophobic solvent (Born, 1920; Bell, 1931). Because ε_{w} is considerably larger than $\varepsilon_{\mathrm{hc}}$, the Born energy is always large and positive, making the partitioning of ions into the membrane highly unfavorable. ΔG_{B} can be regarded as the primary origin of the permeability barrier for charged solutes.

The second term, ΔG_{I}, accounts for the fact that the membrane is not really a bulk phase but a thin slab of hydrophobic matter embedded into an aqueous environment (Neumcke and Läuger, 1969, Parsegian, 1969). ΔG_{I}, which is usually referred to as "image energy," has a negative sign and a maximum at the center of the bilayer. Although the numerical value of ΔG_{I} is only a fraction of ΔG_{B}, its contribution to the overall free energy increases as the thickness of the bilayer is reduced and can have a noticeable effect on the permeabiltiy coefficient.

Because many bilayers possess an internal dipole potential that favors the partitioning of anions over cations into the bilayer, its effect is included by adding ΔG_{D} to ΔG (Flewelling and Hubbell, 1986). The origin of the dipole potential is presumably related to oriented dipole molecules at the bilayer interface, most likely lipid head groups or tightly bound water molecues. The strength of the dipole potential is believed to be in the range of −240 mV for typical membrane lipids.

The hydrophobic term in Eq. (6), ΔG_{H}, accounts for the energy that is required to transfer an uncharged cavity the size of an individual solute atom or molecule from water into the bilayer. ΔG_{H} is always negative and linearly proportional to the surface area of the solute (Uhlig, 1937). Depending on the solute size, ΔG_{H} can make a considerable contribution

TABLE I

Various Contributions to the Gibbs Free Energy of Transfer from Water into the Center of the Hydrophobic Region of a Bilayer[a]

Ion	Dipole
Born energy:	
$\Delta G_{\mathrm{B}} = \frac{N_{\mathrm{a}}\, q^2}{8\, \pi\, \varepsilon_{\mathrm{o}}\, r} \left[\frac{1}{\varepsilon_{\mathrm{hc}}} - \frac{1}{\varepsilon_{\mathrm{w}}}\right]$	$\Delta G_{\mathrm{B}} = \frac{N_a\, p^2}{12\, \pi\, \varepsilon_0\, a^3} \left[\frac{\varepsilon_{\mathrm{w}} - 1}{2\, \varepsilon_{\mathrm{w}} + 1} - \frac{\varepsilon_{\mathrm{hc}} - 1}{2\, \varepsilon_{\mathrm{hc}} + 1}\right]$
Image energy:	
$\Delta G_{\mathrm{I}} = -\frac{N_{\mathrm{a}}\, q^2\, \vartheta}{16\, \pi\, \varepsilon_0\, \varepsilon_{\mathrm{hc}}} \left[\frac{1}{x} + \frac{1}{d} \sum_{n=1}^{\infty} \left[\frac{\vartheta^{2n}}{n + x/d} + \frac{\vartheta^{2n-2}}{n - x/d} - \frac{\vartheta^{2n}}{n + r/d} - \frac{\vartheta^{2n-2}}{n - r/d}\right]\right]$ $\vartheta = \frac{\varepsilon_{\mathrm{w}} - \varepsilon_{\mathrm{hc}}}{\varepsilon_{\mathrm{w}} + \varepsilon_{\mathrm{hc}}}$	$\Delta G_{\mathrm{I(dip)}} = -\frac{N_{\mathrm{a}}\, p^2}{6\, \pi\, \varepsilon_0\, \varepsilon_{\mathrm{hc}}\, d^3} \sum_{i=1}^{\infty} \left[\frac{2\, \vartheta^{2i-1}}{(2\, i - 1)^3}\right]$ $\vartheta = \frac{\varepsilon_{\mathrm{w}} - \varepsilon_{\mathrm{hc}}}{\varepsilon_{w} + \varepsilon_{\mathrm{hc}}}$
Hydrophobic energy:	
$\Delta G_{\mathrm{S}} = -N_{\mathrm{a}}\, 4\, \pi r^2\, \gamma_{\alpha/\beta}$	$\Delta G_{\mathrm{S}} = -N_{\mathrm{a}}\, 4\, \pi r^2\, \gamma$
Energy from dipole potential:	
$\Delta G_{\mathrm{D}} = -N_{\mathrm{a}}\, q\, \Delta\phi$	

[a] N_a, Avogadro's number; q, charge; r, radius, ε_0, dielectric constant of vacuum; ε, dielectric constant of solvent; μ, dipole moment; a, effective dipole size; d, bilayer thickness; x, distance perpendicular from interface; γ, surface energy between water and hydrophobic solvent; $\Delta\phi$, dipole potential

to ΔG, even canceling a significant portion of ΔG_{B} in the case of large molecules.

Finally, if the solute molecule undergoes specific interactions with either the water or the bilayer interior, the corresponding energy contributions are acknowledged in the term ΔG_{SP}. Examples for specific interactions are hydrogen bonds between the solute and water molecules (Paula *et al.*, 1996). Furthermore, the shielding effect of hydrating water molecules reduces the effect of the dipole potential inside the bilayer and is included in this term as well.

After having added all the terms that contribute to ΔG, the partition coefficient K can be obtained according to Eq. (5). If the reference state for the energy calculation is the state of infinite dilution, K must be converted into molarities before P is calculated.

B. Permeability of Nonelectrolytes

The value of any theory is tested when its predictions can be compared with experimental results. If the solubility–diffusion theory is correct, it should be possible to assess the permeability coefficient of any solute given the diffusion coefficient, partition coefficient, and the bilayer thickness. Critical tests of permeability theories can be carried out by measuring the permeability coefficient as a function of a parameter that can be varied experimentally and checking the results against the theoretical prediction. One such parameter is the partition coefficient K. In the past, the solubility–diffusion model has been validated for nonelectrolytes by a large number of measurements in which partition coefficients ranging over many orders of magnitude were graphed against permeability coefficients and the expected straight line was obtained (Fig. 1). Another useful parameter is the thickness of the hydrophobic region of the bilayer. Its effect on P can be obtained directly from Eq. (4), in that varying the bilayer thickness from very thin (20 Å) to very thick (37 Å) should cause P to decrease by a factor of roughly five. As illustrated in Fig. 4, this is exactly what is observed for the nonelectrolytes urea, glycerol, or water itself (Paula *et al.,* 1996), providing additional support for the mechanism.

Overton was the first to recognize that permeability coefficients are proportional to the partition coefficient of the permeant in an oil/water system. Equation (4) verifies the linear relationship between P and K. It should be noted that Overton used measured bulk partition coefficients rather than calculated ones. A drawback of this approach is that the effect of the finite thickness of the bilayer is not accounted for and consequently bilayer-specific effects arising from image forces or internal dipole potentials are ignored. For instance, certain solutes have a higher P value than estimated from Overton's rule, and these exceptions are mostly small molecules with high dipole moments for which the effects of image forces and dipole potentials are strongest.

In summary, Overton's rule provides a good first estimate of P for solutes with known partition coefficients. Furthermore, at a time when there was virtually no knowledge of membrane structure, it offered remarkable insight into the nature of the permeation barrier imposed by membranes.

C. Permeability of Ionic Solutes

Can the same principles that apply to nonelectrolytes be used to estimate the permeation of ions? From the experimental point of view it is noteworthy that permeability coefficients of ions are much smaller than those of

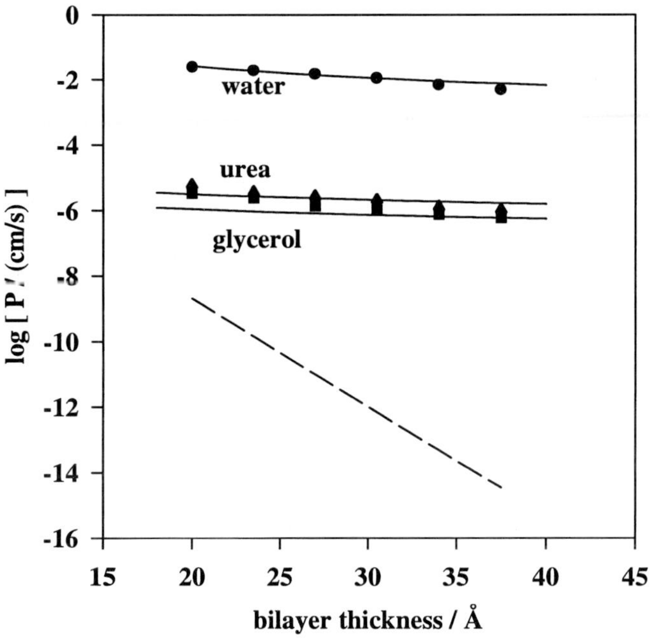

FIGURE 4 Permeability coefficients for water, urea, and glycerol [experimental data from Paula *et al.* (1996)]. Solid lines are calculated values (solubility–diffusion mechanism) and the dashed line is calculated values (pore mechanism for water).

small nonelectrolytes. The higher permeation barrier originates mainly in the Born energy, which is large and positive and causes the solubility of ions in the bilayer interior to be very low (Parsegian, 1969). In contrast, the contribution of the Born energy for nonelectrolytes only plays a role if the molecule is relatively small and possesses a significant dipole moment.

When calculating permeability coefficients according to the solubility–diffusion mechanism, it is crucial to use hydrated ionic radii, as bare radii give values that are orders of magnitude below experimental results. This is most apparent for protons, which give absurd values if the bare radius is used. It has been shown that ionic permeation can be described properly by using hydrated radii in conjunction with the solubility–diffusion mechanism (Paula *et al.,* 1996, 1998). The method works well for halide anions and also for cations, if the bilayer is sufficiently thick. However, discrepancies become apparent for cations in thin bilayers. These are addressed in Section IV.

IV. PERMEATION THROUGH TRANSIENT PORES

Although the solubility–diffusion mechanism is satisfactory for most permeants, there are a few instances in which it clearly fails. These mismatches are encountered in thinner bilayers if the permeants are positively charged ions, such as potassium in liposomes made from phosphatidylcholines. In this case, the permeability coefficients are greater than predicted by the solubility–diffusion model. To account for this discrepancy, transient pores in the bilayer produced by thermal fluctuations have been suggested as an alternative pathway (Fig. 5). Parsegian (1969) showed that the energy of an ion located inside an aqueous pore is significantly smaller than the energy of the same ion in the hydrophobic part of the bilayer. As a result of this lower diffusion barrier, the permeation rate of ions is accelerated. The primary question here is whether a sufficient number of pores is present in the bilayer so that the net rate of ions permeating through pores can exceed that of ions permeating by solubility and diffusion. Intuitively, this pathway becomes an attractive alternative as bilayers tend to become more unstable with decreasing bilayer thickness.

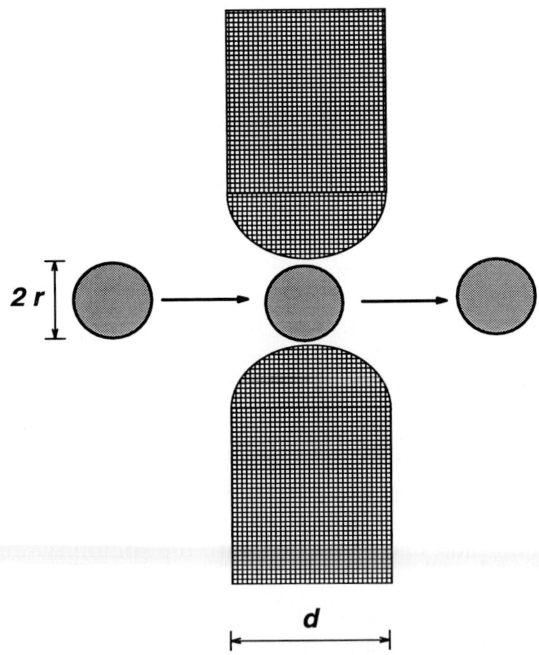

FIGURE 5 Permeation through hydrophilic pores.

A. Theory

The following paragraphs use the potassium ion as an example to illustrate permeation through transient pores. Calculating a permeability coefficient for an ion through a bilayer of a given thickness is mathematically challenging, although attempts have been made to achieve this goal (Markin and Kozlov, 1985). For simplicity, the approach introduced by Hamilton and Kaler (1990) will be used. This model is semiempirical and is based on a number of assumptions, but has the advantage of being easy to approach mathematically without losing its validity.

The model assumes that permeants pass through transient, hydrated pores formed by thermal fluctuations in the bilayer. Hydrophilic pores are assumed in the model rather than unhydrated, hydrophobic pores because their formation is considered to be energetically less costly and therefore more likely. The rate at which permeants cross the bilayer is related to the probability at which pores of sufficient size and depth appear in the bilayer. The probability of pore formation can then be written as an exponential function of the energy necessary to form a pore of given area and depth. Hamilton and Kaler (1990) expressed these ideas in the following equation:

$$P = \frac{D_M \, \sigma \, n_o \, RT}{R_v \, A \, k_1} \left[\pi r^2 + \frac{RT}{k_1} \right] \exp\left[\frac{-k_1 \, \pi r^2}{RT} \right] \exp\left[\frac{-k_2 \, d}{RT} \right], \tag{7}$$

where the parameters k_1 and k_2 are the energies required to create a pore of unity area and of unity depth in the bilayer. They were determined by fitting a set of experimental data to Eq. (7). Alternatively, one could use bending constants of bilayers to calculate the energy that is required to form a pore of given size. Equation (7) can then be used to compare theoretical to measured permeability coefficients as a function of parameters such as bilayer thickness or solute size.

B. Comparison of Pore and Solubility–Diffusion Permeation Models

By using the appropriate parameters for a potassium ion, Eq. (7) yields a more adequate description of potassium permeability in thin bilayers than the corresponding equation for the solubility–diffusion mechanism (Fig. 6). Because of the strong dependence of P on d, the pore mechanism predicts a steep increase in P as the thickness of the bilayer is reduced, which is exactly what is observed in experiments. This strongly suggests that permeation through transient hydrated pores in thin bilayers is the dominating pathway for potassium ions. In thinner bilayers, permeation by

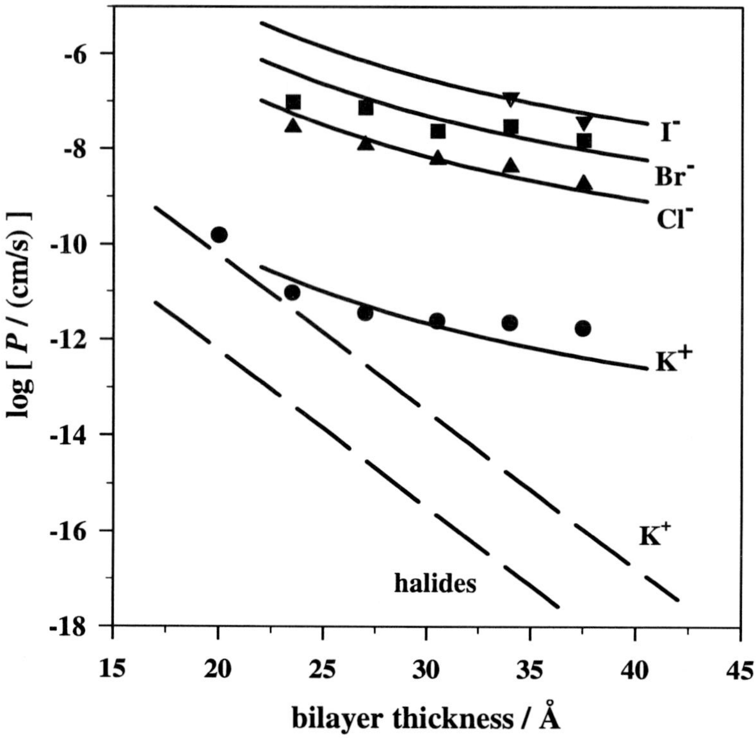

FIGURE 6 Permeability coefficients of potassium ions and halides [experimental data from Paula *et al.* (1998)]. Solid lines are solubility–diffusion mechanism and dashed lines are pore mechanism.

solubility and diffusion contributes insignificantly to the overall ion transport, but becomes increasingly relevant in thicker bilayers.

In general, Eqs. (4) and (7) offer the possibility of comparing the characteristic parameter sensitivity of P for the two mechanisms. For example, as illustrated in Fig. 7, the sensitivity of the pore mechanism to solute size is much less than that of the solubility–diffusion mechanism. Exactly the opposite is true if the bilayer thickness d is varied. Here, the pore mechanism exhibits greater parameter sensitivity (see Fig. 6). These very different characteristics become useful when assigning a permeation mechanism to a certain solute. This point is illustrated by discussing the permeation of the inorganic anions chloride, bromide, and iodide through phospholipid bilayers whose thickness is varied between 32 and 48 carbon atoms (Paula *et al.*, 1998).

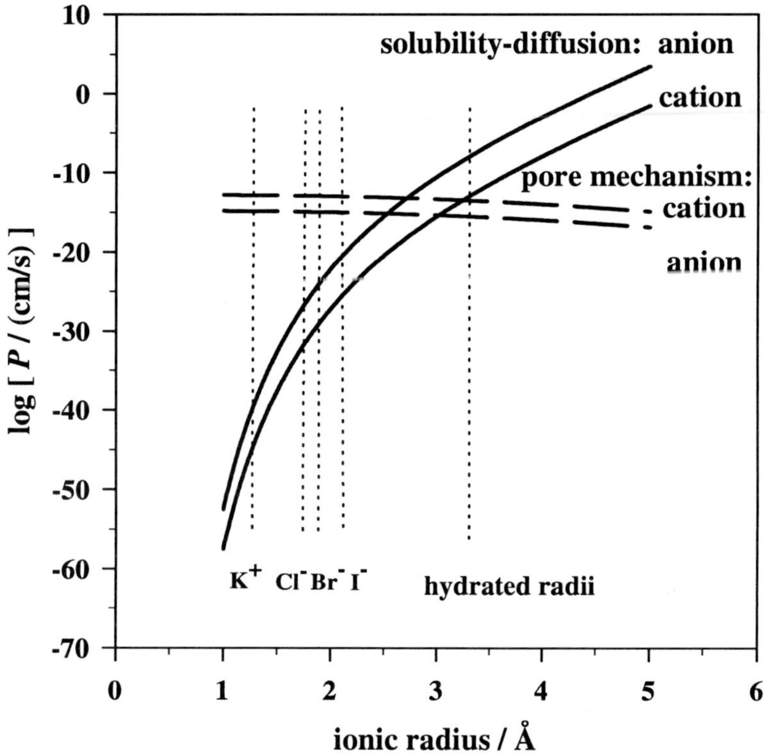

FIGURE 7 Permeability coefficient as a function of solute radius.

Interestingly, the pore mechanism does not provide a satisfactory answer for these halides, despite the fact the their bare radii are in the same size range as that of potassium (K^+ = 1.6 Å; Cl^- = 1.8 Å). If P of a halide is plotted versus d, the resulting graph more closely resembles that expected from the solubility–diffusion mechanism, even in the region of thin bilayers. Furthermore, the permeability coefficient increases by a factor of roughly five in the order $P_{Cl} < P_{Br} < P_{I}$, although the hydrated radii are very similar. This order of increase in P plus its strong radius dependence can only be explained by the solubility–diffusion mechanism using hydrated ionic radii and not by the pore mechanism. Both observations indicate that the solubility–diffusion mechanism is the preferred path for halide permeation. Table II summarizes the contrasting parameter sensitivities of the two mechanisms.

TABLE II

Comparison of the Parameter Sensitivity of the Solubility–Diffusion and the Pore Mechanism

Parameter	Solubility–diffusion mechanism	Pore mechanism
Bilayer thickness	Moderate sensitivity	Pronounced sensitivity
Solute radius	Pronounced sensitivity	Little sensitivity
Sign of solute charge	$P_{anion} > P_{cation}$	$P_{anion} > P_{cation}$[a]

[a] Depends on the nature of the lipid head group.

V. PROTON PERMEATION

Among all the permeants that are of biological interest, protons should be regarded as a special case. Proton transport across cell membranes has many biologically important functions. For example, chemical energy gained by the redox reactions of the respiratory chain is stored in an electrochemical gradient across the mitochondrial membrane, which is then used for ATP synthesis. Obviously, processes like this require membranes that provide a sufficiently large diffusion barrier for protons. Passive proton permeation has therefore been the center of interest of many studies (Nichols and Deamer, 1980; Elamrani and Blume, 1983; Perkins and Cafiso, 1986).

One of the most intriguing experimental observations is the magnitude of the permeability coefficient for protons, which surpasses that expected from other monovalent cations by several orders of magnitude. How might such a large anomaly arise?

We can begin by attempting to calculate the permeability expected from the solubility–diffusion model and the transient pore model. In order to do this, we need to assign an appropriate radius to the permeating hydrogen cation. Clearly, using the radius of a bare proton gives unrealistically small values for P and ignores experimental evidence for the hydration of protons in aqueous solutions. Species such as H_3O^+ or $H_9O_4^+$ seem to be a better choice. In the case of the solubility–diffusion model, the radius of H_3O^+ yields permeability coefficients that are orders of magnitude lower than what is observed experimentally, while the radius of $H_9O_4^+$ gives a P value that is consistent with experimental data (Paula *et al.,* 1996). However, this is an arbitrary choice with no experimental support, and it is therefore worth considering whether other mechanisms could also account for the proton permeability anomaly.

One possibility is that transient defects occur to some extent in virtually all lipid bilayers and that their relative numbers are a function of bilayer thickness. As discussed earlier, thinner bilayers have larger numbers of such defects, which becomes a dominant permeation mechanism for ions such as potassium. It also seems likely that the transient defects exist over a wide range of time scales, with some being so short lived that potassium ions simply do not have sufficient time to diffuse across the bilayer. However, protons can diffuse by a special mechanism that is not available to other cations, which involves hopping along hydrogen-bonded chains of water. One example of this hopping is the translocation that occurs along the single strand of water molecules in the gramicidin channel (Myers and Haydon, 1972; Levitt *et al.*, 1978; Akeson and Deamer, 1991). A more general mechanism for proton transport in biological systems has been explored by Nagle and Tristram-Nagle (1983).

To explain the extraordinarily high proton permeability when it was first observed in liposomes, Nichols and Deamer (1980) postulated that protons might travel along water wires that span the membrane through short-lived defects. A proton could then travel through very short-lived defects that do not exist long enough to accommodate other monovalent inorganic ions.

Molecular dynamics simulations have provided considerable insight into permeation processes across membranes. The strength of this method is that it takes into account inhomogeneities in the bilayer and that it presents molecular details that continuum models such as those described earlier cannot provide. The first solutes whose permeations through bilayers were described by simulations were small molecules such as methane and helium (Sok and Berendsen, 1992). More recently, biologically more interesting species such as water and protons have been addressed as well (Marrink and Berendsen, 1994; Marrink *et al.*, 1996). The result for protons is highly pertinent to the discussion here. That is, Marrink *et al.* (1996) showed that a single strand of water produced within a transient defect in a lipid bilayer could exist for a few picoseconds, just long enough to accommodate the transport of a single proton before being dispersed.

To summarize, we now have two alternative hypotheses that at least qualitatively account for the anomolously high proton permeability of lipid bilayers. If protons can partition into bilayers as $H_9O_4^+$, the solubility–diffusion mechanism can account for the high permeability. However, if short-lived transient defects are present that allow protons, but not other cations, to be translocated along hydrogen-bonded water chains, the anomaly would be explained by a variation of the pore model. It will be interesting to see whether future research approaches will be able to test experimentally the two hypotheses.

VI. SUMMARY

Overton's basic concept of solubility and diffusion can account for a remarkable variety of solute transport across the lipid bilayer moiety of membranes, particularly for the diffusional translocation of small neutral molecules. The solubility–diffusion model can also account for the transport of monovalent cations and halide ions in typical lipid bilayers, but it must take into account other energy barriers, particularly Born energy and hydrophobic energy related to the size of the hydrated ion. Solubility–diffusion fails when one attempts to describe cation transport across thin lipid bilayers such as those composed of 14 carbon phosphatidylcholines, and it becomes necessary to propose that a second process comes into play, which is ion translocation through transient defects in the bilayer. This mechanism becomes most important for explaining proton translocation, which occurs at five to six orders of magnitude greater rates than expected from the permeation of other monovalent cations. Here the best explanation is that very short-lived strands of hydrogen-bonded water in the defects allow protons to cross by hopping along hydrogen-bonded chains of water molecules, a mechanism that is not available to other cations.

References

Akeson, M., and Deamer, D. W. (1991). Proton conductance by the gramicidin water wire. Model for proton conductance in the F1F0 ATPases? *Biophys. J.* **60,** 101–198.

Bangham, A. D., De Gier, J., and Greville, G. D. (1967). Osmotic properties and water permeability of phospholipid liquid crystals. *Chem. Phys. Lipids* **1,** 225–246.

Bell, R. P. (1931). The electrostatic energy of dipole molecules in different media. *Trans. Faraday Soc.* **27,** 797–802.

Born, M. (1920). Volumen und Hydrationswärme der Ionen. *Z. Phys.* **1,** 45–48.

Elamrani, K., and Blume, A. (1983). Effect of the lipid phase transition on the kinetics of H^+/OH^- diffusion across phosphatidic acid bilayers. *Biochim. Biophys. Acta* **727,** 22–30.

Finkelstein, A. (1987). "Water Movement through Lipid Bilayers, Pores, and Plasma Membranes: Theory and Reality." Wiley (Interscience), New York.

Flewelling, R. F., and Hubbell, W. L. (1986). The membrane dipole potential in a total membrane potential model. *Biophys. J.* **49,** 541–552.

Hamilton, R. T., and Kaler, E. W. (1990). Alkali metal ion transport through thin bilayers. *J. Phys. Chem.* **94,** 2560–2566.

Levitt, D. G., Elias, S. R., and Hautman, J. M. (1978). Number of water molecules coupled to the transport of sodium, potassium and hydrogen ions via gramicidin, nanactin or valinomycin. *Biochim. Biophys. Acta* **512,** 436–451.

Lewis, B. A., and Engelman, D. M. (1983). Lipid bilayer thickness varies linearly with acyl chain length in fluid phosphatidylcholine vesicles. *J. Mol. Biol.* **166,** 211–217.

Markin, V. S., and Kozlov, M. M. (1985). Pore statistics in bilayer lipid membranes. *Biol. Mem.* **2,** 404–442.

Marrink, S. J., and Berendsen, H. J. C. (1994). Simulation of water transport through a lipid membrane. *J. Phys. Chem.* **98,** 4155–4168.

Marrink, S. J., Jähnig, F., and Berendsen, H. J. C. (1996). Proton transport across transient single-file water pores in a lipid membrane studied by molecular dynamics simulations. *Biophys. J.* **71,** 632–647.

Myers, V. B., and Haydon, D. A. (1972). Ion transfer across lipid membranes in the presence of gramicidin A. *Biochim. Biophys. Acta* **274,** 313–322.

Nagle, J. F., and Tristram-Nagle, S. (1983). Hydrogen-bonded chain mechanisms for proton conduction and proton pumping. *J. Membr. Biol.* **74,** 1–14.

Neumcke, B., and Läuger, P. (1969). Nonlinear electrical effects in lipid bilayer membranes. II. *Biophys. J.* **9,** 1160–1969.

Nichols, J. W., and Deamer, D. W. (1980). Net proton-hydroxyl permeability of large unilamellar liposomes measured by an acid–base titration technique. *Proc. Natl. Acad. Sci. U.S.A.* **77,** 2038–2042.

Orbach, E., and Finkelstein, A. (1980). The nonelectrolyte permeability of planar lipid bilayer membranes. *J. Gen. Physiol.* **75,** 427–436.

Overton, E. (1895). Über die osmotischen Eigenschaften der lebenden Pflanzen und Tierzelles. *Vierteljahrsschr. Naturforsch. Ges. Zuerich* **40,** 159–201.

Overton, E. (1899). Über die allgemeinen osmotischen Eigenschaften der Zelle, ihre vermutlichen Ursachen und ihre Bedeutung für die Physiologie. *Vierteljahreschrift Naturforsch. Ges.* (*Zürich*) **44,** 88–114.

Parsegian, A. (1969). Energy of an ion crossing a low dielectric membrane: Solutions to four relevant electrostatic problems. *Nature* (*London*) **221,** 844–846.

Paula, S., Volkov, A. G., Van Hoek, A. N., Haines, T. H., and Deamer, D. W. (1996). Permeation of protons, potassium ions, and small polar molecules through phospholipid bilayers as a function of membrane thickness. *Biophys. J.* **70,** 339–348.

Paula, S., Volkov, A. G., and Deamer, D. W. (1998). Permeation of halide anions through phospholipid bilayers occurs by the solubility-diffusion mechanism. *Biophys. J.* **74,** 319–327.

Perkins, W. R., and Cafiso, D. S. (1986). An electrical and structural characterization of H^+/OH^- currents in phospholipid vesicles. *Biochemistry* **25,** 2270–2276.

Sok, R. M., and Berendsen, H. J. C. (1992). Molecular dynamics simulation of the transport of small molecules across a polymer membrane. *J. Chem Phys.* **96,** 4699–4704.

Uhlig, H. H. (1937). The solubilities of gases and surface tension. *J. Phys. Chem.* **41,** 1215–1225.

Walter, A., and Gutknecht, J. (1986). Permeability of small nonelectrolytes through lipid bilayer membranes. *J. Membr. Biol.* **90,** 207–217.

Wiener, M. C., and White, S. H. (1992). Structure of a fluid dioleoylphosphatidylcholine bilayer determined by joint refinement of x-ray and neutron diffraction data. III. Complete structure. *Biophys. J.* **61,** 437–447.

CHAPTER 5

Water Permeation across Membranes

A. S. Verkman
Departments of Medicine and Physiology, Cardiovascular Research Institute, University of California, San Francisco, San Francisco, California 94143

I. INTRODUCTION

One of the early descriptions of water transport appeared in a monograph by Dr. W. Pfeffer (1877), in which many highly perceptive observations are made about how water might move across lipid and cell membranes. There are numerous hints about molecular advances to come more than a century later, such as the existence of "membrane particles" involved in water transport. The underpinnings for much of the recent work were established in the 1950s and 1960s, where water transport was characterized in erythrocytes, kidney tubules, and amphibian urinary bladder. These tissues were found to have remarkably higher water permeability than lipid bilayers, and in the case of amphibian urinary bladder, water permeability could be turned on and off by addition of the hormone vasotocin. In human

1063-5823/99 $30.00

erythrocytes, a constitutively high water permeability could be inhibited by >90% by mercurial sulfhydryl reagents.

Much of the work in the field in the 1970s and 1980s was directed at the characterization of water permeability in different artificial and biological membranes and in the search for the molecules responsible for facilitated water transport. The 1980s were a time of significant advances in defining the cell biology of vasopressin-regulated water permeability in amphibian urinary bladder and kidney-collecting duct epithelia. Elegant analyses by freeze-fracture electron microscopy suggested a membrane-cycling mechanism in which vasopressin induces the exocytic insertion of vesicles containing specific intramembrane particles (putative water channels) into the apical plasma membrane. The resultant increased water permeability is subsequently turned off by hormone removal and endocytic retrieval of the water-transporting particles. The reader is referred to Finkelstein (1987) and Verkman (1989) for further descriptions and references to older work in the field.

The 1990s have been a time of remarkable advances in the molecular identification and structure–function analysis of water transporting proteins in cells. There are now at least 10 water-transporting proteins ("water channels" or "aquaporins") in mammals and many more in plants, bacteria, and other organisms (reviewed in Agre *et al.,* 1993; Chrispeels and Maurel, 1994; Van Os *et al.,* 1994; Knepper, 1994; Verkman *et al.,* 1996). Several different types of membrane transporters have been found to carry water under some conditions. Technical advances in the measurement of water transport and in structure analysis by diffraction methods have yielded key information on water channel structure–function relationships. Transgenic mouse models with specific deletions in water channel proteins have been developed to study the physiological role of water channels. This chapter reviews the basic biophysical description of water transport, modern measurement approaches, current knowledge about how water moves through bilayers and aquaporin water channels, and recent advances in the field. Although significant progress has been made in the past decade, there are major discoveries yet to come, particularly in the realm of translating the advances in basic science into clinically useful diagnostic and therapeutic modalities.

II. BIOPHYSICS OF WATER TRANSPORT

A. *Parameters Defining Water Permeability*

This section briefly reviews the biophysical parameters that characterize water transport. More extensive treatment of this subject can be found in

Finkelstein (1987) and Verkman (1989). Consider a semipermeable membrane separating compartments 1 and 2 (Fig. 1) in which there is one type of impermeant and permeant solute. The volume flow (J_v, cm^3/sec) across the membrane is defined by

$$J_v = P_f\, S\, v_w\, [(c_{i2} - c_{i1}) + \sigma_p(c_{p2} - c_{p1}) + (P_1 - P_2)/RT], \tag{1}$$

where P_f (cm/sec) is the osmotic water permeability coefficient, S (cm^2) is the membrane surface area, v_w (18 cm^3/mol) is the partial molar volume of water, P is the hydrostatic pressure (atm), σ_p is the reflection coefficient of the permeant solute, c_{i1} and c_{i2} are the osmolalities of impermeant solutes on sides 1 and 2, and c_{p1} and c_{p2} are the osmolalities of permeant solutes on sides 1 and 2. J_v is defined as positive in the direction from side 1 to 2. P_f is a phenomenological coefficient that is independent of osmotic gradient size and direction for simple membranes. Equation (1) also provides a accurate description of transport across complex barriers such as epithelia, provided that the details of individual pathways are taken into account. The flux (J_s, mol/sec) of an uncharged permeant solute (c_p) is

$$J_s = P_s\, S\, (c_{p1} - c_{p2}) + J_v\, (1 - \sigma_p)\, <c_p>, \tag{2}$$

where P_s (cm/sec) is the solute permeability coefficient and $<c_p>$ is the mean solute concentration in the pore. Equations (1) and (2) (the Kedem–Katchalsky equations) are formally valid for small osmotic and solute gradients and should be used with caution in real experimental situations. For charged solutes, the equations are modified to include electroosmotic phe-

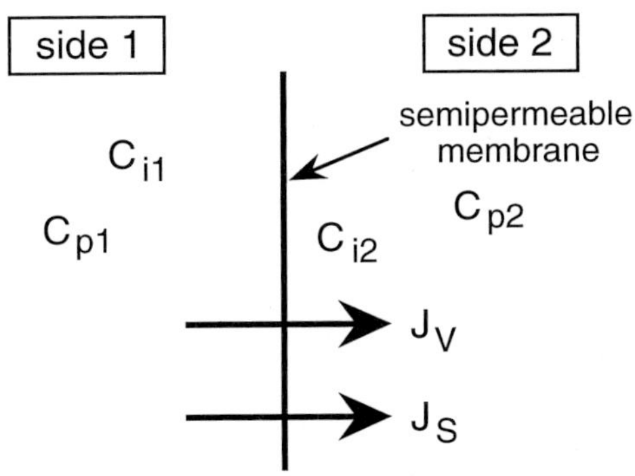

FIGURE 1 Water and solute transport across a semipermeable membrane. A membrane is depicted with specified osmotic water permeability coefficient (P_f), permeant solute diffusion coefficient (P_s), and solute reflection coefficient (σ_p). Impermeant solutes (c_{i1} and c_{12}) and permeant solutes (c_{p1} and c_{p2}) are shown. Transmembrane volume flow (J_v) and solute flux (J_s) are indicated. See Eqs. (1) and (2) for a quantitative description of coupled water and solute transport.

nomena. The second term in Eq. (2) defines the solvent drag of a solute with σ_p less than unity. Equations (1) and (2) have been integrated numerically for specific geometries in various studies involving σ_p determination (Levitt and Mlekoday, 1983; Pearce and Verkman, 1989; Shi *et al.,* 1991; Yang and Verkman, 1998). Values of $\sigma_p << 1$ have been taken as evidence for a common water/solute pathway; however, the experimental determination of σ_p is often challenging and subject to a variety of artifacts.

For water movement through distinct water-transporting units such as aquaporin-type water channels, it is useful to define a single channel osmotic water permeability p_f (cm^3/sec) as $p_f = P_f/n$, where n is the membrane density of functional water channels (number/cm^2). The single channel water permeability has been thought to be useful in describing channel geometry. However, it is becoming evident that water channels have a complex structure in which relating pore geometry to water permeability will require atomic resolution structure information and molecular dynamics computations.

Another useful parameter describing a water-transporting membrane is the Arrhenius activation energy (E_a, kcal/mol), obtained from the slope of a plot of ln P_f vs $1/RT$, where R is the gas constant and T is the absolute temperature. E_a is generally high (> 8 kcal/mol) for water movement through a lipid membrane and low (< 6 kcal/mol) for water movement through aqueous channels (Finkelstein, 1987; Verkman, 1989). The high E_a for water movement through lipid may be related to the formation and breaking of hydrogen bonds as water moves between aqueous and membrane phases. The low E_a for channel-mediated water transport is related to the weak temperature dependence of water self-diffusion and of the diffusion of small solutes in an aqueous phase. It is noted that conclusions about water-transporting mechanisms based on E_a values should be viewed cautiously in biological membranes where multiple water-transporting pathways might be present and where the lipid and protein states may undergo complex temperature-dependent changes. It is also noted that E_a can be low (~5 kcal/mol) if P_f is unstirred-layer limited, even in the absence of water channels.

The ratio of osmotic-to-diffusional water permeability (P_f/P_d) can also provide useful information about the presence of a facilitated water-transporting pathway. The diffusional water permeability coefficient P_d (cm/sec) is defined in terms of the rate of water transport (exchange) across a membrane in the absence of an osmotic gradient,

$$J_{H_2O} = P_d\, S\, ([H_2O^*{}_1] - [H_2O^*{}_2]), \tag{3}$$

where J_{H_2O} is the diffusive water flow and $[H_2O^*]$ is the concentration of "labeled" water (see later). P_f/P_d should equal unity for a simple lipid bilayer membrane that does not contain water channels. P_f/P_d can be greater than unity when water moves through a wide pore or narrow channel or when measured P_d is less than true membrane P_d because of unstirred layers. Various equations have been derived that relate p_f and P_f/P_d to pore geometries (e.g., right cylindrical wide pore and narrow channel, telescope-like pore; Finkelstein, 1987; Hill, 1994); however, as mentioned earlier, it is unlikely that these simplified equations provide useful information for real channels.

B. Measurement of Water Permeability

Of the parameters listed earlier, the osmotic water permeability is the most important parameter in characterizing a water-transporting pathway. Various strategies have been developed for the measurement of osmotic water transport across the limiting membrane in vesicles and cells and across serial membrane barriers in epithelia. In vesicles and cells, the rate of water transport is deduced from the time course of vesicle/cell volume change in response to an imposed osmotic gradient. Transepithelial water transport requires the measurement of net water displacement by direct volume determination or from the dilution of a membrane-impermeant marker. The available methods are reviewed next with emphasis on newer methods for water transport measurement.

1. P_f in Liposomes and Vesicles

Light-scattering and fluorescence-quenching methods are suitable for the measurement of P_f in reconstituted proteoliposomes and in biomembrane vesicles such as plasma membrane vesicles and endosomes. Light scattering is based on the dependence of elastically scattered light on liposome volume. Usually a stopped-flow apparatus is used to establish an osmotic gradient. P_f is computed from the time course of scattered light intensity, vesicle size, and an empirically determined relation between light scattering and vesicle volume (Van Hoek and Verkman, 1992). Fluorescence quenching is based on concentration-dependent self-quenching of certain fluorophores such as FITC–dextran and carboxyfluorescein (Chen *et al.,* 1988). Vesicles or liposomes are loaded with a high concentration of fluorophore and subjected to an inwardly directed osmotic gradient. Water efflux causes vesicle shrinkage and increasing fluorophore concentration, resulting in self-quenching and decreased fluorescence (Fig. 2A). P_f is determined as in the light-scattering method but with an empirical fluorescence vs volume relation. Fluorescence quenching has been used to measure water permeability in plasma membrane vesicles (Chen *et al.,* 1988) and endosomes

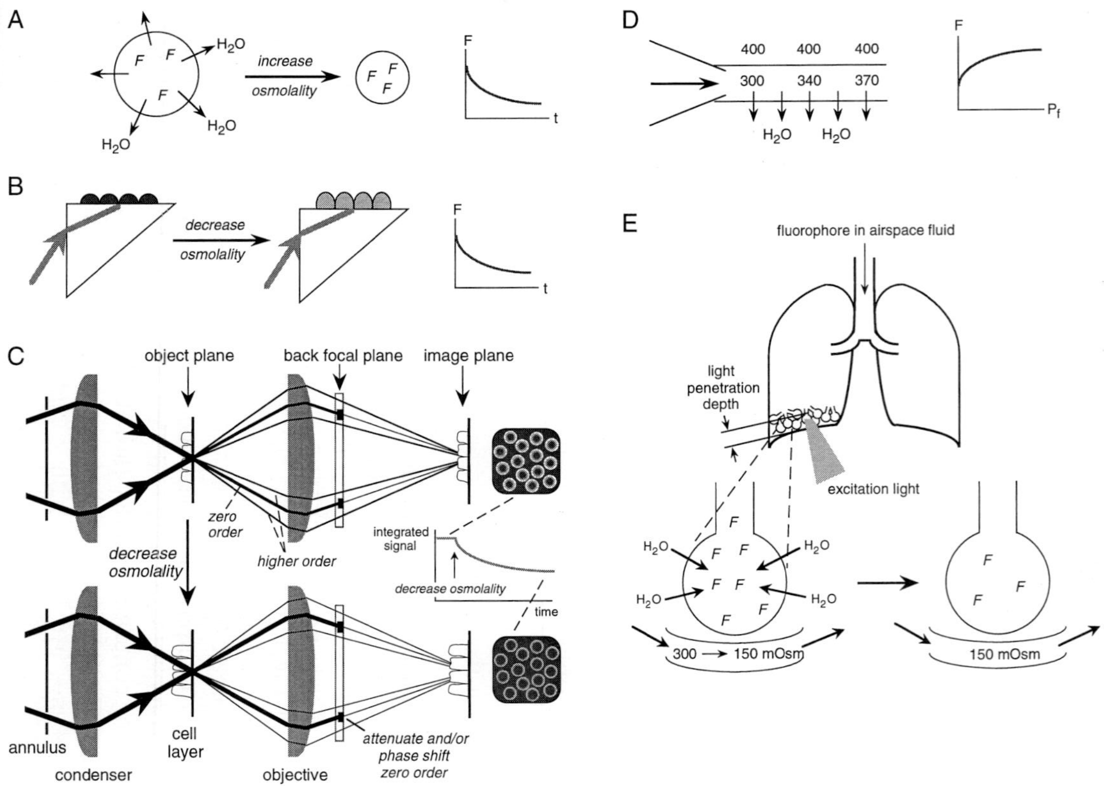
A
H2O
H2O
H2O
increase
osmolality
F
t
B
decrease
osmolality
F
t
C
object plane
back focal plane
image plane
zero
order
higher order
integrated
signal
decrease osmolality
time
decrease
osmolality
annulus
condenser
cell
layer
objective
attenuate and/or
phase shift
zero order
D
400
400
400
300
340
370
H2O
H2O
F
Pf
E
fluorophore in airspace fluid
light
penetration
depth
excitation light
300 → 150 mOsm
150 mOsm

from toad urinary bladder (Shi *et al.*, 1990) and kidney (Verkman *et al.*, 1988) containing the vasopressin-sensitive water channel.

2. P_f in Cells in Suspension

The light-scattering method has been used extensively to measure water permeability in erythrocytes (Levitt and Mlekoday, 1983; Macey, 1984), but has limited applications in larger and more complex cells. Light scattering was used recently to show that immunoisolated type I alveolar epithelial cells from lung had the highest water permeability of any mammalian cell membrane studied to date (Dobbs *et al.*, 1998). Concerns in the use of light scattering in cells include unstirred layers, uncertainties in cell surface area, possible artifacts arising from cell shape changes, and (volume-independent) light scattering from intracellular structures. For this reason, P_f data should be confirmed by measurements in membrane vesicles isolated from the cells, as was done in the type I cell study. For very large cells such as *Xenopus laevis* oocytes, cell volume can be monitored directly by image analysis (Zhang *et al.*, 1990).

FIGURE 2 Optical methods for the measurement of osmotic water permeability across membranes. (A) Self-quenching of an entrapped fluorescent dye. Liposomes or vesicles are loaded with a membrane-impermeant dye. In response to an increase in external osmolality, there is decreased volume, increased fluorophore concentration, and fluorescence self-quenching. P_f is determined from the time course of the decreasing fluorescence. (B) Total internal reflection fluorescence (TIRF) microscopy. Cells are grown or immobilized on a transparent support and the cytoplasm is stained with a fluorescent dye. The cells swell in response to a decrease in extracellular perfusate osmolality, resulting in cytoplasmic dye dilution. The concentration of cytoplasmic dye is measured by TIRF optics using a laser beam to illuminate only 100–200 nm of the cytoplasm beneath the cell surface. P_f is determined from the time course of decreasing fluorescence. (C) Spatial-filtering microscopy. A cell layer is positioned in the object plane of a phase contrast for dark-field microscopy. Cell swelling in response to a decrease in perfusate osmolality results in a redistribution of transmitted light between zero and higher order beams. Spatial filtering in the back focal plane produces a change in integrated signal intensity in the image plane, which is directly related to changes in cell volume. (D) Perfused cylindrical epithelia such as kidney tubules or airways. The tubule is perfused at constant flow with a solution containing a membrane-impermeant fluorescent dye and bathed in a solution of different osmolality. In the steady state, fluorescence changes along the tubule axis because of transepithelial water movement. P_f is determined from the fluorescence at the distal end of the tubule. (E) Airspace-to-capillary water permeability in a perfused lung. The airspace is filled with saline containing a membrane-impermeant fluorescent probe, and pleural surface fluorescence is monitored by epifluorescence microscopy. In response to a decrease in osmolality in fluid perfused into the pulmonary artery, water moves into the airspace and dilutes the fluorophore. Surface fluorescence provides a quantitative measure of intraalveolar fluorophore concentration. See text for references and additional details.

3. Membrane P_f in Adherent Cells and Epithelia

Measurements of cell plasma membrane water permeability are often required in adherent or flat cell sheets such as cultured or transfected cells and in intact epithelia such as amphibian urinary bladder. Although P_f measurements have been made by light scattering in some types of cells cultured on glass coverslips (Fischbarg *et al.,* 1990; Echevarria and Verkman, 1992), the light-scattering method has limited utility in these systems for the reasons mentioned earlier. Several new strategies have been developed to measure P_f in adherent cells and intact epithelia. A total internal reflection fluorescence (TIRF) method was developed for P_f measurement in cells cultured or immobilized on a transparent rigid support (Farinas *et al.,* 1995). Cell cytoplasm was stained with a polar fluorescent indicator such as calcein, and relative cell volume (inversely proportional to indicator concentration) was deduced from the TIR fluorescence from a thin ($<$ 200 nm) layer of cytoplasm near the transparent support (Fig. 2B). TIRF was applied to measure P_f in cells transfected with the vasopressin-regulated water channel AQP2 (Katsura *et al.,* 1995; Valenti *et al.,* 1996). A strategy to measure the volume of nonadherent epithelial cell layers, such as cells cultured on porous supports or epithelial layers, was developed based on the dependence of intracellular refractive index on cell volume. A proof-of-principle study was initially done utilizing laser interferometry (Farinas and Verkman, 1996) in which changes in intracellular refractive index were quantified by measurement of the small changes in optical path length of a laser beam passing through a cell layer. Although membrane P_f could be measured in cultured epithelial cells and an intact epithelium (toad urinary bladder), the interferometry method is impractical for use in most laboratories because of the requirement of specialized optical instrumentation and the need for constant temperature, vibration-free facilities. A simple method has been developed to exploit the dependence of the intracellular refractive index on cell volume utilizing spatial filtering Fourier optics as found in common phase contrast and dark-field microscopes (Fig. 2C; Farinas *et al.,* 1997). With the appropriate arrangement of optics, the intensity of transmitted monochromatic light is a direct measure of cell volume. The spatial filtering method was used to measure apical vs basolateral membrane water permeabilities in MDCK cells, primary cultures of tracheal epithelia, and intact toad urinary bladder.

4. Transepithelial P_f in Intact Tissues

The classical gravimetric method to measure osmotically induced transepithelial water transport in amphibian urinary bladder sacs involves serial measurements of sac weight. For a flat cell layer, such as cells cultured on a porous support or intact epithelia, net volume transport has been measured using Ussing chambers and sensitive capacitance probes (Toriano *et al.*, 1998) or, when water transport rates are high, using a membrane-impermeant volume marker on one side of the chamber (Deen *et al.*, 1997). For a cylindrical epithelial cell layer, such as a kidney tubule, P_f has been measured by *in vitro* perfusion using a membrane-impermeant marker perfused through the lumen. In the presence of a perfusion bath-to-lumen osmotic gradient, transepithelial P_f is determined from perfused vs collected marker concentrations, lumen flow, lumen and bath osmolalities, and tubule length and surface area. A fluorescence method to measure marker concentration (without the need for collection of luminal fluid) was developed in which a membrane-impermeant fluorescent indicator, such as FITC–dextran, was used as the luminal volume marker (Fig. 2D; Kuwahara *et al.*, 1988). This method provided an accurate real-time measurement of P_f in the perfused kidney-collecting duct (Kuwahara *et al.*, 1991) and distal airways (Folkesson *et al.*, 1996). Specialized techniques for P_f measurements in intact organs have been developed. To measurement P_f between the airspace and capillary compartments in lung, the airspace was filled with fluid containing a membrane-impermeant fluorophore at low concentrations and the pulmonary artery was perfused with solutions of specified osmolalities (Fig. 2E; Carter *et al.*, 1996). Pleural surface fluorescence is monitored continuously by epifluorescence microscopy. Because of the finite penetration depth of the excitation light, the surface fluorescence signal is directly proportional to the airspace fluorophore concentration and thus to airspace osmolality. In response to a change in pulmonary artery perfusate osmolality, water flows between the airspace and perfusate compartments. The pleural surface fluorescence method was used to demonstrate high airspace-to-capillary water permeability in mouse lung (Carter *et al.*, 1996) and developing neonatal rabbit lung (Carter *et al.*, 1997). A method has been developed to measure microvascular endothelial P_f in intact lung by filling the airspaces with an inert perfluorocarbon and perfusing the pulmonary artery with solutions of different osmolalities containing identical fluorophore concentrations (Carter *et al.*, 1998).

5. Diffusional Water Permeability

As described earlier, measurements of diffusional water permeability (P_d) are potentially useful if care is taken to avoid or carefully correct for unstirred layer effects. Conventional strategies utilize radioactive water (^3H_2O) or magnetically labeled water [^1H nuclear magnetic resonance (NMR)]. However, ^3H_2O uptake studies are generally impractical because of rapid water diffusional transport rates, and ^1H NMR requires large sample quantities and the use of toxic paramagnetic quenchers to decrease proton relaxation times. A simple optical strategy has been developed to measure P_d based on the dependence of the fluorescence quantum yield of certain fluorophores to solution H_2O/D_2O content. After screening a series of fluorophores, it was found that the fluorescence of aminonapthalene trisulfonic acid (ANTS) increased by more than three-fold in D_2O vs H_2O. D_2O/H_2O exchange was used to measure P_d in liposomes and red cell membranes (Ye and Verkman, 1989), perfused kidney tubules (Kuwahara and Verkman, 1988) and airways (Folkesson *et al.,* 1996), and the intact perfused lung (Carter *et al.,* 1996). Other potentially feasible strategies to measure P_d include exploiting the different infrared absorbances, refractive indices, and densities of D_2O vs H_2O.

III. WATER TRANSPORT ACROSS LIPID MEMBRANES

Lipid bilayers have consistently very low permeability for small ions and polar solutes. However, a wide range of water permeabilities have been found, with P_f values of $< 10^{-5}$ cm/sec for some membranes below their phase transition temperature to values as high as 0.01 cm/sec (reviewed in Fettiplace and Haydon, 1980). Many studies have demonstrated that water permeability through lipid bilayers depends strongly on lipid composition and sterol content, which alter lipid structure and fluid state. In general, maneuvers that decrease membrane fluidity (e.g., cholesterol addition) result in decreased P_f, whereas maneuvers that increase membrane fluidity (e.g., general anesthetics) result in increased P_f. Although a simple solubility–diffusion mechanism (in which permeability depends on water partitioning and diffusion in lipid) has provided crude values (generally underestimates) for P_d in some membranes, such simple considerations are not useful in predicting effects of lipid composition on water permeability. In the older literature, various models involving the diffusion of intramembrane defects have been proposed to describe transport across lipid membranes. Although there has been little recent work in this area, it might be useful to reexamine the mechanistic basis of water movement through lipids using recent biophysical studies of membrane structure by X-ray and neutron-scattering methods and fluorescence spectroscopy.

There has been recent interest in mechanisms that are responsible for making biological membranes relatively water impermeable. The P_f of various biological membranes, such as the ascending limb of Henle in kidney and the mammalian urinary bladder (Chang *et al.,* 1994), are lower than P_f in many artificial lipid bilayers. Of the cultured mammalian cells that have been studied, MDCK canine kidney epithelial cells have relatively low water permeability (Farinas *et al.,* 1995; Timbs and Spring, 1996). It has been proposed that low membrane fluidity in biological membranes (not containing water channels) is a primary determinant of their low water permeabilities (Lande *et al.,* 1995). Additional cellular factors may also be involved, such as effects of membrane proteins on lipid structure, bilayer asymmetry, and effects of skeletal protein–membrane interactions. The maintenance of low water permeability in some membranes is probably important physiologically. In kidney, for example, dilute urine is formed by the transport of salts (without water) out of the lumen of the ascending limb of Henle, with subsequent passage of urine through the lumen of the unstimulated collecting duct.

IV. WATER TRANSPORT ACROSS BIOLOGICAL MEMBRANES: WATER CHANNELS

A. Aquaporin Water Channels

There are at present at least 10 water-transporting proteins identified in mammalian tissues, many more in plants, and a few in other organisms including amphibia and bacteria. Each of the proteins is small (28–31 kDa) and related to the major intrinsic protein (MIP) of lens fiber. Aquaporin-type water channels contain several conserved amino acid sequences, including two NPA motifs, with overall amino acid identities of 20–60% after sequence alignment. Hydropathy plots of the aquaporins are similar and suggest six transmembrane helical segments. In addition, homology in amino acid sequence between the first and second halves of each protein suggests genesis from tandem, intragenic duplication of a three transmembrane segment (Reizer *et al.,* 1993).

Several of the aquaporins are expressed only in one tissue. MIP (also called AQP0) is expressed only in lens, and AQP2 (Fushimi *et al.,* 1993) and AQP6 (Ma *et al.,* 1996, 1997a) only in kidney. The other aquaporins are expressed widely. AQP1 was cloned initially from erythrocytes (Preston and Agre, 1991) and is expressed widely in epithelia (kidney proximal tubule and descending limb of Henle, choroid plexus, ciliary body, gallbladder) and vascular endothelia (Hasegawa *et al.,* 1993, 1994a; Nielsen *et al.,* 1993a,b). AQP3 was cloned from kidney (Ishibashi *et al.,* 1994; Ma *et al.,* 1994; Echevarria *et al.,* 1994) and is expressed in kidney-collecting duct, colonic epithelia, large airways, urinary bladder, conjunctiva, skin, and urinary bladder (Frigeri *et al.,* 1995a,b). AQP4 was cloned initially from lung (Hasegawa *et al.,* 1994a) and is expressed at the basolateral membrane of kidney-collecting duct, ependymal cells lining brain ventricles, astroglial cells in brain and spinal cord, skeletal muscle sarcolemma, and epithelial cells in stomach, trachea, airways, and colon (Frigeri *et al.,* 1995a,b). AQP5

(Raina *et al.*, 1995) is expressed in salivary and lacrimal glands and in lung alveolar epithelia. AQP7 (Ishibashi *et al.*, 1997a; Kuriyama *et al.*, 1997) is expressed in testis, adipose tissue, heart, and kidney. AQP8 (Ishibashi *et al.*, 1997b; Ma *et al.*, 1997b; Koyama *et al.*, 1997) is expressed in gastrointestinal organs (liver, colon, pancreas) and in testis and placenta. AQP9 (Ishibashi *et al.*, 1998) is expressed primarily in peripheral leukocytes and, to a lesser extent, in gastrointestinal organs. Thus, aquaporins are generally expressed in cells involved in fluid transport; however, there are many exceptions such as AQP3 (expressed in skin and urinary bladder), AQP4 (astrocytes and skeletal muscle), AQP8 (hepatocytes), and AQP9 (leukocytes). The expression of water channels in nonfluid-transporting cells raises the possibility that some aquaporins have an as yet unidentified nonwater-transporting role.

Functional studies of aquaporins have been done by the reconstitution of purified protein in liposomes (for AQP1 and AQP4; Van Hoek and Verkman, 1992; Zeidel *et al.*, 1992; Yang *et al.*, 1997) and using heterologous expression systems including *Xenopus* oocytes, transfected mammalian cells and yeast (Ma *et al.*, 1993; Katsura *et al.*, 1995; Laize *et al.*, 1995; Valenti *et al.*, 1996; Yang *et al.*, 1996). When purified and reconstituted in proteoliposomes, AQP1 increases water permeability approximately 100-fold, without increased permeability to protons, urea, and other small solutes (Preston *et al.*, 1992; Van Hoek and Verkman, 1992; Zeidel *et al.*, 1992). In *Xenopus* oocytes, AQPs 1, 2, 4, 5, and 6 appear to form water-selective channels (Fushimi *et al.*, 1993; Hasegawa *et al.*, 1994; Raina *et al.*, 1995; Ma *et al.*, 1996). Water movement through AQP4 is not mercurial sensitive (Hasegawa *et al.*, 1994) because of the absence of a cysteine residue at a position equivalent to cysteine 189 in AQP1 (Shi *et al.*, 1995). Quantitative determination of single channel water permeabilities in *Xenopus* oocytes indicated that AQP4 has the highest intrinsic water permeability, followed by AQP1 and AQP5, with relatively low intrinsic water permeability for MIP, AQP3, and AQP6 (Yang and Verkman, 1997). AQPs 3, 7, 8, and 9 also transport various amounts of urea and glycerol along with water (Ishibashi *et al.*, 1994, 1997a,b, 1998; Ma *et al.*, 1994, 1997b; Echevarria *et al.*, 1994; Kuriyama *et al.*, 1997). The role of solute transport through the aquaporins is not clear nor is it known why many related aquaporins with similar function exist.

B. Aquaporin Water Channel Structure

The majority of information about water channel structure has come from studies on AQP1 because it is purified readily in large quantities from human erythrocytes. Early ultrastructural studies of AQP1 topography were done by freeze-fracture electron microscopy (Verbavatz *et al.*, 1993).

Intramembrane particles of <8 nm diameter and with a tetrameric substructure were found in reconstituted proteoliposomes containing AQP1 and in AQP1-transfected CHO cells. Although AQP1 forms tetramers in membranes, it appears that each of the monomers functions independently as a water channel and probably contains its own water pore. Several heterodimers were constructed containing wild-type and mutated AQP1 monomers (Shi *et al.,* 1994). It was found that the water permeability and mercurial inhibition properties of the heterodimers were equal to the sum of the properties of the individual monomers. Spectroscopy analysis of purified AQP1 by circular dichroism and Fourier transform infrared spectroscopy indicated 40–60% helical content, consistent with multiple membrane-spanning α helices (Van Hoek et al., 1993).

Topological analysis of AQP1 by epitope tagging and vectorial proteolysis suggested that matured AQP1 monomers contain six membrane-spanning domains (Preston *et al.,* 1994a). Interestingly, a study of AQP1 biogenesis indicated that AQP1 monomers are synthesized at the endoplasmic reticulum in an immature form containing four principal membrane-spanning domains (Skach *et al.,* 1994). Water movement through individual AQP1 monomers has been proposed to involve an aqueous channel formed by polar residues of helical domains (Skach *et al.,* 1994) and/or by β-sheet protein folds involving NPA residues (Jung *et al.,* 1994).

A series of AQP1 structural studies were done by three groups by electron crystallography of two-dimensional AQP1 crystals formed in reconstituted proteoliposomes. The Jap group utilized unmodified bovine CHIP29 (Li and Jap, 1997) and the Engel group has used unmodified human AQP1 (Walz *et al.,* 1997). The author's group used enzymatically deglycosylated human AQP1 that is relatively homogeneous in size (Cheng *et al.,* 1997). The former studies were done using glucose-embedded specimens, whereas the Cheng *et al.* (1997) study was done by electron cryocrystallography in ice. The most recent data from each of the groups are in qualitative agreement. As shown in a side view (Fig. 3), each AQP1 monomer probably contains six helical transmembrane domains forming a barrel that surrounds a putative aqueous pore. There is substantial extramembrane density and there may be densities lying above and below the pore; however, at the current resolution (4–6 Å in plane) it is not possible to identify individual chains and residues. Work is in progress to improve the resolution obtainable from two-dimensional crystals and to establish three-dimensional crystals for high-resolution X-ray crystallography.

C. Water Movement through Nonaquaporin Pathways

As described earlier, water can move across a biological membrane through the lipid portion of the membrane and through aquaporin-type

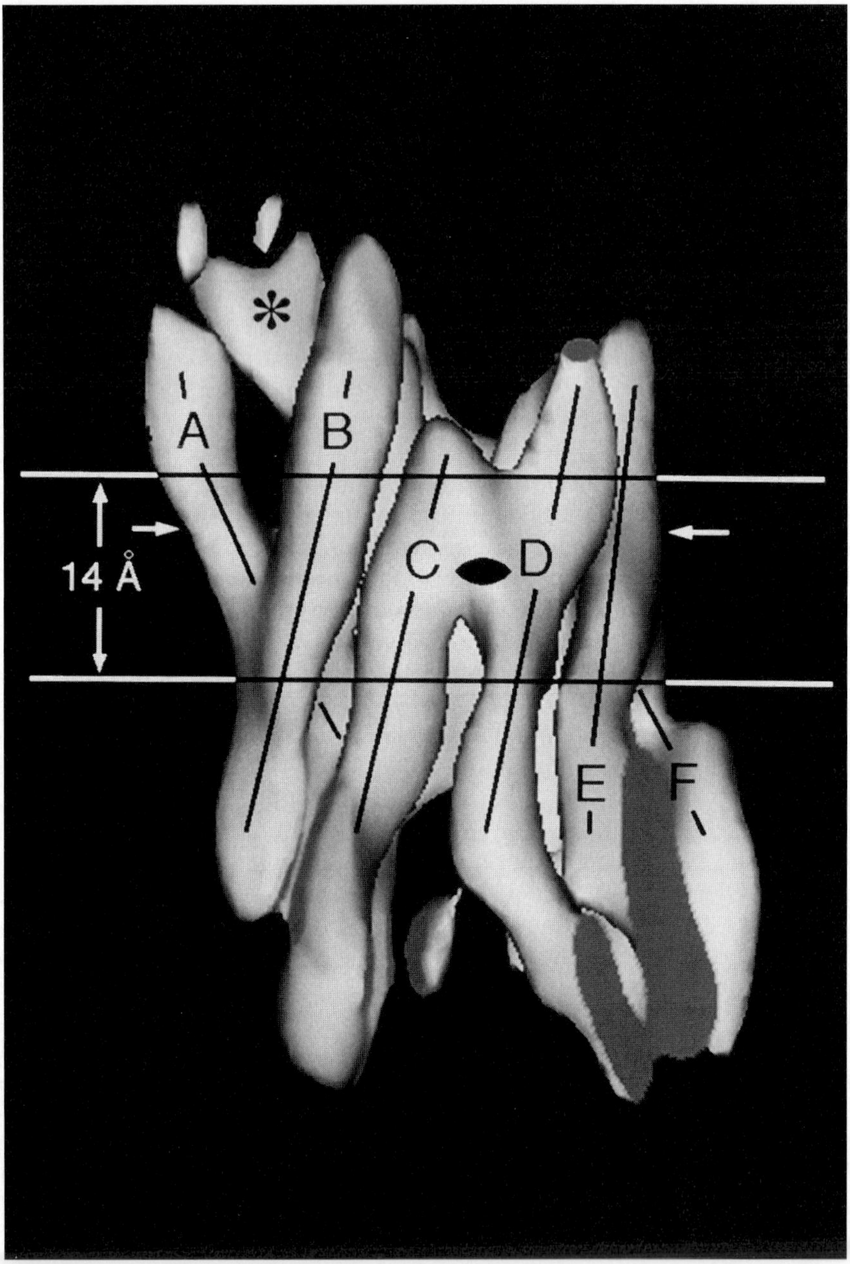

FIGURE 3 Three-dimensional structure of an AQP1 water channel monomer. Reconstruction adapted from Cheng *et al.* (1997) based on electron cryocrystallography of two-dimensional crystals of human AQP1 in reconstituted proteoliposomes. There are six membrane-spanning tiled α helices labeled A–F. A putative aqueous pore is located at the center of the helices.

water channels. Numerous other transmembrane proteins, including ion channels, are presumed to contain an aqueous channel. Demonstration of electroosmotic phenomena supports the idea that ion channels can transport water under some conditions. The CFTR chloride channel, which when mutated causes the disease cystic fibrosis, is able to transport water when activation by cAMP (Hasegawa *et al.,* 1992). However, the membrane densities of CFTR and other ion channels are generally $<1\ \mu m^{-2}$ of membrane. Even if the single channel water permeability is exceptionally high (10^{-12} cm^3/sec), the membrane water permeability conferred by a channel at a density of $1\ \mu m^{-2}$ is quite low (10^{-4} cm/sec). In contrast, aquaporin-type water channels are generally expressed at membrane densities of 10^3–$10^4\ \mu m^{-2}$. It is thus unlikely that ion transporters contribute importantly to net membrane water permeability. Other single-solute and cotransport proteins have been found to be able to transport water when expressed at high densities in *Xenopus* oocytes, including the glucose transporter GLUT1 (Fischbarg *et al.,* 1990; Zhang *et al.,* 1991), the sodium-dependent glucose cotransporter SGLT1 (Loo *et al.,* 1996), and the urea transporter UT3 (Yang and Verkman, 1998). Water transport by GLUT1 and SGLT1 is blocked by glucose transport inhibitors, and water transport by UT3 is blocked by urea transport inhibitors. Demonstration of a low urea reflection coefficient in oocytes expressing UT3 supported the conclusion that water and urea share a common pathway through UT3 (Yang and Verkman, 1998). Although it is likely that water movement through many other membrane proteins will be demonstrated, it is unclear whether nonaquaporin water channels provide a quantitatively important route for water transport in cell membranes. Initial studies in knockout mice lacking specific aquaporins (see later) suggest that aquaporins provide the major route for water permeation in many tissues. Functional studies are needed in knockout mice lacking the nonaquaporin proteins that have nonzero water permeability.

D. Transgenic Aquaporin Knockout Mice

Several lines of indirect evidence have suggested that aquaporins have an important role in mammalian physiology: their strong expression in fluid-transporting tissues, their regulation in development, and the causal relationship of AQP2 mutations with the human disease nephrogenic diabetes insipidus (NDI) (Deen *et al.,* 1995). Humans lacking AQP1 (Colton null blood group) have been reported to be phenotypically normal (Preston *et al.,* 1994b), although formal clinical studies have not been done on these rare individuals. Natural deletions of other aquaporins have not been identified. Unfortunately, nontoxic aquaporin inhibitors are not available at present to study the *in vivo* effect of functional aquaporin deletion. To study the role of aquaporins in mammalian physiology, the author's laboratory has been generating transgenic knockout mice lacking specific aquaporins.

1. AQP1 Knockout Mouse

A transgenic mouse lacking detectable AQP1 protein was generated by targeted gene disruption (Ma *et al.,* 1998). Knockout mice survived and appeared grossly normal, except for 10–15% decreased body weight compared to litter-matched wild-type and heterozygous mice. In kidney, AQP1 is expressed strongly in the plasma membranes of proximal tubule, thin descending limb of Henle, and descending vasa recta. Initial evaluation of renal function indicated a severe urinary concentrating defect in AQP1 knockout mice. After a 36-hr water deprivation, knockout mice became severely dehydrated with a 35% reduction in body weight and an increase in serum osmolality to >500 mOsm. The urine osmolality (~600 mOsm) was no higher than that before water deprivation. In contrast, body weight in wild-type and heterozygous mice decreased by 20–22 %, serum osmolality remained normal (310–330 mOsm), and urine osmolality rose to >2500 mOsm. It was concluded that AQP1 knockout mice are unable to create a hypertonic medullary interstitium by countercurrent multiplication. Experiments indicate that proximal tubule reabsorptive capacity is also defective in AQP1 null mice (Schnermann *et al.,* 1998) and thin descending limb of Henle water permeability is greatly reduced (Chou *et al.,* 1999). In isolated microperfused S2 segments of the proximal tubule, osmotic water permeability was decreased strongly in AQP1 knockout mice, and isosmolar fluid absorption in the absence of an osmotic gradient was reduced by 50%. Free-flow micropuncture confirmed that AQP1 knockout mice manifest defective proximal tubule reabsorption *in vivo.* AQP1 is also expressed strongly in pulmonary microvascular endothelia where it has been proposed to provide the major route for transendothelial water movement. Using the pleural surface fluorescence method described earlier (Carter *et al.,* 1996), it was found that osmotic water permeability between the airspace and capillary compartments is more than five-fold decreased in AQP1 knockout vs wild-type mice (Bai *et al.,* 1999). Phenotype studies are in progress to determine the role of AQP1 in eye, gastrointestinal, and neuromuscular physiology.

2. AQP4 Knockout Mouse

Transgenic knockout mice lacking AQP4 water channels were also generated by targeted gene disruption (Ma *et al.,* 1997c). Their survival, growth, and gross appearance were not different from that in litter-matched wild-type mice. Knockout mice had a mild decrease in maximum urine osmolality after water deprivation. In the vasopressin-treated, microperfused inner medullary-collecting duct, osmotic water permeability was four-fold lower in AQP4 knockout vs wild-type mice (Chou *et al.,* 1998). It was concluded that AQP4 is the major functional basolateral membrane water channel of the inner medullary-collecting duct. The relatively mild defect in the urinary concentrating ability of AQP4 knockout mice is consistent with its selected expression in inner medulla; water channel AQP3 is expressed in more proximal portions of the collecting duct. An interesting result from AQP4 knockout mouse was the demonstration that AQP4 is the "orthogonal array protein" (OAP). OAPs are regular square particle arrays that have been observed by freeze-fracture electron microscopy in many tissues, including the basolateral membrane of kidney-collecting duct epithelia, astrocytes in brain, and skeletal muscle plasmalemma. It was proposed initially that AQP4 is the OAP protein based on its tissue distribution (Frigeri *et al.,* 1995b) and the presence of OAPs in AQP4-transfected CHO cells (Yang *et al.,* 1996). Direct evidence was obtained subsequently from freeze-fracture studies on kidney, skeletal muscle, and brain of wild–type *vs.* Knockout mice. OAPs were identified in every sample from wild-type and heterozygous mice and in no sample from AQP4 knockout mice (Verbavatz *et al.,* 1997). Figure 4 is a freeze-fracture electron micrograph of AQP4-expressing CHO cells, showing E-face and P-face OAPs (A) and a gold label fracture with AQP4 antibody (B).

3. AQP2 and Nephrogenic Diabetes Insipidus

As described earlier, vasopressin increases transepithelial water permeability strongly in kidney-collecting duct and amphibian urinary bladder by a membrane-cycling mechanism. Mutations in AQP2 cause the rare non-X-linked form of hereditary NDI (Deen *et al.,* 1994), in which a defective urinary concentrating ability causes hypernatremia and dehydration if adequate hydration is not provided. At present >15 different mutations in human AQP2 have been identified that are associated with autosomal recessive NDI. Heterologous expression studies in *Xenopus* oocytes suggested that an impaired routing of some AQP2 mutants might contribute to the cellular pathogenesis of NDI (Deen *et al.,* 1995). A mammalian cell transfection model was used to investigate the mechanisms responsible for defective cell water permeability in NDI caused by AQP2 mutations (Tamarappoo and Verkman, 1998). In a *Xenopus* oocyte expression system, it was found that the intrinsic water permeabilities were similar for wild-type AQP2 vs mutants L22V, T126M, and A147T; the R187C and C181W AQP2 mutants were nonfunctional. In transiently transfected CHO cells that were pulselabeled with [^{35}S]methionine, half-times for AQP2 degrada-

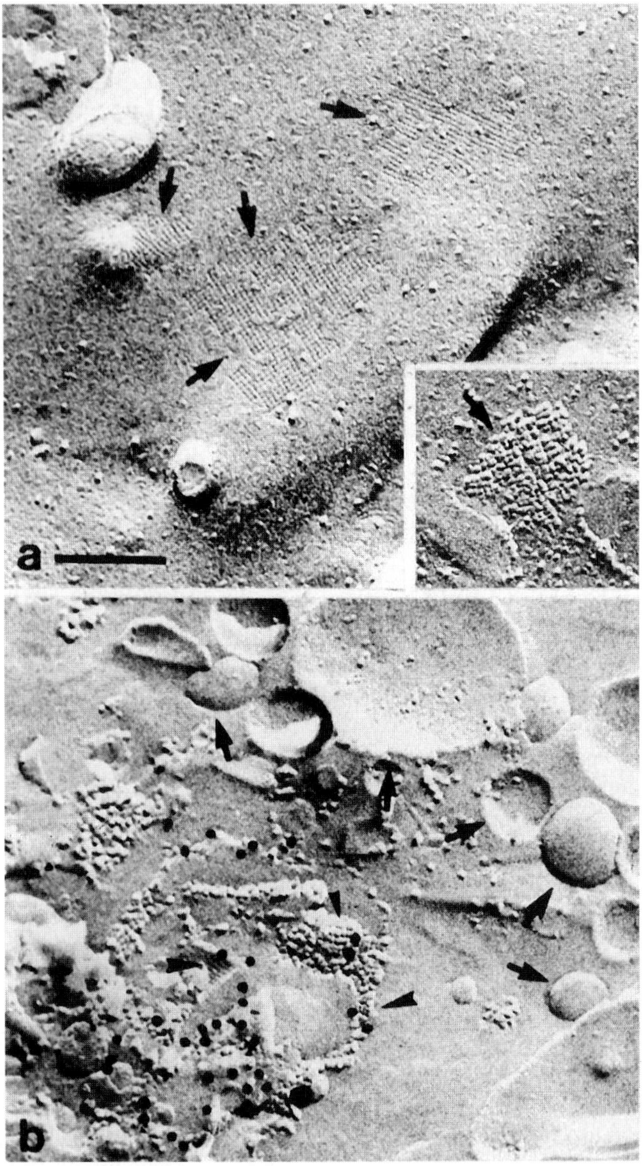

FIGURE 4 Freeze-fracture electron microscopy of CHO cells stably expressing water channel AQP4. (a) E-face and P-face (inset) imprints showing large orthogonal arrays (arrows). (b) Gold label fracture with AQP4 antibody showing gold particles associated with orthogonal arrays. Scale bar: 150 nm.

tion were ~4 hr for wild-type AQP2 and L22V and decreased mildly by up to 50% for NDI-causing mutants. Subcellular localization as determined by immunofluorescence and membrane fractionation indicated that a major reason for defective cell function was mistrafficking of AQP2 mutants with retention at the endoplasmic reticulum.

One strategy used to correct the trafficking defect in NDI involved the use of "chemical chaperones"—small molecules (such as glycerol) that facilitate protein folding *in vitro.* Glycerol has been shown in cell culture models to correct the defective processing of ΔF508 CFTR in cystic fibrosis, the prion protein PrP, the viral oncogene protein $pp60^{src}$, and mutants of the tumor suppressor protein p53 (Tatzel *et al.,* 1996; Sato *et al.,* 1996; Welch and Brown, 1996; Brown *et al.,* 1996, 1997). It was found that the growth of cells in media containing 1 *M* glycerol for 48 hr resulted in the correction of mutant AQP2 mistrafficking as assessed by localization studies and the measurement of cell membrane water permeability (Tamarappoo and Verkman, 1998). The trafficking defect could also be corrected by other chemical chaperones at lower concentrations, including dimethyl sulforide and trimethyamine oxide (50–100 m*M*). Animal studies are in progress to investigate whether chemical chaperones could constitute a clinically useful strategy to treat specific forms of NDI.

V. SUMMARY

Substantial recent progress has been made in understanding the mechanisms of water transport across membranes. Major questions in the field are the detailed molecular mechanisms by which water moves through water channels proteins and the role of water channels in animal and plant physiology. Phenotype analysis of transgenic mouse models of aquaporin deletion and mutation is likely to be highly informative. Screening for aquaporin mutations in human genetic diseases associated with abnormalities of fluid balance may be informative. Finally, the manipulation of aquaporin expression and targeting by pharmacological agents or gene transfer might provide new therapies for water retention, pulmonary edema, glaucoma, and other common clinical problems involving fluid regulation.

Acknowledgments

I thank Drs. Javier Farinas, Tonghui Ma, Michael Matthay, Alok Mitra, B.K. Tamarappoo, Fuminori Umenishi, and Baoxue Yang for collaboration and review of this manuscript. This work was supported by NIH Grants DK35124, HL59198, HL51854, HL60288, and DK43840 and by Grant R613 from the National Cystic Fibrosis Foundation.

References

Agre, P., Preston, G. M., Smith, B. L., Jung, J. S., Raina, S., Moon, C., Guggino, W. B., and Nielsen, S. (1993). Aquaporin CHIP—the archetypal molecular water channel. *Am. J. Physiol.* **265,** F463–F476.

Bai, C. X., Fukuda, N., Song, Y., Ma, T., Matthay, M. A., and Verkman, A. S. (1999). Aquaporin water channels in lung fluid transport: Phenotype analysis of aquaporin-1 and -4 knockout mice. *J. Clin. Invest.,* in press.

Brown, R., Hong-Brown, L., Biwersi, J., Verkman, A. S., and Welch, W. (1996). Chemical chaperones correct the mutant phenotype of the ΔF508 cystic fibrosis transmembrane conductance regulator protein. *Cell Stress Chap.* **1,** 117–125.

Brown, R., Hong-Brown, L., and Welch, W. (1997). Correction of temperature sensitive protein folding defects by chemical chaperones. *J. Clin. Invest.* **99,** 1432–1444.

Carter, E. P., Matthay, M. A., Farinas, J., and Verkman, A. S. (1996). Transalveolar osmotic and diffusional water permeability in intact mouse lung measured by a novel surface fluorescence method. *J. Gen. Physiol.* **108,** 133–142.

Carter, E. P., Umenishi, F., Matthay, M. A., and Verkman, A. S. (1997). Developmental changes in alveolar water permeability in perinatal rabbit lung. *J. Clin. Invest.* **100,** 1071–1078.

Carter, E. P., Ölveczky, B. P., Matthay, M. A., and Verkman, A. S. (1998). High microvascular endothelial water permeability in mouse lung measured by pleural surface fluorescence. *Biophys. J.* **74,** 2121–2128.

Chang, A., Hammond, T. G., Sun, J. T., and Zeidel, M. L. (1994). Permeability properties of the mammalian bladder apical membrane. *Am. J. Physiol.* **267,** C1483–C1492.

Chen, P.-Y., Pearce, D., and Verkman, A. S. (1988). Membrane water and solute permeability determined quantitatively by self-quenching of an entrapped fluorophore. *Biochemistry* **27,** 5713–5719.

Cheng, A., Van Hoek, A. N., Yaeger, M., Verkman, A. S., and Mitra, A. K. (1997). Three-dimensional organization of a human water channel. *Nature* (*London*) **387,** 627–630.

Chrispeels, M. J., and Maurel, C. (1994). Aquaporins—the molecular basis of facilitated water movement through living plant cells. *Plant Physiol.* **105,** 9–13.

Chou, C. L., Knepper, M. A., van Hoek, A. N., Brown, D., Yang, B., Ma, T., and Verkman, A. S. (1999). Reduced water permeability and altered ultrastructure in thin descending limb of Henle in aquaporin-1 null mice. *J. Clin. Invest.,* in press.

Chou, C. L., Ma, T., Yang, B., Knepper, M. A., and Verkman, A. S. (1998). Four-fold reduction in water permeability in inner medullary collecting duct of aquaporin-4 knockout mice. *Am. J. Physiol.* **274,** C549–C554.

Deen, P. M., Verkijk, M. A., Knoers, N. V., Wieringa, B., Monnens, L. A., Van Os, C. H., and Van Oost, B. A. (1994). Requirement of human renal water channel aquaporin-2 for vasopressin-dependent concentration of urine. *Science* **264,** 92–95.

Deen, P. M., Croes, H., Van Aubel, R. A. M. S., Ginsel, L. A., and Van Os, C. H. (1995). Water channels encoded by mutant aquaporin-2 genes in nephrogenic diabetes insipidus are impaired in their cellular routing. *J. Clin. Invest.* **95,** 2291–2296.

Deen, P. M. T., Rijss, J. P. L., Mulders, S. M., Errington, R. J., Van Baal, J., and Van Os, C. H. (1997). Aquaporin-2 transfection of Madin Darby canine kidney cells reconstitutes vasopressin-regulated transcellular osmotic water transport. *J. Am. Soc. Nephrol.* **8,** 1493–1501.

Dobbs, L., Gonzalez, R., Matthay, M. A., Carter, E. P., and Verkman, A. S. (1998). Highly water-permeable type I alveolar epithelial cells confer high water permeability between the airspace and vasculature in rat lung. *Proc. Natl. Acad. Sci. U.S.A.* **95,** in press.

Echevarria, M., and Verkman, A. S. (1992). Optical measurement of osmotic water transport in cultured cells: Evaluation of the role of glucose transporters. *J. Gen. Physiol.* **99,** 573–589.

Echevarria, M., Windhager, E. E., Tate, S. S., and Frindt, G. (1994). Cloning and expression of AQP3, a water channel from the medullary collecting duct of rat kidney. *Proc. Natl. Acad. Sci. U.S.A.* **91,** 10997–11001.

Farinas, J., and Verkman, A. S. (1996). Measurement of cell volume and water permeability in epithelial cell layers by interferometry. *Biophys. J.* **71,** 3511–3522.

Farinas, J., Simenak, V., and Verkman, A. S. (1995). Cell volume measured in adherent cells by total internal reflection microfluorimetry: Application to permeability in cells transfected with water channel homologs. *Biophys. J.* **68,** 1613–1620.

Farinas, J., Kneen, M., Moore, M., and Verkman, A. S. (1997). Plasma membrane water permeability of cultured cells and epithelia measured by light microscopy with spatial filtering. *J. Gen. Physiol.* **110,** 283–296.

Fettiplace, R., and Haydon, D. A. (1980). Water permeability of lipid membranes. *Physiol. Rev.* **60,** 510–550.

Finkelstein, A. (1987). "Water Movement through Lipid Bilayers, Pores, and Plasma Membranes, Theory and Reality." Wiley, New York.

Fischbarg, J., Kunyan, K., Vera, J. C., Arant, S., Silverstein, S., Loike, J., and Rosen, O. M. (1990). Glucose transporters serve as water channels. *Proc. Natl. Acad. Sci. U.S.A.* **87,** 3244–3247.

Folkesson, H., Matthay, M., Frigeri, A., and Verkman, A. S. (1996). High transepithelial water permeability in microperfused distal airways: Evidence for channel-mediated water transport. *J. Clin. Invest.* **97,** 664–671.

Frigeri, A., Gropper, M., Turck, C. W., and Verkman, A. S. (1995a). Immunolocalization of the mercurial-insensitive water channel and glycerol intrinsic protein in epithelial cell plasma membranes. *Proc. Natl. Acad. Sci. U.S.A.* **92,** 4328–4331.

Frigeri, A., Gropper, M., Umenishi, F., Kawashima, M., Brown, D., and Verkman, A. S. (1995b). Localization of MIWC and GLIP water channel homologs in neuromuscular, epithelial and glandular tissues. *J. Cell Sci.* **108,** 2993–3002.

Fushimi, K., Uchida, S., Hara, Y., Hirata, Y., Marumo, F., and Sasaki, S. (1993). Cloning and expression of apical membrane water channel of rat kidney collecting tubule. *Nature (London)* **361,** 549–552.

Hasegawa, H., Skach, W., Baker, O., Calayag, M. C., Lingappa V., and Verkman, A. S. (1992). A multi-functional aqueous channel formed by CFTR. *Science* **258,** 1477–1479.

Hasegawa, H., Zhang, R., Dohrman, A., and Verkman, A. S. (1993). Tissue-specific expression of mRNA encoding the rat kidney water channel CHIP28k by *in situ* hybridization. *Am. J. Physiol.* **264,** C237–C245.

Hasegawa, H., Lian, S. C., Finkbeiner, W. E., and Verkman, A. S. (1994a). Extrarenal tissue distribution of CHIP28 water channels by *in situ* hybridization and antibody staining. *Am. J. Physiol.* **266,** C893–C903.

Hasegawa, H., Ma, T., Skach, W., Matthay, M. A., and Verkman, A. S. (1994b). Molecular cloning of a mercurial-insensitive water channel expressed in selected water-transporting tissues. *J. Biol. Chem.* **269,** 5497–5500.

Hill, A. E. (1994). Osmotic flow in membrane pores of molecular size. *J. Membr. Biol.* **137,** 197–203.

Ishibashi, K., Sasaki, S., Fushimi, K., Uchida, S., Kuwahara, M., Saito, H., Furukawa, T., Nakajima, K., Yamaguchi, Y., Gojobori, T., and Marumo, F. (1994). Molecular cloning and expression of a member of the aquaporin family with permeability to glycerol and urea in addition to water expressed at the basolateral membrane of kidney collecting duct cells. *Proc. Natl. Acad. Sci. U.S.A.* **91,** 6269–6273.

Ishibashi, K., Kuwahara, M., Gu, Y., Kegayama, Y., Tohsaka, A., Marumo, F., and Sasaki, S. (1997a). Cloning and functional expression of a new water channel abundantly expressed in the testis also permeable to glycerol and urea. *J. Biol. Chem.* **272,** 20782–20786.

Ishibashi, K., Kuwahara, M., Kageyama, Y., Tohsaka, A., Marumo, F., and Sasaki, S. (1997b). Cloning and functional expression of a second new aquaporin abundantely expressed in testis. *Biochim. Biophys. Res. Commun.* **237,** 714–718.

Ishibashi, K., Kuwahara, M., Gu, Y., Tanaki, Y., Marumo, F., and Sasaki, S. (1998). Cloning and functional expression of a new aquaporin (AQP9) abundantly expressed in peripheral leukocytes permeable to water and urea, but not to glycerol. *Biochim. Biophys. Res. Commun.* **244,** 268–274.

Jung, J. S., Preston, B. M., Smith, B. L., Guggino, W. B., and Agre, P. (1994). Molecular structure of the water channel through aquaporin CHIP. The hourglass model. *J. Biol. Chem.* **269,** 14648–14654.

Katsura, T., Verbavatz, J. M., Farinas, J., Ma, T., Ausiello, D. A., Verkman, A. S., and Brown, D. (1995). Constitutive and regulated membrane expression of aquaporin-CHIP and aquaporin-2 water channels in stably transfected LLC-PK1 cells. *Proc. Natl. Acad. Sci. U.S.A.* **92,** 7212–7216.

Knepper, M. A. (1994). The aquaporin family of molecular water channels. *Proc. Natl. Acad. Sci. U.S.A.* **91,** 6255–6258.

Koyama, Y., Yamamoto, T., Kondo, D., Funaki, H., Yaoita, E., Kawasaki, K., Sato, N., Hatakeyama, K., and Kihara, I. (1997). Molecular cloning of a new aquaporin from rat pancreas and liver. *J. Biol. Chem.* **272,** 30329–30333.

Kuriyama, H., Kawamoto, S., Ishida, N., Ohno, I., Mita, S., Matsuzawa, Y., Matsubara, K., and Okubo, K. (1997). Molecular cloning and expression of a novel human aquaporin from adipose tissue with glycerol permeability. *Biochim. Biophys. Res. Commun.* **241,** 53–58.

Kuwahara, M., and Verkman, A. S. (1988). Direct fluorescence measurement of diffusional water permeability in the vasopressin-sensitive kidney collecting tubule. *Biophys. J.* **54,** 587–593.

Kuwahara, M., Berry, C. A., and Verkman, A. S. (1988). Rapid development of vasopressin-induced hydroosmosis in kidney collecting tubules measured by a new fluorescence technique. *Biophys. J.* **54,** 595–602.

Kuwahara, M., Shi, L. B., Marumo, F., and Verkman, A. S. (1991). Transcellular water flow modulates water channel exocytosis and endocytosis in kidney collecting tubule. *J. Clin. Invest.* **88,** 423–429.

Laize, V., Rousselet, G., Verbavatz, J. M., Berthonaud, V., Gobin, R., Roudier, N., Abrami, L., Ripoche, P., and Tacnet, F. (1995). Functioinal expression of the human CHIP28 water channel in a yeast secretory mutant. *FEBS Lett.* **373,** 269–274.

Lande, M. B., Donovan, J. M., and Zeidel, M. L. (1995). The relationship between membrane fluidity and permeabilities to water, solutes, ammonia and protons. *J. Gen. Physiol.* **106,** 67–84.

Levitt, D. G., and Mlekoday, H. J. (1983). Reflection coefficient and permeability of urea and ethyleneglycol in the human red cell membrane. *J. Gen. Physiol.* **81,** 239–254.

Li, H. L., Lee, S., and Jap, B. K. (1997). Molecular design of aquaporin-1 water channel as revealed by electron crystallography. *Nat. Struct. Biol.* **4,** 263–265.

Loo, D. D. F., Zeuthen, T., Chandy, G., and Wright, E. M. (1996). Cotransport of water by the Na/glucose cotransporters. *Proc. Natl. Acad. Sci. U.S.A.* **96,** 13367–13370.

Ma, T., Frigeri, A., Tsai, S. T., Verbavatz, J. M., and Verkman, A. S. (1993). Localization and functional analysis of CHIP28k water channels in stably transfected CHO cells. *J. Biol. Chem.* **268,** 22756–22764.

Ma, T., Frigeri, A., Hasegawa, H., and Verkman, A. S. (1994). Cloning of a water channel homolog expressed in brain meningeal cells and kidney collecting duct that functions as a stilbene-sensitive glycerol transporter. *J. Biol. Chem.* **269,** 21845–21849.

Ma, T., Yang, B., Kuo, W. L., and Verkman, A. S. (1996). cDNA cloning and gene structure of a new water channel expressed exclusively in human kidney: Evidence for a gene cluster of aquaporins at chromosome locus 12q13. *Genomics* **35,** 543–550.

Ma, T., Yang, B., Umenishi, F., and Verkman, A. S. (1997a). Closely spaced tandem arrangement of AQP2, AQP5 and AQP6 genes in a 27-kilobase segment at chromosome locus 12q13. *Genomics* **43,** 387–389.

Ma, T., Yang, B., and Verkman, A. S. (1997b). Cloning of a novel water and urea-permeable aquaporin from mouse expressed strongly in colon, placenta, liver and heart. *Biochim. Biophys. Res. Commun.* **240,** 324–328.

Ma, T., Yang, B., Gillespie, A., Carlson, E. J., Epstein, C. J., and Verkman, A. S. (1997c). Generation and phenotype of a transgenic knock-out mouse lacking the mercurial-insensitive water channel aquaporin-4. *J. Clin. Invest.* **100,** 957–962.

Ma, T., Yang, B., Gillespie, A., Carlson, E. J., Epstein, C. J., and Verkman, A. S. (1998). Severely impaired urinary concentrating ability in transgenic mice lacking aquaporin-1 water channels. *J. Biol. Chem.* **273,** 4296–4299.

Macey, R. I. (1984). Transport of water and urea in red blood cells. *Am. J. Physiol.* **246,** C195–C203.

Nielsen, S., Smith, B. L., Christensen, E. I., and Agre, P. (1993a). Distribution of the aquaporin CHIP in secretory and resorptive epithelia and capillary endothelia. *Proc. Natl. Acad. Sci. U.S.A.* **90,** 7275–7279.

Nielsen, S., Smith, B. L., Christensen, E. I., Knepper, M. A., and Agre, P. (1993b). CHIP28 water channels are localized in constitutively water-permeable segments of the nephron. *J. Cell Biol.* **120,** 371–383.

Pearce, D., and Verkman, A. S. (1989). NaCl reflection coefficients in proximal tubule apical and basolateral membrane vesicles measured by induced osmosis and solvent drag. *Biophys. J.* **55,** 1251–1259.

Pfeffer, W. (1877). "Osmotische Untersuchungen." W. Engelmann, Leipzig.

Preston, G. M., and Agre, P. (1991). Isolation of the cDNA for erythrocyte integral membrane protein of 28 kilodaltons: Member of an ancient channel family. *Proc. Natl. Acad. Sci. U.S.A.* **88,** 11110–11114.

Preston, G. M., Carroll, T. P., Guggino, W. B., and Agre, P. (1992). Appearance of water channels in *Xenopus* oocytes expressing red cell CHIP28 protein. *Science* **256,** 385–387.

Preston, G. M., Jung, J. S., Guggino, W. B., and Agre, P. (1994a). Membrane topology of aquaporin CHIP—analysis of function epitope-scanning mutants by vectorial proteolysis. *J. Biol. Chem.* **269,** 1668–1673.

Preston, G. M., Smith, B. L., Zeidel, M. L., Moulds, J. J., and Agre, P. (1994b). Mutations in aquaporin-1 in phenotypically normal humans without functional CHIP water channels. *Science* **265,** 1585–1587.

Raina, S., Preston, B. M., Guggino, W. B., and Agre, P. (1995). Molecular cloning of an aquaporin cDNA from salivary, lacrimal, and respiratory tissues. *J. Biol. Chem.* **270,** 1908–1912.

Reizer, J., Reizer, A., and Saier, M. H. (1993). The MIP family of integral membrane channel proteins: Sequence comparisons, evolutional relationships, reconstructed pathway of evolution, and proposed functional differentiation of two repeated halves of the protein. *Crit. Rev. Biochem. Mol. Biol.* **28,** 235–257.

Sato, S., Ward, C. L., Krouse, E., Wine, J. J., and Kopito, R. R. (1996). Glycerol reverses the misfolding phenotype of the most common cystic fibrosis mutation. *J. Biol. Chem.* **271,** 635–638.

Schnermann, J., Chou, J., Ma, T., Knepper, M. A., and Verkman, A. S. (1998). Defective proximal tubular fluid reabsorption in transgenic aquaporin-1 null mice. *Proc. Natl. Acad. Sci. U.S.A.* **95,** 9660–9664.

Shi, L. B., Brown, D., and Verkman, A. S. (1990). Water, urea and proton transport properties of endosomes containing the vasopressin-sensitive water channel from toad bladder. *J. Gen. Physiol.* **95,** 941–960.

Shi, L. B., Fushimi, K., and Verkman, A. S. (1991). Solvent drag measurement of transcellular and basolateral membrane NaCl reflection coefficient in mammalian proximal tubule. *J. Gen. Physiol.* **98,** 379–398.

Shi, L. B., Skach, W. R., and Verkman, A. S. (1994). Functional independence of monomeric CHIP28 water channels revealed by expression of wild type-mutant heterodimers. *J. Biol. Chem.* **269,** 10417–10422.

Shi, L. B., Skach, W. R., Ma, T., and Verkman, A. S. (1995). Distinct biogenesis mechanisms for water channels MIWC and CHIP28 at the endoplasmic reticulum. *Biochemistry* **34,** 8250–8256.

Skach, W. R., Shi, L. B., Calayag, M. C., Frigeri, A., Lingappa, V. R., and Verkman, A. S. (1994). Topology and biogenesis of the CHIP28 water channel at the endoplasmic reticulum. *J. Cell Biol.* **125,** 803–816.

Tamarappoo, B. K., and Verkman, A. S. (1998). Defective trafficking of AQP2 water channels in Nephrogenic Diabetes Insipidus and correction by chemical chaperones. *J. Clin. Invest.* **101,** 2257–2267.

Tatzel, J., Pruisner, S. B., and Welch, W. J. (1996). Chemical chaperones interfere with the formation of scrapie prion protein. *EMBO J.* **15,** 6363–6373.

Timbs, M. M., and Spring, K. R. (1996). Hydraulic properties of MDCI cell epithelium. *J. Membr. Biol.* **153,** 1–11.

Toriano, R., Ford, P., Rivarola, B., Tamarappoo, B. K., Verkman, A. S., and Parisi, M. (1998). Reconstitution of a regulated transepithelial water pathway in cells transfected with AQP2 and an AQP1/AQP2 hybrid containing the AQP2-C terminus. *J. Membr. Biol.* **161,** 141–149.

Valenti, G., Frigeri, A., Ronco, P. M., Dettorre, C., and Svelto, M. (1996). Expression and functional analysis of water channel in a stably AQP2-transfected human collecting duct cell line. *J. Biol. Chem.* **271,** 24365–24370.

Van Hoek, A. N., and Verkman, A. S. (1992). Functional reconstitution of the isolated erythrocyte water channel CHIP28. *J. Biol. Chem.* **267,** 18267–18269.

Van Hoek, A. N., Wiener, M., Bicknese, S., Miercke, L., Biwersi, J., and Verkman, A. S. (1993). Secondary structure analysis of purified CHIP28 water channels by CD and FTIR spectroscopy. *Biochemistry* **32,** 11847–11856.

Van Os, C. H., Deen, P. M. T., and Dempster, J. A. (1994). Aquaporins—water selective channels in biological membranes—molecular structure and tissue distribution. *Biochim. Biophys. Acta Rev. Biomembr.* **1197,** 291–309.

Verbavatz, J. M., Brown, D., Sabolic, I., Valenti, G., Ausiello, D. A., Van Hoek, A. N., Ma, T., and Verkman, A. S. (1993). Tetrameric assembly of CHIP28 water channels in liposomes and cell membranes. A freeze-fracture study. *J. Cell Biol.* **123,** 605–618.

Verbavatz, J. M., Ma, T., Gobin, R., and Verkman, A. S. (1997). Absence of orthogonal arrays in kidney, brain and muscle from transgenic knockout mice lacking water channel aquaporin-4. *J. Cell Sci.* **110,** 2855–2860.

Verkman, A. S. (1989). Mechanisms and regulation of water permeability in renal epithelia. *Am. J. Physiol.* **257,** C837–C850.

Verkman, A. S., Lencer, W., Brown, D., and Ausiello, A. A. (1988). Endosomes from kidney collecting tubule contain the vasopressin-sensitive water channel. *Nature* (*London*) **333,** 268–269.

Verkman, A. S., Van Hoek, A. N., Ma, T., Frigeri, A., Skach, W. R., Mitra, A., Tamarappoo, B. K., and Farinas, J. (1996). Water transport across mammalian cell membranes. *Am. J. Physiol.* **270,** C12–C30.

Walz, T., Hirai, T., Murata, K., Heymann, J. B., Mitsuoka, K., Fujiyoshi, Y., Smith, B. L., Agre, P., amd Engel, A. (1997). The three-dimensional structure of aquaporin-1. *Nature* (*London*) **387,** 624–627.

Welch, W. J., and Brown, R. (1996). Influence of molecular and chemical chaperones on protein folding. *Cell Stress Chap.* **2,** 109–115.

Yang, B., and Verkman, A. S. (1997). Water and glycerol permeability of aquaporins 1–5 and MIP determined quantitatively by expression of epitope-tagged constructs in *Xenopus* oocytes. *J. Biol. Chem.* **272,** 16140–16146.

Yang, B., and Verkman, A. S. (1998). Urea transporter UT3 functions as an efficient water channel: Direct evidence for a common water/urea pathway. *J. Biol. Chem.* **273,** 9369–9372.

Yang, B., Brown, D., and Verkman, A. S. (1996). The mercurial-insensitive water channel (AQP-4) forms orthogonal arrays in stably transfected CHO cells. *J. Biol. Chem.* **271,** 4577–4580.

Yang, B., Van Hoek, A. N., and Verkman, A. S. (1997). Very high single channel water permeability of aquaporin-4 in baculovirus-infected Sf9 cells and liposomes reconstituted with purified aquaporin-4. *Biochemistry* **36,** 7625–7632.

Ye, R., and Verkman, A. S. (1989). Osmotic and diffusional water permeability measured simultaneously in cells and liposomes. *Biochemistry* **28,** 824–829.

Zeidel, M. L., Ambudkar, S. V., Smith, B. L., and Agre, P. (1992). Reconstitution of functional water channels in liposomes containing purified red cell CHIP28 protein. *Biochemistry* **31,** 7436–7440.

Zhang, R., Logee, K., and Verkman, A. S. (1990). Expression of mRNA coding for kidney and red cell water channels in *Xenopus* oocytes. *J. Biol. Chem.* **265,** 15375–15378.

Zhang, R., Alper, S., Thorens B., and Verkman, A. S. (1991). Evidence from oocyte expression that the erythrocyte water channel is distinct from band 3 and the glucose transporter. *J. Clin. Invest.* **88,** 1553–1558.

CHAPTER 6

Membrane Events Involved in Volume Regulation

Else K. Hoffmann* **and John W. Mills†**

*Department of Biochemistry, August Krogh Institute, University of Copenhagen, Copenhagen, Denmark, and †Department of Biology, Clarkson University, Potsdam, New York 13699

1063-5823/99 $30.00

I. INTRODUCTION: BIOLOGICAL ROLE

Cell membranes of virtually all animal cells are far more permeable to water than to ions, e.g., in Ehrlich ascites tumor (EAT) cells, water permeability exceeds ionic permeability by five orders of magnitude (see Hoffmann, 1978). This high water permeability is caused predominantly by the presence of water channels: aquaporins (Agre *et al.,* 1993; Knepper, 1994; Verkman *et al.,* 1996). Water is therefore in thermodynamic equilibrium across the plasma membrane of animal cells, and cell volume is determined by the relationship between extracellular and intracellular concentrations of osmolytes.

The cytoplasm contains a large number of impermeable anionic macromolecules. The colloid osmotic pressure arising from these impermeable macromolecules must therefore be counteracted by the Na^+,K^+ pump with the constant use of metabolic energy. Thus even at unchanged osmolarity, all cells still have to regulate their volume constantly via this "pump-leak" mechanism (see, e.g., Tosteson, 1964).

In addition to maintaining a constant cell volume under steady-state conditions most cells have developed a multitude of volume regulatory mechanisms that become activated rapidly by the minute and gradual perturbation of cell volume. These mechanisms not only serve to control cell volume but influence a variety of cellular functions, making cell volume control an integral part of many physiological functions.

In mammals, with the exception of the mammalian inner renal medulla, extracellular osmolarity is maintained within narrow limits. However, a variety of clinical conditions can lead to pathological changes in extracellular osmolality, resulting in anisotonic volume disturbances (see McManus and Churchwell, 1994; Hoffmann and Dunham, 1995; Lang *et al.,* 1998a,b).

Conversely, a far more common cause for cell volume pertubations is changes in the cytoplasmic solute concentration, which is associated with a number of physiological events.

i. A wide variety of hormones have been shown to alter cell volume (Häussinger and Lang, 1991; Häussinger *et al.,* 1993; see Lang *et al.,* 1998a,b). Insulin causes hepatocytes to swell by activation of both Na^+/H^+ exchange and Na^+,K^+,$2Cl^-$ cotransport, and this swelling is suggested to be involved in the stimulation of protein and glycogen synthesis and to inhibit proteolysis and glycolysis (see Häussinger *et al.,* 1994; Agius *et al.,* 1994). Glucagon, however, shrinks hepatocytes presumably by activation of K^+ and Cl^- channels and has the opposite effect on the enzyme activity (Vom Dahl *et al.,* 1991). Similar to insulin, several growth factors increase cell volume by stimulating Na^+/H^+ exchange and $Na^+,K^+,2Cl^-$ cotransport (see e.g., Hunziker *et al.,* 1994).

ii. Na^+-dependent uptake of sugars and amino acids leads to cell swelling and has also been shown to activate volume regulatory mechanisms (see MacLeod, 1994). Obviously, the entry of Na^+-coupled nutrients represents a continous challenge to cell volume constancy, but it is also an integral part of the transport function (see Schultz, 1992; Lang and Rehwals, 1992; Harvey, 1994).

iii. Secretagogues enhance ion exit, resulting in cell shrinkage, during the initiation of secretion in salivary glands, pancreatic acinar cells, and shark rectal glands (see, e.g., Hoffmann and Ussing, 1992; Foskett, 1994; Hoffmann and Dunham, 1995).

iv. Exercising muscle may lead to the cellular accumulation of lactate and increased cell volume (Saltin *et al.,* 1987). The intracellular acidosis will increase the cell swelling by activation of Na^+/H^+ exchange.

v. Swelling of astroglial cells is a serious event in trauma, ischemic stroke, hypoxic/anoxic insult, status epilepticus and many other pathological states in the mammalian brain (see Friedman and Haddad, 1993).

For reviews on the fundamentals of short-term cell volume regulation, see Chamberlin and Strange (1989), Hoffmann and Simonsen (1989), Okada and Hazama (1989), Schultz (1989a,b), Lewis and Donaldson (1990), Geck (1990), Lang *et al.* (1990), a series of reviews edited by Gilles *et al.* (1991), Sarkadi and Parker (1991), Hoffmann and Ussing (1992), Spring and Hoffmann (1992), Parker (1993b), Hoffmann *et al.* (1993), Al-Habori (1994), Macknight *et al.* (1993, 1994), a collection of articles edited by Strange (1994a), McManus *et al.* (1995), Sachs (1996), Hoffmann and Dunham (1995), and Lang *et al.* (1998a,b).

One approach to studying these volume regulatory mechanisms is by the experimental manipulation of extracellular solutes (anisosmotic volume regulation). Although there may be some differences in the pathways leading to activation between isosmotic and anisosmotic volume regulation (see, e.g., MacLeod, 1994; Foskett, 1994; Strange, 1994b), the anisosmotic volume regulatory model has produced significant insight into basic regulatory mechanisms and is the major emphasis of this chapter.

Long-time adaptation of volume regulatory mechanisms involves the activation of expression of a series of effector genes. In this way, the synthesis of proteins such as enzymes and membrane transport proteins are induced. These mechanisms are not dealt with here. (For excellent reviews, see, e.g., Yancey, 1994; Burg, 1994; Kwon, 1994; Kwon and Handler, 1995; Burg *et al.,* 1996.)

II. MEMBRANE TRANSPORT MECHANISMS IN REGULATORY VOLUME DECREASE (RVD)

Swelling causes cells to activate separate K^+, Cl^- and organic osmolyte channels, K^+,Cl^- cotransporters, or coupled K^+/H^+ and Cl^-/HCO_3^- exchangers, resulting in a loss of osmolytes and a concomitant loss of cell water (regulatory volume decrease or RVD). In most cell types the majority of the osmotically active solute lost during the RVD response is KCl. In some cell types, a number of organic osmolytes do, however, play a very significant role. In EAT cells, about 70% is KCl and 30% is amino acids, predominantly taurine (Hoffmann and Hendil, 1976). Figure 1 shows the various types of transport sytems involved in the RVD response. That hypotonic swelling (Fig. 1, upper left) results in the activation of separate, but functionally coupled K^+ and Cl^- conductive transport pathways was first proposed in EAT cells (Hoffmann, 1978, 1985; Hoffmann *et al.,* 1984, 1986) and human lymphocytes (Cheung *et al.,* 1982; Grinstein *et al.,* 1982a–c, 1984; Sarkadi *et al.,* 1984a,b, 1985) and has been reported in numerous symetrical cells as well as epithelial cells (for references see Hoffmann and Dunham, 1995; Lang *et al.,* 1998a,b). During the RVD response in EAT cells the Na^+ conductance is decreased (Hoffmann, 1978), whereas the K^+ conductance is increased about 2-fold and the Cl^- conductance is increased about 60-fold (Lambert *et al.,* 1989). At low external (and internal) Ca^{2+}, Kramhøft *et al.* (1986) could demonstrate both K^+,Cl^- cotransport and K^+ and Cl^- conductance channels after swelling of EAT cells (see Fig. 1, upper right).

The cell membrane is depolarized in hypotonically swollen cells because the increase in Cl^- conductance exceeds the increase in the K^+ conductance (see Hoffmann and Dunham, 1995). In EAT cells a depolarization of ~20 mV in a hypotonic solution is observed during RVD (Lang *et al.,* 1987; Lambert *et al.,* 1989).

Usually more cations are lost from cells during RVD than Cl^-, depending on the buffering capacity of the cell (see Freeman *et al.,* 1987; Lew and Bookchin, 1986). In EAT cells the K^+ loss exceeds the Cl^- loss during RVD by about 1.6-fold (Hendil and Hoffmann, 1974); this difference is due to the loss of HCO_3^-, resulting in acidification of the cytoplasm during RVD (Livne and Hoffmann, 1990).

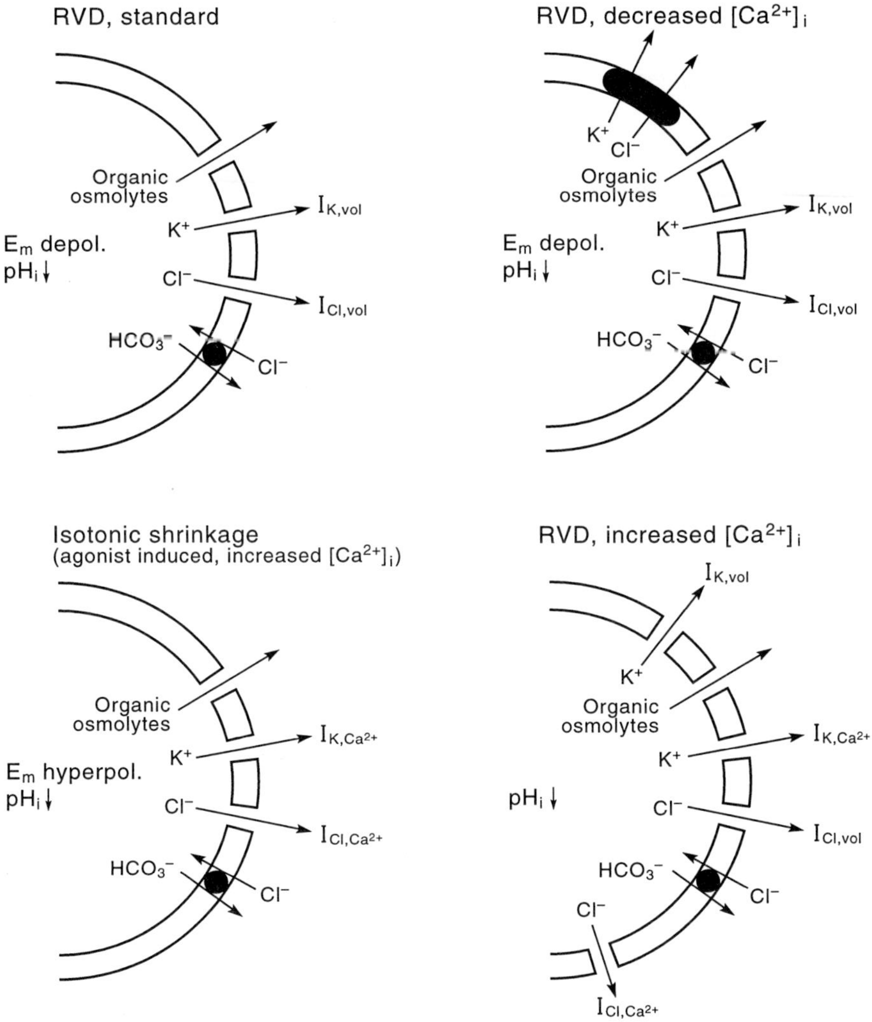

FIGURE 1 Features of RVD and isotonic cell shrinkage after the addition of Ca^{2+}-mobilizing agonists in EAT cells. (*Top left*) During RVD, under standard conditions, a ChTX-insensitive $I_{K,vol}$ is activated together with a tamoxifen-sensitive $I_{Cl,vol}$, a mini Cl^- channel. $I_{Cl,vol}$ increases 30 times more than $I_{K,vol}$ and the membrane potential E_m depolarizes. (*Top right*) In Ca^{2+}-depleted cells, K,Cl cotransport is activated in addition to activation of the channels. (*Bottom left*) Ca^{2+}-mobilizing agonists activate a ChTX-sensitive $I_{K,Ca}$ and a niflumic acid-sensitive $I_{Cl,Ca}$, which is also a mini Cl^- channel. The increase in $I_{K,Ca}$ activity is higher than the increase in $I_{Cl,Ca^{2+}}$, and therefore the cell hyperpolarizes. In the two previous situations, the K^+ loss is partially balanced electrically by HCO_3^- loss through the Cl^-/ HCO_3^- exchanger, resulting in acidification of the cells. After agonist stimulation the initial acidification is, however, followed by an alkalinization caused by shrinkage activation of NHE. (*Bottom right*) An increase in $[Ca^{2+}]_i$ during the period of cell swelling accelerates the RVD response. The ChTX-sensitive $I_{K,Ca}$, the ChTX-insensitive $I_{K,vol}$, the niflumic acid sensitive $I_{Cl,Ca}$, and the tamoxifen sensitive $I_{Cl,vol}$ are all activated. The change in membrane potential depends on whether the increase in K^+ conductance or the increase in Cl^- conductance is highest. Redrawn and extended from Hoffmann (1997).

A. *Swelling-Activated K^+ Channel ($I_{K, vol}$)*

Activation of K^+ leak efflux in response to cell swelling was observed in the early 1970s in lymphocytes (Roti-Roti and Rothstein, 1973) and EAT cells (Hendil and Hoffmann, 1974). Volume-sensitive K^+ channels have later been described in patch-clamp studies in tracheal epithelial cells (Butt *et al.,* 1990), EAT cells (Christensen and Hoffmann, 1992), rat hepatocytes (Sandford *et al.,* 1992), MDCK cells (Banderali and Roy, 1992), and a toad epithelial cell line (Nilius *et al.,*1995a). Directly stretch-activated K^+ channels are located mainly in the basolateral membrane of different kidney cells, where they play a volume regulatory role (see Giebisch, 1995).

Among the cloned K^+ channels shown to be important in volume regulation are the two voltage-gated K^+ channels, the Kv1.3 (n-type) K^+ channel (Deutch and Chen, 1993) and the Kv1.5 channel (Felipe *et al.,* 1993), as well as the "min K^+ channel" (Busch *et al.,* 1992; Busch and Maylie, 1993). The cell membrane is depolarized in hypotonically swollen cells (see earlier), and the depolarization can open voltage-gated K^+ channels. RVD in human lymphocytes involves voltage-gated Kv1.3 (n-type) K^+ channels that are charybdotoxin (ChTX) sensitive as well as a ChTX-insensitive K^+ channel (Lewis and Calahan, 1995). The ChTX-insensitive K^+ channel dominates during RVD in rat thymocytes (Rotin *et al.,* 1991). The molecular identity of this channel is still not clear (see Lewis and Calahan, 1995).

Many of the properties of $I_{K,vol}$ are not explained by the known cloned channels, and additional channels must be operative. In EAT cells, the $I_{K,vol}$ cannot be activated directly by membrane stretch (Christensen and Hoffmann, 1992), and is insensitive to all inhibitors of Kv1.3 tested as well as to ChTX and clotrimazole inhibitors of Ca^{2+}-activated K^+ channels (Jørgensen *et al.,* 1997). The $I_{K,vol}$ in EAT cells is also insensitive to tetraethylammonium (TEA) (Jørgensen *et al.,* 1997), which blocks the voltage-gated type l (Kv3.1) K^+ channel in lymphocytes (Lewis and Calahan, 1995). It can, however, be blocked by Ba^{2+} (Jørgensen *et al.* 1997) and by quinine (Hoffmann *et al.,* 1984). In whole cell studies it has been demonstrated that $I_{K,vol}$ in EAT cells is insensitive to ChTX and apamin and can be activated in the absence of a detectable increase in $[Ca^{2+}]$ (Riquelme *et al.,* 1998). It is likely that this channel and the swelling-activated, ChTX-insensitive, voltage-independent channel seen in thymocytes are the same.

Ca^{2+}-activated K^+ channels will contribute to RVD in cells in which $[Ca^{2+}]_i$ is increased during RVD (see Section IV,H). The Ca^{2+}-activated K^+ channel in EAT cells, described previously as a small inward-rectifier K^+ channel (Christensen and Hoffmann, 1992), was found to have essentially no role in a normal RVD response (Riquelme *et al.,* 1998). Addition of a Ca^{2+}-mobilizing agonist at the time of maximal swelling in hypotonic medium will, however, accelerate the RVD response dramatically (Hoffmann *et al.*, 1993). This is due to the activation of ChTX-sensitive Ca^{2+}-activated K^+ channels (Fig. 1) distinct from and in parallel with the ChTX-insensitive channel activated by cell swelling (Riquelme *et al.*, 1998). In lymphocytes in which cell swelling can induce an increase in $[Ca^{2+}]$, both voltage-sensitive and Ca^{2+}-activated K^+ channels can function simultaneously in volume regulation (Schlichter and Sakellaropoulos, 1994).

B. Swelling-Activated Cl^- Channels

A swelling-activated chloride current ($I_{Cl,vol}$), is probably present in all mammalian cells, where it is responsible for the regulation of membrane potential, intracellular pH, and cell volume. There seems to be cell-type-dependent differences in the voltage-dependent inactivation of $I_{Cl,vol}$, probably reflecting different affinities for divalent cations and protons (Nilius *et al.*, 1997a). The anion permeability sequence of $I_{Cl,vol}$ is, in most cells, including EAT cells: $SCN^- > I^- > NO^- > Br^- > Cl^- > F^-$ corresponding to Eisenmann's sequence I (see Nilius *et al.*, 1994a; Okada, 1997). The inhibitor pharmacology is quite variable between different cells types. The stilbene derivative 4′,4′-diisothiocyano-stilbene-2′,2′-sulfonic acid (DIDS) is a potent inhibitor in some cell types but only inhibits in very high concentrations in others, such as EAT cells (Pedersen, *et al.*, 1998a). This is in accordance with a very moderate inhibition by DIDS of the RVD response in EAT cells (Hoffmann *et al.*, 1986). Niflumic acid has no effect on $I_{Cl,vol}$ in e.g. EAT cells (Pedersen *et al.*, 1998a), whereas it inhibits $I_{Cl,vol}$ strongly in other cells, such as mouse cortical collecting duct cells (Meyer and Korbmacher, 1996). Tamoxifen, an antiestrogen drug, likewise has very different effects on $I_{Cl,vol}$ in different cell types with a strong effect seen in EAT cells (Pedersen *et al.*, 1998a), in T84 cells (Valverde *et al.*, 1993), and in mouse cortical collecting duct cells (Meyer and Korbmacher, 1996). As discussed by Nilius *et al.* (1997a), the varying effect of blockers might as well reflect different pathways for swelling activation as the presence of different channels.

Depletion of intracellular adenosine 5′-triphosphate (ATP) inhibits $I_{Cl,vol}$ in most cell types (see Okada, 1997). It seems that a nonhydrolytic binding of ATP to the channel or to a regulating protein is required for the activation of $I_{Cl,vol}$ in analogy with the regulation of NHE (see later). However, extracellular ATP in a high concentration inhibits $I_{Cl,vol}$ probably by blocking the open channel (see Okada, 1997). For reviews on swelling activated Cl^- channels see, for example, Jentsch (1996), Strange *et al.* (1996), Nilius *et al.* (1997a), and Okada (1997).

1. Single Channel Measurements

On the single channel level the Cl^- channels described to be activated by cell swelling in different cell types can be divided roughly into three major groups based on their single channel conductance Nilius *et al.,* 1997a:

i. Outward rectifiers or nonrectifiers with a small conductance or "mini" Cl^- channels (0.1–12 pS) were described in cell-attached patches from several different cell types (Christensen *et al.,* 1989; Christensen and Hoffmann, 1992; Zhang and Jacob, 1997; Kubitz *et al.,* 1992) and by stationary noise analysis in others (Lewis *et al.,* 1993; Schumacher *et al.,* 1995; Stoddard *et al.,* 1993; Nilius *et al.,* 1994b; Doroshenko and Neher, 1992). From analyses performed on C6 glioma cells, it was suggested that single channel conductances obtained in the previously mentioned studies by stationary noise analysis might be underestimated significantly (Jackson and Strange 1995; Strange *et al.,* 1996). Provided that this is correct, these Cl^- channels might be of the larger conductance type discussed later and different from the EAT cell type mini Cl^- channel, where the small conductance (3–7 pS) is determined directly by single channel current measurements in cell-attached patches (Christensen and Hoffmann, 1992). The volume-sensitive mini Cl^- channel in EAT cells is insensitive to $[Ca^{2+}]_i$ in excised patches and is not activated directly by membrane stretch. The channel activates in an all/or none fashion with few channel closures (Christensen and Hoffmann, 1992).

ii. Outward rectifiers with intermediate unitary conductance (10–100 pS) are described from single channel recordings in cell-attached patches and in the outside-out patch configuration in many cells (Nilius *et al.,* 1997a; Okada, 1997). In several cases, the channels could not be found in cell-attached patches (Boese *et al.,* 1996; Worrell *et al.,* 1989). The relatively few cases in which volume activation of the intermediate conductance channel has been described in the cell-attached patch configuration include osteoblasts (Kelly *et al.,* 1994), rabbit atrial myocytes (Duan *et al.,* 1997a), and nonpigmented ciliary epithelial cells (Zhang and Jacob, 1997). In EAT cells an outwardly rectifying Cl^- channel with a conductance of 34 pS was recorded in isolated inside-out patches, but the channel was never seen in cell-attached patches in either isotonic or hypotonic media (Christensen and Hoffmann, 1992). It is therefore possible but not certain that EAT cells, such as nonpigmented ciliary epithelial cells (Zhang and Jacob, 1997), have both the low conductance and the intermediate conductance of volume-activated channels. Intermediate conductance channels have been suggested to be the channel-mediating efflux of organic osmolytes, the so-called volume-sensitive organic osmolyte/anion channel (VSOAC) (for further description see Strange *et al.,* 1996; Nilius *et al.,* 1997a; Okada, 1997, Kirk and Strange, 1998).

iii. Swelling-activated "maxi channels" with a very large unitary conductance (> 100 pS) are found in renal cortical collecting duct cells (Schwiebert *et al.,* 1994) and bovine ciliary epithelial cells (Zhang and Jacob, 1997).

From these results it seems most reasonable to conclude that there are several types of volume-sensitive anion channels with different volume set points and activation mechanisms. This could be important in the ability of cells to selectively respond to different types of cell swelling (Strange *et al.,* 1996).

2. Molecular Identification of Volume-Regulated Cl^- Channels

Two Cl^- channel proteins expressed from cloned genes seem to be volume-activated Cl^- channels. These are ClC-2 (Thiemann *et al.,* 1992; Gründer *et al.,* 1992; see Jentsch, 1996) and ClC-3 (Coca-Prados *et al.,* 1996; Duan *et al.,* 1997a,b). The properties of expressed ClC-2 channels differ from native volume-activated Cl^- channels in that they are inwardly rectifying and more permeable to Cl^- than to I^- (Jentsch, 1996). Functional expression of ClC-3, however, results in an outwardly rectifying channel more permeable to I^- than to Cl^-, which is modulated strongly by cell volume and exhibits many properties identical to most $I_{Cl,vol}$ in native cells (Duan *et al.,* 1997a,b; Yamazaki *et al.,* 1998). Thus, ClC-3 might encode for $I_{Cl,vol}$ in many mammalian cells. Two other proteins expressed from cloned genes, P-glycoprotein (P-gp) (Valverde *et al.,* 1992) and pI_{Cln} (Paulmichl *et al.,* 1992, 1996) were suggested to be swelling-activated Cl^-channels, but are more likely to be regulators of $I_{Cl,vol}$ (Valverde *et al.,* 1995;1996; Vanoye *et al.,* 1997; Krapivinsky *et al.,* 1994). Even a direct regulatory function for P-gp (Cahalan *et al.* 1994; Rasola *et al.* 1994; Luckie *et al.* 1996) and for pI_{ICln}.(Voets *et al.,* 1996) has, however, been questioned.

3. Separate Swelling- and Ca^{2+}-Activated Anion Currents

Ca^{2+}-activated Cl^- channels have been described in various nonexcitable and excitable cells (for reviews see Valverde *et al.,* 1995; Jentsch, 1996; Nilius *et al.,* 1997b). In most cells, the swelling-induced activation of Cl^- channels does not require an increase in $[Ca^{2+}]_i$ (Nilius *et al.,* 1997a,b). In EAT cells, separate chloride currents, $I_{Cl,Ca}$ and $I_{Cl,vol}$, are activated by an increase in $[Ca^{2+}]_i$ and by cell swelling, respectively. The kinetics of the two currents are different: $I_{Cl,Ca}$ is activated and deactivated at depolarized potentials at hyperpolarized potentials. In contrast, $I_{Cl,vol}$ exhibits time- and voltage-dependent deactivation at depolarized potentials and reactivation at hyperpolarized potentials. $I_{Cl,Ca}$ is inhibited by niflumic acid, 5-nitro-2-(3-phenylpropylamino)benzoic acid (NPPB) and DIDS, with niflumic acid being the most potent inhibitor. In contrast, $I_{Cl,vol}$ was unaffected by niflumic acid (100 μM), but was abolished by tamoxifen (10 μM). Figure 2 demonstrates that RVD in EAT cells is blocked completely by tamoxifen but is unaffected by niflumic acid, indicating that $I_{Cl,Ca}$ does not play any role in the RVD response.

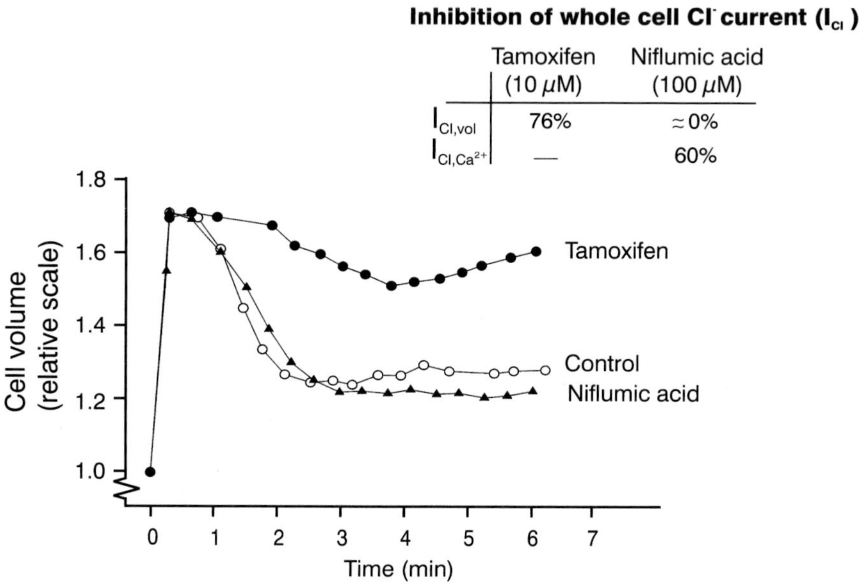

FIGURE 2 Roles of $I_{Cl,vol}$ and $I_{Cl,Ca}$ in the RVD response in Ehrlich ascites tumor cells (EATC). Effect of tamoxifen, an inhibitor of $I_{Cl,vol}$, and niflumic acid, an inhibitor of $I_{Cl,Ca}$, on whole cell Cl^- currents and on RVD in EATC. The whole cell Cl^- current was investigated using the whole cell patch clamp technique. The effect of the inhibitors on $I_{Cl,vol}$ was assesed from the time course of the current activated by exposure to 27% hypotonic solution measured at 100 mV in the presence and absence of tamoxifen (10 μM) or niflumic acid (100 μM); the effect on $I_{Cl,Ca}$ was measured with a pipette solution containing 100 nM [Ca^{2+}]. The percentage inhibition of $I_{Cl,vol}$ and $I_{Cl,Ca}$ by the Cl^- channel blockers was calculated from the current magnitude in nA measured at 100 mV in the presence of the inhibitor in percentage of the mean of the current magnitudes measured before application of the inhibitor. *Cell volume* is monitored over time using a Coulter counter and is given relative to the initial volume in control cells (○), in the presence of tamoxifen (10 μM, ●), or niflumic acid (100 μM, ▲). Inhibitors were added 1 min prior to the hypotonic exposure. At time zero, a sample of the cell suspension is diluted 500-fold in hypotonic (150 mOsm) Ringer solution. The figure is representative of 12 independent experiments for measurements of current inhibition and 3 independent experiments for measurement of RVD. Redrawn from Hoffmann and Pedersen, (1998); values for inhibition of $I_{Cl,vol}$ and $I_{Cl,Ca}$ are from Pedersen *et al.*, (1998a).

C. Swelling-Activated Organic Osmolyte Efflux

The loss of organic osmolytes is mediated via a leak pathway that is separate from the uptake system (channel mediated), as shown first for the swelling-activated taurine efflux in EAT cells (Hoffmann and Lambert, 1983) and astrocytes (Kimelberg *et al.*, 1990). The loss of organic osmolytes has later been shown to be channel mediated in most cells studied (Strange *et al.*, 1996). The high permeability of VSOAC to large organic anions raised the possibility that the channel is also permeable to organic solutes such as taurine (Strange and Jackson, 1995; Strange *et al.*, 1996; Okada, 1997). As described earlier, VSOAC seems to mediate the swelling-activated efflux of organic osmolytes such as myo-inositol, taurine and sorbitol (Okada, 1997; Strange *et al.*, 1996; Boese *et al.*, 1996; Kirk and Strange, 1998).

In EAT cells, however, the major pathway for the efflux of taurine is pharmacologically distinct from $I_{\mathrm{Cl,vol}}$ (Lambert and Hoffmann, 1994). Evidence shows (i) that DIDS is a strong inhibitor of the taurine channel but only inhibits $I_{\mathrm{Cl,vol}}$ marginally and (ii) that arachidonic acid (AA) inhibits $I_{\mathrm{Cl,vol}}$ strongly but activates the swelling-activated taurine efflux (Lambert and Hoffmann, 1994). A possibility for separate Cl^- and taurine channels has also been suggested in rat mammary tissue (Shennan *et al.*, 1994) and in HeLa cells (Stutzin *et al.*, 1997). In renal inner medullary collecting duct (IMCD) cells there seems to be separate channels for betaine and sorbitol loss distinct from the myo-inositol/taurine channel (Ruhfus *et al.*, 1996).

Based on observations obtained with trout erythrocytes and by expression of band 3 anion-exchange protein in oocytes, Motais *et al.* (1997) provided evidence that swelling-induced taurine movements occur via a channel controlled by band 3. However, taurine efflux is not dependent on the presence of band 3 proteins in mammalian cells (Sanchez-Olea *et al.*, 1995), arguing against this as a general phenomenon.

D. K^+,Cl^- Cotransport Involved in RVD

The swelling-activated K^+ efflux in duck erythrocytes was found to occur via a K^+, Cl^- cotransporter with a 1:1 stoichiometry. Volume-sensitive K^+,Cl^- cotransport is very active in red cells from patients with sickle cell anemia (Canessa *et al.*, 1987), resulting in a reduced cell volume and an increased polymerization of hemoglobin. It has been suggested that there is a direct interaction between the mutant hemoglobin and the K^+,Cl^- cotransporter (Oliviera *et al.*, 1992). For reviews dealing specifically with K^+,Cl^- cotransport, see Lauf *et al.* (1992) and Payne *et al.* (1996). For other reviews dealing with K^+,Cl^- cotransport involved in RVD, see Palfrey (1994), Parker (1994), McManus and Churchwell (1994), and Hoffmann and Dunham (1995).

1. Structure of the K^+,Cl^- Cotransporter

The K^+,Cl^- cotransporter was cloned recently (Gillen *et al.,* 1996; Payne *et al.,* 1996). The family of K^+,Cl^- cotransporters is, according to their tissue distribution, divided into two subgroups denoted KCC-1 and KCC-2. Cotransporters of the KCC-1 subgroup are distributed widely, whereas the single KCC-2 gene is found only in brain neurons. Hydropathy profile analyses show that overall structure and transmembrane topology are shared by the cotransporters from the KCC-1 and KCC-2 subgroups: all consist of 12 transmembrane domains, four potential N-linked glycosylation sites, and a large, cytoplasamic COOH-terminal domain (Gillien *et al.,* 1996; Payne *et al.,* 1996).

2. Kinetics of Swelling Activation of the K^+,Cl^- Cotransporter

Applying LK sheep erythrocytes as a model system, Dunham and colleagues (1993) have developed a more advanced model of the two-state model of Jennings and Al-Rohil (1990) with three states A↔B↔C for swelling activation and shrinkage inactivation of K^+,Cl^- cotransport (Dunham *et al.,* 1993; see Hoffmann and Duham, 1995). Swelling causes a relatively slow conversion of cotransporters from state A to state B, followed by a fast conversion of state B to state C. The A to B conversion is dependent on Mg^{2+}, entails an increase in J_{max}, and is controlled by dephosphorylation, whereas the reverse reaction is controlled by a protein kinase. Thus, swelling activation is a consequence of an inhibition of this volume-sensitive kinase (see Hoffman and Dunham, 1995).The B to C conversion, however, entails a decrease in apparent affinity for K^+, and the mechanism of this step has been suggested to involve mechanical changes in the membrane (Kelley and Dunham, 1996). Evidence has been provided that the three-state model for the swelling activation of K^+,Cl^- cotransport also applies to human erythrocytes (Kaji and Gasson, 1995) and it is likely that a similar mechanism may exist for the swelling control of K^+,Cl^- cotransport in other cells.

E. K^+/H^+ Exchange Involved in RVD

Parallel activation of K^+/H^+ exchange and Cl^-/HCO_3^- exchange after cell swelling has been well described in Amphiuma red blood cells (Cala, 1980, 1983a,b; Cala *et al.,* 1986), where Ca^{2+} is found to act as a modulator. K^+/H^+ exchange has been shown to play a role in mitochondrial volume regulation (Garlid, 1996). Evidence indicates that a NHE protein could also be responsible for mediating K^+/H^+ exchange in Amphiuma red blood cells (Cala, 1986). However, the fact that DIDS can specifically label the K^+/H^+ exchanger, whereas it does not seem to have any affinity for NHE, argues against that idea (Cala and Maldonado, 1994).

III. MEMBRANE TRANSPORT SYSTEMS INVOLVED IN VOLUME REGULATORY INCREASE (RVI)

The most common pathways for osmolyte uptake in RVI are the $Na^+,K^+,2Cl^-$ cotransporter as well as the Na^+/H^+ and Cl^-/HCO_3^- exchanger working in parallel. Following exposure to a hypertonic medium and concomitant osmotic cell shrinkage, a regulatory volume increase (hypertonic RVI) can be observed in certain cell types, whereas other cell types show no indication of hypertonic RVI (for references, see Hoffmann and Dunham, 1995). Nearly all cells show, however, a volume recovery in response to reintroduction of an isosmotic medium following a hyposmotic challenge, called the "RVI-after-RVD protocol" or "post-RVD RVI." For a detailed discussion of possible reasons for this failure to elicit hypertonic RVI, see, for example, Hallows and Knauf (1994) and Hoffmann and Dunham (1995). Briefly, the decreased intracellular Cl^- concentration after RVD is suggested to play a permissive role for the activation of $Na^+,K^+,2Cl^-$ cotransport in several cell types (Ussing, 1982; Hoffmann *et al.,* 1983; Grinstein *et al.,* 1983; Breitwieser *et al.,* 1990; Engström *et al.,* 1991; Haas and McBrayer, 1994; Whisenant *et al.,* 1993). In the shark rectal gland, the level of phosphorylation of the NKCC protein is increased by a decline in the intracellular Cl^- concentration under isotonic conditions, suggesting that the dependence of Cl^- might be caused by a Cl^- -sensitive kinase (Lytle and Forbush, 1996). In salivary gland epithelia, an agonist-induced decline in the intracellular Cl^- concentration is required for the activation of Na^+ influx and it is proposed that this reflects an activation of the $Na^+,K^+,2Cl^-$ cotransporter as well as NHE, although the roles of Cl^- on the individual transporters were not assessed directly (Robertson and Foskett, 1994).

A. Shrinkage-Activated Na^+ Channels

There is early evidence that cell shrinkage activates conductive Na^+ leak fluxes (Parker *et al.,* 1975; Hoffmann, 1978) and in a few cells it has been demonstrated that electrolyte accumulation is partly accomplished by a shrinkage-activated Na^+ channel (Civan *et al.,* 1996; Wehner and Tinel, 1998).

B. $Na^+,K^+,2Cl^-$ Cotransport

That an anion–cation cotransport system is activated by cell shrinkage and is an important mechanism in RVI was first described in avian red blood cells (Kregenow, 1973; see Kregenow, 1981), L cells (Gargus and Slayman, 1980), simian virus-transformed 3T3 cells (Bakker-Grunwald *et al.,* 1982), EAT cells (Hoffmann *et al.,* 1983), rat red blood cells (Duhm and Göbel, 1984), and tracheal epithelial cells (Shorofsky *et al.,* 1984) and has later been established in a large number of cell types (for references and/or discussions), see Haas, 1994; Hoffmann and Dunham, 1995; Lang *et al.,* 1998a,b).

In EAT cells, shrinkage activates the bumetanide-inhibitable influxes of K^+ and Cl^- from near zero value at the physiological steady state to about 30 μmol $\times$ g dry wt^{-1} $\times$ min $^{-1}$ after cell shrinkage (Hoffmann *et al.,* 1983). Because $Na^+,K^+,2Cl^-$ cotransport is accompanied by an increased Na^+ efflux via the Na^+/K^+ pump, the net result is K, Cl uptake. The restoration of cell volume is complete within a few minutes, at which time the cotransporter is again inactivated. The bumetanide-sensitive net flux has the stoichiometry 1 : Na, 1 : K, 2 : Cl (when measured in the presence of ouabain and Ba^{2+}) and the K^+ and Cl^- flux ratio (influx/efflux) for the bumetanide-sensitive components is equal to the experimentally determined combined concentration ratio (Jensen *et al.,* 1993). Also, in duck red cells, the stoichiometric ratio of net cotransport is 1 : Na, 1 : K, 2 : Cl, but the ratio of unidirectional fluxes varies, revealing the presence of K^+/K^+ exchange in high K cells and Na^+/Na^+ exchange in high Na cells. A model incorporating these features has been presented (Lytle *et al.,* 1998).

1. Mechanism and Primary Structure

Because $Na^+,K^+,2Cl^-$ cotransport is electroneutral, the transport is independent of and does not affect the membrane potential. Outside the cell, the ions are thought to bind in the order Na^+, Cl^- h, K^+, Cl^- l where Cl^- h and Cl^- l refer to Cl^--binding sites with high and low affinity, respectively, and to be transported in a "first on, first off" manner, known as "glide symmetry" (Stein, 1986; see Hoffmann and Dunham, 1995). The two Cl^- -binding sites seem to have very different affinities in EAT cells, with Cl^-_h having an extremely high Cl^- affinity (Krarup *et al.,* 1996). During the translocation process, the cotransporter passes a stable state in which the ions are trapped (occluded) within the transmembrane domains of the cotransport protein, inaccessible from either side of the membrane (Krarup *et al.,* 1996).

Bumetanide and similar diuretics were suggested to exert their inhibitory effect on the cotransporter by competing for Cl^- at the extracellular face of the membrane (see Haas, 1994). The disruption of actin filaments by cytochalasin B increased the affinity for bumetanide in EAT cells and simultaneously activated the cotransporter (Hoffmann *et al.,* 1994) (see Section V). On a bumetanide–Sepharose affinity gel, two proteins were isolated from EAT cell membranes with molecular masses of 85 and 39 kDa, respectively (Feit *et al.,*1988). Antibodies raised against these proteins inhibited NKCC as well as the RVI response (Dunham *et al.,* 1990). The antibodies also cross-reacted with NKCC from mouse kidney (Haas *et al.,* 1991) and *Xenopus laevis* oocytes (Suvitayavat *et al.,* 1994). Whether these EAT cell proteins are regulatory subunits of NKCC is still not known.

A family of isoforms of NKCCs have been cloned comprising the NKCC-1 and NKCC-2 subgroups. NKCC-1 consists of NKCCs located in the basolateral membrane of epithelia, including the shark rectal gland cotransporter (Xu *et al.,* 1994) and two mammalian homologues (Delpire *et al.,* 1994; Payne *et al.,* 1995). They consist of a polypeptide of about 1200 amino acids, resulting in a molecular mass in the range of 130–132 kDa. They have consensus phosphorylation sites for protein kinase C (PKC) and casein kinase II (Payne and Forbush, 1995) and protein kinase A (PKA) (Reagan, 1995). The NKCC involved in RVI in EAT cells has more than 90% homology with the epithelial NKCC-1 isoform from mouse kidney (B. S. Jensen and E. K. Hoffmann, unpublished results). NKCC-2 consists of NKCCs located in the apical membrane of epithelia. Cloned members include a rabbit kidney NKCC-2 (Payne and Forbush, 1994), a rat kidney NKCC-2 (a protein of ~115 kDa) (Gamba *et al.,* 1994), and a mouse kidney NKCC-2 (Igarashi *et al.,* 1995). The NKCC-2 cotransporters are smaller than the NKCC-1 cotransporters, with molecular masses in the range of 120–121 kDa. Like NKCC-1s, NKCC-2s have consensus sites for PKA, PKC, and casein kinase II (Payne and Forbush, 1995).

C. Parallel Na^+/H^+ and Cl^-/HCO_3^- Exchangers in RVI

The primary event in the RVI response is, in many cells, the activation of NHE. If the intracellular H^+ buffering capacity and/or Cl^-/HCO_3^- exchanger activity is limited, activation of NHE will elicit a sizeable increase in pH_i without an increase in cell volume (Cala and Grinstein, 1988). In contrast, in cells with a large H^+ buffering capacity and a large capacity for Cl^-/HCO_3^- exchange, activation of NHE will result in little change in pH_i but substantial cell swelling because the net result is electroneutral uptake of NaCl followed by osmotically obliged water (Grinstein et al,

1992a). In EAT cells, NHE contributes about 35% to RVI in the presence of HCO_3^-, whereas RVI in EAT cells in nominally HCO_3^- free medium is mediated almost exclusively by NKCC (Pedersen *et al.*, 1996a; see Hoffmann, 1997).

Parallel activation of NHE and a Cl^-/HCO_3^- exchanger has been shown to mediate RVI in many cell types (Cala, 1980, 1983b; Grinstein *et al.*, 1984; Shrode *et al.*, 1995; Pedersen *et al.*, 1996a). For extensive reviews on the functions and regulation of NHE, see Orlowski and Grinstein (1997), Paillard (1997), and Wakabayashi *et al.*, (1997).

Five distict mammalian isoforms of NHE (NHE-1 through NHE-5) have been cloned at this time, and a putative sixth isoform has been cloned (Sardet *et al.*, 1989; Wang *et al.*, 1993; Klanke *et al.*, 1995; see Orlowski and Grinstein, 1997). The "housekeeping" NHE-1 isoform appears to be present in the plasma membrane of nearly all mammalian cells, where it is responsible for pH and volume regulation (Sardet *et al.*, 1989; Grinstein *et al.*, 1992b; Sarkadi and Parker, 1991). It is activated by hyperosmotic cell shrinkage and inhibited by cell swelling (Kapus *et al.*, 1994). In polarized epithelial cells, NHE-1 is localized primarily to the basolateral membrane whereas NHE-3 is localized exclusively in apical membranes (see Wakabayashi *et al.*, 1997; Paillard, 1997). NHE-2 is like NHE-1, both are activated normally by hypertonicity and inhibited by hypotonicity, whereas NHE-3 appears always to be inhibited by hypertonicity (Kapus *et al.*, 1994; Nath *et al.*, 1996). The physiological role of NHE-2 is not clear. However, if transfected stably into AP-1 cells, NHE-2 regulates pH_i and cell volume. NHE-4 may be important for volume regulation in renal inner medulla, as it is activated by relatively strong hypertonicity (Bookstein *et al.*, 1994; 1996).

All NHE isoforms have a membrane topology of 10–12 transmembrane segments. The carboxy-terminal region at the cytoplasmic site has consensus phosphorylation sites for calmodulin mitogen-activated protein (CaM) kinase II and microtubule-associated protein (MAP) kinase but not for PKA and PKC (Fliegel and Fröhlich, 1993).

In adherent fibroblasts, NHE-1 is shown to be localized to focal adhesion sites and with F-actin, vinculin, and talin (see Section V). In fact, the carboxy-terminal cytoplasmic domain of NHE-2 contains two domains, which resemble the SH3-binding domains known from other transporters to mediate interactions between the transport proteins and the cytoskeleton (see Orlowski and Grinstein, 1997).

The cytoplasmic region of NHE-1 contains a high- and low- affinity CaM-binding site (Bertrand *et al.*, 1994). Shrinkage-induced activation of NHE-1 and NHE-2 is inhibited strongly by ATP depletion (see Demaurex and Grinstein, 1994), which appears to cause an acidic shift in the pH_i "set point" in all cell types studied, including EAT cells (Pedersen *et al.*, 1996a).

It seems that the maximal activity of NHE-1 requires ATP hydrolysis-dependent events and hydrolysis-independent binding of ATP (Demaurex *et al.,* 1997). NHE does not have a consensus site for nucleotide binding (Orlowski *et al.,* 1992) and it is suggested that the effect of ATP depletion is related to cytoskeletal rearrangements (Goss *et al.,* 1994) (see also Section V).

D. Organic Osmolytes

High ionic concentrations interfere with the structure and function of macromolecules, including proteins. Thus the possible cellular variation in the concentration of electrolytes is limited. During prolonged exposure to hypertonicity, volume regulation therefore depends mainly on the loss or accumulation of organic osmolytes; during short-term exposure, the release or uptake of organic osmolytes can be quite important. Organic osmolytes are metabolically more expensive to accumulate than ions, but unlike ions, organic osmolytes, even at high concentrations, are non-pertubing to macromolecules (Brown and Simpson, 1972; Somoro, 1986; Winzor *et al.,* 1992). Organic osmolytes are involved in cell volume regulation in many cell types (for reviews see Chamberlin and Strange, 1989; Law, 1991; Law and Burg, 1991; Garcia-Pérez and Burg, 1991; Strange *et al.,* 1996, Kinne 1993; Hoffmann and Dunham, 1995). In some cells the release of organic osmolytes may account for as much of half of the total RVD (Garcia-Romeu *et al.,* 1991).

Four classes of organic osmolytes are used in mammalian cells: free amino acids and their derivatives (e.g., taurine, proline, alanine), sugars and polyols (e.g., sorbitol, myo-inositol), urea, and certain methylated compounds (e.g., betaine, glycerophosphoylcholine) (Yancey, 1994). Organic osmolytes are specifically important for cell volume regulation in the brain, where cell volume alterations and alterations of ionic composition cannot be tolerated (see Strange *et al.,* 1993; Gullans and Verbalis, 1993; Law, 1994), and in the renal medulla, where extracellular osmolarity may become more than fourfold that of isotonicity (Garcia-Pérez and Burg, 1991; Burg *et al.,* 1996).

The intracellular concentration of most individual amino acids is low, and only alanine, glycine, GABA, taurine, glutamate, β-alanine, and proline are quantitatively of any importance in cell volume regulation. The amino-sulfonic acid taurine was the first amino acid shown to be involved in volume regulation in fish erythrocytes (Fugelli and Zachariassen, 1976) and in mammalian cells (Hoffmann and Hendil, 1976). It was later found that taurine plays a role in volume regulation in numerous different cell types (see Law and Burg, 1991; Hoffmann and Dunham, 1995).

Taurine is well suited as an osmolyte. It is the end product of the metabolism of the sulfur-containing amino acids methionine and cysteine, its permeability is very low, and it is present at high concentrations in most invertebrate and vertebrate cells and tissues. In EATC the taurine concentration is 53 mmol/liter cell water, whereas the taurine concentration in the ascites fluid is only 0.01 m*M* (Hoffmann and Lambert, 1983).

An increase in the concentration of organic osmolytes is achieved by increased uptake and synthesis and is decreased by an increased degradation and an increased passive leak. Most of these processes are long-term effects, resulting from an increased expression of genes, and are not dealt with in this chapter. Excellent reviews can be found elsewhere on that subject (see e.g., Kwon, 1994; Kwon and Handler, 1995; Burg *et al.,* 1996).

IV. VOLUME SENSING AND SIGNAL TRANSDUCTION

Signals of volume change can either be mechanical changes, such as deformation of the membrane lipid bilayer or reorganization of the cytoskeleton, or be changes in the concentration of cellular solutes such as small ions or macromolecules (see Hoffmann and Dunham, 1995). These two mechanisms can possibly work together; for instance, in sheep red blood cells, both mechanisms were suggested to be involved as the swelling signal leading to the activation of KCl cotransport (Dunham, 1995).

Sensors of the signals can be stretch-activated channels or other integral membrane proteins responding to the radial strain of the membrane G-proteins and/or enzymes such as phospholipases (PLAs). The sensors of cell volume change initiate the signal transduction pathway, eventually activating the effectors, i.e., volume-sensitive transport proteins. The cytoskeleton may be involved at several levels: (i) in sensing swelling signals as well as generating them, (ii) in transducing the signal, and (iii) in the activation of the effectors. The role of the cytoskeleton in volume regulation will therefore be dealt with in a special section (Section V).

A. Volume Signals

1. Macromolecular Crowding and/or Ionic Strength

The thermodynamic activity of a given protein in a "crowded" solution of macromolecules is greater than the activity of that protein at the same concentration in an "uncrowded" solution, and a slight change in concentration can produce large changes in activity (macromolecular crowding). Several examples indicate that changes in macromolecular crowding can be used as a signal after changes in cell volume. By varying hemoglobin concentrations and volumes of the ghosts and by comparing the set points for the activation of K^+,Cl^- cotransport and NHE, it was demonstrated that the set points expressed as a percentage of dry weight were exactly the same for cells and ghosts, but differed when they were expressed per cell volume (Colclasure and Parker, 1991, 1992). It did not matter whether the dry weight was hemoglobin or serum albumin. The signaling substance was obviously total protein and independent of the specific proteins. Further evidence for this type of mechanism comes from studies with urea, which reduces macromolecular crowding by reducing the thermodynamic activity of the proteins (Collins and Washabaugh, 1985; Minton, 1983; Prakash *et al.,* 1981; Yancey, 1994). Urea strongly activates K^+,Cl^- cotransport in erythrocytes (Parker, 1993a; Dunham, 1995; Kaji and Gasson, 1995).

Based on these results, it seems likely that changes in macromolecular crowding serve as signals of volume change (Minton *et al.,* 1992; Zimmerman and Minton, 1993; Minton, 1994).

The thermodynamic activity of the proteins and thus the macromolecular crowding is reduced by increasing ionic strength. In agreement with this, Parker *et al.* (1995) showed that an increase in ionic strength decreases the set point for both the K^+,Cl^- cotransporter and NHE. Oppositely, in C6 cells (Emma *et al.,* 1997) and in bovine endothelial cells (Nilius *et al.,* 1998), a reduction in intracellular ionic strength is found to decrease the set point for the volume-activated anion channel. Thus, the ion transporters involved in volume regulation may be differentially sensitive to ionic strength and macromolecular crowding.

2. The Intracellular Cl^- Concentration

A Cl^--sensitive protein kinase has been described in human airway epithelium (Treharne *et al.,* 1994), and Cl^- has been found to modulate the function of certain G-proteins (Higashijima *et al.,* 1987). Jackson and co-workers (1996) suggested that a reduction of intracellular Cl^- was the signal for activation of the swelling-activated Cl^- channel in skate hepatocytes. In these experiments, as in others involving changes in cell volume, it is not easy to distinguish between changes in ionic strength and changes in intracellular Cl^-, and the effect seen by Jackson and co-workers might well be caused by the ionic strength (see Section III). As described in Section III there is solid evidence for a decrease in Cl^- playing a permissive role for activation and phosphorylation of a NKCC after cell shrinkage. Conversely, in barnacle muscle, an increase in Cl^- seems to amplify the sensitivity of NHE to cell shrinkage (Hogan *et al.,* 1997). Thus intracellular Cl^-

might be important for determining which mode of osmolyte gain is activated after cell shrinkage; if Cl^- is low, the cotransporter is favored and if Cl^- is high, salt uptake via the coupled exchangers might be favored.

B. Volume Sensors

Only in yeast have osmosensing membrane proteins been demonstrated. These are a transmembrane histidine kinase and another transmembrane protein containing an SH3 domain (Maeda *et al.,* 1996). No homologous protein has as been found yet in mammalian cells (see, however, Section V for a discussion of the possible role of integrins.)

1. G-Proteins

G-proteins are interesting candidates for volume sensors. In cell-free liposomes it has been demonstrated that a heterotrimeric G-protein can be activated by shear stress (Frangos and Gudi, 1997), and G-proteins seem to be involved in the activation of RVD as well as RVI responses. Thus, G-proteins appear to play a role in the swelling-induced activation of the 85-kDa cytosolic phospholipase A_2 (PLA_2) (see later) and of $I_{Cl,vol}$ in a number of cell types (Nilius *et al.,* 1997a), and G-proteins are involved in the shrinkage-induced activation of NHE in barnacle muscle fibers (Hogan *et al.,* 1997).

2. Phospholipase A_2

PLA_2s are a large group of acylhydrolases that catalyze the hydrolysis of the sn-2 fatty ester bond of phospholipid, forming arachidonic acid and lysophospolipid. Nine groups of PLAs have been identified (for reviews, see Dennis, 1994, 1997). Of these, the 85-kDa cytosolic PLA_2 ($cPLA_2$), a group IV phospholipase A_2, seems to be the one of interest for cell volume regulation. The $cPLA_2$ is phosphorylated by PKC and a MAP kinase and this event is dependent on transducing G-proteins and a rise in $[Ca^{2+}]_i$ (see, e.g., Piomelli, 1996). Cell swelling was first suggested to activate phospholipase A_2 in human platelets (Margalit *et al.,* 1993a), and inhibitors of phospholipase A_2 interfere with the RVD process, the activation of $I_{Cl,vol}$, and/or an organic osmolyte channel in several cell types (Lambert and Hoffmann, 1994; Mitchell *et al.,* 1997; Nilius *et al.,* 1994c; Basavappa *et al.,* 1998). Swelling-induced release of arachidonic acid, a result of PLA_2 activation, is demonstrated in EAT cells (Thoroed *et al.,* 1997) and CHP-100 neuroblastoma cells (Basavappa *et al.,* 1998)(see Fig. 3, top and Fig. 5, top). The swelling-induced arachidonic acid release, $I_{Cl,vol}$, and RVD was inhibited by $AACOCF_3$, an inhibitor of $cPLA_2$ in both cell types (Basavappa *et al.,*

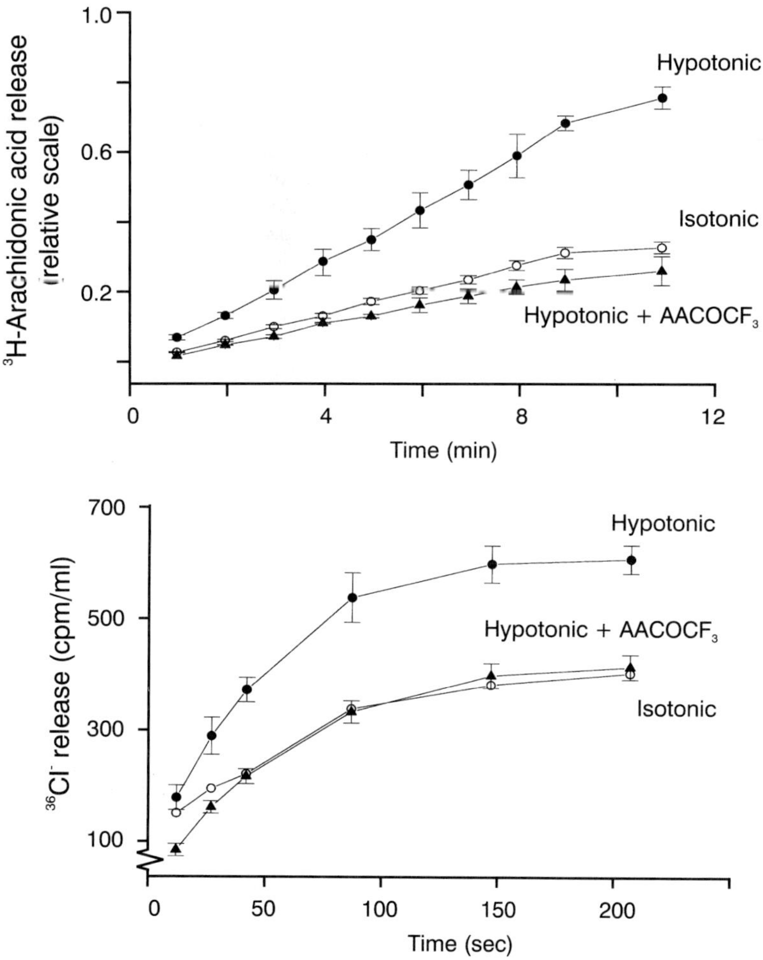

FIGURE 3 Effect of an inhibitor of the cytosolic 85-kDa $cPLA_2$ on the swelling-activated [^{3}H]-arachidonic acid release and on activation of $I_{Cl,vol}$ in CHP-100 neuroblastoma cells. (Top) Effect of hypoosmotic solution on release of [^{3}H]-AA in CHP-100 cells. Cells were loaded with [^{3}H]-AA (1 μCi/ml) for 18 hr, washed, and subsequently exposed to isotonic [290 mOsm (○)] or hypotonic [190 mOsm (●)] medium, respectively, both containing 1.0% (w/v) BSA to scavenge extracellular fatty acids. In studies with the $cPLA_2$ blocker, cells were pretreated for 4 hr with $AACOCF_3$ (10 μM)(▲). Samples of the extracellular medium were collected every 2 min. Accumulated [^{3}H]-AA cpm values are relative to the total amount of radioactivity. The total amount of radioactivity was calculated as the sum of radioactivity in all time samples plus radioactivity in the lysed monolayers at the end of the experiment. (Bottom) Cells were preloaded with ^{36}Cl (2 μCi/ml) for 2 hr, washed, and exposed to the same treatments. Redrawn from Figs. 2 and 5 in Basavappa *et al.,* (1998).

1998; Thoroed *et al.,* 1997) (see Fig. 3). In EAT cells, the swelling-induced release of arachidonic acid was inhibited by guanosine-5′-*O*-(2-thiodiphosphate) (GDPβS), consistent with the involvement of a G-protein (Thoroed *et al.,* 1997). Moreover, from inhibitor studies and from a phospholipase D (PLD) assay it is concluded that neither PLD nor PLC is involved in the increased release of arachidonic acid during RVD and that events associated with RVD are mediated exclusively by the activation of $cPLA_2$ (see Fig. 4). Because there is no change in the release of [^{14}C]stearic acid from EAT cells after the reduction of osmolarity (Thoroed *et al.,* 1997), this argues against PLC or PLA_1 being involved in the release of AA (Fig. 5, top). In endothelial cells it was similarly found that shear stress activated $cPLA_2$ (Oike *et al.,* 1994; Nilius *et al.,* 1994c; Pearce *et al.,* 1996). Finally, osmotic swelling can increase PLA_2 activity directly in large membrane vesicles (Lehtonen and Kinnunen, 1995).

A number of different mechanisms leading to the swelling-induced activation of $cPLA_2$ can now be suggested: (i) changes in the lipid bilayer may act as a mechanosignal activating a heterotrimeric G-protein or by direct

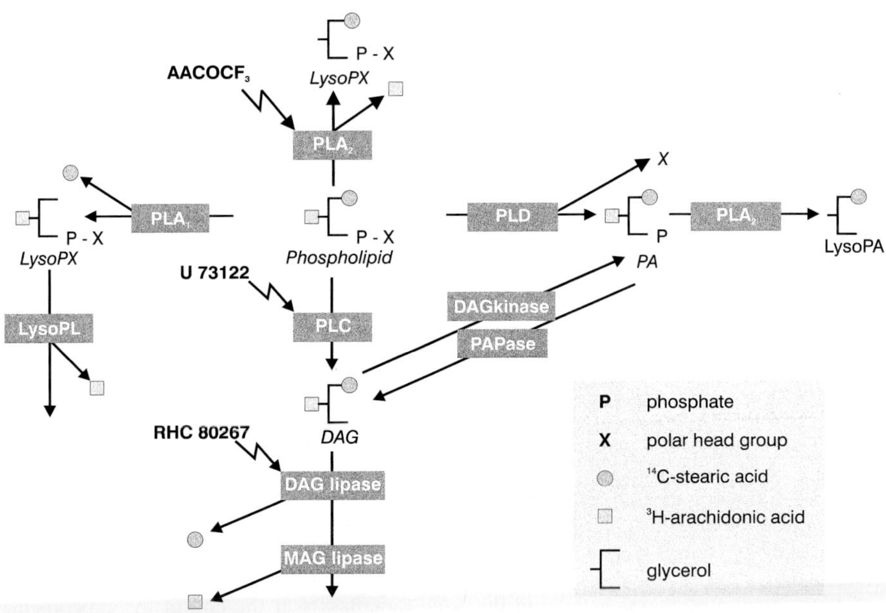

FIGURE 4 Enzymatic pathways involved in the release of phosphatidic acid (PA), arachidonic acid, and stearic acid from membrane phospholipids. X: the polar head group can be choline, ethanolamine, serine, or inositol. The site of action of inhibitors used in Thoroed *et al.* (1997) is indicated. Redrawn and extended from Thoroed *et al.,* (1997).

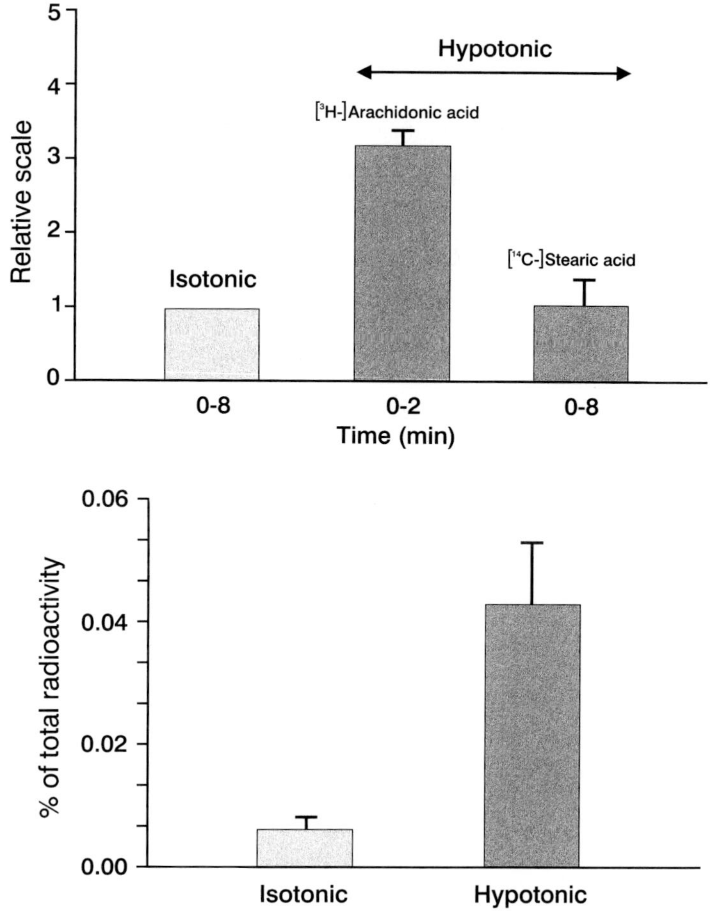

FIGURE 5 (Top) Effect of cell swelling on the rate of [^{3}H]-arachidonic acid and [^{14}C]stearic acid release from Ehrlich cells. Cells are preincubated with the respective isotop for 2 hr, washed, and exposed to a hypotonic medium. The hypotonic incubation medium contains 0.5% (w/v) BSA in order to scavenge the fatty acids released from the cells. The release of [^{3}H]-arachidonic acid is linear for 2 min after which the rate of release again decreases; the release of [^{14}C]stearic acid is linear in the whole 8-min period. Values for the rate of release in hypotonic medium is given relative to the release in isotonic medium. Values are given ± SEM (n = 4). In 50 additional experiments the [^{3}H]-arachidonic acid release after cell swelling has been determined (as a control for various inhibitors) with similar results. Redrawn from Figs. 2 and 3 in Thoroed *et al.,* (1997). (Bottom) Formation of [^{14}C]stearic acid-labeled lysophosphatidyl choline ([^{14}C]-lysoPC) during the first 145 sec of a RVD response or during the same time period in isotonic medium. Cells are preincubated with [^{14}C]stearic acid as described. The [^{14}C]stearic acid activity in the lysoPC spot was estimated as a percentage of the total cell associated radioactivity. The various radioactive lipid spots on the thin-layer plates were localized by autoradiography, scraped off the plate, and measured by liquid scintillation spectrometry. The formation of [^{14}C]-lysoPC was significantly higher during RVD that under isotonic conditions. Redrawn from Fig. 4 in Thoroed *et al.* (1997).

activation of $cPLA_2$ (Fig. 6); (ii) swelling-induced changes in the cytoskeleton could alter $cPLA_2$ activity (Akiba *et al.,* 1993); (iii) swelling-induced activation of SA-cat channels (Christensen and Hoffmann, 1992) could lead to localized increases in $[Ca^{2+}]$, which would also be expected to stimulate PLA_2 activity; and (iv) a swelling-activated tyrosin kinase could be upstream of the activation PLA_2 (Craig and Johnson, 1996) as it is known that certain tyrosine kinases, via Ras or Rac, activate PLA_2 (Peppelenbosch *et al.,* 1995).

C. Role of Arachidonic Acid and Lysophosphatidic Acid

Arachidonic acid in itself can affect ion channels and can influence $[Ca^{2+}]_i$ in cells. Extracelllular application of AA ihhibits $I_{Cl,vol}$ directly in many cell types (Lambert and Hoffmann, 1994; Basavappa *et al.,* 1998; for further references, see Okada, 1997). In contrast, a micromolar concentration of AA is well known to activate K^+ channels in several cell types (Petrou *et al.,* 1994; Meves, 1994). Arachidonic acid has been shown directly to release Ca^{2+} from internal stores in EAT cells (Jørgensen *et al.,* 1998) and to induce an increase in $[Ca^{2+}]_i$ in B cells from pancreas (Ramanadham *et al.,* 1992). This in turn can activate $I_{Cl,Ca}$ and $I_{K,Ca}$. In rat inner medullary

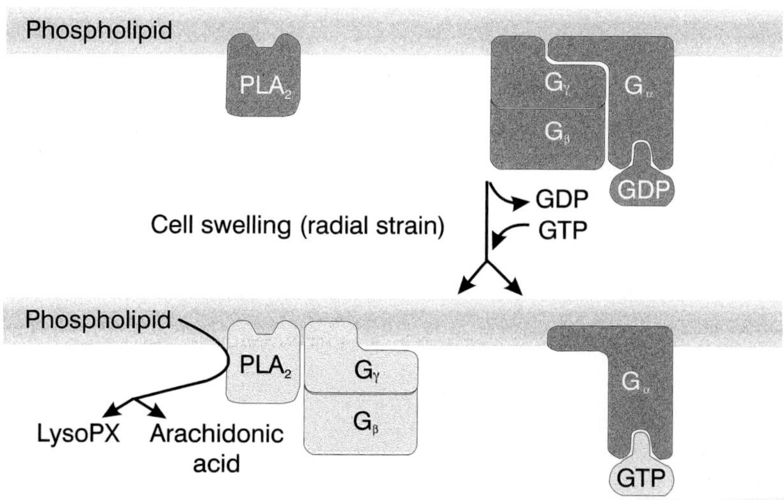

FIGURE 6. PLA_2 and/or a heterotrimeric G-protein as the volume sensor. Model showing how changes in the lipid bilayer may act as a mechanosignal by activating heterotrimeric G-proteins directly or by direct activation of $cPLA_2$. Cell swelling results in some kind of radial strain, and it is suggested that this can activate $cPLA_2$ and/or heterotrimeric G-proteins (see text and Figure 4 for discussion and references).

collecting duct cells, cell swelling activates PLA_2, resulting in a release of arachidonic acid, which in turn leads to an increase in $[Ca^{2+}]_i$ (Tinel *et al.,* 1997).

As seen from Fig. 4, phosphatidic acid may also be hydrolyzed by PLA_2 with the two products arachidonic acid and lysophosphatidic acid (LPA). LPA is known to elicit numerous biological functions (Moolenaar *et al.,* 1997). The potential role of LPA in cell volume regulation has not yet been investigated but it is an obvious candidate because its downstream signal cascade includes the Ras-Raf-MAP kinase cascade and Rho, which has been implicated in RVD (Tilly *et al.,* 1996) and in cytoskeletal rearrangement (see Section V).

D. Eicosanoids as Messengers

Three main classes of oxygenase enzymes are implicated in the metabolism of arachidonic acid to eicosanoids (see, e.g., Piomelli, 1996):

1. Prostaglandin H-synthase (cyclooxygenase and peroxidase) producing prostaglandins, prostacyclins, and thromboxanes.
2. Cytochrome P450 produces a family of epoxyeicosatrienoic acids (EETs).
3. A variety of lipoxygenases, producing leukotrienes (LTs), hepoxilins, and hydroxy eicosatetraenoic acid (HETEs). The lipoxygenase (LIP) 5-LIP catalyzes the oxygenation of AA to 5-HPETE, which is converted via a series of steps to the peptidoleukotrienes, including leukotriene D_4 (LTD_4) (see, e.g., Ford-Hutchinson *et al.,* 1994).

1. LTD_4 as Messenger

In several cell types, inhibitors of the 5-LIP pathway or the leukotriene receptors inhibit RVD and/or $I_{Cl,vol}$ (Diener and Scharrer, 1993; Doroshenko and Neher, 1992; Mastrocola *et al.,* 1993; Thoroed and Fugelli, 1994a,b; Fatherazi *et al.,* 1994; Jørgensen *et al.,* 1996; Light *et al.,* 1997). However in C6 glioma cells, LIP inhibitors are found to affect $I_{Cl,vol}$ directly (Strange *et al.,* 1996). It is therefore important to measure swelling-induced changes in levels of arachidonic acid metabolites and to directly test the effect of these on the activation of the channel before a role in volume regulation can be assigned. Studies on EAT cells have provided evidence for an involvement of LTD_4 in regulating swelling-activated channels. The main lines of evidence are:

1. A swelling-induced increase in release of LTD_4 and/or C_4 (LTC_4) has been demonstrated (Lambert *et al.,* 1987).

2. LTD_4 accelerates RVD (Lambert *et al.,* 1987; Jørgensen *et al.,* 1996).
3. Activation of taurine efflux after the addition of LTD_4, apparently via a Ca^{2+}-independent organic osmolyte channel, has been demonstrated (Lambert and Hoffmann, 1993; Lambert, 1998).
4. The RVD response, as well as the LTD_4-induced acceleration of the RVD response, is inhibited strongly by the LTD_4 receptor antagonist L649,123 (1 μM) (Jørgensen *et al.,* 1996).
5. Desensitization of the LTD_4 receptor by preincubation with 100 nM LTD_4 inhibits a subsequent RVD response (Jørgensen *et al.,* 1996).

LTD_4 is also demonstrated to be the messenger for swelling activation of the Cl^- channel during RVD in isolated crypt cells from rat colonic epithelium (Diener and Scharrer, 1993). Moreover, in hepatocytes from lipopolysaccharide-treated rats, LTD_4 appears to activate Ca^{2+}-dependent and -independent anion conductances (Meng *et al.,* 1997). In contrast, in some other cells, LTD_4 does not seem to play any role in the activation of $I_{Cl,vol}$ (Basavappa *et al.,* 1998; Fatherazi *et al.,* 1994; Hainsworth *et al.,* 1996; Ackerman *et al.,* 1994).

Further evidence for the role of LTD_4 comes from studies where the addition of omega-3 FA (EPA C20:5, n-3) to the diet of the mouse hosts of EAT cells altered the fatty acid composition of the cellular membrane phospholipid pool. This resulted in an accelerated RVD response and a net Cl^- loss following hypotonic exposure in the EAT cells (Lauritzen *et al.,* 1993). The total leukotriene synthesis (LTD_4 plus LTD_5) in cells enriched with EPA was enhanced compared to the total leukotriene synthesis in control cells. In agreement with this, evidence shows that EPA is a poor substrate or even an inhibitor of the cyclooxygenase (Needleman *et al.,* 1986), whereas it may be a better substrate than AA for the lipoxygenase (Careaga-Houck and Sprecher, 1990). In cells where the membrane is enriched with EPA, the swelling-induced shift from prostaglandin to leukotriene synthesis (Lambert *et al.,* 1987) could thus be augmented and the RVD response accelerated (see Lauritzen *et al.,* 1993). This might be of significance for humans on an omega-3 FA diet.

2. Lipoxygenase Products and K^+ Channels

Lipoxygenase products can modulate K^+ channels, but only in a few cases has this been described to be important in the swelling activation of K^+ channels: The 12-LIP product, hepoxilin A_3, and the 5-LIP metabolite, LTD_4 activate the swelling-activated K^+ channel in human platelets (Margalit *et al.,* 1993b) and in EAT cells (Hoffmann, 1998), respectively. In red blood cells from the mud puppy, a 5-LIP metabolite is likewise responsible for the swelling-induced activation of K^+ channels (Light *et al.,* 1997). In

EAT cells, LTD_4 activates a ChTX-sensitive, Ca^{2+}-activated K^+ efflux with an EC_{50} of about 10 n*M* as well as a ChTX-insensitive, Ca^{2+}-independent K^+ efflux with an EC_{50} of about 2 n*M* (Hoffmann, 1998). As described earlier, the volume-regulating K^+ channel is ChTx insensitive and Ca^{2+} independent. Because the addition of 3 n*M* LTD_4 to EAT cells in hypotonic, Ca^{2+}-free medium accelerates the RVD response in the absence of a measurable increase in $[Ca^{2+}]_i$ (Jørgensen *et al.,* 1996), this indicates that LTD_4 interacts with the volume-activated channel. An interesting idea would be that the swelling-activated K^+ channel could be a LTD_4 receptor-gated channel.

E. ATP as an Autocrine Messenger

ATP is released and appears to play important roles in autocrine signaling under stress situations in various cell types, including mechanical stress in an epithelial cell line (Grygorczyk and Hanrahan, 1997), endothelial cells (Hamada *et al.,* 1998), and EAT cells (Pedersen *et al.,* 1999); hydrostatic pressure in rabbit urinary bladder (Ferguson *et al.,* 1997); and osmotic cell swelling in rat hepatocytes (Wang *et al.,* 1996a). Extracellular ATP exerts its effects on ion transport via P2X receptors, which are ligand-gated cation channels, and P2Y receptors, which are G-protein coupled (see Burnstock, 1996, 1997). In EAT cells, as in many other cells types, the main P2 receptors are the $P2Y_1$ and $P2Y_2$ subtype (Pedersen *et al.,* 1999). Stimulation of these receptors with micromolar concentrations of ATP results in the release of inositol 1,4,5-trisphosphate $[Ins(1,4,5)P_3]$ (Dubyak, 1986) and an increase in $[Ca^{2+}]_i$ (Artalejo and García-Sancho, 1988).

1. K^+ Channels

ATP activates Ca^{2+}-dependent ChTX-sensitive K^+ channels in EAT cells (Hoffmann, 1998) and in many other cell types (Nilius *et al.,* 1995b; Strøbaek *et al.,* 1996), accelerates the RVD response in EAT cells (Hoffmann, 1998), and induces cell shrinkage when added in isotonic steady state (Pedersen *et al.,* 1998b). However, as discussed earlier, separate swelling- and Ca^{2+}-activated K^+ currents have been demonstrated in EAT cells, and ATP is found to only activate the ChTX-sensitive, Ca^{2+}-activated K^+ efflux, not the ChTX-insensitive, volume-activated K^+ efflux (Hoffmann, 1998). Thus, at least in EAT cells, ATP is not the message for activating $I_{K,vol}$.

2. $I_{Cl,vol}$

Separate $I_{Cl,vol}$ and $I_{Cl,Ca}$ were also demonstrated in many cells, including EAT cells (see earlier). $I_{Cl,Ca}$ is activated by ATP in many cell types (see

Nilius *et al.*, 1995b; Strøbaek *et al.*, 1996), and activation of Ca^{2+}-independent Cl^- channels by ATP has actually been reported (Guo *et al.*, 1997). It awaits further analysis whether ATP via P2Y receptors might activate both $I_{Cl,vol}$ and $I_{Cl,Ca}$.

F. The Role of Phosphatidyl Inositol 4,5-bisphosphate

Osmotic swelling and shrinkage in EAT cells are associated with a decrease and an increase in phosphatidyl insositol 4,5-bisphosphate [PtdIns(4,5)P_2], respectively (Christensen *et al.*, 1988; Harbak *et al.*, 1995). In other cells, an increase in volume results in a transient increase in Ins(1,4,5)P_3 (Dahl *et al.*, 1991; Suzuki *et al.*, 1990; Baquet *et al.*, 1991), indicating that there would be a concomitant decrease in PtdIns(4,5)P_2. Because a decrease in PtdIns(4,5)P_2 regulates actin polymerization (Schafer and Cooper, 1995), the change in this phospholipid during RVD and RVI could be important for the volume-associated changes in the F-actin cytoskeleton that may be part of the signaling mechanism. (see Section V).

G. Ca^{2+}/Calmodulin in Signal Transduction

Ca^{2+} is an ubiquitous intracellular messenger involved in the regulation of numerous cellular processes. Ca^{2+} has also been implicated as having either a direct or a participatory role in the RVD response following cell swelling in most cells investigated, but although a certain level of $[Ca^{2+}]$ is often required, an increase in $[Ca^{2+}]_i$ is not necessary in many of the cells investigated. For reviews of the role of Ca^{2+} in cell volume regulation, see Pierce and Politis (1990), McCarty and O'Neil (1992), Hoffmann *et al.*, (1993), Foskett (1994), and Hoffmann and Dunham (1995).

1. Ca^+ Influx and/or a Change in $[Ca^{2+}]_i$?

An increase in $[Ca^{2+}]_i$ following a hypotonic exposure has been demonstrated in several cell types, especially epithelial cells, fibroblasts, and astrocytes, and in most of these cells Ca^{2+} influx plays a key role in the RVD response (for references, see Foskett, 1994; Hoffman and Dunham, 1995). For example, volume recovery is blocked by the removal of external Ca^{2+} and by Ca^{2+} channel blockers. A biphasic $[Ca^{2+}]_i$ increase is induced by hypotonicity in the case of intestine 407 cells (Hazama and Okada, 1990). Both phases were dependent on the influx of extracellular Ca^{2+}, but the second rise was due to Ca^{2+} release from internal stores, a Ca^{2+} -induced

Ca^{2+} release (CICR). Such a mechanism has also been proposed in renal A6 cells (Urbach *et al.,* 1995) and HeLa cells (Sardini *et al.,* 1995).

In other cell types, the volume response is essentially unaffected by the removal of external Ca^{2+} (see Foskett, 1994; Hoffmann and Dunham, 1995). In proximal convoluted tubule cells and salivary acinar cells, there is still a swelling-induced increase in $[Ca^{2+}]_i$ in the Ca^{2+}-free medium, which is mediated by the release from internal stores (Beck *et al.,* 1991; Breton *et al.,* 1992; Foskett *et al.,* 1994). However, in these cases, it was concluded that the RVD response was not dependent on this increase in $[Ca^{2+}]_i$. Finally, no change in $[Ca^{2+}]_i$ could be detected in human lymphocytes during RVD (Grinstein and Smith, 1990; Ross and Cahalan, 1995) and no evidence of increased $[Ca^{2+}]$ was recorded in a cell suspension of fura-2-loaded EAT cells (Jensen *et al.,* 1993; Thomas-Young *et al.,* 1993; Harbak and Simonsen, 1995). When $[Ca^{2+}]_i$ was measured in single EAT cells, a transient increase in $[Ca^{2+}]_i$ was observed in only 6% of the cells at 1 m*M* external Ca^{2+} and almost never in media with low or zero extracellular Ca^{2+} (Jørgensen *et al.,* 1997). Similar results were obtained in single cell analyses of human lymphocytes during RVD (Schlichter and Sakellaropoulos, 1994). Thus, a measurable increase in $[Ca^{2+}]_i$ is not necessary for a RVD response in EAT cells; in the few cells in which it is seen, it is probably an epiphenomenon as was suggested in salivary acinar cells and proximal tubule cells (Breton *et al.,* 1992; Foskett *et al.,* 1994).

Although the RVD response of EAT cells in hypotonic medium thus takes place without a detectable increase in $[Ca^{2+}]_i$, an increase in $[Ca^{2+}]_i$ induced by the addition of Ca^{2+}-mobilizing agonists at the time of maximal swelling will accelerate the rate of the RVD response dramatically. A transient increase in $[Ca^{2+}]_i$ can be induced by the addition of ATP, uridin-triphosphate (UTP), bradykinin, thrombin, or histamine in EAT cells (see Hoffmann *et al.,* 1993; Pedersen *et al.,* 1998b,c), which is shown to accelerate the RVD response significantly (Hoffmann *et al.,* 1993; Hoffmann, 1998). As shown in Fig. 1, (lower right), this is due to the activation of Ca^{2+}-dependent, ChTX-sensitive K^+ channels, in parallel with the swelling-activated, ChTX-insensitive K^+ channels (Jørgensen *et al.,* 1997; Riquelme *et al.,* 1998; Hoffmann, 1998). In summary, the RVD response can take place without an increase in $[Ca^{2+}]_i$, but any increase will accelerate the response (for similar conclusions, see Schlichter and Sakellaropoulos, 1994; Ross and Cahalan, 1995).

2. Is a Certain $[Ca^{2+}]_i$ a Prerequisite?

An unperturbed resting $[Ca^{2+}]_i$ seems to play a permissive role and to be necesssary for RVD in rabbit medullary thick ascending limb cells (Montrose-Rafizadeh and Guggino, 1991) and in vascular endothelial cells

(Szücs *et al.,* 1996). In initial studies with EAT cells, there was a complete block of the RVD response and of the swelling-induced efflux of K^+ and Cl^- after chelating $[Ca^{2+}]_i$ by 1,2-bis(*O*-aminophenoxy) ethane-*N,N,N′,N′*-tetraacetic acid (BAPTA) (Jørgensen *et al.,* 1996). Subsequently, it was shown that the RVD response in cells where $[Ca^{2+}]_i$ is buffered by BAPTA is not inhibited significantly when BAPTA loading is performed in a K^+ equilibrium medium, thus avoiding the BAPTA-induced KCl loss (Jørgensen *et al.,* 1997). In line with this, depletion of intracellular Ca^{2+} or chelation of cytosolic Ca^{2+} does not affect the RVD response in lymphocytes (Grinstein and Smith, 1990). Thus the role of an unperturbed resting level of $[Ca^{2+}]$ in the RVD response is still an open question and may vary between cell types.

3. Role of $[Ca^{2+}]_i$ and Calmodulin (CaM) in RVI

An increase in $[Ca^{2+}]_i$ is generally not observed during RVI and does not seem to play a role in the shrinkage-induced activation of NKCC or in the activation of NHE (Hoffmann and Dunham, 1995). However, a localized increase in $[Ca^{2+}]_i$ cannot of course be excluded (Marsault *et al.,* 1997). CaM, however, seems to play an important role. CaM antagonists inhibit the shrinkage-induced activation of NKCC (Jensen *et al.,* 1993; Krarup *et al.,* 1998) and of NHE (Pedersen *et al.,* 1996a) in EAT cells as well as in many other cell types (O'Donnell *et al.,* 1995; Lytle and Forbush, 1992b; Tanimura *et al.,* 1995; Pewitt *et al.,* 1990; Wakabayashi *et al.,* 1997). CaM-binding motifs are located in the cytoplasmic tail of NHE-1, and if the high-affinity CaM-binding site is deleted, NHE is activated (Wakabayashi *et al.,* 1994) and cannot be activated further by hypertonicity (Bertrand *et al.,* 1994), which led to the suggestion that cell shrinkage could remove an autoinhibitory effect of the CaM-binding regions.

4. Isotonic Cell Shrinkage

Many agonists, secretagogues, several hormones, growth factors and neurotransmitters exert their effect via plasma membrane receptors and elicit a transient rise in $[Ca^{2+}]_i$ by the activation of PLC, which hydrolyzes PtdIns(4,5)P_2, resulting in a production of Ins(1,4,5)P_3 and diacylglycerol (DAG). The increase in $[Ca^{2+}]_i$ evokes $I_{Cl,Ca}$ and $I_{K,Ca}$, resulting in KCl loss and cell shrinkage. This is a very important part of the stimulated fluid and electrolyte secretion by secretory epithelia in airways, salivary glands, and pancreas (see Petersen, 1992; Hoffmann and Ussing, 1992; Frizzell and Morris, 1994; Greger, 1996) but also occurs in nonsecretory, nonpolarized cells. For example in EAT cells (see Hoffmann *et al.,* 1993; Pedersen *et al.,* 1998c) and other cell types (for references, see Lang *et al.,*1998a,b), ATP, bradykinin, and thrombin elicit an increase in Ins(1,4,5)P_3 and an increase in $[Ca^{2+}]_i$ followed by a transient shrinkage due to the activation of $I_{Cl,Ca}$ and $I_{K,Ca}$ (see Hoffmann *et al.,* 1993; Pedersen *et al.,* 1998c). Recovery is caused primarily by the activation of NKCC, but NHE is also activated (Pedersen *et al.,* 1998c). A similar effect has been described in rat parotid acini (Manganel and Turner, 1991) and in Ha-ras-transformed NIH-3T3 fibroblasts (Wöll *et al.,* 1993).

H. Protein Phosphorylation

Protein kinases can be tranducers of the volume signal. An obvious idea was that a protein kinase could be inhibited by cell swelling and activated by cell shrinkage, which seems to apply to human, duck, and sheep red blood cells (Parker, 1994; Palfrey, 1994). In general, the picture is more complicated, however, and several kinases seem to be involved.

1. RVD and Ion Channels

Activation of tyrosine kinases during RVD was demonstrated in several cell types (Lepple-Wienhues *et al.,* 1998; Schliess *et al.,* 1995; Sadoshima *et al.,* 1996; Tilly *et al.,* 1996a). In Jurkat lymphocytes, the activity of the tyrosine kinase (p56lck) correlated with the time course of RVD and appeared to be involved in the activation of $I_{Cl,vol}$ (Lepple-Wienhues *et al.,* 1998). In intestine 407 cells, inhibition of the focal adhesion kinase, p125FAK, inhibited $I_{Cl,vol}$ (Tilly *et al.,* 1996a) (see Section V,D.2 for further discussion).

In contrast, protein kinase C inhibits $I_{Cl,vol}$ in some cell types (Duan *et al.,* 1997a; Hardy *et al.,* 1995; Coca-Prados *et al.,* 1996). PKC is activated after cell shrinkage and in the late phases of RVD in EAT cells (Larsen *et al.,* 1994). This might play a role in keeping $I_{Cl,vol}$ low during RVI and in the termination of the RVD response (Hoffmann and Dunham, 1995).

2. K^+,Cl^- Cotransporter

K^+,Cl^- cotransport is deactivated by phosphorylation and activated by dephosphorylation, thus K^+,Cl^- cotransport in rabbit, human, and sheep red blood cells is blocked by inhibitors of the serine/threonine protein phosphatases 1 and 2A (Jennings and Schulz, 1991; Starke and Jennings, 1993; Kaji and Tsukitani, 1991; Bize and Dunham, 1994). That protein phosphatase 1 is indeed responsible for the dephosphorylation step was demonstrated by Krarup and Dunham (1996) on resealed ghosts from dog erythrocytes.

As mentioned earlier, kinase activity is volume sensitive, being inhibited by cell swelling (Jennings and Al-Rohil, 1990; Dunham *et al.,* 1993; Kaji and Tsukitani, 1991). This volume-sensitive kinase is still unidentified. However, sequence analyses from cloned $Na^+,2Cl^-$ and K^+,Cl^- cotransporters have identified consensus sites for casein kinase II and PKC (Gillen *et al.*, 1996; Payne *et al.*, 1996). Because there seems to be a coordinated activation/inhibition of the $Na^+,K^+,2Cl^-$ cotransporter and the $K^+,2Cl^-$ cotransporter (see Parker, 1994; Palfrey, 1994), it is tempting to speculate that PKC and/or casein kinase II could be responsible for the phosphorylation of both cotransporters. It should be noted, however, that the coordinated regulation might as well involve phosphorylation/dephosphorylation of a regulator.

3. RVI

Activation of tyrosine kinases has been measured during RVI in several cells. Thus cell shrinkage activated the tyrosine kinases p59fgr and p56/59hck rapidly in human neutrophils. The signaling pathway was proposed to involve the activation of tyrosine kinases followed by the activation of a small G-protein RhoA and increased phosphorylation of myosin light chain (MLC) (Krump *et al.*, 1997), which may be the result of RhoA's inhibition of myosin phosphatases (Kimura *et al.*, 1996). In several cell types, shrinkage-induced phosphorylation of MLC is due to the activation of myosin light chain kinase (MLCK) (Shrode *et al.*, 1995; Klein and O'Neill, 1995; Takeda *et al.*, 1993). Finally, PKC, which is activated by cell shrinkage in EAT cells (Larsen *et al.*,1994), will also phosphorylate MLC (Takeda *et al.*, 1993). A likely hypothesis therefore is that a crucial event in the activation of NHE and NKCC may be the increased phosphorylation of MLC, which in turn leads to changes in the cytoskeleton (see Section V).

Extracellular signal-regulated protein kinases (ERKs) are serine/threonine protein kinases, including ERK1 and ERK2, that activate stress-activated protein kinases (SAPKs), and p38 kinases. Excessive osmotic shrinkage triggers several proteins involved in this ERK cascade, including SAPKs and p38 in several mammalian cell types (Sheikh-Hamad *et al.*, 1998; Matsuda *et al.*, 1995). In MDCK cells, p38 activity is essential for the shrinkage-induced induction of the betaine transporter (Sheikh-Hamad *et al.*, 1998), and preliminary results in EAT cells have indicated that the inhibition of p38 inhibits the upregulation of NKCC after 30 min of hypertonicity (E. K. Hoffmann, unpublished results). In yeast, the activation of a p38-like kinase by hypertonicity seems to involve two independent osmosensors: a transmembrane histidine kinase and a transmembrane protein that contains an SH3 domain (Maeda *et al.*, 1996). In CHO cells, inhibition of the Janus kinase (JAK), which is upstream from ERK, inhibits the shrinkage-induced activation of NHE (Garnovskaya *et al.*, 1997). It has also been demonstrated that the tyrosin kinase, Pyk2, is activated by hyperosmolarity and can activate SAPK through the small G-protein Rho (Tokiwa *et al.*, 1996). However, even though the ERK superfamily appears to be activated by cell shrinkage in many cell types, in other cell types, such as human neutrophils, ERK1/2, SAPK, or p38 are not activated upon hypertonic cell shrinkage (Krump *et al.,* 1997). Finally, it must be noted that the activation of SAPK and p38 by cell swelling has also been reported in a few studies (Sadoshima *et al.,* 1996; Tilly *et al.,* 1996b), thus the picture is not very clear at the moment.

4. NHEs

Even though NHE is activated by phosphorylation, cell shrinkage does not result in phosphorylation of the NHE protein (Grinstein *et al.,* 1992b), and shrinkage-induced activation of NHE is not prevented by the deletion of the phosphorylation site on NHE (Bianchini *et al.,* 1995). Thus, the phosphorylation-dependent events probably involve accessory proteins rather than NHE itself. Candidates could be components of the cytoskeleton (see Wakabayashi *et al.,* 1997; Orlowski and Grinstein, 1997). Activation of NHE by hypertonicity in astrocytes seems to involve the myosin light chain, which is phosphorylated upon cell shrinkage. ML-7 inhibits the shrinkage-induced activation of NHE, although it should be noted that the IC_{50} value was 56 μM which is a very high compared to the IC_{50} value for MLCK, which is 0.3 μM, making it possible that the observed effect of ML-7 was nonspecific. In EAT cells, MLCK inhibition inhibits NKCC and RVI strongly (Krarup *et al.,* 1998) but does not inhibit NHE-induced alkalinization seen after exposure to hypertonicity (Varming *et al.,* 1998). As mentioned earlier, PKC can phosphorylate MLC as well. In EAT cells it is suggested that a PKC isoform is involved in the shrinkage activation of NHE, as chelerythine, a specific inhibitor of PKC, inhibits the shrinkage-induced activation of NHE at concentrations where the drug is specific for PKC (Pedersen *et al.,* 1996a; Varming *et al.,* 1998). Thus a role for MLC in NHE activity in EAT cells cannot be ruled out. It should also be noted that PKC does not appear to be involved in the shrinkage-induced activation of NHE-1 in other cell types (Wakabayashi *et al.,* 1997).

5. NKCC

A number of studies have demonstrated that NKCC is activated via protein kinases (Pewitt *et al.,* 1990; Torchia *et al.,* 1992; Jensen *et al.,* 1993; Lytle and Forbush, 1992b; Haas *et al.,* 1995; Klein and O'Neill, 1995; O'Donnell *et al.,* 1995; Krarup *et al.,* 1998) and is inhibited by phosphatases (Pewitt *et al.,* 1990; Palfrey and Pewitt, 1993; Paulais and Turner, 1992; Vigne *et al.,* 1994; O'Donnell *et al.,* 1995; Krarup *et al.,* 1998) and direct phosphorylation of NKCC proteins after cell shrinkage has been demonstrated (Lytte and Forbush, 1992b; Xu *et al.,* 1994; Torchia *et al.,* 1992; O'Donnell *et al.,* 1995).

a. Protein Kinase A. After agonist-mediated increases in cAMP (Lytle and Forbush, 1992b; Tanimura *et al.,* 1995), PKA is found to play an important role in the activation of NKCC, but PKA is apparently not involved in the shrinkage-induced activation (Pewitt *et al.,* 1990; Hoffmann and Dunham, 1995; O'Donnell *et al.,* 1995; Krarup *et al.,* 1998).

b. Ca^{2+}-Dependent Kinases. A permissive $[Ca^{2+}]_i$ is necessary for the activation of NKCC as the buffering of intracellular Ca^{2+} prevents shrinkage activation of NKCC (O'Donnell *et al.,* 1995; Ikeda *et al.,* 1995; Krarup *et al.,* 1998), pointing to an involvement of Ca^{2+}-dependent protein kinases. The role of PKC in the shrinkage activation of NKCC differs between cell types from a weak stimulatory role in EAT cells to no effect in bovine endothelial cells (O'Donnell *et al.,* 1995) and to downregulation of shrinkage-induced NKCC activity in sweat glands (Toyomoto *et al.,* 1997). Anti-CaM drugs are very effective inhibitors of shrinkage-induced activation of NKCC (O'Donnell, 1991; Jensen *et al.,* 1993; Krarup *et al.,* 1998) pointing to the involvement of CaM-dependent kinases. The CaM-dependent protein kinase MLCK seems to play a central role in the shrinkage-induced activation of NKCC in EATC (Krarup *et al.,* 1998) and in bovine endothelial cells (Klein and O'Neill, 1995; O'Donnel *et al.,* 1995), whereas inhibitors of CaM-dependent kinase II are uneffective as inhibitors of the acetylcholine-induced or shrinkage-activated NKCC (Ikeda *et al.,* 1995; Krarup *et al.,* 1998).

c. Steady State. Under isotonic conditions, NKCC activity is low in many cell types. Inhibition of serine/threonine phosphatases activates NKCC dramatically (Hoffmann and Dunham, 1995; Haas, 1994; Vigne *et al.,* 1994; O'Donnell *et al.,* 1995; Krarup *et al.,* 1998), indicating a continuous phosphorylation/dephosphorylation of the transporter in the steady state. Three kinases seem to be involved in the steady-state phosphorylation of NKCC in Ehrlich cells, i.e., PKA, PKC, and MLCK, whereas only PKC and MLCK are involved in the shrinkage-induced activation of NKCC (Krarup *et al.,* 1998).

In conclusion, increased phosphorylation of MLC seems to be an important step in the activation of volume regulatory transporters after cell shrinkage. This phosphorylation could be mediated by MLCK and/or by PKC or by an inhibition of myosin phosphatase, which is downstream of the small G-protein, RhoA. Phosphorylation of MLC could then in turn promote coordinated myosin–actin interactions with associated proteins, resulting in activation of NKCC and possible NHE.

There is also evidence for long-term (5 hr) effects of cell shrinkage on the amount of NKCC expressed in EAT cells (Jensen and Hoffmann,

1997). On such a longer time scale, it is likely that the ERK family kinases, p38 and SAPKs play important roles.

V. CYTOSKELETON AND CELL VOLUME REGULATION

For the purposes of this chapter, the term cytoskeleton will refer to the actin-associated cytoskeletal system. The broad use of the term will also include filament arrays composed of other distinct cytoskeletal proteins such as myosin and spectrin, realizing that these could be discussed as a separate functional group (Bennett and Lambert, 1991; Baehler, 1996). This chapter does not discuss the microtubule or intermediate filament portion of the cytoskeleton. For reviews on isotypes, polymerization kinetics, and general cellular distribution and function of all three of these systems, see elsewhere (Herman, 1993; Carlier, 1991; Hasson and Mooseker, 1995; Klymkowsky, 1995; Mandelkow and Mandelkow, 1995; Sun *et al.*, 1995; Cowin and Burke, 1996; Viel and Branton, 1996; Chou *et al.*, 1997; Wade and Hyman, 1997).

A. Cytoskeleton and the Cell Volume Regulatory Response

This section discusses the evidence that specifically addresses the role that the cytoskeleton may play in volume regulatory mechanisms. This discussion focuses entirely on the potential role of the actin cytoskeleton as there is an abundance of evidence that microtubules and intermediate filaments do not appear critical to any of the volume regulatory mechanisms (for exceptions, see Downey *et al.*, 1995; Linshaw *et al.*, 1992).

Numerous studies have attempted to correlate changes in the actin-based cytoskeleton with volume changes or to determine if experimental alteration of the cytoskeleton affects the volume regulatory process. There is no consensus on what role, if any, actin filaments may play in volume regulation, although there is some indication that disruption of the steady-state organization of actin filaments does affect volume regulatory processes. For example, in EAT cells, cell swelling results in a loss of cellular actin staining, and treatment with high concentrations of cytochalasin, which lowered F-actin levels, partially inhibits RVD (Cornet *et al.*, 1993a). In shark rectal gland cells there is a reversible reduction in basolateral F-actin, staining and a rearrangement of F-actin bundles is seen during hypotonic challenge (Ziyadeh *et al.*, 1992; Henson *et al.*, 1997). This rearrangement is also seen in RCCT-28A cells (Fig. 7) and in T84 cells (Matthews *et al.*, 1998). In PC12 cells there is a rearrangement of actin that correlates with a reduction in F-actin staining (Cornet *et al.*, 1993b, 1994). In HSG cells, 1 μM cytochalasin D inhibits the RVD response completely (Fatherazi *et al.*, 1994). This effect of cytochalasin treatment is also seen in osteosarcoma cells (Lo *et al.*, 1995). In human fibroblasts, exposure to hypotonic medium results in a reduction in F-actin content per cell followed by recovery (Bibby and McCulloch, 1994). In B lymphocyte myeloma cells, the cortical actin ring is disrupted by hypotonic challenge and treatment with cytochalasin increases the rate of chloride channel activation under mild (10%) hypotonic challenge (Levitan *et al.*, 1995). In human neutrophils, treatment with cytochalasins did not alter RVD and did not alter the F-actin content (Downey *et al.*, 1995). The same authors did report, however, that cytochalasin treatment in Jurkat cells caused a reduction in F-actin and completely inhibited the RVD response in dilute media. Conversely, in cardiac myocytes, F-actin does not change during hypotonic swelling or RVD (Zhang *et al.*, 1997). Treatment with phalloidin to stabilize the actin filament system decreases the degree of swelling as compared to untreated controls (Zhang *et al.*, 1997).

Although the evidence just presented indicates that the actin filament system may be involved in the RVD response, it also suggests that this role may be highly cell specific or take part in only a part of the regulatory response. In a detailed study in HL-60 cells, Hallows and co-workers (1991, 1996) reported that the F-actin content dropped in cells exposed to hypotonic challenge and rose in those undergoing RVI in hypertonic medium. When the cells were exposed to cytochalasin, which lowered F-actin levels, the cells exhibited normal RVD and RVI, indicating that F-actin levels, and normal organization of filaments, were not critical to the regulatory process (Hallows *et al.*, 1991). A subsequent study from these authors pointed out that the portion of F-actin present in the Triton-insoluble fraction ("cytoskeletal actin") declined during cell swelling but did not recover during RVD (Hallows *et al.*, 1996). After treatment with cytochalasin there was a dramatic drop in the Triton-insoluble fraction of F-actin. During subsequent RVD and RVI there was no further change in this pool of F-actin, but the cells exhibited normal regulatory responses. In agreement with Hallows *et al.* (1991, 1996), Moran *et al.*, (1996) reported that there is a reversible decrease in F-actin during cell swelling in cultured astrocytes; however, disruption of the cytoskeleton with cytochalasins does not affect RVD. These results support the hypothesis that the "cytoskeletal actin," especially those aggregated and cross-linked filaments visualized by phalloidin staining, may not be involved in regulatory mechanisms. That portion of F-actin present as possibly short oligomeric filaments and not cross-

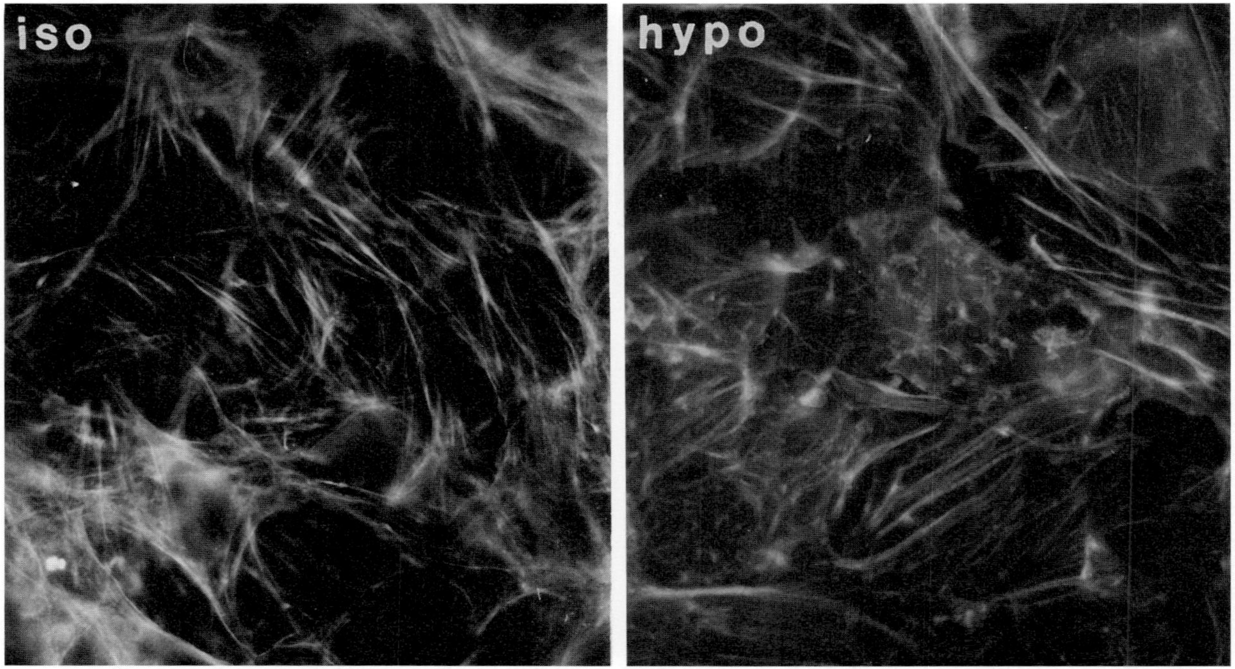

FIGURE 7 Localization of F-actin in RCCT-28A cells. Plane of focus is at base of cells. (Left) Control conditions. Actin filaments can be seen as stress fibers and as membrane-associated actin at the cell periphery. (Right) Five minutes in hypotonic medium. Notice that the F-actin organization is disrupted. (J. W. Mills, unpublished results). Results are similar to the results reported in T84 cells after 30 min in hypotonic buffer (Matthews *et al.,* 1998).

linked tightly into the structural actin may instead be the pool susceptible to volume perturbation and therefore linked to the volume regulatory process. A decrease in F-actin content after cell swelling has, however, not been observed universally. In human intestine 407 cells exposed to hypotonicity, the F-actin content increases approximately 40% during the time course of swelling and returns to control levels within 5 min (Tilly *et al.,* 1996). Correlated with this change in F–actin content was a domain-specific change in F-actin organization, with an increase in F-actin associated with stress fibers, and a decrease in F-actin in the apical membrane.

In summary, volume perturbations have been shown to alter the organization and content of F-actin in many cell types. These changes are, for the most part, reversible and coincide with the time course of the volume changes. In some cases, perturbation of the actin filaments by treatment with cytochalasins, followed by osmotic challenge, results in an inhibition of the volume recovery process, indicating that the alteration in the F-actin domain that occurs during volume changes may be a necessary part of the RVD response. However, in other cases, disruption of F-actin has no effect on volume recovery, even when the perturbing effect of the cytochalasin has been verified. Thus it is not yet clear what role the actin filaments play in RVD, and whether a given role is purely cell specific or related to the type of transporter(s) activated during the volume recovery phase. The specific transporters whose activity may be linked to cytoskeletal alterations will be discussed next.

B. Actin and Volume-Regulated Ion Channels

1. $I_{Cl,vol}$

Various lines of evidence indicate that the actin filament system may play a role in regulating the activity of this channel. In each case, data rely on the use of cytochalasins or phalloidin to disrupt or stabilize the actin filament system. In HSG cells, cytochalasin treatment inhibits $I_{Cl,vol}$ during RVD (Fatherazi *et al.,* 1994). In cardiac myocytes, both cytochalasin and phalloidin inhibited $I_{Cl,vol}$ (Zhang *et al.,* 1997). In membrane patches of RCCT-28A cells, cytochalasin stimulated $I_{Cl,vol}$ and phalloidin prevented channel activation (Schwiebert *et al.,* 1994). Potentiation of a swelling-induced Cl^- efflux pathway in intestinal epithelial cells by prior treatment with cytochalasin has also been mentioned by Tilly *et al.* (1996a).

In myeloma cells cytochalasin enhanced activation of $I_{Cl,vol}$ and phalloidin treatment inhibited it (Levitan *et al.,* 1995), in agreement with data on RCCT-28A cells. Interestingly, when medium osmolality is reduced to 25% of control, neither cytochalasin nor phalloidin affected the Cl^--channel response. The results were interpreted with regard to a membrane tension model of channel activation. Under mild hypotonic shock (10%), the swelling is sensed through a change in membrane tension and a linkage to the actin filament system. If the actin filaments are disrupted further, channel activation is enhanced. If the stimulus is great enough (25%), the tension–actin mechanism is activated maximally, and further disruption of the filament system has no effect. The fact that phalloidin treatment inhibited the response under mild hypotonic conditions indicates that the swelling and therefore the stretch of membrane is not sufficient in itself, but some interaction with the dynamic state of the actin filaments is required (Levitan *et al.,* 1995).

One of the studies arguing against this possible role of actin changes on the RVD mechanism are the results in astrocytes that show that cytochalasin treatment, which clearly alters F-actin, does not affect RVD or Cl^- efflux (Moran *et al.,* 1996).

2. $I_{K,vol}$

$I_{K,vol}$ involved in the RVD response may also be regulated via an interaction with the actin cytoskeleton (Cantiello *et al.,* 1993). In actin-binding protein (ABP)-deficient melanoma cells, hypotonic challenge does not result in any RVD response, which is associated with a failure of K^+ channels to be activated. When the expression of ABP was recovered by transfection with the cDNA for ABP, both K^+ channel activation and a RVD response were recovered (Cantiello et al., 1993). In ABP-deficient cells, a higher baseline permeability to K^+ was found, indicating that in the absence of ABP the K^+ channel was in an unregulated, chronically active state. Thus ABP may be acting, via a link between the channel and the actin cytoskeleton, to modulate channel activity. It is not clear why, however, that when inside-out patches of ABP-positive cells are exposed to actin on the cytoplasmic side, under conditions that should support the formation of actin filaments, cation-selective channel activity is enhanced (Cantiello, 1997).

3. Swelling-Activated Ca^{2+} Current

In guinea pig gastric myocytes, hypotonicity induces the activation of voltage-dependent Ca^{2+} channels (Xu *et al.,* 1997). When this preparation is treated with high doses of cytochalasin, base line Ca^{2+} currents are unaffected. However, Ca^{2+} currents activated by the addition of a hypotonic medium are inhibited markedly. Conversely, treatment with phalloidin leads to an enhanced current, which can be inhibited by cytochalasin treatment. This indicates that the state of actin filament organization plays a role in the activation of a volume-sensitive Ca^{2+} channel and is supported by findings in other cell types. Cytochalasin treatment inhibits neurotransmitter-stimulated Ca^{2+} influx in hippocampal neurons and accelerates run-down of the NMDA channel in these cells (Rosenmund and Westbrook, 1993; Furukawa and Mattson, 1995). A model for the regulation of these channels by actin, proposed by Rosenmund and Westbrook (1993), is shown in Fig. 8. Similarly, cytochalasin accelerates and phalloidin delays the inactivation of voltage-dependent Ca^{2+} channels in *Lymnae* neurons (Johnson and Byerly, 1993).

C. Actin, the Na^+,K^+,2Cl^- Cotransporter, and NHE

1. Na^+,K^+,2Cl^- Cotransporter

A significant body of data obtained from studies of T84 cells indicates that the actin filament system may play a role in the cAMP-induced activation of the Na^+,K^+,2Cl^- cotransporter (Shapiro *et al.,* 1991; Matthews *et al.,* 1992, 1994, 1997). This potential regulatory link between the Na^+,K^+, 2Cl^- cotransporter and the actin filament system is supported by data from studies of volume regulation in Ehrlich cells (Jessen and Hoffman, 1992; Hoffmann

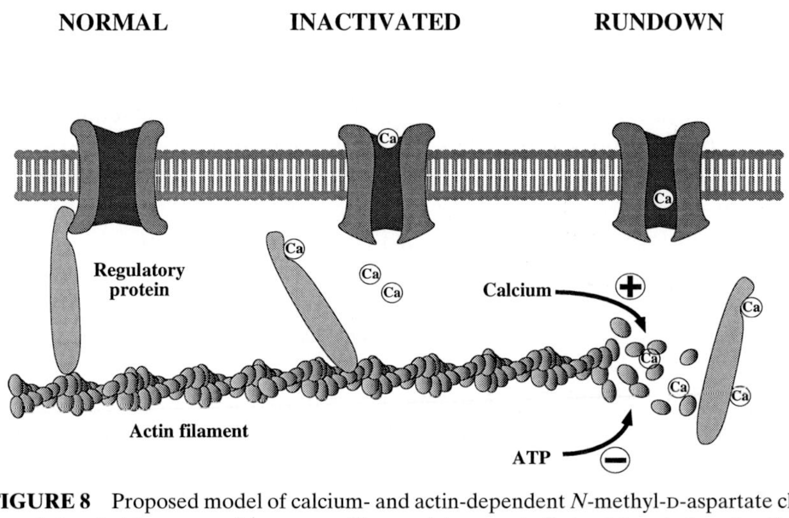

FIGURE 8 Proposed model of calcium- and actin-dependent *N*-methyl-D-aspartate channel regulation in hippocampal neurones. The model suggests that calcium-induced actin depolymerization underlies rundown of the channel. The model includes a regulatory protein that dissociates from the channel in a calcium-dependent manner, leading to channel inactivation. However, unless calcium-dependent depolymerization of actin filaments occurs, the regulatory protein can reassociate quickly. Actin depolymerization results in removal of the soluble regulatory protein, and the channel runs down. Reproduced from Rosenmund and Westbrook (1993), with permission from Cell Press.

et al., 1994). In isotonic medium, pretreatment with cytochalasin stimulates the activity of the transporter. This activation occurs in the absence of Ca^{2+}, in agreement with data on T84 cells. When EAT cells are exposed to hypotonic medium, a Ca^{2+}-dependent disruption of the cytoskeleton and a parallel partial activation of the cotransporter are seen. However, upon exposure to hypertonic medium, cytochalasin treatment inhibits the shrinkage activation of the cotransporter.

When EAT cells are exposed to high concentrations of cytochalasin in isotonic medium, they form numerous extensive membrane blebs. In these blebs, flux data demonstrate that the $Na^+/K^+/2Cl$ cotransporter is in a permanently activated state (Hoffmann *et al.,* 1994) and cannot be activated further by cell shrinkage (E. K. Hoffmann, unpublished observation). Thus, it is possible that a change in association of actin filaments with the cytoplast membrane is responsible for inducing the partial activated state of the cotransporter.

The paradoxical stimulation of the cotransporter seen in EAT cells in both hypotonic and hypertonic medium has also been demonstrated in T84 cells (Matthews *et al.,* 1998). In hypertonic medium, phalloidin pretreatment had no effect on activation of the cotransporter in T84 cells, whereas cytochalasin inhibited activation, similar to what occurred in EAT cells.

Further support for an actin filament dependent regulation of the cotransporter comes from studies of RVI in cultured endothelial cells (Klein and O'Neill, 1995) and EAT cells (Krarup *et al.,* 1998). After exposure to hypertonic medium, phosphorylation of MLCK is increased in endothelial cells, an effect inhibited by the MLCK inhibitor ML-7 (Klein and O'Neill, 1995). In endothelial cells and EAT cells, ML-7 inhibits the swelling-induced activation of the cotransporter. A similar effect of ML-7 on shrinkage-induced activation of NHE in astrocytes has also been reported (Shrode *et al.,* 1995). Because the myosin light chain kinase is involved, this implies a critical interaction between myosin II and the actin filament system. As discussed in Section V,D,3, MLCK activity appears critical for exocytotic events, implying that the RVI response leading to the activation of cotransporters or NHEs in the cell membrane may involve the insertion of transporters.

2. NHE

As described in Section IV,H,4, direct phosphorylation of NHE does not occur during osmotic challenge (Grinstein *et al.,* 1992b). The hypothesis has been presented that the shrinkage-induced phosphorylation events that do occur are associated with accessory proteins. These could include cytoskeletal proteins interacting with the cytoplasmic domain of NHE (Noëll and Pouysségur, 1995). In fibroblasts, NHE has been shown to localize to the adhesion complex (Grinstein *et al.,* 1993). However, the significance of this association is not clear, as there was no indication that the cytoskeletal-associated NHEs were any more active than those not associated with the cytoskeleton. An alternative role for the cytoskeleton in regulating NHE could be via the vesicle cycling mechanism. A reduction in NHE activity in serum-deprived Caco-2 cells has been demonstrated (Watson *et al.,* 1992). Serum deprivation is known to alter the cytoskeleton. When Caco-2 cells are treated with cytochalasin, the serum deprivation effect is inhibited. Thus, under these conditions, NHEs are retained in the membrane even in the absence of serum. Because endocytosis is dependent on the actin filament system, the results imply that one aspect of NHE regulation is via a vesicle shuttling mechanism

As discussed in Section III,B., shrinkage-induced activation of NHE-1 and NHE-2 is inhibited strongly by ATP depletion, suggesting that this is related to cytoskeletal rearrangement. ATP is needed for actin polymerization *in vitro* and it is known that the depletion of ATP in intact cells leads to the depolymerization of cortical actin (Bacallao *et al.,* 1994). As discussed later, vesicle insertion mechanisms require intact cortical actin filaments.

D. Potential Regulatory Sites

Initially, the idea that the cytoskeleton could be involved in cell volume regulation was based on the concept of an actin–myosin-based contractile mechanism acting either to resist the swelling pressure induced by the influx of water or to prevent the entry of water by a sol-gel transformation (reviewed in Kleinzeller and Ziyadeh, 1990; Mills *et al.,* 1993). It was later proposed that the cytoskeleton may regulate transport processes directly, either those involved in the maintenance of steady-state volume or those that were stimulated by the volume perturbation (Mills, 1987). This suggestion was based on data showing that cytoskeletal proteins are associated directly with membrane transport proteins; that perturbation of cytoskeletal filament systems was associated with altered membrane transport properties; that membrane receptors and their characteristics were altered by cytoskeletal interaction; and that the state of organization of the cytoskeleton played a role in the responsiveness of second messenger systems that altered membrane transport events (Forte and Lee, 1977; Fukuda *et al.,* 1981; Matsumoto and Sakai, 1979a,b; Matsumoto *et al.,* 1984; Insel and Koachman, 1982; Melmed *et al.,* 1981; Siman *et al.,* 1985).

The sites where elements of the cytoskeleton may interact with the volume regulatory mechanism include the sensor of volume, the transduction mechanism for this initial signal, and the actual effectors (transporters involved in the movement of solutes and water).

E. Volume Sensors

As mentioned in Section IV, membrane-associated sensors of the volume signal could be integral membrane proteins, G-proteins, or phospholipases responding to a radial strain in the cell membrane. Some of the possible interactions of these with the cytoskeleton are discussed next.

1. Integrins

There is abundant evidence demonstrating that elements of the cytoskeleton interact with membrane proteins. Two of the best studied and elucidated models are those of the neurotransmitter receptors for acetylcholine and glutamate (Whatley and Harris, 1996). In some cases, the interaction of the receptor with the cytoskeleton appears to be functionally important for aggregation and positioning of the receptors at appropriate sites (Froehner, 1991), and the interaction with the cytoskeleton has been shown to alter functional characteristics such as binding affinity and responsivenes (for reviews, see Whatley and Harris, 1996; Bechade and Triller, 1996).

Other types of membrane receptors have also been shown to interact in a functionally relevant way with elements of the cytoskeleton. This includes the epidermal growth factor (EGF) receptor (EGFR), which has been shown to colocalize with actin filaments. Importantly, it was shown in vivo that the high-affinity class of the receptor is associated with the actin cytoskeleton (Defize *et al.,* 1989). An increase in the association with actin filaments was demonstrated after exposure to EGF (Rijken *et al.,* 1991; den Hartigh *et al.,* 1992), suggesting that the state of the cytoskeleton may be part of the signaling mechanism or play a critical role as a scaffold to provide the correct topology for propagation of the signal.

If there is a receptor or volume sensor in the membrane of mammalian cells, separate and distinct from the transporter, and it interacts with the cytoskeleton, what are the candidates? One ubiquitous class of integral membrane proteins that has been demonstrated to interact with the cytoskeleton are the integrins. These are present as heterodimers in the cell membrane and have a variable association with the actin filament system, partially dependent on whether the receptor is occupied by its ligand (outside-in signaling). Once occupied, a cascade of events result in assembly of many cytoskeletal proteins at the ligand–receptor site. Also recruited and associated closely with the cytoskeleton are many (as much as 20) proteins known to take part in signal transduction, including PKC, protein phosphatases, and the focal adhesion kinase, p125FAK (Shattil *et al.,* 1994; Miyamoto *et al.,* 1995). When cytoskeletal assembly is disrupted by treatment with cytochalasin, the signaling molecules fail to assemble at the integrin site (Miyamoto *et al.,* 1995).

Integrin occupation also alters ion transport in the target cell. This includes a rise in Ca^{2+}, presumably by influx into the cell, and a rise in intracellular pH due to the activation of NHE (Schwartz *et al.,* 1989, 1990; Schwartz, 1993). As mentioned in Sections III,B and V,C, an increase in the association of NHE with integrin receptors and actin has been demonstrated in fibroblasts after adherence to a substrate (Grinstein, *et al.,* 1993). Thus, integrin receptor activation results in cytoskeletal changes, recruitment of the antiporter to the integrin site, and a rise in intracellular pH mediated by the activation of NHE.

Because the integrin receptor clearly plays a role in adherence and interacts with the actin cytoskeleton, what is the evidence that this could play a role in volume sensing? Evidence that could suggest such a role comes from studies employing magnetic beads coated with the amino acid sequence that mediates integrin–receptor binding. When shear stress was applied to the cell via the beads bound to membrane receptors, an increase in the stiffness of the entire cytoskeleton system was seen (Ingber, 1997). The increase in stiffness upon application of the twisting force was eliminated in cells treated with agents that depolymerize actin filaments, microtubules, and intermediate filaments. Thus application of a tensile force to integrin receptors, attached to beads, leads to a response in the cytoskeletal elements in the cytoplasm (Ingber, 1997). Whether this response to tension could also occur via an integrin receptor that is not occupied or not attached to an external substrate is not known. However, association of integrin subunits with the cytoskeleton could be induced in hepatocytes in suspension when treated with anti-integrin antibodies and then cross-linked with a secondary antibody indicating that the outside-in signaling pathway is responsive and independent of substrate adhesion (Nebe *et al.,* 1996). Furthermore, some evidence shows that the signaling apparatus known to be associated with the integrin receptor is altered by cell swelling (membrane strain) and this stimulation does not appear to be mediated by receptor occupancy (see later).

2. G-Proteins and Phospholipase A_2

G-proteins involved in signaling pathways are known to associate with the cell membrane. In most cases, this association has been shown to be via a direct interaction with the membrane and membrane receptor proteins. Recent evidence, however, indicates that several classes of G-protein interact with components of the cytoskeleton. For example in several cell types,

G-proteins colocalize with actin, vinculin, and spectrin (Bourguignon *et al.*, 1990; Hansen *et al.*, 1994), as well as with intermediate filaments (Chiba *et al.*, 1995). The association with the actin cytoskeleton can be disrupted by cytochalasin treatment (Heringdorf *et al.*, 1996), and activation of the G-proteins involves a dissociation from the cytoskeleton (Sarndahl *et al.*, 1993).

As discussed in Section IV,B,2, $cPLA_2$ was proposed as a volume sensor. Stimulation of arachidonic acid production via $cPLA_2$ was shown to be associated with a translocation of $cPLA_2$ to the nuclear membrane. Cytochalasin B inhibits collagen-induced $cPLA_2$ activation in human platelets. In rabbit platelets, thrombin stimulation of $cPLA_2$ results in increased association with the Triton-insoluble cytoskeletal fraction (Akiba *et al.*, 1993). Thus, this potential volume sensor has a variable and functionally critical interaction with the actin cytoskeleton, which could be part of the volume-sensing mechanism.

F. Transduction of the Signal

There are many components of the "signaling" systems, such as protein kinases (discussed in Section IV,H), phopholipids and G-proteins, that are known to be associated with elements of the cytoskeleton (Mochly-Rosen, 1995; Sohn and Goldschmidt-Clermont, 1994). In most cases, these associations have been interpreted to serve a role in efficient association of the enzyme with its target substrate. However, in some cases, the activity of the signal protein is altered when its association with cytoskeletal proteins is altered, indicating a potential regulatory role. For example, in synaptosomes, actin serves as a site for the activated form of the PKC isozyme and stabilizes its activity, acting as a regulator for the activity of the enzyme in the exocytotic pathway. Therefore the cytoskeletal–signal molecule interaction is both permissive (serving to mediate association of enzyme and substrate) and required (preserving activity when bound).

A signaling system associated closely with the cytoskeleton has been assigned a specific role in cell volume regulation. In a series of papers, Tilly and co-workers (Tilly *et al.*, 1993, 1994, 1996a,b) reported on the RVD response in intestine epithelial cells. As mentioned in Section IV,H,1, activation of tyrosine kinases is required for the stimulation of $I_{Cl,vol}$ in these cells. This kinase activity results in the phosphorylation, among several proteins, of the focal adhesion kinase p125FAK that binds to the cytoplasmic side of the integrin receptor. Treatment of the cells with the C3 exoenzyme of *Clostridium botulinum* blocked the phosphorylation of p125FAK and inhibited the $I_{Cl,vol}$. The toxin inhibits the small G-protein Rho, which plays a key role in actin filament assembly and is also known to stimulate tyrosine phosphorylation of p125FAK (Ridley, 1994). Because assembly of p125FAK at the integrin site is also associated with the assembly of RhoA at the same macromolecular complex (Yamada and Miyamoto, 1995), Tilly *et al.*, (1996a) proposed that the p21rho/p125FAK signaling cascade played a key role in activating the $I_{\mathrm{Cl,vol}}$ and thus was linked to the volume sensor. The exact nature of the "sensor" is still not known. Because the signaling cascade appears to involve RhoA, we must look for those things that stimulate Rho that may be activated by swelling. A well-established stimulatory agent for Rho is lysophosphatidic acid (LPA), a product of $cPLA_2$ (see Fig. 4), the activity of which is stimulated by cell swelling (see Fig. 6). LPA binds to a G-protein-coupled receptor in the cell membrane and activates Rho (Moolenar *et al.*, 1997). Activation of Rho leads to the reorganization of actin and integrin assembly. This integrin/cytoskeletal-associated signal cascade involving p125FAK then leads to $I_{\mathrm{Cl,vol}}$ activation. The actin assembly may be serving a critical role to position the signaling molecules in the orientation needed to propagate the signals efficiently to the correct effectors. That the signal pathway involving LPA can lead to ion channel activation and Rho-mediated cytoskeletal changes has been demonstrated in fibroblasts (Postma *et al.*, 1996).

These above results indicate that the actin cytoskeleton, through both a structural and a regulatory interaction with signaling molecules, could play a key role in the transmission of a volume regulatory stimulus from the sensor through the second messenger cascade to the effector responsible for osmolyte flux.

G. Regulation of Membrane Transporters

1. Vesicle Insertion

The next step along the pathway from cell volume perturbation to membrane transport activation could involve a change in the number and/or position of the transporters in the cell membrane. Onc way in which this regulatory component could be manifested is by a vesicle insertion and retrieval process. There is abundant information supporting the hypothesis that transport proteins are delivered to the cell membrane by a cytoskeletal-dependent vesicle shuttle mechanism (Hays *et al.*, 1994; Forte and Yao, 1996). The models can be described as basically two general schemes: (i) a permissive role in which the cortical actin filaments, initially forming a barrier to vesicle fusion with the cell membrane, are broken down or rearranged, thus allowing vesicle–cell membrane interaction, and (ii) a more active role, in which the actin filaments, in conjunction with motor proteins, are involved in the movement or fusion of the vesicle with the cell membrane.

Evidence for the permissive role of the cortical actin barrier in exocytosis is the demonstration in chromaffin cells that there is a transient depolymerization of the actin filaments during the release of cortical ganules and that this depolymerization precedes the granule release (Vitale *et al.,* 1991; Trifaro *et al.,* 1992). A similar depolymerization of actin filaments is seen in the toad urinary bladder after stimulation with vasopressin or cAMP and this loss of filaments precedes the increase in osmotic water flow. Moreover, as seen in the chromaffin cell, the breakdown of the actin web occurred in patches and was not continuous across the entire apical actin filament domain (Ding *et al.,* 1991; Holmgren *et al.,* 1992). This partial and site-specific disruption of the cortical actin barrier appears critical to the exocytotic process. However, it is not clear whether this actin breakdown, in itself, is enough to allow exocytosis as extensive breakdown, versus only a partial disruption actually inhibits the exocytotic release of granules (Muallem *et al.,* 1995).

Instead of being simply a barrier, the actin filaments may also serve as the scaffold on which the machinery for vesicle tranlocation is assembled and moves. This is suggested from studies investigating the role of myosin in the exocytotic release of granules. For example, exocytosis is associated with MLCK activity leading to the diphosphorylation of MLC (Choi *et al.,* 1994) and inhibition of MLCK inhibits agonist-induced exocytosis (Rao *et al.,* 1997). Inhibition of MLCK in chromaffin cells prevents priming of docked vesicles (Kumakura *et al.,* 1994). Priming is the term used to describe the processes that already docked (vesicle-cell membrane in physical contact as viewed by electron microscopy) vesicles undergo before they can fuse with the cell membrane (Augustine *et al.,* 1996). MLCK regulates the actin–myosin interaction, and myosin II has been shown to be located in association with the membranes of cortical granules in fish eggs (see Fig. 9) (Becker and Hart, 1996).

In addition to the role of myosin II in exocytosis, there is some indication that unconventional myosins may play a role (Goodson *et al.,* 1997). For example, myosin I has been found in association with the membranes of Golgi-derived vesicles and has been implicated in the movement of vesicles from the microtubule transport system to the cortical cytoplasm (Fath and Burgess, 1994). Myosin V has been shown to be involved in the movement of beads in an *in vitro* motility assay (Cheney *et al.,* 1993). Thus the movement of vesicles, possibly containing membrane transport proteins, could be mediated by several components of the actin–myosin cytoskeletal apparatus. Vesicles moving along microtubule tracks would be passed to the unconventional myosins that would serve as motor proteins to move the

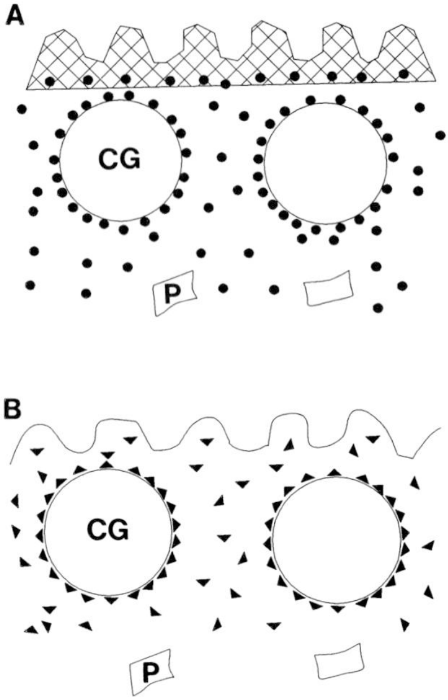

FIGURE 9 Schematic representation of the spatial distribution of filamentous actin (A, crosshatch), nonfilamentous actin (A, ●), and myosin-II (B, ▲) in the unactivated zebrafish egg based on results of light microscopic observations. CG, cortical granule; P, yolk platelet. Reproduced from "The cortical actin cytoskelton of unactivated zebrafish eggs: Spatial organization and distribution of filamentous actin, nonfilamentous actin and myosin-II," Becker and Hart, *Mol. Reprod. Dev.,* Copyright © 1996. Reprinted by permission of Wiley-Liss, Inc., a subsidiary of John Wiley & Sons, Inc.

vesicle through the cortical cytoplasm (either through the cortical web or through gaps in it formed by the breakdown of actin filaments). The transport by these proteins would require the presence of at least some intact actin filaments to serve as the scaffold. The final delivery/insertion of proteins into the cell membrane would require an interaction between actin filaments and myosin II to complete insertion and establishment of the functional domain in the cell membrane.

2. Direct Interaction with Membrane Transporters

A final site for a regulatory role of the cytoskeleton in cell volume control is directly at the membrane transporter mediating the changes in osmolyte fluxes. As originally proposed (Mills, 1987), this was thought to be via a direct interaction of a cytoskeletal protein with a membrane transport protein. Alteration of this interaction, via conformational changes or changes in affinity mediated by phosphorylation, could lead to changes in activity or number of functional units. A role for components of the actin cytoskeletal system and actin itself in regulating Na^+ channels and K_{ATP} channels is especially convincing (Cantiello and Prat, 1996; Terzic and Kurachi, 1996; Brady *et al.*, 1996). For example, a preparation of purified Na^+ channels in planar lipid bilayers shows altered conductance levels and a doubling of channel open probability (P_o) when exposed to actin on the cytoplasmic side (Berdiev *et al.*, 1996). Importantly, the change in P_o occurred only when conditions were suitable for the polymerization of actin. When the Cl^- channel protein CFTR was added to the planar lipid bilayer containing the Na^+ channel, P_o was reduced. This effect was enhanced dramatically when the transporters were present along with actin (Ismailov *et al.*, 1997). This result implies that a regulatory complex is established between CFTR and the Na^+ channel that involves a linkage mediated by actin filaments.

A similar actin-dependent regulatory complex appears to be involved in controlling ATP-inhibitable K^+ channels (K_{ATP}). The K_{ATP} channel is complexed with and inhibited by a sulfonylurea-binding protein (SUR), which is also the site of ATP binding (Inagaki *et al.*, 1995). When actin filaments are depolymerized by exposure to DNase I or cytochalasin, inhibition of the K_{ATP} channel by sulfonylureas or ATP is abolished. In addition, treatment with DNase I or cytochalasin alone stimulates K_{ATP} channel activity (Brady *et al.*, 1996; Terzic and Kurachi; Yokoshiki *et al.*, 1997,). These results indicate that actin filament organization mediates the regulatory action of SUR. This "tonic inhibitory" effect is in agreement with the data on CFTR and the Na^+ channel discussed earlier.

There are no data indicating that such direct interaction and regulatory function occurs between the actin cytoskeleton and volume-sensitive membrane transport proteins. However, the fact that the volume regulatory process and transport proteins involved in affecting volume recovery are altered by changes in the steady-state organization of actin filaments indicates that this is an important question to pursue.

VI. SUMMARY

Evidence suggests that the sensor of cell swelling could be the G-protein-activated, Ca^{2+}-dependent, cytosolic $cPLA_2$. The G-protein and/or the $cPLA_2$ may be activated directly as a consequence of the radial strain in the membrane, but there is also strong evidence that the actin cytoskeleton can be involved in the activation of G-proteins and PLA_2. Activation of PLA_2 might also be downstream to activation of a tyrosine kinase, which in turn could be activated via a putative, volume-sensing integral membrane protein, similar to that described in yeast cells.

Activation of $cPLA_2$ can cause the release of LPA and AA. LPA can, via a LPA receptor, activate Rho, leading to the reorganization of actin and integrin assembly. An integrin/cytoskeletal-associated signal cascade involving p125FAK can lead to the activation of $I_{Cl,vol}$. AA gives rise to the production of various eicosanoids that act as autocrine messengers regulating different volume-sensitive channels. Thus, the 12-LIP metabolite hepoxilin A_3 and the 5-LIP metabolite LTD_4 have both been found to activate swelling-activated K^+ channels. In some cell types, ATP seems to play a similar autocrine role in the swelling-induced activation of channels.

Swelling-activated channels are, in many cells, distinct from Ca^{2+}-activated channels. Some swelling-activated channels have been cloned, but many of the properties of the volume regulatory channels are not in agreement with those of the known cloned channels and additional channels must be operative.

Cell shrinkage seems to involve the phosphorylation of MLC either by the activation of MLCK or PKC or by the inhibition of myosin phosphatase. This in turn leads to an interaction between myosin and actin filaments and to the activation of NKCC and NHE. MLCK activity appears critical to exocytotic events, implying that the RVI response leading to the increased activity of both NKCC and NHEs might possibly involve insertion of these transporters in the membrane. Moreover, in several cell systems, shrinkage-induced activation of the cotransporter is also associated with direct serine/threonine phosphorylation of the NKCC protein itself by a volume-sensitive protein kinase. The relationship between MLCK and this volume-sensitive kinase is not yet known.

References

Ackerman, M. J., Wickman, K. D., and Clapham, D. E. (1994). Hypotonicity activates a native chloride current in *Xenopus* oocytes. *J. Gen. Physiol.* **103,** 153–179.

Agre, P., Preston, G. M., Smith, B. L., Jung, J. S., Raina, S., Moon, C., Guggino, W. B., and Nielsen, S. (1993). Aquaporin CHIP; the archetypal molecular water channel. *Am. J. Physiol.* **265,** F463–F476.

Agius, L., Peak, M., Beresford, G., Al-Habori, M., and Thomas, T. H. (1994). The role of ion content and cell volume in insulin action. *Biochem. Soc. Trans.* **22,** 516–522.

Akiba, S., Sato, T., and Fujii, T. (1993). Evidence for an increase in the association of cytosolic phospholipase A_2 with the cytoskeleton of stimulated rabbit platelets. *J. Biochem.* (*Tokyo*) **113,** 4–6.

Al-Habori, M. (1994). Cell volume and ion transport regulation. *Int. J. Biochem.* **26,** 319–334.

Artalejo, A. R., and García-Sancho, J. (1988). Mobilization of intracellular calcium by extracellular ATP and by calcium ionophores in the Ehrlich ascites tumor cell. *Biochim. Biophys. Acta* **941,** 48–54.

Augustine, G. J., Burns, M. E., DeBello, W. M., Pettit, D. L., and Schweizer, F. E. (1996). Exocytosis: Proteins and perturbations. *Annu. Rev. Pharmacol. Toxicol.* **36,** 659–701.

Bacallao, R., Garfinkel, A., Monke, S., Zampighi, G., and Mandel, L. J. (1994). ATP depletion: A novel method to study junctional properties in epithelial tissues. *J. Cell Sci.* **107,** 3301–3313.

Baehler, M. (1996). Myosins on the move to signal transduction. *Curr. Opin. Cell Biol.* **8,** 18–22.

Bakker-Grunwald, T., Ogden, P., and Lamb, J. F. (1982). Effects of ouabain and osmolarity on bumetanide-sensitive potassium transport in Simian virus-transformed 3T3 cells. *Biochim. Biophys. Acta* **687,** 333–336.

Banderali, U., and Roy, G. (1992). Anion channels for amino acids in MDCK cells. *Am. J. Physiol.* **263,** C1200–C1207.

Baquet, A., Meijer, A. J., and Hue, L. (1991). Hepatocyte swelling increases inositol 1,4,5-trisphosphate, calcium and cyclic AMP concentration but antagonizes phosphorylase activation by Ca^{2+}-dependent hormones. *FEBS Lett.* **278,** 103–106.

Basavappa, S., Pedersen, S. F., Jørgensen, N. K., Ellory, J. C., and Hoffmann, E. K. (1998). Swelling-induced arachidonic acid release via 85 kDa $cPLA_2$ in human neuroblastoma cell. *J. Neurophysiol.* **79,** 1441–1449.

Bechade, C., and Triller, A. (1996). The distribution of glycine receptors and interactions with the cytoskeleton. *Cell Dev. Biol.* **7,** 717–724.

Beck, J. S., Breton, S., Laprade, R., and Giebisch, G. (1991). Volume regulation and intracellular calcium in the rabbit proximal convoluted tubule. *Am. J. Physiol.* **260,** F861–F867.

Becker, K., and Hart, N. H. (1996). The cortical actin cytoskelton of unactivated zebrafish eggs: Spatial organization and distribution of filamntous actin, nonfilamentous actin and myosin-II. *Mol. Reprod. Dev.* **43,** 536–547.

Bennett, V., and Lambert, S. (1991). The spectrin skeleton: From red cells to brain. *J. Clin. Invest.* **87,** 1483–1489.

Berdiev, B. K., Prat, A. G., Cantiello, H. F., Ausiello, D. A., Fuller, C. M., Jovov, B., Benos, D. J., and Ismailov, I. I. (1996). Regulation of epithelial sodium channels by short actin filaments. *J. Biol. Chem.* **271,** 17704–17710.

Bertrand, B., Wakabayashi, S., Ikeda, T., Poussegur, J., and Shigekawa, M. (1994). The Na^+/H^+ exchanger isoform 1 (NHE1) is a novel member of the calmodulin-binding proteins. *J. Biol. Chem.* **269,** 13703–13709.

Bianchini, L., Kapus, A., Lukacs, G., Wasan, S., Wakabayashi, S., Pouysségur, J., Yu, F. H., Orlowski, J., and Grinstein, S. (1995). Responsiveness of mutants of NHE-1 isoform of Na^+/H^+ antiport to osmotic stress. *Am. J. Physiol.* **269,** C998–C1007.

Bibby, K. J., and McCulloch, C. A. G. (1994). Regulation of cell volume and $[Ca^{2+}]_i$ in attached human fibroblasts responding to anisosmotic buffers. *Am. J. Physiol.* **266,** C1639–C1649.

Bize, I., and Dunham, P. B. (1994). Staurosporine, a protein kinase inhibitor, activates K-Cl cotransport in LK sheep erythrocytes. *Am. J. Physiol.* **266,** C759–C770.

Boese, S. H., Wehner, F., and Kinne, R. K. H. (1996). Taurine permeation through swelling-activated anion conductance in rat IMCD cells in primary culture. *Am. J. Physiol.* **271,** F498–F507.

Bookstein, C., Musch, M. W., DePaoli, A., Villereal, M., Rao, M. C., and Chang, E. B. (1994). A unique sodium-hydrogen exchange isoform (NHE-4) of the inner medulla of the rat kidney is induced by hyperosmolarity. *J. Biol. Chem.* **269,** 29704–29709.

Bookstein, C., Musch, M. W., DePaoli, A., Xie, Y., Rabenau, K., Villereal, M., Rao, M. C., and Chang, E. B. (1996). Characterization of the rat Na^+/H^+ exchanger isoform NHE-4 and localization in rat hippocampus. *Am. J. Physiol.* **271,** C1629–C1638.

Bourguignon, L. Y. W., Walker, G., and Huang, H. S. (1990). Interactions between a lymphoma membrane-associayted guanosine 5′-triphosphate binding protein and the cytoskeleton during receptor patching and capping. *J. Immunol.* **144,** 2242–2252.

Brady, P. A., Alekseev, A. E., Aleksandrova, L. A., Gomez, L. A., and Terzic, A. (1996). A disrupter of actin microfilaments impairs sulphonylurea-inhibitory gating of cardiac K_{ATP} channels. *Am. J. Physiol.* **271,** H2710–H2716.

Breitwieser, G. E., Altamirano, A. A., and Russell, J. M. (1990). Osmotic stimulation of Na^+-K^+-Cl^- cotransport in squid giant axon is $[Cl^-]_i$ dependent. *Am. J. Physiol.* **258,** C749–C753.

Breton, S., Beck, J. S., Cardinal, J., Giebisch, G., and Laprade, R. (1992). Involvement and source of calcium in volume regulatory decrease of collapsed proximal convoluted tubule. *Am. J. Physiol.* **263**(No. 4, Part 2), F656–F664.

Brown, A., and Simpson, J. (1972). Water relations of sugar-tolerant yeasts: The role of intracellular polyols. *J. Gen. Microbiol.* **72,** 589–591

Burg, M. B. (1994). Molecular basis for osmoregulation of organic osmolytes in renal medullary cells. *J. Exp. Zool.* **268**(2), 171–175.

Burg, M. B., Kwon, E. D., and Kültz, D. (1996). Osmotic regulation of gene expression. *FASEB J.* **10,** 1598–1606.

Burnstock, G. (1996). P2 purinoreceptors: Historical perspective and classification. *In*: "P2 Purinoreceptors: Localization, Function and Transduction Mechanisms." Ciba Foundation Symposium **198,** 1–34.

Burnstock, G. (1997). The past, present and future of purine nucleotides as signaling molecules. *Neuropharmacology* **36,** 1127–1139.

Busch, A. E., and Maylie, J. (1993). MinK channels: A minimal channel protein with a maximal impact. *Cell. Physiol. Biochem.* **3,** 270–276.

Busch, A. E., Varnum, M., Adelman, J. P., and North, R. A. (1992). Hypotonic solution increases the slowly activating potassium current ISK expressed in *Xenopus* oocytes. *Biochem. Biophys. Res. Commun.* **184,** 804–810.

Butt, A. G., Clapp, W. L., and Frizzel, R. A. (1990). Potassium conductances in tracheal epithelium activated by secretion and cell swelling. *Am. J. Physiol.* **258,** C630–C638.

Cahalan, M. D., and Lewis, R. S. (1988). Role of potassium and chloride channels in volume regulation by thymphocytes. *In* "Cell Physiology of Blood" (R. B. Gunn and J. C. Parker, eds.), pp. 281–301. Rockefeller Univ. Press, New York.

Cahalan, M. D., Ehring, G. R., Osipchuk, Y. V., and Ross, P. E. (1994). Volume-sensitive Cl^- channels in lymphocytes and multidrug-resistant cell lines. *J. Gen. Physiol.* **104,** 43a–44a.

Cala, P. M. (1980). Volume regulation by Amphiuma red blood cells. The membrane potential and its implications regarding the nature of the ion-flux pathways. *J. Gen. Physiol.* **76,** 683–708.

Cala, P. M. (1983a). Volume regulation by red blood cells: Mechanisms of ion transport. *Mol. Physiol.* **4,** 33–52.

Cala, P. M. (1983b). Cell volume regulation by Amphiuma red blood cells. The role of Ca^{+2} as a modulator of alkali metal/H^+ exchange. *J. Gen. Physiol.* **82,** 761–784.

Cala, P. M. (1986). Volume-sensitive alkali metal-H transport in Amphiuma red blood cells. *In* "Na^+-H^+ Exchange, Intracellular pH, and Cell Function" (P. S. Aronson and W. F. Boron, eds.) *Curr. Top. Membr. Transp.* **26,** 79–99.

Cala, P. M., and Grinstein, S. (1988). Coupling between Na^+/H^+ and Cl^-/HCO^-_3 exchange in pH and volume regulation. *In* "Na^+/H^+ Exchange" (S. Grinstein, ed.), pp. 201–208. CRC Press, Boca Raton, Florida.

Cala, P. M., and Maldonado, H. M. (1994). Regulatory Na/H exchange by Amphiuma red blood cells. *J. Gen. Physiol.* **105,** 1035–1054.

Cala, P. M., Mandel, L. J., and Murphy, E. (1986). Volume regulation by Amphiuma red blood cells: Cytosolic free Ca and alkali metal-H exchange. *Am. J. Physiol.* **250,** C423–C429.

Canessa, M., Fabry, M. E., Blumenfeld, N., and Nagel, R. L. (1987). Volume-stimulated, Cl^- dependent K^+ efflux is highly expressed in young human red cells containing normal hemoglobin or HbS. *J. Membr. Biol.* **97,** 97–105.

Cantiello, H. F. (1997). Role of actin filament organization in cell volume and ion channel regulation. *J. Exp. Zool.* **279,** 425–435

Cantiello, H. F., and Prat, A. G. (1996). Role of actin filament organization in ion channel activity and cell volume regulation. *In* "Current Topics in Membranes" (W. J. Nelson, eds.), Vol. 43, pp. 373–396. Academic Press, New York.

Cantiello, H. F., Prat, A. G., Bonventre, J. V., Cunningham, C. C., Hartwig, J. H., and Ausiello, D. A. (1993). Actin-binding protein contributes to cell volume regulatory ion channel activation in melanoma cells. *J. Biol. Chem.* **268,** 4596–4599.

Careaga-Houck, M., and Sprecher, H. (1990). Effects of a fish oil diet on the metabolism of endogenous (n-6) and (n-3) fatty acids in rat neutrophils. *Biochim. Biophys. Acta* **1947,** 29–34.

Carlier, M.-F. (1991). Nucleotide hydrolysis in cytoskeletal assembly. *Curr. Opin. Cell Biol.* **3,** 12–17.

Chamberlin, M. E., and Strange, K. (1989). Anisosmotic cell volume regulation: A comparative view. *Am. J. Physiol.* **257,** C159–C173.

Cheney, R. E., Oshea, M. K., Heuser, J. E., Coelho, M. V., Wolenski, J. S., Espreafico, E. M., Forscher, P., Larson, R. E., and Mooseker, M. S. (1993). Brain myosin-V is a two-headed unconventional myosin with motor activity. *Cell (Cambridge, Mass.)* **75,** 13–23.

Cheung, R. K., Grinstein, S., Dosch, H.-M., and Gelfand, E. W. (1982). Volume regulation by human lymphocytes: Characterization of the ionic basis for regulatory volume decrease. *J. Cell. Physiol.* **112,** 189–196.

Chiba, K., Longo, F. J., Kontani, K., Katada, T., and Hoshi, M. (1995). A periodic network of G protein $\beta\gamma$ subunit coexisting with cytokeratin filaments in starfish oocytes. *Dev. Biol.* **169,** 415–420.

Choi, O. H., Adelstein, R. S., and Beaven, M. A. (1994). Secretion from rat basophilic cell RBL-2H3 cells is associated with diphosphorylation of myosin light chains by myosin light chain kinase as well as phosphorylation by protein kinase C. *J. Biol. Chem.* **269,** 536–541.

Chou, Y.-H., Skalli, O., and Goldman, R. D. (1997). Intermediate filaments and cytoplasmic networking: New connections and more functions. *Curr. Opin. Cell Biol.* **9,** 49–53.

Christensen, O., and Hoffmann, E. K. (1992). Cell swelling activates K^+- and Cl^--channels as well as nonselective stretch-activated cation channels in Ehrlich ascites tumor cells. *J. Membr. Biol.* **129,** 13–36.

Christensen, O., Simon, M., and Randlev, T. (1989). Anion channels in a leaky epithelium. A patch clamp study of choroid plexus. *Pfluegers Arch.* **415,** 37–46.

Christensen, S., Hoffmann, E. K., Saermark, T., and Simonsen, L. O. (1988). Inositol trisphosphate may be a second messenger in regulatory volume decrease in Ehrlich mouse ascites-tumour cells. *J. Physiol. (London)* **403,** 109P.

Civan, M. M., Coca–Prados, M., and Peterson–Yantorno, K. (1996). Regulatory volume increase of human non-pigmented ciliary epithelial cells. *Exp. Eye Res.* **62,** 627–640.

Coca-Prados, M., Sanches-Torres, J., Peterson-Yantorno, K., and Civan, M. M. (1996). Association of ClC-3 channel with Cl^- transport by human nonpigmented ciliary epithelial cells. *J. Membr. Biol.* **150,** 197–208.

Colclasure, G. C., and Parker, J. C. (1991). Cytosolic protein concentration is the primary volume signal in dog red cells. *J. Gen. Physiol.* **98,** 881–892.

Colclasure, G. C., and Parker, J. C. (1992). Cytoplasmic protein concentration is the primary volume signal for swelling induced [K-Cl] cotransport in dog red cells. *J. Gen. Physiol.* **100,** 1–10.

Collins, K. D., and Washabaugh, M. W. (1985). The Hofmeister effect and the behaviour of water at interfaces. *Q. Rev. Biophys.* **18,** 323–422.

Cornet, M., Lambert, I. H., and Hoffmann, E. K. (1993a). Relation between cytoskeleton, hypo-osmotic treatment and volume regulation in Ehrlich ascites tumor cells. *J. Membr. Biol.* **131,** 55–66.

Cornet, M., Ubl, J., and Kolb, H. A. (1993b). Cytoskeleton and ion movements during volume regulation in cultured PC12 cells. *J. Membr. Biol.* **133,** 161–170.

Cornet, M., Isobe, Y., and Lemanski, L. F. (1994). Effects of anisosmotic conditions on the cytoskeletal architecture of cultured PC12 cells. *J. Morphol.* **222,** 269–286.

Cowin, P., and Burke, B. (1996). Cytoskeleton–membrane interactions. *Curr. Opin. Cell Biol.* **8,** 56–85.

Craig, S. W., and Johnson, R. P. (1996). Assembly of focal adhesions: Progress, paradigms, and portents. *Curr. Opin. Cell Biol.* **8,** 74–85.

Dahl, S. V., Hallbrucker, C., Lang, F., and Häussinger, D. (1991). Role of eicosanoids, inositol phosphates and extracellular Ca^{2+} in cell-volume regulation of rat liver. *Eur. J. Biochem.* **198,** 73–83.

Defize, L. H. K., Boonstra, J., Meisenhelder, J., Kruyer, W., Tertoolen, L. G. J., Tilly, B. C., Hunter, T., van Bergen en Henegouwen, P. M. P., Moolenaar, W. H., and de Laat, S. W. (1989). Signal transduction by epidermal growth factor through the subclass of high affinity receptors. *J. Cell Biol.,* **109,** 2495–2507.

Delpire, E., Rauchman, M. I., Beier, D. R., Hebert, S. C., and Gullans, S. R. (1994). Molecular cloning and chromosome localization of a putative basolateral Na^+-K^+-$2Cl^-$ cotransoprter from mouse innner medullary collecting duct (mIMCD-3) cells. *J. Biol. Chem.* **269,** 25677–25683.

Demaurex, N., and Grinstein, S. (1994). Na^+/H^+ antiport: Modulation by ATP and role in cell volume regulation. *J. Exp. Biol.* **196,** 389–404.

Demaurex, N., Romanek, R. R., Orlowski, J., and Grinstein, S. (1997). ATP dependence of Na^+/H^+ exchange. *J. Gen. Physiol.* **109,** 117–128.

Dennis, E. A. (1994). Diversity of group types, regulation, and function of phospholipase A_2. *J. Biol. Chem.* **269,** 13057–13060.

Dennis, E. A. (1997). The growing phospholipase A_2 superfamily of signal transduction enzymes. *Trends Biochem. Sci.* **22,** 1–2.

den Hartigh, J. C., van Bergen en Henegouwen, P. M. P., Verkleij, A. K., and Boonstra, J. (1992). The EGF receptor is an actin-binding protein. *J. Cell Biol.* **119,** 349–356.

Deutsch, C., and Chen, L. Q. (1993). Heterologous expression of specific K^+ channels in T lymphocytes: Functional consequences for volume regulation. *Proc. Natl. Acad. Sci. U.S.A.* **90,** 10036–10040.

Diener, M., and Scharrer, E. (1993). The leukotriene D_4 receptor blocker, SK&F 104353, inhibits volume regulation in isolated crypts from the rat distal colon. *Eur. J. Pharmacol.* **238,** 217–222.

Ding, G., Franki, N., Condeelis, J., and Hays, R. M. (1991). Vasopressin depolymerizes F-actin in toad bladder epithelial cells. *Am. J. Physiol.* **260,** C9–C16.

Doroshenko, P., and Neher, E. (1992). Volume-sensitive chloride conductance in bovine chromaffin cell membrane. *J. Physiol.* (*London*) **449,** 197–218.

Downey, G. P., Grinstein, S., Sue-A-Quan, A., Czaban, B., and Chan, C. K. (1995). Volume regulation in leukocytes: Requirement for an intact cytoskeleton. *J. Cell. Physiol.* **163,** 96–104.

Duan, D., Hume, J. R., and Nattel, S. (1997a). Evidence that outwardly rectifying Cl^- channels underlie volume-regulated Cl^- currents in the heart. *Circ. Res.* **80,** 103–113.

Duan, D., Winter, C., Cowley, S., Hume, J. R., and Horowitz, B. (1997b). Molecular identification of a volume-regulated chloride channel. *Nature* (*London*) **390,** 417–421

Dubyak, G. R. (1986). Extracellular ATP activates polyphosphoinositide breakdown and Ca^{2+} mobilization in Ehrlich ascites tumor cells. *Arch. Biochem. Biophys.* **245,** 84–95.

Duhm, J., and Göbel, B. O. (1984). Na^+-K^+ transport and volume of rat erythrocytes under dietary K^+ deficiency. *Am. J. Physiol.* **246,** C20–C29.

Dunham, P. B. (1995). Effects of urea on K-Cl cotransport in sheep red blood cells. Evidence for two signals of swelling. *Am. J. Physiol.* 265, C447–C452.

Dunham, P. B., Jessen, F., and Hoffmann, E. K. (1990). Inhibition of Na-K-Cl cotransport in Ehrlich ascites cells by antiserum against purified proteins of the cotransporter. *Proc. Natl. Acad. Sci. U.S.A.* **87,** 6828–6832.

Dunham, P. B., Klimczak, J., and Logue, P. J. (1993). Swelling activation of K-Cl cotransport in LK sheep erythrocytes: A three-state process. *J. Gen. Physiol.* **101,** 733–765.

Emma, F., McManus, M., and Strange, K. (1997). Intracellular electrolytes regulate the volume set point of the organic osmolyte/anion channel VSOAC. *Am. J. Physiol.* **272,** C1766–C1775.

Engström, K. G., Sandström, P.-E., and Sehlin, J. (1991). Volume regulation in mouse pancreatic beta-cells is mediated by a furose-mide-sensitive mechanism. *Biochim. Biophys. Acta* **1091,** 145–150

Fath, K. R., and Burgess, D. R. (1994). Membrane motility mediated by unconventional myosin. *Curr. Opin. Cell Biol.* **6,** 131–135.

Fatherazi, S., Izutsu, K. T., Wellner, R. B., and Belton, C. M. (1994). Hypotonically activated chloride current in HSG cells. *J. Membr. Biol.* **142,**181–193.

Feit, P. W., Hoffmann, E. K., Schiødt, M., Kristensen, P., Jessen, F., and Dunham, P. B. (1988). Purification of proteins of the Na/Cl cotransporter from membranes of Ehrlich ascites cells using a bumetanice-Sepharose affinity column. *J. Membr. Biol.* **103,** 135–147.

Felipe, A., Snyders, D. J., Deal, K. K., and Tamkun, M. M. (1993). Influence of cloned voltage-gated K^+ channel expression on alanine transport, Rb^+ uptake, and cell volume. *Am. J. Physiol.* **265** (Cell Physiol. 34), C1230–C1238.

Ferguson, D. R., Kennedy, I., and Burton T. J. (1997). ATP is released from rabbit urinary bladder epithelial cells by hydrostatic pressure changes—a possible sensory mechanism? *J. Physiol.* (*London*) **505,** 503–511

Fliegel, L., and Fröhlich, O. (1993). The Na^+/H^+ exchanger: An update on structure, regulation and cardiac physiology. *Biochem. J.* **296,** 273–285.

Ford-Hutchinson, A. W., Gesser, M., and Young, R. N. (1994). 5-Lipoxygenase. *Annu. Rev. Biochem.* **63,** 383–417.

Forte, J. G., and Lee, H. C. (1977). Gastric adenosine triphosphatases: A review of their possible role in HCL secretion. *Gastroenterology* **73,** 921–926.

Forte, J. G., and Yao, X. (1996). The membrane-recruitment-and-recycling hypothesis of gastric HCl secretion. *Trends Cell Biol.* **6,** 45–48.

Foskett, J. K. (1994). The role of calcium in the control of volume regulatory transport pathways. *In* "Cellular and Molecular Physiology of Cell Volume Regulation" (K. Strange, ed.). pp. 259–277. CRC Press, Boca Raton, Florida.

Foskett, J. K., Wong, M. M. M., Sue-A-Quan, G., and Robertson, M. A. (1994). Isosmotic modulation of cell volume and intracellular ion activities during stimulation of single exocrine cells. *J. Exp. Zool.* **268,** 104–110.

Frangos, J. A., and Gudi, S. R. P. (1997). The phospholipid bilayer as a mechanoreceptor in shear stress-induced G protein activation. *XXXIII IUPS Congress* L079. 03.

Freeman, C. J., Bookchin, R. M., Ortiz, O. E., and Lew, V. L. (1987). K-permeabilized human red cells lose an alkaline, hypertonic fluid containing excess K^+ over diffusible anions. *J. Membr. Biol.* **96,** 235–241.

Friedman, J. E., and Haddad, G. G. (1993). Major differences in $Ca^{2+}{}_i$ response to anoxia between neonatal and adult rat Ca1 neurons: Role of $Ca^{2+}{}_o$ and $Na^+{}_o$. *J. Neurosci.* **13,** 63–72.

Frizzell, R. A., and Morris, A. P. (1994). Chloride conductances of salt-secreting epithelia. *In* "Current Topics in Membranes" (W. B. Guggino, ed.), Vol 42, pp. 173–214. Academic Press, San Diego.

Froehner, S. (1991). The submembrane machinery for nicotinic acetylcholine receptor clustering. *J. Cell Biol.* **114,** 1–7.

Fugelli, K. and Zachariassen, K. E. (1976). The distribution of taurine, gamma-aminobutyric acid and inorganic ions between plasma and erythrocytes in flounder (*Platichthys flesus*) at different plasma osmolarities. *Comp. Biochem. Physiol.* **55A,** 173–177.

Fukuda, J., Kameyama, M., and Yamaguchi, K. (1981). Breakdown of cytoskeletal filaments selectively reduces Na and Ca spikes in cultured mammal neurones. *Nature* (*London*) **294,** 82–85.

Furukawa, K. and Mattson, M. P. (1995). Cytochalsins protect hippocampal neurons against amyloid β-peptide toxicity: Evidence that actin depolymerization suppresses Ca^{2+} influx. *J. Neurochem.* **65,** 1061–1068.

Gamba, G., Miyanoshita, A., Lombardi, M., Lytton, J., Lee, W.-S., Hedinger, M. A., and Hebert, S. C. (1994). Molecular cloning, primary structure and characterization of two members of the mammalian electroneutral sodium-(potassium)-chloride cotransporter family expressed in kidney. *J. Biol. Chem.* **269,** 17713–17722.

Garcia-Pérez, A., and Burg, M. B. (1991). Renal medullary organic osmolytes. *Physiol. Rev.* **71,** 1081.

Garcia-Romeu, F., Cossins, A. R., and Motais, R. (1991). Cell volume regulation by trout erythrocytes: Characteristics of the transport systems activated by hypotonic swelling. *J. Physiol.* (*London*). **440,** 547–567.

Gargus, J. J., and Slayman, C. W. (1980). Mechanism and role of furosemide-sensitive K^+ transport in L cells: A genetic approach. *J. Membr. Biol.* **52,** 245–256.

Garlid, K. D. (1996). Cation transport in mitochondria: The potassium cycle. *Biochim Biophys. Acta.* **1275**(1–2), 123–126.

Garnovskaya, M. N., Mukhin, Y. V., and Raymond, J. R. (1997). Shrinkage-induced activation of Na^+/H^+ exchange: In search of signalling intermediates involved in the regulatory volume increase in fibroblasts. *XXXIII IUPS Congress* P025. 02.

Geck, P. (1990). Volume regulation in Ehrlich cells. *In* "Cell Volume Regulation" (K. W. Beyenbach, ed.), pp. 26–59, Karger, Basel.

Giebisch, G. (1995). Renal potassium channels: An overview. *Kidney Int.* **48,** 1004–1009.

Gillen, C. M., Brill, S., Payne, J. A., and Forbush III, B. (1996). Molecular cloning and functional expression of the Kcl cotransporter from rabbit, rat and human. *J. Biol. Chem.* **271,** 16273–16244.

Gilles, R., Hoffmann, E. K., and Bolis, L. (1991). "Comparative and Environmental physiology 9," pp. 1–249. Springer-Verlag, Berlin.

Goodson, H. V., Valetti, C., and Kreis, T. E. (1997). Motors and membrane traffic. *Curr. Opin. Cell Biol.* **9,** 18–28.

Goss, G. G., Woodside, M., Wakabayashi, S., Pouysségur, J., Waddel, T., Downey, G. P., and Grinstein, S. (1994). ATP dependence of NHE-1, the ubiquitous isoform of the Na^+/H^+ antiporter. *J. Biol. Chem.* **269,** 8741–8748.

Greger, R. (1996). The membrane transporters regulating epithelial NaCl secretion. *Pfluegers Arch.* **432,** 579–588

Grinstein, S., and Smith, J. D. (1990). Calcium-independent cell volume regulation in human lymphocytes. Inhibition by charybdotoxin. *J. Gen. Physiol.* **95,** 97–120.

Grinstein, S., Clarke, C. A., DuPre, A., and Rothstein, A. (1982a). Volume-induced increase of anion permeability in human lymphocytes. *J. Gen. Physiol.* **80,** 801–823.

Grinstein, S., Clarke, C. A., and Rothstein, A. (1982b). Increased anion permeability during volume regulation in human lymphocytes. *Philos. Trans. R. Soc. London B* **299,** 509–518.

Grinstein, S., DuPre, A., and Rothstein, A. (1982c). Volume regulation by human lymphocytes. Role of calcium. *J. Gen. Physiol.* **79,** 849–868.

Grinstein, S., Cohen, S., Sarkadi, B., and Rothstein, A. (1983). Induction of 86^{Rb} fluxes by Ca^{2+} and volume changes in thymocytes and their isolated membranes. *J. Cell. Physiol.* **116,** 352–362.

Grinstein, S., Rothstein, A., Sarkadi, B., and Gelfand, E. W. (1984). Responses of lymphocytes to anisotonic media: Volume regulating behavior. *Am. J. Physiol.* **246,** C204–C215.

Grinstein, S., Furuya, W., and Bianchini, L. (1992a). Protein kinases, phosphatases, and the control of cell volume. *News Physiol. Sci.* **7,** 232–237.

Grinstein, S., Woodside, M., Sardet, C., Pouysségur, J., and Rotin, D. (1992b). Activation of the Na^+/H^+ antiporter during cell volume regulation. Evidence for a phosphorylation-independent mechanism. *J. Biol. Chem.* **267,** 23823–23828.

Grinstein, S., Woodside, M., Waddell, T. K., Downey, G. P., Orlowski, J., Pouysségur, J., Wong, D. C., and Foskett, J. K. (1993). Focal localization of the NHE-1 isoform of the Na^+/H^+ antiport: Assessment of effects on intracellular pH. *EMBO J.* **12,** 5209–5218.

Grinstein, S., Woodside, M., Goss, G. G., and Kapus, A. (1994). Osmotic activation of the Na^+/H^+ antiporter during volume regulation. *Biochem. Soc. Trans.* **22,** 512–516.

Gründer, S., Thiemann, A., Pusch, M., and Jentsch, T. J. (1992). Regions involved in the opening of the CIC-2 chloride channel by voltage and cell volume. *Nature* (*London*) **360,** 759–762.

Grygorczyk, R., and Hanrahan, J. W. (1997). CFTR-independent ATP release from epithelial cells triggered by mechanical stimuli. *Am. J. Physiol.* **272,** C1058–C1066.

Gullans, S. R., and Verbalis, J. G. (1993). Control of brain volume during hyperosmolar and hypoosmolar conditions. *Annu. Rev. Med.* **44,** 289–301.

Guo, X., Merlin, D., Harvey, R. D., Laboisse, C., and Hopfer, U. (1997). Pharmacological evidence that calcium is not required for P_2 receptorstimulated Cl^- secretion in HT29-Cl. 16E. *J. Membr. Biol.* **155,** 239–246

Haas, M. (1994). The Na-K-Cl cotransporters. *Am. J. Physiol.* **267,** C869–C885.

Haas, M., and McBrayer, D. G. (1994). Na-K-Cl cotransport in nystatin treated tracheal cells: Regulation by isoproterenol, apical UTP, and $[Cl]_i$. *J. Am. Physiol.* **266,** C440–C452.

Haas, M., Dunham, P. B. and Forbush, B. (1991). $[^3H]$ bumetanide binding to mouse kidney membranes: Identification of corresponding membrane proteins. *Am. J. Physiol.* **260,** C791–C804.

Haas, M., McBrayer,D. G., and Lytle, C. (1995). $[Cl^-]_2$ -dependent phosphorylation of the Na-K-Cl cotransport protein of dog tracheal epithelial cells. *J. Biol. Chem.* **270,** 28955– 289621.

Hainsworth, A. H., Barrand, M. A., and Hladky, S. B. (1996). Evidence against the involvement of leukotrienes LTC_4 and LTD_4 in transduction of volume-activated anion currents in COR- L23R cells. *J. Physiol.* (*London*) **495,** 122P.

Hallows, K. R., and Knauf, P. A. (1994). Principles of cell volume regulation. *In* "Cellular and Molecular Physiology of Cell Volume Regulation" (K. Strange, ed.), pp. 3–29. CRC Press, Boca Raton, Florida.

Hallows, K. R., Packman, C. H., and Knauf, P. A. (1991). Acute cell volume changes in anisotonic media affect F-actin content of HL-60 cells. *Am. J. Physiol.* **261,** C1154–C1161.

Hallows, K. R., Law, F.-Y., Packman, C. H., and Knauf, P. A. (1996). Changes in cytoskeletal actin content, F-actin distribution, and surface morphology during HL-60 cell volume regulation. *J. Cell. Physiol.* **167,** 60–71.

Hamada, K., Takuwa, N., Yokoyama, K., and Takuwa, Y. (1998). Stretch activates jun N-terminal kinase/stress-activated protein kinase in vascular smooth muscle cells through mechanisms involving autocrine ATP stimulation of purinoceptors. *J. Biol. Chem.* **273,** 6334–6340.

Hansen, C. A., Schroering, A. G., Carey, D. J., and Robishaw, J. D. (1994). Localization of a heterotrimeric G-protein γ subunit to focal adhesions and associated stress fibers. *J. Cell Biol.* **126,** 811–819.

Harbak, H., and Simonsen, L. O. (1995). The K^+ channels activated during regulatory volume decrease (RVD) are distinct from those activated by Ca^{2+}-mobilizing agonists in Ehrlich mouse ascites tumour cells. *J. Physiol.* (*London*). **482,** 12P.

Harbak, H., Nielsen, D., Christensen, S., and Simonsen, L. O. (1995). Cell volume-induced changes in phosphatidyl-inositol(4,5)bisphosphate in Ehrlich mouse ascites tumor cells. *J. Physiol.* **489,** 12P.

Hardy, S. P., Goodfellow, H. R., Valverde, M. A., Gill, D. R., Sepulveda, F. V., and Higgins, C. F. (1995). Protein kinase C-mediated phosphorylation of the human multidrug resistance P-glycoprotein regulates cell volume-activated chloride channels. *EMBO J.* **14,** 68–75.

Harvey, B. J. (1994). Crosstalk and epithelial ion transport. *Curr. Opin. Nephrol. Hypertens.* **3,** 523–528.

Hasson, T., and Mooseker, M. S. (1995). Molecular motors, membrane movements and physiology: Emerging roles for myosins. *Curr. Opin. Cell Biol.* **7,** 587–594.

Häussinger, D., and Lang, F. (1991). Regulation of cell volume in the hepatic function: A mechanism for metabolic control. *Biochim. Biophys. Acta* **1071,** 331–350.

Häussinger, D., Gerok, W., and Lang, F. (1993). Cell volume and hepertic metabolism. *In* "Interaction of Cell Volume and Cell Function" (F. Lang and D. Häussinger, eds.), Vol. 14., pp. 33–67. Comp. Env. Physiol. Springer-Verlag, Berlin.

Häussinger, D., Newsome, W., Von Dahl, S., Stoll, B., Noe, B., Schreiber, R., Wettstein, M., and Lang, F. (1994). Control of liver cell function by the hydration state. *Biochem. Soc. Trans.* 22, 497–502.

Hays, R. M., Franki, N., Simon, H., and Gao, Y. (1994). Antidiuretic hormone and exocytosis: Lessons from neurosecretion. *Am. J. Physiol.* **267,** C1507–C1524.

Hazama, A., and Okada, Y. (1990). Biphasic rises in cytosolic free Ca^{2+} in association with activation of K^+ and Cl^- conductance during the regulatory volume decrease in cultured human epithelial cells. *Pfluegers Arch.* **416,** 710–714.

Hendil, K. B., and Hoffmann, E. K. (1974). Cell volume regulation in Ehrlich ascites tumor cells. *J. Cell. Physiol.* **84,** 115–125.

Henson, J. H., Roesener, C. D., Gaetano, C. J., Mendola, R. J., Forrest, J. N., Holy, J., and Kleinzeller, A. (1997). Confocal microscopic observation of cytoskeletal reorganizations in cultured shark rectal gland cells following treatment with hypotonic shock and high external K^+. *J. Exp. Zool.* **279,** 415–424.

Heringdorf, D. M. Z., Liedel, K., Kaldenbergstasch, S., Michel, M. C., Jacobs, K. H., and Wieland, T. (1996). Translocation of microfilament-associated inhibitory guanine-nucleotide-binding proteins to the plasma membrane in myeloma differentiated human leukemia (HL-60) cell. *Eur. J. Biochem.* **235,** 670–676.

Herman, I. M. (1993). Actin isoforms. *Curr. Opin. Cell Biol.* **5,** 48–55.

Higashijima, T., Ferguson, K. M., and Sternweis, P. C. (1987). Regulation of hormone-sensitive GTP-dependent regulatory proteins by chloride. *J. Biol. Chem.* **262,** 3597–3602.

Hoffmann, E. K. (1978). Regulation of cell volume by selective changes in the leak permeabilities of Ehrlich ascites tumor cells. *In* "Alfred Benzon Symposium XI. Osmotic and Volume

Regulation" (C. B. Jørgensen and E. Skadhauge, eds.), pp. 397–417. Munksgaard, Copenhagen.

Hoffmann, E. K. (1985). Role of separate K^+ and Cl^- channels and of Na^+/Cl^- cotransport in volume regulation in Ehrlich cells. *Fed. Proc.* **44,** 2513–2519.

Hoffmann, E. K. (1997). Intracellular transmission in cell volume regulation in Ehrlich ascites tumor cells. *J. Exp. Zool.* **279,** 398–414.

Hoffmann, E. K. (1998). LTD_4 activates ChTX-sensitive and -insensitive K^+ channels in Ehrlich ascites tumor cells. *Eur. J. Physiol.* (in press).

Hoffmann, E. K., and Dunham, P. B. (1995). Membrane mechanisms and intracellular signalling in cell volume regulation. *Int. Rev. Cytol.* **161,** 173–262.

Hoffmann, E. K., and Hendil, K. B. (1976). The role of amino acids and taurine in isosmotic intracellular regulation in Ehrlich ascites mouse tumour cells. *J. Comp. Physiol.* **108,** 279–286.

Hoffmann, E. K., and Lambert, I. H. (1983). Amino acid transport and cell volume regulation in Ehrlich ascites tumour cells. *J. Physiol. (London)* **338,** 613–625.

Hoffmann, E. K., and Pedersen, S. F. (1998). Sensors and signal transduction in the activation of cell volume regulatory ion transport systems. *In* "Cell Volume Regulation," (F. Lang, ed.), p. 50–78. Basel, Karger.

Hoffmann, E. K., and Simonsen, L. O. (1989). Membrane mechanisms in volume and pH regulation in vertebrate cells. *Physiol. Rev.* **69,** 315–382.

Hoffmann, E. K., and Ussing, H. H. (1992). Membrane mechanisms in volume regulation in vertebrate cells and epithelia. *In* "Membrane Transport in Biology", (G. H. Giebisch, J. A. Schafer, H. H. Ussing, and P. Kristensen, eds.), Vol. 5, pp. 317–399. Springer-Verlag, Heidelberg.

Hoffmann, E. K., Sjöholm, C., and Simonsen, L. O. (1983). Na^+,Cl^- co-transport in Ehrlich ascites tumor cells activated during volume regulation (regulatory volume increase). *J. Membr. Biol.* **76,** 269–280.

Hoffmann, E. K., Simonsen, L. O., and Lambert, I. H. (1984). Volume-induced increase of K^+ and Cl^- permeabilities in Ehrlich ascites tumor cells. Role of internal Ca^{2+}. *J. Membr. Biol.* **78,** 211–222.

Hoffmann, E. K., Lambert, I. H., and Simonsen, L. O. (1986). Separate, Ca^{2+}-activated K^+ and Cl^- transport pathways in Ehrlich ascites tumor cells. *J. Membr. Biol.* **91,** 227–244.

Hoffmann, E. K., Simonsen, L. O., and Lambert, I. H. (1993). Cell volume regulation: Intracellular transmission. *In* "Advances in Comparative and Environmental Physiology, Interaction, Cell Volume, Cell Function" (F. Lang and D. Häussinger, eds.), Vol. 14, pp. 187–248. Springer-Verlag, Berlin.

Hoffmann, E. K., Jessen, F., and Dunham, P. B. (1994). The Na-K-2Cl cotransporter is in a permanently activated state in cytoplasts from Ehrlich ascites tumor cells. *J. Membr. Biol.* **138,** 229–239.

Hogan, E. M., Davis, B. A., and Boron, W. F. (1997). Intracellular Cl^- dependence of Na-H exchange in barnacle muscle fibers under normotonic and hypertonic conditions. *J. Gen. Physiol.* **110,** 629–639.

Holmgren, K., Magnusson, K. E., Franki, N., and Hays, R. M. (1992). ADH-induced depolymerization of F-actin in the toad bladder granular cell: A confocal microscope study. *Am. J. Physiol.* **262,** C672–C677.

Hunziker, E. B., Wagner, J., and Zapf, J. (1994). Differential effects of insulin-like growth factor I and growth hormone on developmental stages of rat growth plate chondrocytes *in vivo. J. Clin. Invest.* **93,** 1078–1086.

Igarashi, P., Vanden-Heuvel, G. B., Payne, J. A., and Forbush III, B. (1995). Cloning, embryonic expression, and alternative splicing of a murine kidney-specific Na-K-Cl cotransporter. *Am. J. Physiol.* **269,** F405–F418.

Ikeda, K., Wu, D., and Takasaka, T. (1995). Cellular mechanisms in activation of Na-K-C cotransport in nasal gland acinar cells of guinea pigs. *J. Membr. Biol.* **146,** 307–314.

Inagaki, N., Gonoi, T., Clement, J. P., Namba, N., Inazawa, J., Gonzalez, G., Aguilar-Bryan, L., and Bryan, J. (1995). Reconstitution of the IKATP: An inward rectifier subunit plus the sulfonylurea receptor. *Science,* **270,** 1166–1170.

Ingber, D. E. (1997). Tensegrity: The architectural basis of cellular mechanotransduction. *Annu. Rev. Physiol.* **59,** 575–599.

Insel, P. A., and Koachman, A. M. (1982). Cytochalasin B enhances hormone and cholera toxin-stimulated cyclic AMP accumulation in S49 lymphoma cells. *J. Biol. Chem.* **257,** 9717–9723.

Ismailov, I. I., Berdiev, B. K., Shlyonsky, V. G., Fuller, C. M., Prat, A. G., Jovov, B., Cantiello, H. F., Ausiello, D. A., and Benos, D. J. (1997). Role of actin in regulation of epithelial sodium channels by CFTR. *Am. J. Physiol.* **272,** C1077–C1086.

Jackson, P. S., and Strange, K. (1995). Single channel properties of volume sensitive anion conductance. Current activation occurs by abrupt switching of closed channels to an open state. *J. Gen. Physiol.* **105,** 643–660.

Jackson, P. S., Churchwell, K., Ballatori, N., Boyer, J. L., and Strange, K. (1996). Swelling-activated anion conductance in skate hepatocytes: Regulation by cell Cl^- and ATP. *Am. J. Physiol.* **270,** C57–C66

Jennings, M. L., and Al-Rohil, N. (1990). Kinetics of activation and inactivation of swelling-stimulated K^+/Cl^- transport. The volume-sensitive parameter is the rate constant for inactivation. *J. Gen. Physiol.* **95,** 1021–1040.

Jennings, M. L., and Schulz, R. K. (1991). Okadaic acid inhibition of KCl cotransport. Evidence that protein dephosphorylation is necessary for activation of transport by either cell swelling or *N*-ethylmaleimide. *J. Gen. Physiol.* **97,** 799–818.

Jensen, B. S., and Hoffmann, E. K. (1997). Hypertonicity enhances expression of functional $Na^+,K^+,2Cl^-$ cotransporters in Ehrlich ascites tumour cells. *Biochim. Biophys. Acta* **1329,** 1–6.

Jensen, B. S., Jessen, F., and Hoffmann, E. K. (1993). Na^+,K^+,Cl^- cotransport and its regulation in Ehrlich ascites tumor cells. Ca^{2+}/calmedulin and protein kinase C dependent pathways. *J. Membr. Biol.* **131,** 161–178.

Jentsch, T. J. (1996). Chloride channels: A molecular perspective. *Curr. Opin. Neurobiol.* **6,** 303–310

Jessen, F., and Hoffmann, E. K. (1992). Activation of the Na-K-2Cl cotransport system by reorganization of the actin filaments in Ehrlich ascites tumor cells. *Biochim. Biophys. Acta* **1110,** 199–201.

Johnson, B. D., and Byerly, L. (1993). A cytoskeletal mechanism for Ca^{2+} channel metabolic dependence and inactivation by intracellular Ca^{2+}. *Neuron* **10,** 797–804.

Jørgensen, N. K., Lambert, I. H., and Hoffmann, E. K. (1996). Role of LTD_4 in the regulatory volume decrease response in Ehrlich ascites tumor cells. *J. Membr. Biol.* **151,** 159–173.

Jørgensen, N. K., Christensen, S., Harbak, H., Brown, A. M., Lambert, I. H., Hoffmann, E. K., and Simonsen, L. O. (1997). On the role of Ca^{2+} in regulatory volume decrease (RVD) response in Ehrlich mouse ascites tumor cells . *J. Membr. Biol.* **157,** 281–299.

Jørgensen, N. K., Pedersen,S. F., and Hoffmann, E. K. (1998). Thrombin-bradykinin-and arachidonic acid-induced Ca^{2+}- signalling in Ehrlich acites tumor cells. *Am. J. Physiol.* **276,** C26–C37.

Kaji, D. M. and Gasson, C. (1995). Urea activation of KCl transport in human erythrocytes. *Am. J. Physiol.* **268,** C1018–C1025.

Kaji, D. M., and Tsukitani, Y. (1991). Role of protein phosphatase in activation of KCl cotransport in human erythrocytes. *Am. J. Physiol.* **260,** C176–C180.

Kapus, A., Grinstein, S., Wasan, S., Kandasamy, R., and Orlowski, J. (1994). Functional characterization of three isoforms of the Na^+/H^+ exchanger stably expressed in Chinese hamster ovary cells. *J. Biol. Chem.* **269,** 23544–23552.

Kelley, S. J., and Dunham,P. B. (1996). Mechanism of swelling activation of K-Cl cotransport in inside-out vesicles of LK sheep erythrocyte membranes. *Am. J. Physiol.* **270,** C1122–C1130.

Kelly, M. E. M., Dixon, S. J., and Sims, S. M. (1994). Outwardly rectifying chloride current in rabbit osteiclasts is activated by hyposmotic stimulation. *J. Physiol.* (*London*) **475.3,** 377–389.

Kimelberg, H. K., Goderie, S. K., Higman, S., Pang, S., and Wanievski, R. A. (1990). Swelling-induced release of glutamate, aspartate, and taurine from astrocyte cultures. *J. Neurosci.* **10,** 1583–1591.

Kimura, K., Ito, M., Amano, M., Chihara, K., Fukata, Y., Nakafuku, M., Yamamori, B., Feng, J., Nakano, T., Okawa, K., Iwamatsu, A., and Kaibuchi, K. (1996). Regulation of myosin phosphatase by Rho and Rho-associated kinase (Rho-kinase). *Science* **273,** 245–248.

Kinne, R. K. H. (1993). The role of organic osmolytes in osmoregulation: From bacteria to mammals. *J. Exp. Zool.* **265,** 346–355.

Kirk, K. and Strange, K. (1998). Functional properties and physiological roles of organic solute channels. *Annu. Rev. Physiol.* **60,** 719–739.

Klanke, C. A., Su, Y. R., Callen, D. F., Wang, Z., Meneton, P., Baird, N., Kandasamy, R. A., Orlowski, J., Otterud, B. E., Leppert, M., Shull, G. E., and Menon, A. G. (1995). Molecular cloning and physical and genetic mapping of a novel human Na^+/H^+ exchanger (NHE-5/SLC9A5) to chromosome 16q22. 1. *Genomics* **25,** 615–622.

Klein, J. D., and O'Neill, W. C. (1995). Volume sensitive myosin phosphorylation in vascular endothelial cells: Correlation with Na-K-Cl cotransport. *Am. J. Physiol.* **269,** C15524–C15532.

Kleinzeller, A., and Ziyadeh, F. N. (1990). Cell volume regulation in epithelia—with emphasis on the role of osmolytes and the cytoskeleton. In "Cell Volume Regulation" (R. K. H. Kinne, E. Kinne-Saffran and K. W. Beyenbach, eds.), Vol. 4, pp. 59–86. Karger, Basel.

Klymkowsky, M. W. (1995). Intermediate filaments: New proteins, some answers, more questions. *Curr. Opin. Cell Biol.* **7,** 46–54.

Knepper, M. A. (1994). The aquaporin family of molecular water channels. *Proc. Natl. Acad. Sci. U.S.A.* **91,** 6255–6258.

Kramhøft, B., Lambert, I. H., Hoffmann, E. K., and Jørgensen, F. (1986). Activation of Cl-dependent K transport in Ehrlich ascites tumor cells. *Am. J. Physiol.* **251,** C369–C379.

Krapivinsky, G. B., Ackerman, M. J., Gordon, E. A., Krapivinsky, M. J., and Clapham, D. E. (1994). Molecular characterization of a swelling-induced chloride conductance regulatory protein, pI_{Cln}. *Cell.* (*Cambridge, Mass.*) **76,** 439–448.

Krarup, T., and Dunham, P. B. (1996). Reconstitution of calyculin-inhibited K^+/Cl^- cotransport in dog erythrocyte ghosts by exogenous PP-1. *Am. J. Physiol.* **270,** C898–C902.

Krarup, T., Jensen, B. S., and Hoffmann, E. K. (1996). Occlusion of K^+ in the Na^+, K^+, Cl^- cotransporter of Ehrlich ascites tumor cells. B*iochim. Biophys. Acta* **1284,** 97–108.

Krarup, T., Jakobsen, L. D., Jensen, B. S., and Hoffmann, E. K. (1998). $Na^+/K^+/2Cl^-$ cotransport in Ehrlich cells: Regulation by protein phosphatases and kinases. *Am. J. Physiol.* **275,** C239–C250.

Kregenow, F. M. (1973). The response of duck erythrocytes to norepinephrine and an elevated extracellular potassium: Volume regulation in anisotonic media. *J. Gen. Physiol.* **61,** 509–527.

Kregenow, F. M. (1981). Osmoregulatory salt transporting mechanisms: Control of cell volume in anisotonic media. *Annu. Rev. Physiol.* **43,** 493–505.

Krump, E., Nikitas, K., and Grinstein, S. (1997). Induction of tyrosine phosphorylation and Na^+/H^+ exchanger activation during shrinkage of human neutrophils. *J. Biol. Chem.* **272,** 17303–17311.

Kubitz, R., Warth, R., Allert, N., Kunzelman, K., and Greger, R. (1992). Small-conductance chloride channels induced by cAMP, Ca^{2+}, and hypotonicity in HT_{29} cells. Ion selectivity, additivity and stilbene sensitivity. *Pfluegers Arch.* **421,** 447–454.

Kumakura, K., Sasaki, K., Sakuri, T., Ohara-Imaizumi, M., Misonu, H., Nakamura, S., Matsuda, Y., and Nonomura, Y. (1994). Essential role of myosin light chain kinase in the mechanism for MgATP-dependent priming of exocytosis in adrenal chromaffin cells. *J. Neurosci.* **14,** 7695–7703.

Kwon, H. M. (1994). Molecular regulation of mammalian osmolyte transporters. *In* "Cellular and Molecular Physiology of Cell Volume Regulation" (K. Strange, ed.), pp. 383–394. CRC Press, Boca Raton, Florida.

Kwon, H. M., and Handler, J. S. (1995). Cell volume regulated transporters of compatible osmolytes. *Curr. Opin. Cell Biol.* **7,** 465–471.

Lambert, I. H. (1998). Regulation of the taurine content in Ehrlich ascites tumor cells; *In:* "Taurine 3" (J. B. Lombardini, S. Schaeffer, and R. J. Huxtable, eds.), pp. 269–276. Plenum Publ. New York.

Lambert, I. H., and Hoffmann, E. K. (1993). Regulation of taurine transport in Ehrlich ascites tumor cells. *J. Membr. Biol.* **131,** 67–79.

Lambert, I. H., and Hoffmann, E. K. (1994). Cell swelling activates separate taurine and chloride channels in Ehrlich mouse ascites tumor cells. *J. Membr. Biol.* 142, 289–298.

Lambert I. H., Hoffmann, E. K., and Christensen, P. (1987). Role of prostaglandins and leukotrienes in volume regulation by Ehrlich ascites tumor cells. *J. Membr. Biol.* **98,** 247–256.

Lambert, I. H., Hoffmann, E. K., and Jørgensen, F. (1989). Membrane potential, anion and cation conductances in Ehrlich asites tumor cells. *J. Membr. Biol.* **111,** 113–132.

Lang, F., and Rehwald, W. (1992). Potassium channels in renal epithelial transport regulation. *Physiol. Rev.* **72,** 1–32.

Lang, F., Paulmichl, M., Völkl, H., Gstrein, E., and Friedrich, F. (1987). Electrophysiology of cell volume regulation. *In* "Molecular Nephrology. Biochemical Aspects of Kidney Function" (Z. Kovacevic and W. G. Guder, eds.), pp. 133–139. Berlin, New York.

Lang, F., Völkl, H., and Häussinger, D. (1990). General principles in cell volume regulation. *In "Cell Volume Regulation"* (K. W. Beyenbach, ed.), pp. 1–26. Karger, Basel.

Lang, F., Busch, G. L., and Völkl, H. (1998a). The diversity of volume regulatory mechanisms. *Cell Physiol. Biochem.* **8,** 1–45

Lang, F., Busch, G. L., Ritter, M., Völkl, H., Waldegger, S., Gulbins, E., and Häussinger, D. (1998b). Functional significance of cell volume regulatory mechanisms. *Physiol. Rev.* **78,** 1–60.

Larsen, A. K., Jensen, B. S., and Hoffmann, E. K. (1994). Activation of protein kinase C during cell volume regulation in Ehrlich mouse ascites tumor cells. *Biochim. Biophys. Acta* **1222,** 477–482.

Lauf, P. K., Bauer, J., Adragna, N. C., Fujise, H., Zade-Oppen, A. M. M., Ryu, K. H., and Delpire, E. (1992). Erythrocyte K-Cl cotransport: Properties and regulation. *Am. J. Physiol.* **263,** C917–C932.

Lauritzen, L., Hoffmann, E. K., Hansen H. S., and Jensen, B. (1993). Dietary (n-3) and (n-6) fatty acids are equipotent in stimulating volume regulation in Ehrlich ascites tumor cells. *Am. J. Physiol.* **264,** C109–C117.

Law, R. O. (1991). Amino acids as volume-regulatory osmolytes in mammalian cells. *Comp. Biochem. Physiol.* **99A,** 263–277.

Law, R. O. (1994). Regulation of mammalian brain cell volume. *J. Exp. Zool.* **268,** 90–96.

Law, R. O., and Burg, M. B. (1991). The role of organic osmolytes in the regulation of mammalian cell volume. *In* "Advances in Comparative and Enviromental Physiology", (R. Gilles, E. K. Hoffmann, and L. Bolis, eds.), Vol. 9, pp. 189–226. Springer-Verlag, Berlin and Heidelberg.

Lehtonen, J. Y. A., and Kinnunen, P. K. J. (1995). Phospholipase A_2 as a mechanosensor. *Biophys. J.* **68,**1888–1894.

Lepple-Wienhues, A., Szabó, I., Laun, T., Kaba, N. K., Gulbins, E., and Lang, F. (1998). The tyrosine kinase $p56^{lck}$ mediates activation of swelling-induced chloride channels in lymphocytes. *J. Biol. Chem.* **141,** 281–286.

Levitan, I., Almonte, C., Mollard, P., and Garber, S. S. (1995). Modulation of a volume-regulated chloride current by F-actin. *J. Membr. Biol.* **147,** 283–294.

Lew, V. L., and Bookchin, R. M. (1986). Volume, pH, and ion-content regulation in human red cells: Analysis of transient behavior with an integrated model. *J. Membr. Biol.* **92,** 57–74.

Lewis, R. S., and Cahalan, M. D. (1995). Potassium and calcium channels in lymphocytes. *Annu. Rev. Immunol.* **13,** 623–653.

Lewis, R. S., Ross, P. E., and Cahalan, M. D. (1993). Chloride channels activated by osmotic stress in T lymphocytes. *J. Gen. Physiol.* **101,** 801–826.

Lewis, S. A., and Donaldson, P. (1990). Ion channels and cell volume regulation: Chaos in an organized system. *News Physiol. Sci.* **5,** 112–119.

Light, D. B., Mertins, T. M., Belongia, J. A., and Witt, C. A. (1997). 5-Lipoxygenase metabolites of arachidonic acid regulate volume decrease by mudpuppy red blood cells. *J. Membr. Biol.* **158,** 229–239

Linshaw, M. A., Fogel, C. A., Downey, G. P., Koo, E. W. Y., and Gotlieb, A. I. (1992). Role of cytoskeleton in volume regulation of rabbit proximal tubule in dilute medium *Am. J. Physiol.* **262,** F144–F150.

Livne, A., and Hoffmann, E. K. (1990). Cytoplasmic acidification and activation of Na^+/H^+ exchange during regulatory volume decrease in Ehrlich ascites tumour cells. *J. Membr. Biol.* **114,** 153–157.

Lo, C., Ferrier, J., Tenenbaum, H. C., and McCulloch, C. A. G. (1995). Regulation of cell volume and intracellular pH in hypoosmotically swollen rat osteosarcoma cells. *Biochem. Cell Biol.* **73,** 535–544.

Luckie, D. B., Krause, M. E., Law, T. C., Sikic, B. I., and Wine, J. (1996). Doxorubicin selection for MDR 1/P-glycoprotein reduces swelling-activated K^+ and Cl^- currents in MES-Sa cells. *Am. J. Physiol.* **270,** C1029–C1036.

Lytle, C., and Forbush III, B. (1992a). Is $[Cl]_i$ the switch controlling Na-K-2Cl cotransport in shark rectal gland? *Biophys. J.* **61,** A384.

Lytle, C., and Forbush III, B. (1992b). The Na-K-Cl cotransport protein in shark rectal gland. II. Regulation by direct phosphorylation. *J. Biol. Chem.* **267,** 25438–25443.

Lytle, C., and Forbush III, B. (1996). Regulatory phosphorylation of the secretory Na-K-Cl cotransporter: modulation by cytoplasmic Cl. *Am. J. Physiol.* **270,** C437–C448

Lytle, C., *et al.* (1998). A model of Na-K-2Cl cotransport based on ordered ion binding and glide symmetry. *Am. J. Physiol.* **274,** C299–309.

Macknight, A. D. C., Grantham, J, and Leaf, A. (1993). Physiological responses to changes in extracellular osmolarity. *In* "Clinical Disturbances of Water Metabolism" (D. W. Seldin and G. Giebisch, Eds.) pp. 31–49. Raven, New York.

Macknight, A. D. C., Gordon,L. G. M. and Purves, R. D. (1994). Problems in understanding of cell volume regulation. *J. Exp. Zool.* **268,** 80–89.

MacLeod, R. J. (1994). How an epithelial cell swells in a determinant of the signaling pathways that activate RVD. *In* "Cellular and Molecular Physiology of Cell Volume Regulation" (K. Strange, ed.), pp. 191–200. CRC Press, Boca Raton.

McCarty, N. A. and O'Neil, R. G. (1992). Calcium signaling in cell volume regulation. *Physiol. Rev.* **72,** 1037–1061.

McManus, M. L., and Churchwell, K. B. (1994). Clinical significance of cellular osmoregulation. *In* "Cellular and Molecular Physiology of Cell Volume Regulation" (K. Strange, ed.), pp. 63–77. CRC Press, Boca Raton, Florida.

McManus, M. L., Churchwell, K. B., and Strange, K. (1995). Regulation of cell volume in health and disease. *N. Engl. J. Med.* **333,** 1260–1266.

Maeda, T., Takekawa, M., and Saito, H. (1996). Activation of yeat PBS2 MAPKK by MAPKKKs or by binding of an SH3-containing osmosensor. *Science* **269,** 554–558.

Mandelkow, E., and Mandelkow, E.-M. (1995). Microtubules and microtubule-associated proteins. *Curr. Opin. Cell Biol.* **7,** 72–81.

Manganel, M., and Turner, R. J. (1991). Rapid secretagogue-induced activation of Na^+/H^+ exchange in rat parotid acinar cells. Possible interrelationship between volume regulation and stimulus-secretion coupling. *J. Biol. Chem.* **266,** 10182–10188.

Margalit, A., Livne, A. A., Funder, J., and Granot, Y. (1993a). Initiation of RVD response in human platelets: Mechanical–biochemical transduction involves petussis-toxin-sensitive G protein and phospholipase A. *J. Membr. Biol.* **136,** 303–311.

Margalit, A., Sofer, Y., Grossman, S., Reynaud, D., Pace-Asciak, C. R., and Livne, A. A. (1993b). Hepoxilin A_3 is the endogenous lipid mediator opposing hypotonic swelling of intact human platelets. *Proc. Natl. Acad. Sci. U.S.A.* **90,** 2589–2592.

Marsault, R., Murgia, M., Pozzan, T., and Rizzuto, R. (1997). Domains of high Ca^{2+} beneath the plasma membrane of living A7r5 cells. *EMBO J.* **16,** 1575–1581

Mastrocola, T., Lambert, I. H., Kramhöft, B., Rugolo, M., and Hoffmann, E. K. (1993). Volume regulation in human fibroblasts: Role of Ca^{2+} and 5-lipoxygenase products. *J. Membr. Biol.* **136,** 55–62.

Matsuda, S., Kawasaki, H., Moriguchi, T., Gotoh, Y., and Nishida, E. (1995). Activation of protein kinase cascades by osmotic shock. *J. Biol. Chem.* **270,** 12781–12786.

Matsumoto, G., and Sakai, H. (1979a). Microtubules inside the plasma membrane of squid giant axons and their possible physiological function. *J. Membr. Biol.* **50,** 1–14.

Matsumoto, G., and Sakai, H. (1979b). Restoration of membrane excitability of squid giant axons by reagents activating tyrosine-tubulin ligase. *J. Membr. Biol.,* **50,** 15–22.

Matsumoto, G., Ichikawa, M., and Tasaki, A. (1984). Axonal microtubules necessary for generation of sodium current in squid giant axons: II. Effect of colchicine upon asymmetrical displacement current. *J. Membr. Biol.,* **77,** 93–99.

Matthews, J. B., Awtrey, C. S., and Madara, J. L. (1992). Microfilament-dependent activation of $Na^+/K^+/2Cl^-$ cotransport by cAMP in intestinal epithelial monolayers. *J. Clin. Invest.* **90,** 1608–1613.

Matthews, J. B., Tally, K. J., Smith, J. A., and Awtrey, C. S. (1994). F-actin differentially alters epithelial transport and barrier function. *J. Surg. Res.* **56,** 505–509.

Matthews, J. B., Smith, J. A., and Hrnjez, B. J. (1997). Effect of F-actin stabilization or disassembly on epithelial Cl^- secretion and Na-K-2Cl cotransport. *Am. J. Physiol.* **272,** C254–C262.

Matthews, J. B., Smith, J. A., Mun, E. C., and Sicklick, J. K. (1998). Osmotic regulation of intestinal epithelial Na^+-K^+-Cl^- cotransport: Role of Cl^- and F-actin. *Am. J. Physiol.,* **274,** C697–C706.

Melmed, R. N., Karanian, P. J., and Berlin, R. D. (1981). Control of cell volume in the J774 macrophage by microtubule disassembly and cyclic AMP. *J. Cell Biol.* **90,** 761.

Meng, X.-J., Carruth, M. W., and Weinman, S. A. (1997). Leulotriene D_4 activates a chloride conductance in hepatocytes from lipopolysaccharide-treated rats. *J. Clin. Invest.* **99,** 2915–2922.

Meves, H. (1994). Modulation of ion channels by arachidonic acid. *Prog. Neurobiol.* **43,** 175–186.

Meyer, K., and Korbmacher, C. (1996). Cell swelling activates ATP-dependent voltage-gated chloride channels in M-1 mouse cortical collecting duct cells. *J. Gen. Physiol.* **108,** 177–193.

Mills, J. W. (1987). The cell cytoskeleton: Possible role in volume control. *In* "Current Topics in Membranes and Transport" (R. Gilles, A. Kleinzeller, and L. Bolis, eds.), Vol. 30, pp. 75–101. Academic Press, New York.

Mills, J. W., Schwiebert, E. M., and Stanton, B. A. (1993). Evidence for the role of actin filaments in regulating cell swelling. *J. Exp. Zool.* **268,** 111–120.

Minton, A. P. (1983). The effect of volume occupancy upon the thermodynamic activity of proteins: Some biochemical consequences. *Mol. Cell Biochem.* **55,** 119–140.

Minton, A. P. (1994). Influence of macromolecular crowding on intracellular association reactions. Possible role in volume regulation. *In* "Cellular and Molecular Physiology of Cell Volume Regulation" (K. Strange, ed.), pp. 181–190. CRC Press, Boca Raton, Florida.

Minton, A. P., Colclasure, G. C., and Parker, J. C. (1992). Model for the role of macromolecular crowding in regulation of cellular volume. *Proc. Natl. Acad. Sci. U.S.A.* **89,** 10504–10506.

Mitchell, C. H., Zhang, J. J., Wang, L., and Jacob, T. J. C. (1997). Volume-sensitive chloride current in pigmented ciliary epithelial cells: Role of phospholipases. *Am. J. Physiol.* **272,** C212–C222.

Miyamoto, S., Teramoto, H., Coso, O. A., Gutkind, J. S., Burbelo, P. D., Akiyama, S. K., and Yamada, K. M. (1995). Integrin function: Molecular hierachies of cytoskeletal and signaling molecules. *J. Cell Biol.* **131,** 791–805.

Mochly-Rosen, D. (1995). Localization of protein kinases by anchoring proteins: A theme in signal transduction. *Science* **268,** 247–251

Montrose-Rafizadeh, C., and Guggino, W. B. (1991). Role of intracellular calcium in volume regulation by rabbit medullary thick ascending limb cells. *Am. J. Physiol.* **260,** F402–F409.

Moolenaar, W. H., Kranenburg, O., Postma, F. R., and Zondag, G. C. M. (1997). Lysophosphatidic acid: G-protein signalling and cellular responses. *Curr. Opin. Cell Biol.* **9,** 168–173.

Moran, J., Sabanero, M., Meza, I., and Pasantes-Morales, H. (1996). Changes of actin cytoskeleton during swelling and regulatoy volume decrease in cultured astrocytes. *Am. J. Physiol.* **271,** C1901–C1907.

Motais, R., Fiévet, B., Borgese, F., and Garcia-Romeu, F. (1997). Association of the band 3 protein with a volume-activated, anion and amino acid channel: A molecular approach. *J. Exp. Biol.* **200,** 361–367.

Muallem, S., Kwiatkowska, K., Xu, X., and Yin, H. L. (1995). Actin filament disassembly is a sufficient final trigger for exocytosis in nonexcitable cells. *J. Cell Biol.* **128,** 589–598.

Nath, S. K., Hang, Y., Levine, S. A., Yun, C. H. C., Montrose, M. H., Donowitz, M., and Tsé, C.-M. (1996). Hyperosmolarity inhibits the Na^+/H^+ exchanger isoforms NHE-2 and NHE-3: An effect opposite to that on NHE-1. *Am. J. Physiol.* **270,** G431–G441.

Nebe, B., Bohn, W., Sanftleben, H., and Rychly, J. (1996). Induction of a physical linkage between integrins and the cytoskeleton depends on intracellular calcium in an epithelial cell line. *Exp. Cell Res.* **229,** 100–110.

Needleman, P., Turk, J., Jakschik, B. A., Morrison, A. R., and Lefkowith, J. B. (1986). Arachidonic acid metabolism. *Annu. Rev. Biochem.* **55,** 69–102.

Nilius, B., Sehrer, J., and Droogmans, G. (1994a). Permeation properties and modulation of volume-activated Cl^- channels in human endothelial cells. *Br. J. Pharmacol.* **112,** 1049–1056.

Nilius, B., Sehrer, J., Viana, F., De Greef, C., Raeymaekers, L., Eggermont, J., and Droogmans, G. (1994b). Volume-activated Cl^- currents in different mammalian non-excitable cell types. *Pfluegers Arch.* **428,** 364–371.

Nilius, B., Oike, M., Zahradnik, I., and Droogmans, G. (1994c). Activation of a Cl^- current by hypotonic volume increase in human endothelial cells. *J. Gen. Physiol.* **103,** 787–805.

Nilius, B., Sehrer, J., De Smet, P., Van Driessche, W., and Droogmans, G. (1995a). Volume regulation in a toad epithelial cell line: role of coactivation of K^+ and Cl^- channels. *J. Physiol. (London)* **487,** 367–378.

Nilius, B., Sehrer, J., Heinke, S., and Droogmans, G. (1995b). Ca^{2+} release and activation of K^+ and Cl^- currents by extracellular ATP in distal nephron epithelial cells. *Am. J. Physiol.* **269,** C376–C384.

Nilius, B., Eggermont, J., Voets, T., Buyse, G., Manopoulos, V., and Droogmans, G. (1997a). Properties of volume-regulated anion channels in mammalian cells. *Prog. Biophys. Mol. Biol.* **64,** 69–119.

Nilius, B., Prenen, J., Szucs, G., Wei, L., Tanzi, F., Voets, T., and Droogmans, G. (1997b). Calcium-activated chloride channels in bovine pulmonary artery endothelial cells. *J. Physiol. (London)* **498,** 381–396.

Nilius, B., Prenen, J., Voets, T., Eggermont, J., and Droogmans, G. (1998). Activation of volume-regulated chloride currents by reduction of intracellular ionic strength in bovine endothelial cells. *J. Physiol. (London)* **506,** 353–361.

Noël, J., and Pouysségur, J. (1995). Hormonal regulation, pharmacology and membrane sorting of vertebrate Na^+/H^+ exchanger isoforms. *Am. J. Physiol.* **268,** C283–C296.

O'Donnell, M. E. (1991). Endothelia cell sodium-potassium-chloride cotransport. Evidence of regulation by Ca^{2+} and protein kinase. *C. J. Biol. Chem.* **266(18),** 11559–11566.

O'Donnell, M. E., Martinez, A., and Sun, D. (1995). Endothelial Na-K-Cl cotransport regulation by tonicity and hormones: Phosphorylation of cotransport protein. *Am. J. Physiol.* **269,** C1513–C1523.

Oike, M., Schwarz, G., Sehrer, J., Jost, M., Gerke, V., Weber, K., Droogmans, G., and Nilius, B. (1994). Cytoskeletal modulation of the response to mechanical stimulation in human vascular endothelial cells. *Pfluegers Arch.* **428,** 569–576.

Okada, Y. (1997). Volume expansion-sensing outward-rectifier Cl^- channel: Fresh start to the molecular identity and volume sensor. *Am. J. Physiol.* **273,** C755–C789.

Okada, Y., and Hazama, A. (1989). Volume-regulatory ion channels in epithelial cells. *NIPS* **4,** 238–242.

Olivieri, O., Vitoux, D., Galacteros, F., Bachir, D., Bloquit, Y., Beuzard, Y., and Brugnara, C. (1992). Hemoglobin variants and activity of the (K^+Cl^-) cotransport system in human erythrocytes. *Blood* **79,** 793–797.

Orlowski, J., and Grinstein, S. (1997). Na^+/H^+ exchangers of mammalian cells. *J. Biol. Chem.* **272,** 22373–22376.

Orlowski, J., Kandasamy, R. A., and Shull, G. E. (1992). Molecular cloning of putative members of the Na/H exchanger gene family. cDNA cloning, deduced amino acid sequence, and mRNA tissue expression of the rat Na/H exchanger NHE-1 and two structurally related proteins. *J. Biol. Chem.* **267,** 9331–9339.

Paillard, M. (1997). Na^+/H^+ exchanger subtypes in the renal tubule: Function and regulation in physiology and disease. *Exp. Nephrol.* **5,** 277–284.

Palfrey, H. C. (1994). Protein phosphorylation control in the activity of volume-sensitive transport systems. *In* "Cellular and Molecular Physiology in Cell Volume Regulation" (K. Strange, ed.), pp. 201–214. CRC Press, Boca Raton, Florida.

Palfrey, H. C., and Pewitt, E. B. (1993). The ATP and Mg2+ dependence of Na-K-2Cl cotransport reflects a requirement for protein phosphorylation: Studies using calyculin A. *Pfluegers Arch.* **425,** 321–328.

Parker, J. C. (1993a). Urea alters set point volume for K-Cl cotransport, Na–H proton exchange, and Ca–Na exchange in dog red blood cells. *Am. J. Physiol.* **265,** C447–C452.

Parker, J. C. (1993b). In defense of cell volume? *Am. J. Physiol.* **265,** C1191–C1200.

Parker, J. C. (1994). Coordinated regulation of volume-activated transport pathways. *In* "Cellular and Molecular Physiology of Cell Volume Regulation" (K. Strange, ed.), pp. 311–321. CRC Press, Boca Raton, Florida.

Parker, J. C., Gitelman, H. J., Glosson, P. S., and Leonard, D. L. (1975). Role of calcium in volume regulation by dog red blood cells. *J. Gen. Physiol.* **65,** 84–96.

Parker, J. C., Dunham, P. B., and Minton, A. P. (1995). Effects of ionic strength on the regulation of Na/H exchange and K-Cl cotransport in dog red blood cells. *J. Gen. Physiol.* **105,** 677–699.

Paulais, M., and Turner, R. J. (1992). Activation of the Na^+,K^+,$2Cl^-$ cotransporter in rat parotid acinar cells by aluminum flouride and phosphatase inhibitors. *J. Biol. Chem.* **267.** 21558–21563.

Paulmichl, M., Li, Y., Wickman, K., Ackerman, M., Peralta, E., and Clapham, D. (1992). New mammalian chloride channel identified by expression cloning. *Nature (London)* **356,** 238–241.

Paulmichl, M., Laich, A., Fürst, H., Gschwenter, M., Nagl, U. O., Hittmair, A., and Ritter, M. (1996). Transposition of the chloride channel I_{CLN} from the cytosol into the cell membrane after volume stress. *Biophys. J.* **70,** A9.

Payne, J. A., and Forbush III, B. (1994). Alternatively spliced isoforms of the putative renal Na-K-Cl cotransporter are differentially distributed within the rabbit kidney. *Proc. Natl. Acad. Sci. U.S.A.* **91,** 4544–4548.

Payne, J. A., and Forbush III, B. (1995). Molecular characterization of the epithelial Na-K-Cl cotransporter isoforms. *Curr. Opin. Cell Biol.* **7,** 493–503.

Payne, J. A., Stevenson, T. J., and Donaldson, L. F. (1996). Molecular characterization of a putative KCl cotransporter in rat brain. *J. Biol. Chem.* **271,** 16245–16252.

Payne, J. A., Xu, J.-C., Haas, M., Lytle, C. Y., Ward, D. and Forbush, B., III (1995). Primary structure, functional expression, and chromosomal localization of the bumetanide-sensitive Na-K-Cl cotransporter in human colon. *J. Biol. Chem.* **270,** 17977–17985.

Pearce, M. J., McIntyre, T. M., Prescott, S. M., Zimmerman, G. A., and Whatley, R. E. (1996). Shear stress activates phospholipase A_2 ($cPLA_2$) and MAP kinase in human endothelial cells. *Biochem. Biophys. Res. Commun.* **218,** 500–504.

Pedersen, S. F., Kramhøft, B., Jørgensen, N. K., and Hoffmann, E. K. (1996a). Shrinkage induced activation of the Na^+/H^+ exchanger in Ehrlich ascites tumor cells: Mechanisms involved in the activation and a role for the exchanger in cell volume regulation. *J. Membr. Biol.* **149,** 141–159.

Pedersen, S., Hoffmann, E. K., Hougaard, C., Jørgensen, N. K. Wybrandt, G. B., and Lambert, I. H. (1996b). Leukotriene D_4-induced Ca^{2+} mobilization in Ehrlich ascites tumor cells. *J. Membr. Biol.* **155,** 61–73.

Pedersen, S. F., Prenen, J., Droogmans, G., Hoffmann, E. K., and Nilius, B. (1998a). Separate swelling- and Ca^{2+}-activated anion currents in Ehrlich ascites tumor cells. *J. Membr. Biol.* **163,** 97–110.

Pedersen, S., Pedersen, S. F., Lambert, I. H., Nileus, B., and Hoffmann, E. K. (1999). Mechanical stress-induced release of nucleotides from Ehrlcih ascites tumor cells. Activation of two distinct P2Y2 receptors. *Biochim. Biophys. Acta* **1416,** 271–284.

Pedersen, S. F., Jørgensen,N. K., and Hoffmann, E. K. (1998c). Dynamics of $Ca^{2+}{}_i$ and pH_i in Ehrlich ascites tumor cells after Ca^{2+} mobilizing agonists or hypertonic exposure. *Pfluegers Arch.* **436,** 199–210.

Pedersen, S. F., Pedersen, S., Lambert, I. H., and Hoffmann, E. K. (1998b). P2-receptor-mediated signal transduction in Ehrlich ascites tumor cells. *Biochim. Biophys. Acta.* **1374,** 94–106.

Peppelenbosch, M. P., Qiu, R.-G., deVries-Smits, A. M. M., Tertoolen, L. G. J., deLaat, S. W., McCormick, F., Hall, A., Symons, M. H., and Bos, J. L. (1995). Rac mediates growth factor induced arachidonic acid release. *Cell* **81,** 8490–8546.

Petersen, O. H. (1992). Stimulus-secretion coupling: Cytoplasmic calcium signals and the control of ion channels in exocrine acinar cells. *J. Physiol.* (*London*) **448,** 1–51.

Petrou,S., Ordway, R. W., Hamilton, J. A., Walsh, J. W., Jr., and Singer, J. J. (1994). Structual requirements for draged lipid molecules to directly increase or suppress K^+ channel activity in smooth muscle cells. Effects of fatty acid, lysophosphatidate, acylcoenzyme A and sphingosine. *J. Gen. Physiol.* **103,** 471–486.

Pewitt, E. B., Hegde, R. S., Haas, M., and Palfrey, H. C. (1990). The regulation of Na/K/2Cl cotransport and bumetanide binding in avian erythrocytes by protein phosphorylation and dephosphorylation. *J. Biol. Chem.* **265,** 20747–20756.

Pierce, S. K., and Politis, A. D. (1990). Ca^{2+}-activated cell volume recovery mechanisms. *Annu. Rev. Physiol.* **52,** 27–42.

Pierce, S. K., Politis, A. D., Cronkite, D. H., Rowland, L. M., and Smith, L. H., Jr. (1989). Evidence of calmodulin involvement in cell volume recovery following hypo-osmotic stress. *Cell Calcium* **10,** 159–169.

Piomelli, D. (1996). "A rachidonic Acid in Cell Signalling. Molecular Biology Intelligence Unit." Springer-Verlag, Berlin.

Postma, F. R., Jalink, K., Hengeveld, T., Bot, A., Alblas, J., De Jonge, H., and Moolenar, W. H. (1996). Serum-induced membrane depolarization in quiescent fibroblasts: Activation of a chloride conductance through the G protein-coupled receptor. *EMBO J.* **15,** 63–72.

Prakash, V., Loucheux, C., Scheufele, S., Gorbunoff, M. J., and Timasheff, S. N. (1981). Interactions of proteins with solvent components in 8 *M* urea. *Arch. Biochem. Biophys.* **210,** 455–464.

Ramanadham, S., Gross, R. W. and Turk, J. (1992). Arachidonic acid induces an increase in the cytosolic calcium concentration in single pancreatic islet β-cells. *Biochem. Biophys. Res. Commun.* **184,** 647–653.

Rao, K., Paik, W.-Y., Zheng, L., Jobin, R. M., Tomic, M., Jiang, H., Nakanishi, S., and Stojilkovic, S. S. (1997). Wortmannin-sensitive and-insensitive steps in calcium-controlled exocytosis in pituitary gonadotrophs: Evidence that myosin light chain kinase mediates calcium-dependent and wortmannin-sensitive gonadotropin secretion. *Endocrinology* (*Baltimore*) **138,** 1440–1449.

Rasola, A., Galieta, L. J., Gruenert, D. C., and Romeo, G. (1994). Volume-sensitive chloride currents in four epithelial cell lines are not directly correlated to the expression of the MDR-1 gene. *J. Biol. Chem.* **269,** 1432–1436.

Reagan, J. D. (1995). Molecular cloning of a putative Na^+,K^+,$2Cl^-$ cotransporter from the malpighian tubules of the tobacco hornworm, *Manduca sexta. Insect Biochem. Mol. Biol.* **25,** 875–880.

Ridley, A. J. (1994). Signal transduction through the GTP-binding proteins Rac and Rho. *J. Cell Sci.,* **Suppl. 18,** 127–131.

Rijken, P. J., Hage, W. J., van Bergen en Henegouwen, P. M. P., Verkleij, A. J., and Boonstra, J. (1991). Epidermal growth factor induces rapid reorganization of the actin microfilament system in human A431 cells. *J. Cell Sci.* **100,** 491–499.

Riquelme, G., Sepúlveda, F. V., Jørgensen, F., Pedersen, S., and Hoffmann, E. K. (1998). Swelling-activated potassium currents of Ehrlich ascites tumor cells. *Biochim. Biophys. Acta* **1371,** 101–106.

Robertson, M. A., and Foskett, J. K. (1994). Na^+ transport pathways in secretory acinar cells: Membrane cross talk mediated by $[Cl^-]_i$. *Am. J. Physiol.* **267,** C146–C156.

Rosenmund, C., and Westbrook, G. L. (1993). Calcium-induced actin depolymerization reduces NMDA channel activity. *Neuron* **10,** 805–814.

Ross, P. E., and Cahalan, M. D. (1995). Ca^{2+} influx pathways mediated by swelling or stores depletion in mouse thymocytes. *J. Gen. Physiol.* **106,** 415–444.

Roti-Roti, L. W., and Rothstein, A. (1973). Adaption of mouse leukemic cells (L51784) to anisosmotic media. *Exp. Cell Res.* **79,** 295–310.

Rotin, D., Mason, M. J., and Grinstein, S. (1991). Channels, antiports, and regulation of cell volume in lymphoid cells. *In* "Advances in Comparative and Enviromental Physiology" (R. Gilles, E. K. Hoffmann, and L. Bolis, eds.), Vol. 9, Chap. 5, pp. 118–139. Springer-Verlag, Berlin and Heidelberg.

Ruhfus, B., Tinel, H., and Kinne, R. K. (1996). Role of G-proteins in the regulation of organic osmolyte efflux from isolated rat renal inner medullary collecting duct cells. *Pflugers-Arch.* **433**(1–2), 35–41.

Sachs, J. R. (1996). Cell volume regulation. *In* "Molecular Biology of Membrane Transport Disorders" (Schultz, S. G., *et al.,* eds.) pp. 379–406. Plenum, New York.

Sadoshima, J., Qui, Z., Morgan, J. P., and Izumo, S. (1996). Tyrosine kinase activation is an immediate and essential step in hypotonic swelling-induced ERK activation and c-fos gene expression in cardiac mysocytes. *EMBO J.* **15,** 5535–5546.

Saltin, B., Sjögaard, G., Strange, S., and Juel, C. (1987). Redistribution of K^+ in the human body during muscular exercise; its role to maintain whole body homeostasis. *In* "Man in Stressful Environments. Thermal and Work Physiology" (K. Shiraki and M. K. Yousef, eds.), pp. 247–267. Thomas, Springfield, Illinois.

Sandford, C., Sweiry, J. H., and Jenkinson, D. H. (1992). Properties of cell volume-sensitive potassium conductance in isolated guinea-pig and rat hepatocytes. *J. Physiol.* (*London*) **447,** 133–148.

Sanchez-Olea, R., Fuller, C., Benos, D., and Pasantes-Morales, H. (1995). Volume-associated osmolyte fluxes in cell lines with and without the anion exchanger. *Am. J. Physiol.* **269,** C1280–C1286.

Sardet, C., Franchi, A., and Pousségur, J. (1989). Molecular cloning, primary structure and expression of the human growth factor-activatable Na^+/H^+ antiporter. *Cell* (*Cambridge, Mass.*) **56,** 271–280.

Sardini, A., Sepulveda, F. V., and McNaughton, P. A. (1995). Intracellular calcium increase in response to swelling in HeLa cells. *J. Physiol.* (*London*) **482,** 11P.

Sarkadi, B., and Parker, J. C. (1991). Activation of ion transport pathways by changes in cell volume. *Biochim. Biophys. Acta.* **1071,** 407–427.

Sarkadi, B., Mack, E., and Rothstein, A. (1984a). Ionic events during the volume response of human peripheral blood lymphocytes to hypotonic media. I. Distinctions between volume-activated Cl^- and K^+ conductance pathways. *J. Gen. Physiol.* **83,** 497–512.

Sarkadi, B., Mack, E., and Rothstein, A. (1984b). Ionic events during the volume response of human peripheral blood lymphocytes to hypotonic media. II. Volume- and time-dependent activation and inactivation of ion transport pathways. *J. Gen. Physiol.* **83,** 513–527.

Sarkadi, B., Cheung, R., Mack, E., Grinstein, S., Gelfand, E. W., and Rothstein, A. (1985). Cation and anion transport pathways in volume regulatory response of human lymphocytes to hyposmotic media. *Am. J. Physiol.* **248,** C480–C487.

Sarndahl, E., Bokoch, G. M., Stendahl, O., and Andersson, T. (1993). Stimulus-induced association of α subunits of heterotrimeric GTP-binding proteins from the cytoskeleton of human neutrophils. *Proc. Natl. Acad. Sci., U.S.A.* **90,** 6552–6556.

Schafer, D. A., and Cooper, J. A. (1995). Control of actin filament assembly. *Annu. Rev. Cell Dev. Biol.* **11,** 497–518

Schlichter, L. C., and Sakellaropoulos, G. (1994). Intracellular Ca^{2+} signaling induced by osmotic shock in human T lymphocytes. *Exp. Cell Res.* **215,** 211–222.

Schliess, F., Schreiber, R., and Häussinger, D. (1995). Activation of extracellular signal-regulated kinases Erk-1 and Erk-2 by cell swelling in H4IIE hepatoma cells. *Biochem. J.* **309,** 13–17.

Schultz, S. G. (1989a). Intracellular sodium activities and basolateral membrane potassium conductances of sodium-absorbing epithelial cells. *In* "Cellular and Molecular Biology of Sodium Transport". (S. G. Schultz, ed.), Curr. Top. Membr. Vol. 34, pp. 21–44. Academic Press, San Diego.

Schultz, S. G. (1989b). Volume preservation. Then and now. *News Physiol. Sci.* **4,** 169–172.

Schultz, S. G. (1992). Membrane cross-talk in sodium-absorbing epithelial cells. *In* "The Kidney: Physiology and Pathophysiology (2nd ed.)," (D. W. Seldin and G. Giebisch, eds.), Vol. 1, pp. 287–299. Raven Press, New York.

Schumacher, P. A., Sakellaropoulos, G., Phipps, D. J., and Schlichter, L. C. (1995). Small-conductance chloride channels in human peripheral T lymphocytes. *J. Membr. Biol.* **145,** 217–232.

Schwartz, M. A. (1993). Spreading of human endothelial cells on fibronectin or vitronectin triggers elevation of intracellular free calcium. *J. Cell Biol.* **120** (4), 1003–1010.

Schwartz, M. A., Both, G., and Lechene, C. (1989). Effect of cell spreading on cytoplasmic pH in normal and transformed fibroblasts. *Proc. Natl. Acad. Sci. U.S.A.* **86,** 4525–4529.

Schwartz, M. A., Cragoe, E. J., and Lechene, C. P. (1990). pH regulation in spread cells and round cells. *J. Biol. Chem.* **265,** 1327–1332.

Schwiebert, E. M., Mills, J. W., and Stanton, B. A. (1994). Actin-based cytoskeleton regulates a chloride channel and cell volume in a renal cortical collecting duct cell line. *J. Biol. Chem.* **269,** 7081–7089

Shapiro, M., Matthews, J., Hecht, G., Delp, C., and Madara, J. l. (1991). Stabilization of F-actin prevents cAMP-elicited Cl^- secretion in T84 cells. *J. Clin. Invest.* **87,** 1903–1909.

Shattil, S. J., Ginsberg, M. H., and Brugge, J. S. (1994). Adhesive signaling in platelets. *Curr. Opin. Cell Biol.* **6,** 695–704.

Sheikh-Hamad, D., Di Mari, J., Suki, W. D., Safirstein, R., Watts III, B. A., and Rouse, D. (1998). p38 Activity is essential for osmotic induction of HSP70 and transporter for organic solute betaine in Madin-Darby canine kidney cells. *J. Biol. Chem.* **273,** 1832–1837.

Shennan, D. B., McNeillie, S. A., and Curran, D. E. (1994). The effect of a hyposmotic shock on amino acid efflux from lactating rat mammary tissue: Stimulation of taurine and glycine efflux via a pathway distinct from anion exchange and volume-activated anion channels. *Exp. Physiol.* **79,** 797–808.

Shorofsky, S. R., Field, M., and Fozzard, H. A. (1984). Mechanism of Cl secretion in canine trachea: Changes in intracellular chloride activity with secretion. *J. Membr. Biol.* **81,** 1–8.

Shrode, L. D., Klein, J. D., O'Neill, W. C., and Putnam, R. W. (1995). Shrinkage-induced activation of Na^+/H^+ exchange in primary rat astrocytes: Role of myosin light chain kinase. *Am. J. Physiol.* **269,** C257–C266.

Siman, R., Baudry, M., and Lynch, G. (1985). Regulation of glutamate receptor binding by the cytoskeletal protein fodrin. *Nature* (*London*) **313,** 225–228.

Sohn, R. H., and Goldschmidt-Clermont, P. J. (1994). Profilin: At the crossroads of signal transduction and the actin cytoskeleton. *BioEssays* **16,** 465–472.

Somero, G. N. (1986). Protons, osmolytes and fitness of internal milieu for protein function. *Am. J. Physiol.* **251,** R197–R213.

Spring, K. R., and Hoffmann, E. K. (1992). Cell volume control. *In* "The Kidney: Physiology and Pathophysiology". (D. W. Seldin and G. Giebisch, eds.), pp. 147–169. Raven, New York.

Starke, L. C., and Jennings, M. L. (1993). K-Cl cotransport in rabbit red cells: Further evidence for regulation by protein phosphatase type 1. *Am. J. Physiol.* **264,** C118–C124.

Stein, W. D. (1986). Intrinsic, apparent, and effective affinities of co- and countertransport systems. *Am. J. Physiol.* **250,** C523–C533.

Stoddard, J. S., Steinbach, J. H., and Simchowitz, L. (1993). Whole cell Cl^- current in human neutrophils induced by cell swelling. *Am. J. Physiol.* **265,** C156–C165.

Strange, K. (1994a). "Cellular and Molecular Physiology of Cell Volume Regulation." CRC Press, Boca Raton, Florida.

Strange, K. (1994b). Are all cell volume changes the same? *News Physiol. Sci.* **9,** 223–228.

Strange, K., and Jackson, P. S. (1995). Swelling-activated organic osmolyte efflux: A new role for anion channels. *Kidney Int.* **48,** 994–1003.

Strange, K., Morrison, R., Shrode, L., and Putnam, R. (1993). Mechanism and regulation of swelling-activated inositol efflux in brain glial cells. *Am. J. Physiol.* **265,** C244–C256.

Strange, K., Emma, F., and Jackson, P. S. (1996). Cellular and molecular physiology of volume-sensitive anion channels. *Am. J. Physiol.* **270,** C711–C730.

Strøbaek, D., Christophersen, P., Dissing, S., and Olesen, S.-P. (1996). ATP activates K and Cl channels via purinoreceptor-mediated release of Ca^{2+} in human coronary artery smooth muscle. *Am. J. Physiol.* **271,** C1463–C1471.

Stutzin, A., Eguiguren, A. L., Cid, L. P., and Sepulveda, F. V. (1997). Modulation by extracellular Cl^- of volume-activated organic osmolyte and halide permeabilities in HeLa cells. *Am. J. Physiol.* **273,** C999–C1007.

Sun, H.-Q., Kwiatkowska, K., and Yin, H. L. (1995). Actin monomer binding proteins. *Curr. Opin. Cell Biol.* **7,** 102–110.

Suvitayavat, W., Palfrey, H. C., Haas, P. B., Dunham, P. B., Kalmar, F., and Rao, M. C. (1994). Characterization of the endogeneous Na^+-K^+-$2Cl^-$ cotransporter in *Xenopus* oocytes. *Am. J. Physiol.* **266,** C284–C292

Suzuki, M., Kawahara, K., Ogawa, A., Morita, T., Kawaguchi, Y., Kurihara, S., and Sakai, O. (1990). $[Ca^{2+}]_i$ rises via G protein during regulatory volume decrease in rabbit proximal tubule cells. *Am. J. Physiol.* **258,** F690–696.

Szücs, G., Heinke, S., Droogmans, G., and Nilius, B. (1996). Activation of the volume sensitive chloride current in vascular endothelial cells requires a permissive intracellular Ca^{2+} concentration. *Pfluegers Arch.* **431,** 467–469

Takeda, M., Homma, T., Breyer, M. D., Horiba, N., Hoover, R. L, Kawamoto, S., Ichikawa, I., and Kon, V. (1993). Volume- and agonist-induced regulation of myosin-light chain phosphorylation in glomerular mesangial cells. *Am. J. Physiol.* **264,** F421–F426.

Tanimura, A., Kurihara, K., Reshkin, S. J., and Turner, R. J. (1995). Involvement of direct phosphorylation in the regulation of the rat parotid $Na^+,K^+,2Cl^-$ cotransporter. *J. Biol. Chem.* **270,** 25252–25258.

Terzic, A., and Kurachi, Y. (1996). Actin microfilament disrupters enhance K_{ATP} channel opening in patches from guinea-pig cardiomyocytes. *J. Physiol. (London)* **492,** 395–404.

Thiemann, A., Grunder, S., Pusch, M., and Jentsch, T. J. (1992). A chloride channel widely expressed in epithelial and non-epithelial cells. *Nature (London)* **356,** 57–60.

Thomas-Young, R. J., Smith, T. C., and Levinson, C. (1993). Regulatory volume decrease in Ehrlich ascites tumor cells is not mediated by a rise in intracellular calcium. *Biochim. Biophys. Acta* **1146,** 81–86.

Thoroed, S. M., and Fugelli, K. (1994a). The Na^+-independent taurine influx in flounder erythrocytes and its association with the volume regulatory taurine efflux. *J. Exp. Biol.* **186,** 245–268.

Thoroed, S. M., and Fugelli, K. (1994b). Free amino compounds and cell volume regulation in erythrocytes from different marine fish species under hypoosmotic conditions: The role of a taurine channel. *J. Comp. Physiol. B* **164,** 1–10.

Thoroed, S. M., Lauritzen, L., Lambert, I., Hansen, H. S., and Hoffmann, E. K. (1997). Cell swelling activates phospholipasse A_2 in Ehrlich ascites tumor cells. *J. Membr. Biol.* **160,** 47–58.

Tilly, B. C., van den Berghe, N., Tertoolen, L. G. J., Edixhoven, M. J., and de Jonge, H. R. (1993). Protein tyrosine phosporylation is involved in osmo regulation of ionic conductances. *J. Biol. Chem.,* **268,** 19919–19922.

Tilly, B. C., Edixhoven, M. J., van den Berghe, N., Bot, A. G. M., and de Jonge, H. R. (1994). Ca^{2+} -mobilizing hormones potentiate hypotonicity-induced activation of ionic conductances in intestine 407 cells. *Am. J. Physiol.,* **267,** C1271–C1278.

Tilly, B. C., Edixhoven, M. J., Tertoolen, L. G. J., Morii, N., Saitoh, Y., Narumiya, S., and deJonge, H. R. (1996a). Activation of the osmosensitive chloride conductance involves p21rho and is accompanied by a transient reorganization of the F-actin cytoskeleton. *Mol. Biol. Cell* **7,** 1419–1427.

Tilly, B. C., Gaestel, M., Engel, K., Edixhoven, M. J., and Jorge, H. R. D. (1996b). Hypo-osmotic cell swelling activates the p38 MAP kinase signalling cascade. *FEBS Lett.* **395,** 133–136.

Tinel, H., Wehner, F., and Kinne, R. K. H. (1997). Arachidonic acid as a second messenger for hypotonicity-induced calcium transients in IMDC cells. *Pfluegers Arch.* **433,** 245–253.

Tokiwa, G., Dikic, I., Lev, S., and Schlessinger, J. (1996). Activation of Pyk2 by stress signals and coupling with JNK signaling pathway. *Science* **273,** 792–794.

Torchia, J., Lytle, D. J., Pon, D. J., Forbush III, B., and Sen, A. K. (1992). The Na-K-Cl cotransporter of avian salt gland. Phosphorylation in response to cAMP-dependent and calcium-dependent secretagogues. *J. Biol. Chem.* **267,** 25444–25450.

Tosteson, D. C. (1964). Regulation of cell volume by sodium and potassium transport . *In* "The Cellular Functions of Membrane Transport" (J. F. Hoffmann and J. F. Prentice, eds.), pp. 3–23. Hall, New York.

Toyomoto, T., Knutsen, D., Soos, G., and Sato, K. (1997). Na-K-2Cl cotransporters are present and regulated in simian eccrine clear cells. *Am. J. Physiol.* **273,** R270–R277.

Treharne, K. J., Marshall, L. J., and Mehta, A. (1994). A novel chloride-dependent GTP-utilizing protein kinase in plasma membranes from human respiratory epithelium. *Am. J. Physiol.* **267,** L592–L601.

Trifaro, J.-M., Del castillo, A. R., and Vitale, M. L. (1992). Dynamic changes in chromaffin cell cytoskeleton as prelude to exocytosis. *Mol. Neurobiol.* **6,** 339–358.

Urbach,V., Ó'Cróinín, D., and Harvey, B. J. (1995). Mechanosensitive calcium entry in renal A6 cells. *J. Physiol.* (*London*) **482,** 21P.

Ussing, H. H. (1982). Pathways for transport in epithelia. *In* "Functional Regulation at the Cellular and Molecular Levels" (R. A. Corradino, ed.), pp. 285–297. Elsevier/North-Holland, Amsterdam.

Valverde, M. A., Diaz, M., Sepulveda, F. V., Gill, D. R., Hyde, S. C., and Higgins, C. F. (1992). Volume-regulated chloride channel associated with the human multidrug resistant P-glycoprotein. *Nature* (*London*) **355,** 830–833.

Valverde, M. A., Mintenig, G. M., and Sepúlveda, F. V. (1993). Differential effects of tamoxifen and I^- on three distinguishable chloride currents activated in T84 cells. *Pfluegers Arch.* **425,** 552–554.

Valverde, M. A., Hardy, S. P., and Sepúlveda, F. V. (1995). Chloride channels: A state of flux. *FASEB J.* **9,** 509–515.

Valverde, M. A., Bond, T. D., Hardy, S. P., Taylor, J. C., Higgins, C. F., Altamirano, J., and Alvarez-Leefmans, F. J. (1996). The multidrug-resistance-P-glycoprotein modulates cell volume regulatory decrease. *EMBO J.* **15,** 4460–4468.

Vanoye, C. G., Altenberg, G. A., and Reuss, L. (1997). P-glycoprotein is not a swelling-activated Cl^- channel: Possible role as a Cl^- channel regulator. *J. Physiol.* (*London*) **502,** 249–258.

Varming, C., Pedersen, S. F., and Hoffmann, E. K. (1998). Protein kinases, phosphatases, and the regulation of the Na^+/H^+exchanger in Ehrlich ascites tumor cells. *Acta Physiol. Scand.* **163,** A24.

Verkman, A. S., Van Hoek, A. N., Ma, T., Frigeri, A., Skach, W. R., Mitra, A., Tamarappoo, B. K., and Farinas, J. (1996). Water transport across mammalian cell membranes. *Am. J. Physiol.* **270,** C12–C30.

Viel, A., and Branton, D. (1996). Spectrin: On the path from structure to function. *Curr. Opin. Cell Biol.* **8,** 49–55

Vigne, P., Farré, A. L., and Frelin, C. (1994). The $Na^+,K^+,2Cl^-$ cotransporter of brain capillary endothelial cells. Properties and regulation by endothelins, hyperosmolar solutions, calyculin A and interleukin. *J. Biol. Chem.* **269,** 19925–19930.

Vitale, M. L., Del Castillo, A. R., Tchakarov, L., and Trifaro, J.-M. (1991). Cortical filamentous actin disassembly and scinderin redistribution during chromaffin cell stimulation precede exocytosis, a phenomenon not exhibited by gelsolin. *J. Cell Biol.,* **113,** 1057–1067.

Voets, T., Buyse, G., Tytgat, J., Droogmans, G., Eggermont, J., and Nilius, B. (1996). The chloride current induced by expression of the protein pI_{Cln} in Xenopus oocytes differs from the endogenous volume-sensitive chloride current. *J. Physiol.* (*London*) **495.2,** 441–447.

Vom Dahl, S., Hallbrucker, C., Lang, F., and Häussinger, D. (1991). Regulation of liver cell volume and proteolysis by glucagon and insulin. *Biochem. J.* **278,** 771–777.

Wade, R. H., and Hyman, A. A. (1997). Microtubule structure and dynamics. *Curr. Opin. Cell Biol.* **9,** 12–17.

Wakabayashi, S., Bertrand, B., Ikeda, T., and Pousségur, J. (1994). Mutation of calmodulin-binding site renders the Na^+/H^+ exchanger (NHE-1) highly H^+ sensitive and Ca^{2+} regulation-defective. *J. Biol. Chem.* **269,** 13710–13715.

Wakabayashi, S., Shigekawa, M., and Pousségur, J. (1997). Molecular physiology of vertebrate Na^+/H^+ exchangers. *Physiol. Rev.* **77,** 51–74.

Wang, Z., Orlowski J., and Shull G. E. (1993). Primary structure and functional expression of a novel gastrointestinal isoform of the rat Na^+/H^+ exchanger. *J. Biol. Chem.* **268,** 11925–11928.

Wang, Y., Roman, R., Lidofsky, S. D., and Fitz, J. G. (1996). Autocrine signalling through ATP release represents a novel mechanism for cell volume regulation. *Proc. Natl. Acad. Sci. U.S.A.* **93,** 12020–12025

Watson, A. J. M., Levine, S., Donowitz, M., and Montrose, M. H. (1992). Serum regulates Na^+/H^+ exchange in Caco-2 cells by a mechanism which is dependent on F-actin. *J. Biol. Chem.* **267,** 956–962.

Wehner, F., and Tinel, H. (1998). Role of Na^+ conductance, Na^+/H^+ exchange, and Na^+, K^+, $2Cl^-$ symport in the regulatory volume increase of rat hepatocytes. *J. Physiol.* (*London*) **506,** 127–142.

Whatley, V. J., and Harris, R. A. (1996). The cytoskeleton and neurotransmitter receptors. *Int. Rev. Neurobiol.* **39,** 113–143.

Whisenant, N., Khademazad, M., and Muallem, S. (1993). Regulatory interaction of ATP, Na^+, and Cl^- in the turnover cycle of the NaK2Cl cotransporter. *J. Gen. Physiol.* **101,** 889–908.

Winzov, D. J., Paleg, L. G., Jones, G. P., and Naidu, B. P. (1992). Rationalization of the effects of compatible solutes on protein stability in terms of thermodynamic nonideality. *Arch. Biochem. Biophys.* **296,** 102–107.

Wöll, E., Ritter, M., Scholz, W., Häussinger, D., and Lang, F. (1993). The role of calcium in cell shrinkage and intracellular alkalinization by bradykinin in Ha-Ras oncogene expressing cells. *FEBS Lett.* **322,** 261–265.

Worrell, R. T., Butt, A. G., Cliff, W. H., and Frizzell, R. A. (1989). A volume-sensitive chloride conductance in human colonic cell line T84. *Amer. J. Physiol.* **256**(6 part 1), C1111–C1119.

Xu, J.-C., Lytle, C., Zhu, T. T., Payne, J. A., Benz, E., Jr., and Forbush III, B. (1994). Molecular characterization and functional expression of the bumetanide-sensitive Na-K-Cl cotransporter. *Proc. Natl. Acad. Sci. U.S.A.* **91,** 2201–2205.

Xu, W. X., Kim, S. J., So, I., and Kim, K. W. (1997). Role of actin microfilament in osmotic stretch-induced increase of voltage-operated calcium channnel current in guinea-pig gastric myocytes. *Pfluegers Arch.,* **434,** 502–504.

Yamada, K. M., and Miyamoto, S. (1995). Integrin transmembrane signaling and cytoskeletal control. *Curr. Opin. Cell Biol.* **7,** 681–689.

Yamazaki, J., Duan, D., Janiak, R., Kuenzli, K., Horowitz, B., and Hume, J. R. (1998). Functional and molecular expression of volume-regulated chloride channels in canine vascular smooth muscle cells. *J. Physiol. (London)* **507,** 729–736.

Yancey, P. H. (1994). Compatible and counteracting solutes. *In* "Cellular and Molecular Physiology of Cell Volume Regulation" (K. Strange, ed.), pp. 81–109. CRC Press, Boca Raton, Florida.

Yokoshiki, H., Katsube, Y., Sunagawa, M., Seki, T., and Sperelakis, N. (1997). Disruption of actin cytoskeleton attenuates sulphonylurea inhibition of cardiac ATP-sensitive K^+ channels. *Pfluegers Arch. (Eur. J. Physiol.)* **434,** 203–205.

Zhang, J. J., and Jacob, T. J. C. (1997). Three different Cl^- channels in bovine ciliary epithelium activated by hypotonic stress. *J. Physiol. (London)* **499,** 379–389.

Zhang, J., Larsen, T. H., and Lieberman, M. (1997). F-actin modulates swelling-activated chloride current in cultured chick cardiac myocytes. *Am. J. Physiol.* **273,** C1215–C1224.

Zimmerman, S. B., and Minton, A. P. (1993). Macromolecular crowding: Biochemical, biophysical, and physiological consequences. *Annu. Rev. Biophys. Biomol. Struct.* **22,** 27–65.

Ziyadeh, F. N., Mills, J. W., and Kleinzeller, A. (1992). Hypotonicity and cell volume regulation in shark rectal gland: Role of organic osmolytes and F-actin. *Am. J. Physiol.* **262,** F468–F480.

CHAPTER 7

Interaction of Natural and Model Peptides with Membranes

David S. Cafiso
Department of Chemistry and Biophysics Program, University of Virginia
Charlottesville, Virginia 22901

I. INTRODUCTION

The study of membrane–peptide interactions can elucidate a wide range of biologically relevant questions. For example, a number of naturally occurring peptides that interact strongly with membranes are known to have important antibiotic activities. Some of these peptides promote ion channel activity or bilayer destabilization, and information on the peptide structures and the interactions they make with membranes provides insight into the mechanism of their antibiotic activity. Membrane–peptide interactions also provide information on critical protein–membrane electrostatic interactions. Highly positively charged peptides interact strongly with acidic

1063-5823/99 $30.00

lipid surfaces, and these same interactions bind water-soluble proteins that are involved in cell-signaling pathways to membrane interfaces. The association of highly charged peptides with membrane surfaces also facilitates the formation of lateral domains of acidic lipids, a process that may be important in the timing of cell-signaling events.

High-resolution structural techniques that have worked well for water-soluble proteins frequently fail when applied to membrane proteins; as a result, membrane protein structure and molecular function presently represent one of the greatest challenges in the field of structural biology. Studies aimed at the investigation of peptide–membrane interactions are central to understanding the structure and molecular function of membrane proteins. Work on peptide–membrane interactions provides information on the free energy contributions to the association of protein domains with membranes. A number of proteins, such as the bacterial toxin colicin (Cramer *et al.,* 1995), insert spontaneously into membranes and are driven by the free energy contributions to their binding. In addition, these energies are likely to be important in determining the association of signal sequences and other protein domains with membranes.

Peptide–membrane interactions are clearly important to characterize, but these interactions, as well as the behavior of bilayers, are influenced strongly by a large number of relatively weak intermolecular forces. As a result, developing a molecular understanding of biological membranes and membrane–protein function has been quite challenging. Only recently have computational approaches begun to elaborate the complexity of peptide–membrane interactions, and a number of these approaches have had some success in reproducing experimental data.

This chapter discusses recent work that is aimed at investigating the energetics of peptide–membrane interactions, focusing in particular on interactions that are mediated by electrostatic interactions between peptides and the membrane interface. It also discusses factors that determine the position and conformation of peptides on membrane surfaces and briefly discusses several classes of membrane-active peptides. There is vast literature detailing the fascinating biology and phenomenology of peptide–membrane interactions. Because of the limited scope of this chapter, the reader will be referred to several more comprehensive review articles detailing this work.

II. FREE ENERGY OF PEPTIDE BINDING TO MEMBRANES

A. The Hydrophobic Effect and Peptide Solvation

There has been considerable interest in understanding the factors that contribute to the free energy of binding peptides to membranes. A number of experimental measurements of the free energy required to move an amino acid side chain from water to a hydrocarbon *are used in* estimates of protein stability, as well as to predict transmembrane topology based on the protein primary sequence. One important component in these prediction methods is the use of a hydrophobicity scale, which allows the estimate of the partition-free energy for an amino acid segment sufficiently long to traverse the bilayer as an α-helix (see reviews by Eisenberg, 1984; Engleman *et al.,* 1986). There has been considerable discussion over the types of hydrophobicity scale that should be used in the prediction of protein topology, and many of these are based on experimental studies of the partitioning of amino acids between bulk phases as well as statistical analysis of the location of amino acids within proteins.

In order to describe the free energy contributions to membrane binding made by a peptide sequence, the contributions that specific amino acids make to the partitioning have been evaluated. Many measurements have been made of the partition coefficients for amino acids (e.g., acetyl amino acid amides, Ac-X-amide) from water into less polar solvents (Eisenberg and McLachlan; 1986, Shirley *et al.,* 1992) such as octanol. A general observation that emerges from these types of measurements is that the measured hydrophobic free energy is proportional to the total surface area in contact with water. The solvation parameter, C_s, is used frequently to describe the proportionality so that $\Delta G_t = C_s A$, where A is the solvent-accessible surface area of the group being considered. Typically, values for C_s are about 24 cal mol^{-1}Å^{-2} for the free energy of transfer per solvent-accessible surface area. This is an important quantity in estimating the free energy of binding of peptides to membranes and in determining the energetics of protein folding.

There has been considerable debate over the evaluation of transfer-free energies in terms of molecule partition coefficients (White and Wimley, 1994). Several articles suggest that Flory–Huggins-corrected volume fraction units may be more appropriate than mole fraction partition coefficients (De Young and Dill, 1990; Sharp *et al.,* 1991). The assumption that the molecular volumes of solvent and solute are the same is implicit in the use of mole fraction partition coefficients; as a result, the free energy to transfer a solute from water into other solvents yields an unrealistically high energy cost on a mole fraction basis. As a result, the value of C_s reported earlier based on volume-corrected units yields a value of about 42 cal mol^{-1}Å^{-2}. The general use of Flory–Huggins-corrected volume units is controversial, in part because the theory can not be tested directly, and because relatively small corrections should result for most partitioning measurements (Holtzer, 1992, 1994).

More elaborate measurements of amino acid partitioning have been made in order to estimate the contribution that an amino acid makes in the context of a peptide structure, as well as to approximate the environments that proteins encounter in a lipid bilayer. For example, using neutron diffraction, Jacobs and White (1989) measured the partitioning of tripeptides from water into lipid bilayers and investigated the positions of these tripeptides. This work demonstrated that the absorption of peptides to interfaces is very favorable, particularly for peptides containing aromatic amino acids, and an interfacial hydrophobicity scale to evaluate the interaction and folding of proteins within membranes was developed. To extend the examination of hydrophobicity in the context of a larger peptide, the partitioning from water to either octanol or bilayers of host–guest pentapeptides was examined, where the guest position included each of the 20 natural amino acids (Wimley *et al.,* 1996, Wimley and White, 1996).

In contrast to the measurements made on acyl amino acid amides, measurements on pentapeptides yielded solvation energies for the uncharged amino acids that were largely positive, a result that may have reflected differences due to peptide conformation, hydrogen bonding, or the presence of flanking amino acids. Partitioning of tripeptides and pentapeptides between water and membrane yielded transfer-free energies for amino acids that were similar to each other but about one-half that found for these peptides partitioning to octanol (Wimley and White, 1996). It has been suggested that this is largely a result of differences between octanol and the complex interfacial membrane region where these peptides bind.

A different approach was employed by Thorgeirsson *et al.* (1996) who used a spin-labeled 25 amino acid peptide derived from the cytochrome c oxidase subunit IV signal sequence (COX IV) and produced 14 different substitutions of uncharged amino acids within this sequence. The COX IV peptide appears to be in an extended structure, bound within the membrane interior in the presence of membranes containing acidic lipids (Yu *et al.,* 1994). These peptides were spin labeled so that the membrane–aqueous partitioning could be determined directly from the electron paramagnetic resonance (EPR) spectrum of the peptide and the partitioning of the 14 different peptides to negatively charged membranes containing palmitogloleoylphosphatidylcholine and palmitoyloleoylphosphatidylglycerol. These data produced transfer-free energies in agreement with previous partitioning data based on acetyl amino acid amides, but in contrast with data obtained using the host–guest pentapeptides described earlier. The differences between these two sets of measurements is not entirely clear, but a number of possibilities have been suggested, including differences in their localization within the bilayer or the presence of electrostatic interactions with acidic lipids (Wimley and White, 1996).

The partitioning of hydrophobic or amphipathic molecules from a water phase into a bulk hydrocarbon phase is often associated with a large increase in enthalpy, usually attributed to changes in water enthalpy (Tanford, 1978).

This classical hydrophobic interaction is, however, often absent when a hydrophobic or amphipathic molecule is in equilibrium between water and a bilayer phase (Huang and Charlton, 1972). For example, the partitioning of *N*-methylindole (a tryptophan side chain analog) from water to cyclohexane yields a ΔG, ΔH, and ΔS of -7.2 kcal/mole, 1.1 kcal/mole, and 27.7 cal $mol^{-1}K^{-1}$, respectively, whereas its partitioning from water to a bilayer yields values of 8.1 kcal/mole, -3.4 kcal/mole, and 15.5 cal $mol^{-1}K^{-1}$ for ΔG, ΔH, and ΔS, respectively (Wimley and White, 1992, 1993). Instead of entropy-driven partitioning, this enthalpy has been referred to as a nonclassical hydrophobic effect (Seelig and Ganz, 1991). Although it has been suggested that the classical hydrophobic effect may not be operative in these cases, evaluation of the heat capacity changes associated with the membrane binding of nonpolar solutes indicates that the hydrophobic effect is operative (Wimley and White, 1993). It is important to note that the assembly of bilayers is the result of large enthalpy and entropy contributions. As a result, perturbations of the bilayer due to the binding of peptides or other solutes may lead to changes in energy or entropy that are not seen in bulk phases.

B. Backbone Electrostatic Effects

The hydrophobicity of the amino acid side chains is only one component in determining the water–membrane partition coefficient of a peptide, and making an estimate of the free energy of binding for a particular sequence to lipid bilayers can be quite challenging. For a simple 20 amino acid polymer of alanine in an α-helical conformation, the free energy of partitioning can be estimated (based primarily on the hydrophobic free energy contribution made by alanine side chains). This value is found to be approximately -30 kcal/mole (Engleman *et al.,* 1986) and is significantly more negative than experimentally determined values, which are approximately -5 kcal/mol (Moll and Thompson, 1994). Two additional terms can be included in this estimate: one due to the entropy loss experienced by the peptide in binding to the membrane and a second due to a reduction in the conformational entropy experienced by the lipid acyl chains on the insertion of a rigid peptide (Jahnig, 1983). For a 20-mer of polyalanine, these numbers yield values that are on the order of -20 kcal/mole (Moll and Thompson, 1994), but still -15 kcal/mole more negative than the experimental value.

It has been argued that the discrepancy between these predictions, which yield a binding free energy far too negative, and experimental data is largely a result of underestimating the cost of inserting a hydrogen bond into a low dielectric medium (Ben-Tal *et al.,* 1996a). A determination of the electrostatic energy penalty resulting from the insertion of a peptide hydrogen bond into a low dielectric medium was made using finite-difference solutions to the Poisson–Boltzmann equation. It was determined that the cost for inserting a C=O ••• H–N pair into the membrane is 2.1 kcal/mole, and the combination of electrostatic, immobilization, and hydrophobic interaction terms was found to lead to estimates for peptide binding that were in close agreement with experimental-binding data.

For the simple polyalanine just described, two insertion modes into the membrane were examined for the α-helical structure: one parallel and another perpendicular to the membrane interface (see Fig. 1). This calculation utilized a simple hydrocarbon slab for the bilayer and did not include complexity that would be associated with the bilayer interface (Ben-Tal *et al.,* 1996a). Calculations suggest that even quite hydrophobic peptides should bind parallel to the membrane interface before inserting and aligning along the bilayer normal. As a peptide is *first* inserted into a bilayer, the energetics initially favor binding of the peptide in an orientation parallel

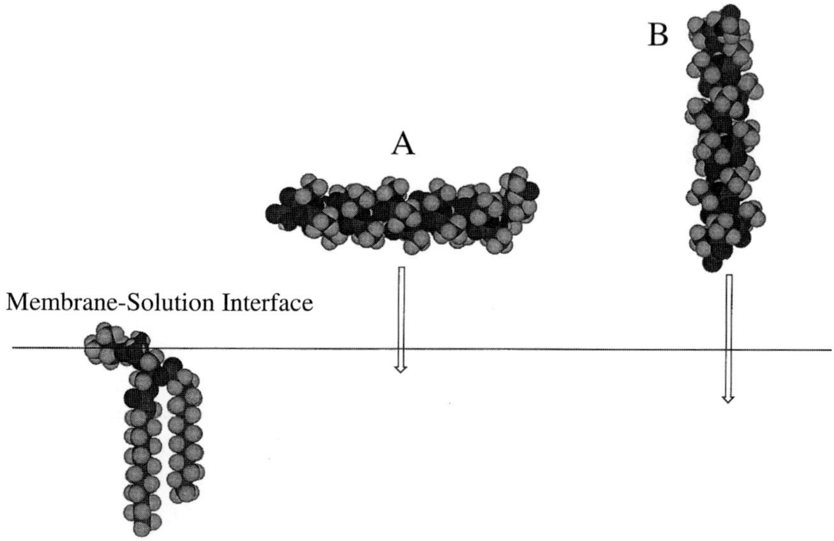

FIGURE 1 Two modes (**A,** horizontal, and **B,** vertical) of membrane insertion for a membrane-spanning alanine-containing helix (after Ben-Tal *et al.,* 1996a). Detailed calculations of the energy terms that control binding indicate that the interaction of capped helices is preferred and that the energetically favored insertion mode is the horizontal insertion indicated in A.

to the membrane interface (Fig. 1A). The perpendicular insertion requires the insertion of a helix end with the loss of solvent hydrogen bonding and no significant gain in hydrophobic (solvent exclusion) free energy. An initial parallel insertion of the peptide maximizes the hydrophobic contact that is needed to overcome the unfavorable insertion of the helix ends. This study also demonstrates that there are significant differences in energy for the insertion of capped and uncapped helices in either orientation.

In many cases, peptides do not assume a secondary structure in solution, presumably because of the entropy reduction associated with the formation of a secondary structure. Frequently, these peptides can be induced to form secondary structure such as helices on membrane binding (Kaiser and Kezdy, 1983). In addition to promoting the formation of backbone hydrogen bonds in a low dielectric environment, the interface may also reduce the entropy penalty by restricting the peptide flexibility (Chan *et al.,* 1991, Wattenbarger *et al.,* 1990). It has been proposed that transbilayer helices first associate as extended structures on the membrane interface and then develop secondary structure (Jacobs and White, 1989). In the case of larger peptides and proteins, this may then be followed by the insertion of a helical loop (Engleman and Steiz, 1981).

From the previous discussion, it can readily be appreciated that the partitioning of a peptide from an aqueous to a membrane environment must in general take into account a number of free energy terms. Several that have been discussed in some detail are represented in Eq. (1) (for a discussion of these terms, see Ben Tal *et al.,* 1996a; Engleman and Steiz, 1981; Jacobs and White, 1989; Jahnig, 1983):

$$\Delta G_{\text{total}} = \Delta G_{\text{nonpolar}} + \Delta G_{\text{elect}} = \Delta G_{\text{lipid}} = \Delta G_{\text{immobil}} + \Delta G_{\text{conf}}. \quad (1)$$

Where $\Delta G_{\text{nonpolar}}$ is the contribution due to the hydrophobic or solvent excluding the free energy change and ΔG_{elect} is the electrostatic contribution to the binding, which includes primarily the free energy associated with moving polar or charged groups to a region of low dielectric. This also includes any effects due to the movement of polar or charged groups through electrostatic fields that are known to be associated with lipid bilayers (surface, dipole, or transmembrane fields and their effect on peptide binding are discussed later); ΔG_{lipid} includes any effects that the peptide has, for example, on the conformation and entropy of the lipid; ΔG_{immob} includes the free energy change due to the loss of entropy by the peptide on membrane binding; and ΔG_{conf} includes any free energy changes associated with conformational changes in the peptide that might occur on membrane binding.

C. Effect of Membrane Electrostatic Fields on Peptide–Membrane Interactions

Electrostatic fields can arise from several different sources in biological membranes. Charge separation across the bilayer can give rise to a transmembrane potential difference, $\Delta\Psi$, which represents a potential difference between the two bulk aqueous phases. The electric field from this potential drops across the membrane hydrocarbon and is known to regulate the activity of voltage-gated channels. Biological membranes typically contain some percentage of negatively charged lipid, such as phosphatidylserine (PS) or phosphatidylinositol (PI). For these charged interfaces the balance of counterion diffusion and electrostatic attraction results in an electrostatic surface potential, Ψ_s, where the membrane surface is negative with respect to the adjacent aqueous phase. The surface potential for most biological membranes under normal ionic strength conditions has a magnitude that is on the order of a few tens of millivolts. This potential can be described accurately for charged surfaces containing monovalent lipid by the Gouy–Chapmann–Stern theory (GCS) (Cafiso *et al.,* 1989; Hartsel and Cafiso, 1986; McLauglin, 1989; Winiski *et al.,* 1986); however, more sophisticated analyses are needed to account for the potentials that are observed in the presence of multivalent lipids such as phosphatidylinositol 4,5-bisphosphate (PIP_2) (Langner *et al.,* 1990). In addition to surface and transmembrane potentials, biological membranes also possess a large internal dipole potential, Ψ_d. The dipole potential is thought to arise from dipole moments of the carbonyl oxygens and/or water that is exhibiting some net orientation within the membrane interface (Ellena *et al.*, 1987; Flewelling and Hubbell, 1986; Simon and McIntosh, 1989). This potential is quite large, it is on the order of 300 mV in egg phosphatidylcholine, and it results in the hydrocarbon being positive with respect to either aqueous phase (for a review, see Brockman, 1994).

1. Role of Membrane Surface Potentials

For any charged peptide, its binding to the membrane or the interface should be strongly affected by electrostatic fields. As indicated in Fig. 2, the interface of a membrane containing negatively charged lipids is at a negative electrostatic potential relative to the bulk aqueous phase. Each charged species in solution will be in equilibrium with this electrostatic potential and its distribution between the membrane interface and the bulk phase is described by a Boltzmann distribution. Assuming an ideal behavior so that activities can be approximated by concentrations yields

$$[P]_{\text{surface}} = [P]_{\text{bulk}} e^{-zF\Psi_s/RT}, \tag{2}$$

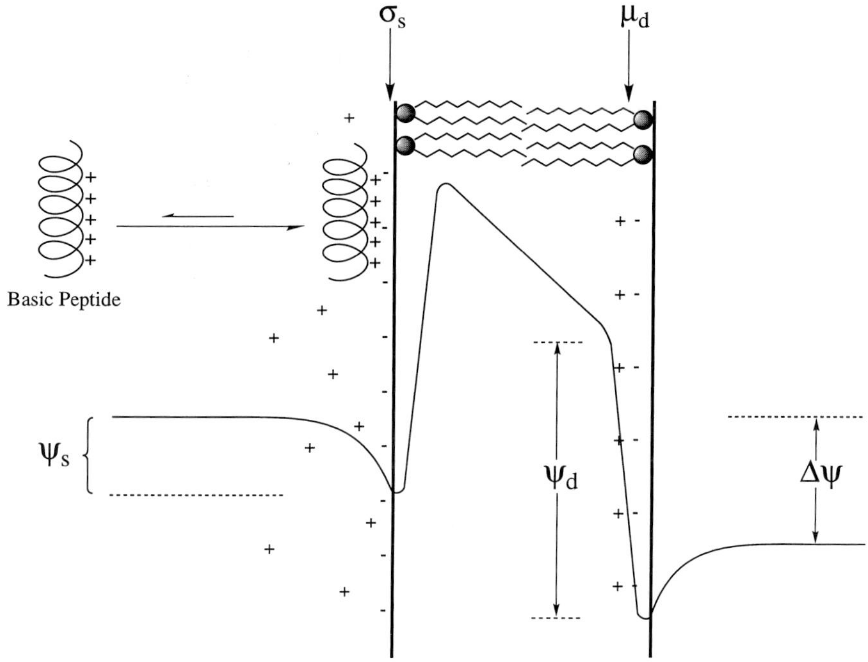

FIGURE 2 Membrane electrostatic fields. The membrane surface charge density σ_s gives rise to the electrostatic surface potential Ψ_s. Dipole moments, μ_d, within the membrane interface give rise to the large internal membrane dipole potential, Ψ_d. Transmembrane potentials, $\Delta\Psi$, result from the net separation of charge across the membrane. In the presence of membranes containing acidic lipids, basic multivalent peptides are highly concentrated at the membrane interface relative to the bulk aqueous phase and can associate strongly with the interface, even though they lack nonpolar residues.

where $[P]_{\text{surface}}$ and $[P]_{\text{bulk}}$ represent the concentrations of the peptide at the membrane surface and in the bulk aqueous phase and z is the valence of the charged species. Equation (2) predicts that the membrane interface will be at a slightly lower pH than the bulk phase and that simple anions such as chloride will be depleted at the interface. For a multivalent species, such as a charged peptide or charged protein domain, this effect will be much more significant. For example, for a membrane having a surface potential of -60 mV and for a peptide with a valence of 4 the surface concentration of the peptide would be expected to be 10,000 times that in the bulk phase.

As a result of this interaction, basic peptides and basic domains within proteins interact strongly with negatively charged surfaces, even when they are highly polar. This type of interaction was observed a number of years ago and has been characterized widely (de Kruijff *et al.,* 1985; Dufourcq *et al.,* 1981). For example, the interactions of signal sequences with membranes are influenced strongly by surface electrostatics (Wang and Weiner, 1994), and evidence shows that surface electrostatics can change the position and structures of peptides on the membrane interface (see e.g., Frey and Tamm, 1990; Qin and Cafiso, 1996).

A number of model basic peptides have been used to investigate the energetics of electrostatic peptide–membrane interactions. For example, simple peptides such as pentalysine (Lys_5) have been observed to interact strongly with membrane surfaces containing phosphatidylserine or phosphatidylglycerol (PS or PG) and addition of a lysine residue strengthens the binding affinity such that each lysine residue appears to bind to PS with an apparent binding free energy of 1.4 kcal/mole (Kim *et al.,* 1991). Positively charged model peptides containing either arginine or lysine also bind to acidic lipid surfaces, although arginine-containing peptides seem to bind with slightly higher affinity (Mosior and McLaughlin, 1992a). When alanines are inserted into model peptides containing arginine or lysine, they reduce the strength of the binding. Because the charges are spaced further apart, this effect may be due in part to a reduction of the electrostatic potential at each positively charged residue. In addition, it may simply reflect the increased entropy loss that the larger alanine-containing peptides experience when they bind to the membrane interface (Mosior and McLaughlin, 1992a).

When the binding of basic peptides to membranes is investigated as a function of acidic lipid content, the binding is found to be highly cooperative. It is possible to account for this type of binding, at least qualitatively, by a combination of the GCS theory and mass action. Although the GCS theory can reproduce the shapes of the binding curves, this type of simple analysis frequently requires the use of a valence that is significantly less than the known charge on the peptide (Mosior and McLaughlin, 1992b), although in the case of the cytochrome oxidase signal sequence, binding data agree quite well with the predictions of the GCS theory (Thorgeirsson *et al.,* 1995). The quantitative failure of this simple GCS theory for some peptides is not surprising given that it completely ignores the peptide structure and the placement of charges away from the membrane interface. Attempts, have been made to more realistically estimate the binding of charged peptides to membrane interfaces by taking into account both bilayer and peptide structures. For example, Murray *et al.* (1998) calculated the electrostatic interaction between the myristoylated N-terminal peptide derived from the src tryrosine kinase with a membrane surface constructed of 2:1 PC:PS. This calculation utilized solutions to the nonlinear Poisson–Boltzmann equation to *estimate* the electrostatic interaction between the

N terminus of src and interface and produced binding *estimates* that agreed well with experimental data obtained for the binding of the peptide. In addition to a long-range electrostatic attraction, these calculations indicate that highly charged peptides can also experience a Born repulsion as they get close to the membrane surface. This repulsion results from the interaction of the electric field from the peptide charges with the low dielectric of the membrane interior. Structural work on these types of charged peptides is discussed in Section III and indicates that this Born repulsion and the peptide hydrophobicity play important roles in determining the position of these peptides on the membrane surface. Generally, these types of electrostatic calculations reproduce the binding constants of charged proteins and peptides, within an order of magnitude, and accurately predict the dependence of the binding constant on surface charge density and ionic strength (Ben-Tal *et al.,* 1996b, 1997).

2. Biological Significance

Electrostatic interactions between membranes and peptides appear to play important roles in a wide range of processes. For example, many water-soluble proteins that are involved in cell signaling become attached to membrane surfaces as a result of acylation and/or electrostatic interactions with the membrane interface (McLaughlin and Aderem, 1995; Towler *et al.,* 1988). For example, the myristoylated alanine-rich C kinase substrate (MARCKS) resides in brain and several other tissues and interacts with both actin and calmodulin (CaM) (Rosen *et al.,* 1990; Taniguchi and Manenti, 1993). Although its role is not entirely understood, MARCKS has been proposed to act as a protein kinase C-regulated CaM buffer (McIlroy *et al.,* 1991; Taniguchi and Manenti, 1993). In addition to its N-terminal myristoylation, the MARCKS protein contains a highly basic, 25 amino acid domain (151–175), incorporating 12 lysines and 1 arginine residue, which electrostatically associates with negatively charged membranes (Kim *et al.,* 1994). Both the acylation and the basic domain are necessary to attach MARCKS to the membrane interface (Peitzsch and McLaughlin, 1993). As a result, the phosphorylation of MARCKS effectively eliminates the favorable electrostatic interaction and results in the membrane dissociation of MARCKS. This event functions as an electrostatic "switch" that can be used to control the association of MARCKS or other similarly structured proteins with the membrane interface (McLaughlin and Aderem, 1995).

The interaction of the src nonreceptor tyrosine kinase pp60c-src is similarly dependent on acylation as well as an electrostatic interaction for membrane attachment. In the case of the oncogenic form of src ($pp60^{v\text{-}src}$), the removal of residues responsible for either the charge interaction or the acylation site eliminates the membrane attachment of $pp60^{v\text{-}src}$ and its ability to transform cells (Cross *et al.,* 1984; Kamps *et al.,* 1985; Silverman and Resh, 1992).

A peptide derived from the N-terminal segment of src having a charge of 5 was found to interact with PC/PG (2:1) containing membranes with a molar partition coefficient of 10^7 M^{-1} (Buser *et al.,* 1994). The molar partition coefficient of the peptide is given as $K_1 = [P_m]/([L][P])$, where $[P_m]$ is the membrane concentration of the peptide, $[L]$ is the concentration of lipid accessible to the peptide, and $[P]$ is the concentration of free peptide (Peitzsch and McLaughlin, 1993). Binding of the myristoylated src peptide to neutral membranes was 10^4 M^{-1}, whereas the binding of the nonmyristoylated form of the peptide to PC:PG (2:1) membranes yielded a partition coefficient of 10^3 M^{-1}; thus, the product of the two individual binding constants is approximately the binding constant for the myristoylated peptide the charged membrane surface.

3. Lateral Segregation of Charged Lipids Induced by Peptide and Protein Binding

In addition to binding to membranes, domains from peptides such as MARCKS induce the lateral segregation of negatively charged lipids. Using fluorescence microscopy, a peptide corresponding to the membrane and CaM-binding domain of MARCKS (151–175) has been observed to promote the formation of lateral domains rich in the acidic lipid phosphatidylserine (Denisov *et al.,* 1998; Glaser *et al.,* 1996). This type of liquid–liquid domain formation occurs in general when basic peptides bind to membranes containing acidic lipids and can be induced by positively charged multivalent molecules such pentalysine or spermine. In these cases, the formation of the domain is driven by the reduction in electrostatic free energy within the double layer, but is opposed by the reduction in entropy that accompanies domain formation (Denisov *et al.,* 1998). At higher ionic strengths, the electrostatic energy gain associated with domain formation is reduced, and domains formed from pentalysine are observed to dissipate. This is not the case for MARCKS (151–175), which is observed to form domains at high ionic strength (Yang and Glaser, 1995). It has been proposed that the ability of MARCKS (151–175) to penetrate the membrane interface accounts for its altered behavior compared to pentalysine (Denisov *et al.,* 1998).

The intact MARCKS protein and the MARCKS-derived peptide both inhibit phospholipase C (PLC)-catalyzed hydrolysis of PIP_2 in phospholipid vesicles (Glaser *et al.,* 1996). The highly charged lipid PIP_2 is also sequestered within the lateral PS-rich domains that are induced by MARCKS, and the inhibition of PLC by MARCKS appears to be a result from the removal of the substrate from PLC. Why should the substrate PIP_2 be sequestered away from PLC in this type of signaling system? Although the biological role of this lateral domain formation is not entirely understood, it may provide a mechanism for positive cooperativity in the timing of the cell-signaling response (Glaser *et al.,* 1996).

4. Role of Transmembrane and Dipole Potentials

A number of peptides are known to exhibit behavior that is dependent on transmembrane electric fields. For example, peptides such as alamethicin induce a voltage-dependent conductance in lipid bilayers that is likely the result of a voltage-dependent conformational change. Several other examples of interactions of peptides with transmembrane fields have also been described. For example, signal sequences can be imported into lipid vesicles in a potential-dependent manner (Maduke and Roise, 1993), and it has been reported that short peptides that possess a single positive charge at their N terminus can be transported on application of a membrane potential (de Kroon *et al.,* 1989). Subsequent work demonstrated that transport of these short peptides is actually a result of a pH gradient that is present in these membrane systems (de Kroon *et al.,* 1991).

As indicated earlier, membranes also possess an internal dipole potential that is large and falls over a relatively short distance. It is conceivable that this type of potential might modulate the binding or transport of peptides (Cafiso, 1991). For example, as a result of the dipole potential, a peptide that is negatively charged would be more likely to become buried within the interface and transit more rapidly compared with a positively charged peptide. However, at the present time there is relatively little evidence for the role of this type of potential in these interactions.

Membrane electric fields also appear to be important in determining the folding of membrane proteins. The membrane protein structure prediction routines described earlier typically include an analysis of the positions of basic residues within the protein sequence. Clusters of basic residues are typically found on the cytoplasmic surface of membrane proteins, an observation that is referred to as the inside positive rule (Dalbey, 1990; von Heijne, 1992, 1995; von Heijne and Gavel, 1988). Surface potentials, transmembrane potentials, or dipole potentials might play a role in this effect, and there have been reports that transmembrane potentials (Andersson and von Heijne, 1994), and surface charge density (van Klompenburg *et al.,* 1997) affect the passage of positively charged protein segments through membranes.

D. Membrane Lipid Composition and Spontaneous Curvature

A wide body of literature discusses the effects of membrane composition on peptide–membrane interactions. As indicated earlier, hydrophobicity and electrostatic interactions between acidic lipids and basic peptides have strong and easily recognizable effects on peptide–membrane interactions. The biological importance of a membrane property referred to as the spontaneous curvature has become more apparent. Biological membranes typically contain a fraction of lipid such as PE that in pure form tends to form nonbilayer phases. As a result, bilayer phase membranes formed from PC and PE will have an intrinsic strain or curvature stress that is not present when the membrane is composed of PC alone. This strain is sometimes referred to as the spontaneous curvature of a bilayer, which represents the intrinsic curvature that the one monolayer of the bilayer would assume if it were freed from its constraints in the bilayer configuration (Gruner, 1989). This curvature appears to be highly regulated in some organisms (Morein *et al.,* 1996; Osterberg *et al.,* 1995) and is believed to affect the activity of a number of membrane proteins such as rhodopsin (see, e.g., Brown, 1994).

Peptides can both modulate this curvature and respond to it, and such interactions may be involved in a number of processes (Batenburg and de Kruijff, 1988). For example, the single channel properties of the peptide channel alamethicin, which exhibits a voltage-dependent gating in lipid bilayers, have been observed to correlate with the membrane spontaneous curvature (Keller *et al.,* 1993). It has been shown that an increase in the mole fraction of dioleoylphosphatidylethanolamine versus dioleoylphosphatidylcholine in the bilayer shifts alamethicin channels to higher conductance levels in a manner that is directly related to the spontaneous curvature of these lipid mixtures. The peptide itself, when added to the membrane, alters the bilayer curvature and changes the transition temperature from bilayer to nonbilayer phase (Keller *et al.,* 1996). In addition, the free energy for alamethicin binding to bilayers is affected by the spontaneous curvature (J. R. Lewis and D. S. Cafiso, 1999). For another channel-forming peptide, gramicidin, the equilibrium between gramicidin monomer and dimer (the channel-conducting form) can be modulated by altering the spontaneous curvature (Lundbaek *et al.,* 1997). In this case, dioleoylphosphatidylserine and the ionic composition were varied to alter the curvature stress. It was found that channel (dimer) formation requires the bilayer to assume a localized positive curvature. Furthermore, it was demonstrated that a reduction of the positive curvature stress in the bilayer decreased channel stability.

There are many other examples of work that connect membrane curvature stress with peptide and protein binding. For example, polyunsaturates alter curvature stress and are important for the activity of protein kinase C (Slater *et al.,* 1996). Gramicidin, acylated gramicidin analogs, signal sequences, and membrane fusion inhibitors have all been observed to modulate lipid phase behavior and or curvature (Epand *et al.,* 1997; Killian *et al.,* 1990; Vogt *et al.,* 1994). This work provides a clear example of how lipid composition and the intrinsic properties of fluid phase bilayers can modulate the energies of peptide–membrane interactions.

E. Kinetics of Peptide Binding

The free energy of peptide binding to a membrane, K, is directly related to its association and dissociation rates, $K = k_{on}/k_{off}$. If the binding of the peptide occurs at a diffusion limited rate, changes in the binding constant should reflect changes in the dissociation rate of the peptide. For many peptides, such as those from the highly charged effector domain of MARCKS, the binding constant for the peptide is exceedingly strong to membranes containing acidic lipids; as a result, if the peptide binding is diffusion limited, the membrane dissociation rate for the peptide should be quite slow. This may have implications for the proposed role of MARCKS as a PKC–regulated calmodulin buffer. For example, if the MARCKS effector domain must first dissociate from the membrane before interacting with calmodulin, its response to changing levels of CaM within the cell may be quite slow and limited by its membrane dissociation rate.

To investigate the kinetics of binding of peptides derived from the MARCKS effector domain, stopped flow fluorescence was employed to study the kinetics of binding for several acrylodan-labeled MARCKS (151–175) derivatives (Arbuzova *et al.,* 1997). These investigators found that the MARCKS-derived peptide bound to 100-nm-diameter PC:PS liposomes at a diffusion limited rate, which was approximately 10^{11} M^{-1} $\sec^{-1}$. As expected, the binding rate was independent of factors that alter the binding constant, such as the surface charge density and ionic strength. The dissociation rates could be quite slow and were approximately 1 $\sec^{-1}$ in PC:PS membranes (5:1). In this system, CaM was found to dramatically decrease the lifetime of the peptide on the membrane interface. A likely interpretation of kinetic data is that CaM catalyzes the dissociation of MARCKS (151–175) by colliding with it on the membrane surface. This type of interaction by CaM may play a role in the interactions of CaM with other important proteins such as CaM kinase II and myosin light chain kinase.

Other peptides, such as amphipathic α-helical peptides that interact electrostatically with membranes, have been observed by tryptophan fluorescence or energy transfer and were found to bind to membranes at a diffusion-limited rate (Polozov *et al.,* 1998). In addition, the protein annexin VI, which is surface associated, associates with membranes at a diffusion limited rate (Lu *et al.,* 1995); however, melittin and alamethicin bind with rates that are approximately 30- and 6-fold slower than the diffusion limit, respectively (Schwarz and Beschiaschvili, 1989; Schwarz *et al.,* 1987). The slower rate of association of alamethicin may be a result of the fact that this peptide binds deeply within the hydrophobic core of the membrane (Barranger-Mathys and Cafiso, 1996). Although melittin lies in the plane of the membrane under fully hydrated conditions, it likely penetrates deeply into the membrane interior (see, e.g., Dempsey and Butler, 1992; Frey and Tamm, 1991); in addition, it may also dissociate first from a helical tetrameric structure in solution.

The kinetics of binding of the signal sequence from subunit IV of cytochrome c oxidase to model membranes has also been examined (Golding *et al.,* 1996). This binding was monitored by examining the fluorescence of a surface-associated probe that was sensitive to changes in membrane charge density. Kinetic data indicate that the binding occurs in two steps, which can be interpreted as membrane surface binding followed by the formation of an α helix and the insertion of the peptide.

III. FACTORS INFLUENCING PEPTIDE STRUCTURE AND ORIENTATION IN MEMBRANES

A. Basic Domains of Membrane Proteins

As indicated earlier, positively charged peptides interact electrostatically with acidic lipid interfaces, and this electrostatic interaction plays an important role in attaching proteins to membrane interfaces. The structures of a number of basic peptides on the membrane interface have been investigated using several experimental approaches. Simple peptides such as pentalysine, which associate electrostatically with membrane surfaces, do not appear to penetrate and remain on the aqueous side of the interface. Deuterium nuclear magnetic resonance (NMR) was used to investigate the effect of pentalysine (Lys_5) on the residual quadrupolar coupling of head group-labeled phosphatidylcholine bilayers. Unlike charged molecules, which penetrate the surface, Lys_5 produced no change in the residual quadrupolar splitting on binding (Roux *et al.,* 1988). Peptides that penetrate the interface are also expected to produce changes in monolayer surface pressure when added to the subphase of a Langmuir trough, and indeed, for peptides such as Lys_5, no alteration in monolayer surface pressure can be detected when they are added to subphase below PC:PS monolayers (Ben-Tal *et al.,* 1996b). Thus, available evidence suggests that highly charged peptides such as Lys_5 reside in the aqueous phase and do not come in physical contact with the lipid head group domain.

Model basic peptides containing alanine and lysine have been studied with amino acids selectively inserted into guest positions (Liu and Deber, 1997). These peptides were largely random in buffer but formed α-helical structures in the presence of acidic lipids. In the presence of neutral lipids only, peptides containing a more hydrophobic guest formed helical structures, whereas the less hydrophobic peptides could assume helical structures, provided that negatively charged lipids were added. Fluorescence measurements also provided an indication that a population of these peptides could assume a transmembrane configuration.

Peptides derived from the charged domains of MARCKS and neuromodulin were examined by site-directed spin labeling and EPR spectroscopy to determine both their conformation and position on the membrane surface (Qin and Cafiso, 1996; Wertz *et al.,* 1996). The highly charged domain from MARCKS appears to assume the conformation shown below with its hydrophobic phenylalanine residues positioned within the membrane interface and its charged residues off the membrane surface in the aqueous phase. In contrast to Lys_5, EPR data suggest that charged peptides from MARCKS and neuromodulin penetrate the surface of the lipid interface, and MARCKS (151–175) does produce changes in surface pressure when added to the subphase of PC:PS monolayers. For this peptide, the change in surface pressure is less than that expected if the entire peptide were buried, but is consistent with the five phenylalanine residues penetrating the membrane interface (Glaser *et al.,* 1996). For MARCKS (151–175) it is also striking that the N-terminal domain also turns away from the membrane surface and does not penetrate the membrane interface.

As indicated earlier, modeling the interactions between basic peptides and negatively charged membrane surfaces indicates that these highly charged peptides experience a Born repulsion near the membrane interface. This is probably the case for the N–terminal region of MARCKS, which should resemble Lys_5 in terms of its energetics of interaction with the membrane interface. The peptide is attracted to the membrane surface as a result of a long-range electrostatic attraction with the negatively charged lipid surface, but as the peptide approaches close (within a few angstroms) to the surface it experiences a strong repulsion due to the interaction of charge on the peptide with the low dielectric of the membrane interior. The distance between the highly charged N terminus of MARCKS (151–175) and the membrane surface can apparently be modulated by altering the long-range attractive force, and an increase in the concentration of negatively charged lipid brings this end of the peptide closer to the membrane interface (Qin and Cafiso, 1996).

Like Lys_5, MARCKS and neuromodulin-derived peptides do not interact with neutral zwitterionic PC bilayers. However, such an observation should not be used to conclude that peptides, such as that from MARCKS, do not interact hydrophobically with membranes. In the presence of neutral membranes, such as those formed from PC, peptides such as MARCKS (151–175) fail to bind because there is an insufficient hydrophobic free energy to overcome the Born repulsion that these peptides experience as they approach the membrane interface.

Site-directed spin-labeling and EPR spectroscopy have been used to examine the structure and position of the myristoylated N-terminal domain of src at membrane interfaces containing PC or PC:PS (Victor and Cafiso, 1998). Except for the N-terminal myristolyation, this peptide is much more polar than the peptides derived from MARCKS of neuromodulin. EPR spectroscopy shows that although this peptide lies close to the interface near its N terminus, it does not penetrate the membrane interface as does MARCKS (see Fig. 3). This picture of src combined with the computational methods described earlier predict accurately the binding free energy for the src-derived peptide to membranes (Murray *et al.,* 1998).

B. Channel-Forming Peptides

A large number of naturally occurring peptides that are made by organisms as diverse as fungi, insects, and humans exhibit antibiotic activity as

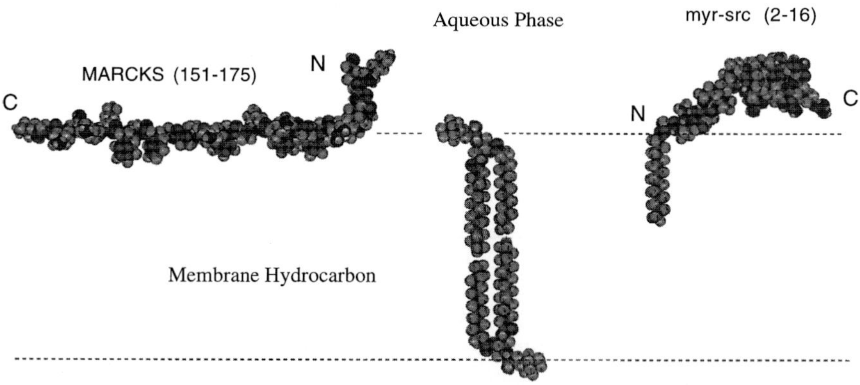

FIGURE 3 Membrane position and structures of the membrane, CaM, and PKC-binding domain of MARCKS (Qin and Cafiso, 1996) and the N-terminal membrane-binding segment of the src tyrosine kinase (Victor and Cafiso, 1998). MARCKS (151–175) is in an extended structure with its phenylalanine residues inserted into the membrane interface. The peptide myr–src (2–16) is also extended, but it is much more polar than the MARCKS-derived peptide and does not penetrate the membrane interface.

a result of their strong interactions with cell membranes. Because of their relatively small size and because they can often be synthesized in large quantities, these peptides are readily amenable to structural techniques such as NMR spectroscopy. However, these membrane active peptides are quite challenging to study. For example, it is not clear for many of these peptides whether their biological activity lies in their ability to form discrete ion channels across membranes or whether they act by destabilizing the membrane bilayer structure. Many of these membrane-active peptides exhibit surface to transmembrane orientational equilibria that depend on the peptide concentration and lipid hydration. In addition, a number of these peptides appear to exhibit their channel-forming activity through the formation of amphipathic helical aggregates within the membrane. In the case of alamethicin (see later), this equilibria appears to strongly favor the monomeric state, which makes direct spectroscopic studies designed to define their active structures difficult if not impossible.

Defensins and cecropins are peptides that are employed by humans and insects, respectively, as rapid-response acting defense mechanisms. Defensins are 29 to 35 amino acids in length and contain six cysteines forming three intramolecular disulfide bonds (for a review, see Kagan *et al.,* 1994). They are found in high concentrations in neutrophils and have antibiotic activity against a range of organisms and cell types, including bacteria, fungi, viruses, and malignant cells. These peptides exhibit voltage-dependent activity but, unlike many channel-forming peptides, they appear to be largely β sheet in structure. Cecropins vary in length from 35 to 39 residues, exhibit activity against bacteria, and possess two amphipathic helical regions connected by a flexible hinge region where the N-terminal region is charged and the C-terminal region is more hydrophobic (for a review, see Bechinger, 1997). Although evidence shows that these peptide may act by forming tail-to-tail dimers, which then aggregate to form channels (Wade *et al.,* 1992), evidence from infrared spectroscopy also indicates that the peptide lies along the bilayer surface and that the antibiotic mechanism involves the formation of a surface aggregate that destabilizes the bilayer (Gazit *et al.,* 1996).

Magainins are a series of peptides produced in the skin and tissues of certain frogs that exhibit antibacterial, fungicidal, and antivirual activity (for a review, see Maloy and Kari, 1995). These peptides range in length from 21 to 26 residues and promote channel formation in lipid bilayers and cell lysis. Magainins appear to be in random coil configurations in solution, but are α helical when associated with membranes. From solid-state NMR experiments and fluorescence measurements (Bechinger, 1997; Matsuzaki *et al.,* 1994), the helix formed by these peptides lies in the plane of the bilayer at low surface concentrations, with several hydrophobic phenylalanines buried at the interface, and charged lysine residues facing the aqueous phase. In this configuration, modeling indicates that magainins may distort the bilayer and cause a decrease in the bilayer thickness; indeed these peptides are observed to induce a destabilization of bilayer phases at concentrations that are physiologically relevant. At higher surface concentrations, magainins are reported to take up an opposite orientation along the bilayer normal and the presence of water-filled pores is observed (Ludtke *et al.,* 1996).

Electrophysiological recordings indicate a variety of channel-like activities for magainins, but it is not clear whether this channel activity or the general bilayer destabilizing activity of the peptide is directly responsible for its biological activity. The presence of magainin concentrations less than 3 mole % have been observed to uncouple oxidative phosphorylation in a number of systems, and it is proposed that this effect is responsible for its cytotoxic activity (Westerhoff *et al.,* 1995).

Melittin is a basic α-helical and surface active peptide that is a major component of bee venom (for reviews, see Bechinger, 1997; Dempsey, 1990; Matsuzaki, 1997). This peptide exhibits bilayer destabilizing tendencies for some lipid mixtures but it can also promote ion channel conductances upon the application of a transmembrane membrane potential. Depending on the conditions, the conductances can be discrete or poorly defined. Although the peptide exhibits a monomer–tetramer equilibrium in solution (Brown *et al.,* 1980; Vogel and Jahnig, 1986), it appears to be monomeric in liquid crystalline membranes at membrane concentrations less than 0.5 mole % (John and Jahnig, 1988).

Alamethicin is a 20 amino acid peptide that is isolated from the fungus *Trichoderma viride* (for reviews, see Bechinger, 1997; Cafiso, 1994; Sansom, 1993; Wooley and Wallace, 1992). This peptide has been widely studied in part because it produces strong voltage-dependent conductances in bilayers. Alamethicin belongs to a family of related peptides called peptaibols that contain reduced C termini rich in the amino acid α-methylalanine (also referred to as AiB or α-aminoisobutyric acid). This peptide produces very elaborate channel conductances that appear to be the result of the formation of a conducting aggregate. The association and dissociation of monomers from this aggregate are dynamic processes that can account for discrete channel conductance states, and structure–function studies performed on dimers of alamethicin provide strong evidence for this view (You *et al.,* 1996). Although neutron diffraction has detected water presumably located within alamethicin pores (He *et al.,* 1996b), no direct structural studies on a well-defined channel aggregate have been made thus far.

The nature of the voltage-dependent event for alamethicin that leads to the observed channel conductances has not been resolved. This process likely involves the interaction of dipole moments of the peptide with the membrane electric field as analogs of alamethicin that lack any formal charge exhibit voltage gating. Changes in orientation from a surface-bound state to a transmembrane orientation are one mechanism by which alamethicin may act, and such orientational changes are in fact observed as a function of peptide:lipid ratio and lipid hydration (He *et al.,* 1996a). However, under conditions that lead to channel conductance, the majority of the population of alamethicin in bilayers is oriented normal to the bilayer (Barranger-Mathys and Cafiso, 1996; North *et al.,* 1995).

Peptides such as alamethicin demonstrate some of the difficulties that are encountered frequently with this general class of membrane-active peptides. As indicated in Fig. 4, alamethicin likely undergoes a monomer–multimer equilibrium where the oligomer is of varying size. In addition, under some conditions, alamethicin also undergoes a transition between surface and inserted states. Magnetic resonance studies in the liquid crystalline phase indicate that the peptide is largely monomeric, so that the channel-forming state may represent only a small fraction of the total membrane-bound population of peptide (Archer *et al.,* 1991; Barranger and Cafiso, 1994).

Of the small channel-forming peptides, the linear gramicidins are by far the best defined. Gramicidin A is a short 15 amino acid peptide made up of alternating L and D amino where the conductive form is a single-stranded β_6 helical dimer in an end-to-end orientation. The high-resolution membrane-bound structure is known as a result of solid-state NMR studies (Cross, 1997) and the highly defined nature of the peptide has allowed detailed structure–function studies. These studies have produced, among other properties, voltage dependence in the gramicidin channel (Koeppe and Anderson, 1996). This peptide has also provided evidence for the importance of typtophan interactions in the interfacial regions of membrane proteins (Wallace, 1996) and for the role of dynamics in ion channel conduction (Roux and Karplus, 1994). Gramicidin can occur in several alternate double helical configurations (Veatach *et al.,* 1974), which are known to interconvert slowly in organic solvents (Arumugam *et al.,* 1996). In addition, these double helical conformations can become trapped kinetically in a membrane environment and are converted to the active end-to-end dimer with heating (Arumugam *et al.,* 1996). This work demonstrates that the preparation procedure and solvent history of the peptide–membrane mixture can be critical determinants in the structure and activity of a membrane-associated peptide.

C. *De Novo* Design of Membrane-Active Peptides

The *de novo* design of membrane-active peptides is a powerful approach to directly access the importance of peptide sequence and structure in

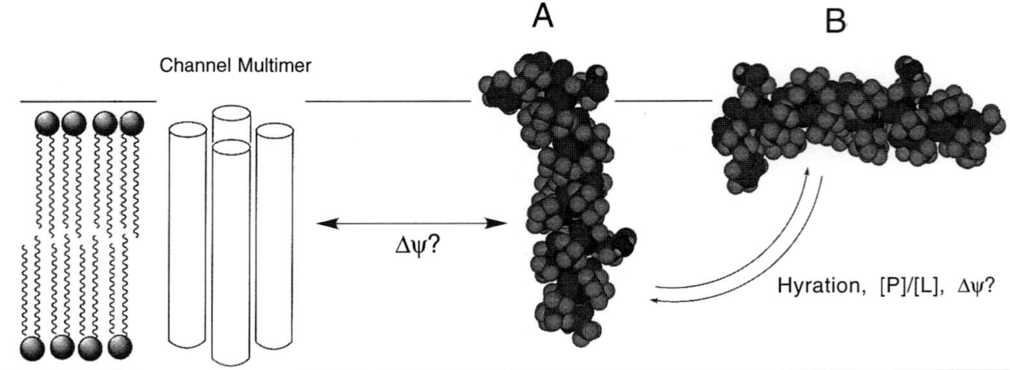

FIGURE 4 Alamethicin can exist in both perpendicular, (A), and parallel, (B), orientations relative to the membrane interface. The parallel orientation is found primarily under conditions of limited hydration (He *et al.,* 1996a). The perpendicular, inserted orientation is found at full hydration for most lipids under conditions where alamethicin exhibits voltage gating. The model shown here is based on EPR depth measurements of spin-labeled alamethicin analogs (Barranger-Mathys and Cafiso, 1996). Alamethicin channels form by the assembly of a channel multimer that can vary in size. The tetramer depicted is likely a minimal size for the conductive aggregate, and alamethicin monomers are in equilibrium with the channel aggregate. The step in this equilibrium at which the voltage-dependent event takes place has not yet been elucidated.

membrane–peptide interactions (Degrado *et al.,* 1989). *De novo* peptides have been synthesized to address fundamental questions regarding the length and hydrophobicity of peptides necessary to achieve transmembrane insertion of a peptide helix. For example, designed membrane-active peptides have been shown to orient as a function of the peptide length, membrane thickness, and cholesterol content (Ren *et al.,* 1997). Water-soluble membrane-inserting peptides have been produced based on stretches of alanine having a more hydrophobic N terminus and a charged C terminus (Chung and Thompson, 1996). Designed transmembrane peptides have also been produced from highly hydrophobic peptides having a central 24 central leucine stretch and charged residues on either end (Axelsen *et al.,* 1995; Zhang *et al.,* 1992, 1995). Amphiphathic model peptides have been produced with a varied hydrophobic–hydrophilic balance (Kiyota *et al.,* 1996), and it was found that peptides having a larger hydrophobic balance not only bound to membranes more strongly, but also exhibit a greater tendency toward aggregation.

Based on the idea that aggregates of laterally amphipathic helices form the structural basis for at least some membrane ion channels, model peptides based on leucine and serine were synthesized and shown to exhibit voltage-dependent, Na^+/K^+ or H^+ conductances in planar bilayers (DeGrado and Lear, 1990; Lear *et al.,* 1988; Kienker *et al.,* 1994). Peptides that are highly amphipathic have also been synthesized and found to permeabilize membranes in a pH-dependent fashion (Fattal *et al.,* 1994; Parente *et al.,* 1990). This appears to be a result of the ability of these peptides to form an α-helical structure at specific pH ranges. Positively charged analogs have also been designed to bind to DNA and have been found to assist in the delivery of DNA into cells (Wyman *et al.,* 1997). Insertion has been observed for a peptide fragment from the C-helix of bacteriorhodopsin (Hunt *et al.,* 1997). It is anticipated that this type of peptide–membrane interaction will permit an investigation of the sequence of folding and insertion of proteins into membranes.

To facilitate the study of channel peptides in their active state, both native and *de novo* peptides have been covalently joined or templated in order to preform the channel aggregate and eliminate the monomer–multimer equilibrium. For example, in the case of the membrane-active peptide melittin, templating the peptide as a four-helix bundle enhanced membrane binding dramatically and led to a peptide that was much more active than native melittin (Pawlak *et al.,* 1994). Measurements of this four-helix aggregate on monolayers also yielded changes in surface pressure that were consistent with the expected size of the bundle. Model peptides based on leucine and serine have also been templated in order to improve their channel activity (Akerfeldt *et al.,* 1992). To mimic the channel-forming domains of native ion channels, peptides made from the putative channel-forming helices of Ca^{2+} channels, the nicotinic acetylcholine receptor and glycine receptor, have been templated and studied in planar bilayers (Iwamoto *et al.,* 1994; Oblatt-Montal *et al.,* 1993; Reddy *et al.,* 1993).

IV. SUMMARY

Biological membranes are liquids with properties determined by noncovalent forces. This is also true for membrane-active peptides, where peptide structure, orientation, and their effects on membranes can be a balance between a number of weak forces. This balance of forces has several important consequences. First, it leads to a remarkably diverse and interesting behavior for membrane-active interacting peptides. Even among a single type of peptide, quite varied behavior can be observed as a function of the lipid composition, the peptide–lipid ratio, and membrane hydration. A second consequence of the nature of membranes is that determination of the structures of membrane-associated peptides and the important forces that determine their behavior is quite challenging.

This chapter provided a summary of the interaction of a number of membrane-active peptides and model peptides with bilayers. It should be noted that work on a number of interesting membrane-active peptides derived from larger proteins has not been covered in this chapter. These include protein signal sequences (Tamm, 1991; Zheng and Gierasch, 1996) and peptides derived from proteins that direct membrane fusion (Chernomordik *et al.,* 1998; Skehel *et al.,* 1995).

Work on peptide–membrane interactions has progressed rapidly in recent years, partly as a result of the success of a number of experimental approaches, including magnetic resonance (NMR and EPR spectroscopy), vibrational and optical spectroscopies, and the continued development of methodologies for peptide synthesis and purification. In addition, computational methods are beginning to provide important insight into the nature of the forces governing peptide–membrane interactions, and they have reached a level of sophistication that makes them an important complement to experimental work.

Acknowledgments

The published and unpublished work alamethicin discussed from our laboratory was supported by NIH Grant GM35215. Structural studies on electrostatic peptide–membrane interactions were supported by NSF Grant MCB 9728083.

References

Akerfeldt, K. S., Kim, R. M., Camac, D., Groves, J. T., Lear, J. D., and Degrado, W. F. (1992). Tetraphilin: A four-helix proton channel built on a tetraphenylporphyrin framework. *J. Am. Chem. Soc.* **114,** 9656–9657.

Andersson, H., and von Heijne, G. (1994). Membrane protein topology Effects of delta mu H^+ on the translocation of charged residues explain the 'positive inside' rule. *EMBO J.* **13,** 2267–2272.

Arbuzova, A., Wang, J., Murray D., Jacob, J. Cafiso, D. S., and McLaughlin, S. (1997). Kinetics of interaction of the myristoylated alanine-rich C kinase substrate, membranes, and calmodulin. *J. Biol. Chem.* **272,** 27167–27177.

Archer, S. J., Ellena, J. F., and Cafiso, D. S. (1991). Dynamics and aggregation of the peptide ion channel alamethicin. *Biophys. J.* **60,** 389–398.

Arumugam, S., Pascal, S., North, C. L., Hu, W. Lee, K.-C., Cotten, M., Ketchem, R. R., Xu, F., Brenneman, M., Kovacs, F., Tian, F., Wany, A., Huo, S., and Cross, T. A. (1996). Conformational trapping in a membrane environment: A regulatory mechanism for protein activity. *Proc. Natl. Acad. Sci. U.S.A.* **93,** 5872–5876.

Axelsen, P. H., Kaufman, B. K., McElhaney, R. N., and Lewis, R. N. (1995). The infrared dichroism of transmembrane helical polypeptides. *Biophys. J.* **69,** 2770–2781.

Barranger-Mathys, M., and Cafiso, D. S. (1994). Collisions between helical peptides in membranes monitored using electron paramagnetic resonance. Evidence that alamethicin is monomeric in the absence of a membrane potential. *Biophys. J.* **67,** 172–176.

Barranger-Mathys, M., and Cafiso, D. S. (1996). Membrane structure of voltage-gated channel forming peptides revealed by site-directed spin labeling. *Biochemistry* **35,** 498–505.

Batenburg, A. M., and de Kruijff, B. (1988). Modulation of membrane surface curvature by peptide–lipid interactions. *Biosci. Rep.* **8,** 299–307.

Bechinger, B. (1997). Structure and functions of channel-forming peptides: Magainins, cecropins, mellitin and alamethicin. *J. Membr. Biol.* **156,** 197–211.

Ben-Tal, N., Honig, B., Peitzsch, R. M., Denisov, G., and McLaughin, S. (1996b). Binding of small basic peptides to membranes containing acid lipids: Theoretical models and experimental results. *Biophys. J.* **71,** 561–575.

Ben-Tal, N., Ben-Shaul, A., Nicholls, A., and Honig, B. (1996a). Free-energy determinants of alpha-helix insertion into lipid bilayers. *Biophys J.* **70,** 1803–1812.

Ben-Tal, N., Honig, B., Miller, C., and McLaughlin, S. (1997). Electrostatic binding of proteins to membranes. Theoretical predictions and experimental results with charybdotoxin and phospholipid vesicles. *Biophys. J.* **73,** 1717–1727.

Brockman, H. (1994). Dipole potential of lipid membranes. *Chem. Phys. Lipids* **73,** 57–79.

Brown, L. R., Lauterwein, J., and Wuthrich, K. (1980). High-resolution ^{1}H-NMR studies of self-aggregation of melittin in aqueous solution. *Biochim. Biophys. Acta.* **622,** 231–244.

Brown, M. F. (1994). Modulation of rhodopsin function by properties of the membrane bilayer. *Chem. Phys. Lipids* **73,** 159–180.

Buser, C. A., Sigal, C. T., Resh, M. D., and McLaughlin, S. A. (1994). Membrane binding of myristylated peptides corresponding to the NH_2 terminus of Src. *Biochemistry* **33,** 13093–13101.

Cafiso, D. S. (1991). Lipid bilayers: Membrane–protein electrostatic interactions. *Curr. Opin. Struct. Biol.* **1,** 185–190.

Cafiso, D. S. (1994). Alamethicin: A peptide model for voltage-gating and protein membrane electrostatic interactions. *Annu. Rev. Biophys. Biomol. Struct.* **23,** 141–165.

Cafiso, D., McLaughlin, A., McLaughlin, S., and Winiski, A. (1989). Measuring electrostatic potentials adjacent to membranes. *In* "Methods in Enzymology" (S. Fleischer and B. Fleischer, eds.), Vol. 171, pp. 342–364. Academic Press, San Diego.

Chan, H. S., Wattenbarger, M. R., Evans, D. F., Bloomfield, V. A., and Dill, K. A. (1991). Enhanced structure of polymers at interfaces. *J. Chem. Phys.* **94,** 8542–8557.

Chernomordik, L. V., Frolov, V. A., Leikina, E., Bronk, P., and Zimmerberg, J. (1998). The pathway of membrane fusion catalyzed by influenza hemagglutinin: Restriction of lipids, hemifusion, and lipidic fusion pore formation. *J. Cell Biol.* **140,** 1369–1382.

Chung, L. A., and Thompson, T. E. (1996). Design of membrane inserting peptides: Spectroscopic characterization with and without lipid bilayers. *Biochemistry* **35,** 11343–11354.

Cramer, W. A., Heymann, J. B., Schendel, S. L., Deriy, B. N., Cohen, F. S., Elkins, P. A., and Stauffacher, C. V. (1995). Structure-function of the channel-forming colicins. *Annu. Rev. Biophys. Biomol. Struct.* **24,** 611–641.

Cross, F. R., Garber, E. A., Pellman, D., and Hanafusa, H. (1984). A short sequence in the pp60src N-terminus is required for pp60src myristylation and membrane association and for cell transformation. *Mol. Cell. Biol.* **4,** 1834–1842.

Cross, T. A. (1997). Solid-state nuclear magnetic resonance characterization of gramicidin channel structure. *In* "Methods in Enzymology" (G. B. Fields, ed.), Vol. 289, pp. 672–696. Academic Press, San Diego.

Dalbey, R. E. (1990). Positively charged residues are important determinants of membrane protein topology. *Trends Biochem. Sci.* **15,** 253–257.

DeGrado, W. F., Wasserman, Z. R., and Lear, J. D. (1989). Protein design, a minimalist approach. *Science* **243,** 622–628.

DeGrado, W. F., and Lear, J. D. (1990). Conformationally constrained alpha-helical peptide models for protein ion channels. *Biopolymers* **29,** 205–213.

de Kroon, A. I., de Gier, J., and de Kruijff, B. (1989). Association of synthetic model peptides with phospholipid vesicles induced by a membrane potential. *Biochim. Biophys. Acta.* **981,** 371–373.

de Kroon, A. I., Vogt, B., van't Hof, R., de Kruijff, B., and de Gier, J. (1991). Ion gradient-induced membrane translocation of model peptides. *Biophys. J.* **60,** 525–537.

de Kruijff, B., Rietveld, A., Telders, N., and Vaandrager, B. (1985). Molecular aspects of bilayer stabilization induced by poly(L-lysines) of varying size in cardiolipin liposomes. *Biochim. Biophys. Acta* **820,** 295–304.

Dempsey, C. E. (1990). The actions of melittin on membranes. *Biochim. Biophys. Acta* **1031,** 143–161.

Dempsey, C. E., and Butler, G. S. (1992). Helical structure and orientation of melittin in dispersed phospholipid membranes from amide exchange analysis *in situ. Biochemistry* **31,** 11973–11977.

Denisov, G., Wanaski, S., Luan, P., Glaser, M., and McLaughlin, S. (1998). Binding of basic peptides to membranes produces lateral domains enriched in the acid lipids phosphatidylserine and phosphatidylinositol 4,5-bisphosphate: An electrostatic model and experimental results. *Biophys. J.* **74,** 731–744.

De Young, L. R., and Dill, K. A. (1990). Partitioning of non-polar solutes into lipid bilayers and amorphous normal alkanes. *J. Phys. Chem.* **94,** 801–809.

Dufourcq, J., Faucon, J. F., Maget-Dana, R., Pileni, M. P., and Helene, C. (1981). A fluroescence study of the binding of oligopeptides containing aromatic and basic residues to phospholipid vesicles. *Biochem. Biophys. Acta* **649,** 67–75.

Eisenberg, D. (1984). Three-dimensional structure of membrane and surface proteins. *Annu. Rev. Biochem.* **53,** 595–623.

Eisenberg, D., and McLachlan, A. D. (1986). Solvation energy in protein folding and binding. *Nature* (*London*) **319,** 199–203.

Ellena, J. F., Dominey, R. N., Archer, S. J., Zu, Z.-C., and Cafiso, D. S. (1987). The localization of hydrophobic ions in phospholipid bilayers using ^{1}H nuclear Overhauser effect spectroscopy. *Biochemistry* **26,** 4584–4592.

Engleman, D. M., and Steiz, T. A. (1981). The spontaneous insertion of proteins into and across membranes: The helical hairpin hypothesis. *Cell* (*Cambridge, Mass.*) **23,** 411–422.

Engleman, D., Steitz, T. A., and Goldman, A. (1986). Identifying non-polar transbilayer helices in amino acid sequences of membrane proteins. *Annu. Rev. Biophys. Biophys. Chem.* **15,** 321–353.

Epand, R. F., Moroder, L., Lutz, J., Flanagan, T. D., Nir, S., and Epand, R. M. (1997). Lipogastrins as potent inhibitors of viral fusion. *Biochim. Biophys. Acta* **1327,** 259–268.

Fattal, E., Nir, S., Parente, R. A., and Szoka, F. C., Jr. (1994). Pore-forming peptides induce rapid phospholipid flip-flop in membranes. *Biochemistry* **33,** 6721–6731.

Flewelling, R. F., and Hubbell, W. L. (1986). The membrane dipole potential in a total membrane potential model. *Biophys. J.* **49,** 541–552.

Frey, S., and Tamm, L. K. (1990). Membrane insertion and lateral diffusion of fluorescence-labelled cytochrome c oxidase subunit IV signal peptide in charged and uncharged phospholipid bilayers. *Biochem. J.* **272,** 713–719.

Frey, S., and Tamm, L. K. (1991). Orientation of melittin in phospholipid bilayers. A polarized attenuated total reflection infrared study. *Biophys. J.* **60,** 922–930.

Gazit, E., Miller, I. R., Biggin, P. C., Sansom, M. S., and Shai, Y. (1996). Structure and orientation of the mammalian antibacterial peptide cecropin P1 within phospholipid membranes. *J. Mole. Biol.* **258,** 860–870.

Glaser, M., Wanaski, S., Buser, C. A., Boguslavsky, V., Rashidzada, W., Morris, A., Rebecchi, M., Scarlata, S. F., Runnels, L. W., Prestwich, G. D., Chen, J., Aderem, A., Ahn, J., and McLaughlin, S. (1996). Myristoylated alanine-rich C kinase substrate (MARCKS) produces reversible inhibition of phospholipase C by sequestering phosphatidylinositol 4,5-bisphosphate in lateral domains. *J. Biol. Chem.* **271,** 26187–26193.

Golding, C., Senior, S., Wilson, M. T., and O'Shea, P. (1996). Time resolution of binding and membrane insertion of a mitochondrial signal peptide: Correlation with structural changes and evidence for cooperativity. *Biochemistry* **35,** 10931–10937.

Gruner, S. M. (1989). Stability of lyotropic phases with curved interfaces. *J. Phys. Chem.* **93,** 7562–7570.

Hartsel, S. C., and Cafiso, D. S. (1986). A test of discreteness-of-charge effects in phospholipid vesicles: Measurements using paramagnetic amphiphiles. *Biochemistry* **25,** 8214–8219.

He, K., Ludtke, S. J., Heller, W. T., and Huang, H. W. (1996a). Mechanism of alamethicin insertion into lipid bilayers. *Biophys. J.* **71,** 2669–2679.

He, K., Ludtke, S. J., Worcester, D. L., and Huang, H. W. (1996b). Neutron scattering in the plane of membranes: Structure of alamethicin pores. *Biophys. J.* **70,** 2659–2666.

Holtzer, A. (1992). The use of Flory-Huggins theory in interpreting partitioning of solutes between organic liquids and water. *Biopolymers* **32,** 711–715.

Holtzer, A. (1994). Does Flory-Huggins theory help in interpreting solute partitioning experiments? *Biopolymers* **34,** 315–320.

Huang, C.-H., and Charlton, J. P. (1972). Interactions of phosphatidylcholine vesicles with 2-*p*-toluidinylnaphthalene-6-sulfonate. *Biochemistry* **11,** 735–740.

Hunt, J. F., Rath, P., Rothschild, K. J., and Engelman, D. M. (1997). Spontaneous, pH-dependent membrane insertion of a transbilayer alpha-helix. *Biochemistry* **36,** 15177–15192.

Iwamoto, T., Grove, A., Montal, M. O., Montal, M., and Tomich, J. M. (1994). Chemical synthesis and characterization of peptides and oligomeric proteins designed to form transmembrane ion channels. *Int. J. Pept. Protein Res.* **43,** 597–607.

Jacobs, R. E., and White, S. H. (1989). The nature of the hydrophobic binding of small peptides at the bilayer interface: Implications for the insertion of transbilayer helices. *Biochemistry* **28,** 3421–3437.

Jahnig, F. (1983). Thermodynamics and kinetics of protein incorporation into membranes. *Proc. Natl. Acad. Sci. U.S.A.* **80,** 3691–3695.

John, E., and Jahnig, F. (1988). Dynamics of melittin in water and membranes as determined by fluorescence anisotropy decay. *Biophys. J.* **54,** 817–827.

Kagan, B. L., Ganz, T., and Lehrer, R. I. (1994). Defensins: A family of antimicrobial and cytotoxic peptides. *Toxicology* **87,** 131–149.

Kaiser, E. T., and Kezdy, F. J. (1983). Secondary structures of proteins and peptides in amphiphilic environments. *Proc. Natl. Acad. Sci. U.S.A.* **80,** 1137–1143.

Kamps, M. P., Buss, J. E, and Sefton, B. M. (1985). Mutation of NH_2-terminal glycine of p60src prevents both myristoylation and morphological transformation. *Proc. Natl. Acad. Sci. U.S.A.* **82,** 4625–4628.

Keller, S. L., Bezrukov, S. M., Gruner, S. M., Tate, M. W., Vodyanoy, I., and Parsegian, V. A. (1993). Probablilty of alamethicin conductance states varies with nonlamellar tendency of bilayer phospholipids. *Biophys. J.* **65,** 23–27.

Keller, S. L., Gruner, S. M., and Gawrisch, K. (1996). Small concentrations of alamethicin induce a cubic phase in bulk phosphatidylethanolamine mixtures. *Biochim. Biophys. Acta* **1278,** 241–246.

Kienker, P. K., DeGrado, W. F., and Lear, J. D. (1994). A helical-dipole model describes the single-channel current rectification of an uncharged peptide ion channel. *Proc. Natl. Acad. Sci. U.S.A.* **91,** 4859–4863.

Killian, J. A., de Jong, A. M., Bijvelt, J., Verkleij, A. J., and de Kruijff, B. (1990). Induction of non-bilayer lipid structures by functional signal peptides. *EMBO J.* **9,** 815–819.

Kim, J., Mosior, M., Chung, L. A., Wu, H., and McLaughlin, S. (1991). Binding of peptides with basic residues to membranes containing acidic phospholipids. *Biophys. J.* **60,** 135–148.

Kim, J., Blackshear, P. J., Johnson, J. D., and McLaughlin, S. (1994). Phosphorylation reverses the membrane association of peptides that correspond to the basic domains of MARCKS and neuromodulin. *Biophys J.* **67,** 227–237.

Kiyota, T., Lee, S., and Sugihara, G. (1996). Design and synthesis of amphiphilic-helical model peptides with systematically varied hydrophobic–hydrophilic balance and their interaction with lipid- and bio-membranes. *Biochemistry* **35,** 13196–13204.

Koeppe, R. E. N., and Anderson, O. S. (1996). Engineering the gramicidin channel. *Annu. Rev. Biophys. Biomol. Struct.* **25,** 231–258.

Langner, M., Cafiso, D., Marcelja, S., and McLaughlin, S. (1990). Electrostatics of phosphoinositide bilayer membranes. Theoretical and experimental results. *Biophys. J.* **57,** 335–349.

Lear, J. D., Wasserman, Z. R., and DeGrado, W. F. (1988). Synthetic amphiphilic peptide models for protein ion channels. *Science* **240,** 1177–1181.

Lewis, J. R. and Cafiso, D. S. (1999). Binding energy of the channel forming peptide alamethicin varies with membrane spontaneous curvature. *Biophys. J.* **76,** A18.

Liu, L. P., and Deber, C. M. (1997). Anionic phospholipids modulate peptide insertion into membranes. *Biochemistry* **36,** 5476–5482.

Lu, Y., Bazzi, M. D., and Nelsestuen, G. L. (1995). Kinetics of annexin VI, calcium, and phospholipid association and dissociation. *Biochemistry* **34,** 10777–10785.

Ludtke, S. J., He, K., Heller, W. T., Harroun, T. A., Yang, L., and Huang, H. W. (1996). Membrane pores induced by magainin. *Biochemistry* **35,** 13723–13728.

Lundbaek, J. A., Maer, A. M., and Andersen, O. S. (1997). Lipid bilayer electrostatic energy, curvature stress, and assembly of gramicidin channels. *Biochemistry* **36,** 5695–5701.

McIlroy, B. K., Walters, J. D., Blackshear, P. J., and Johnson, J. D. (1991). Phosphorylation-dependent binding of a synthetic MARCKS peptide to calmodulin. *J. Biol. Chem.* **266,** 4959–4964.

McLaughlin, S. A. (1989). The electrostatic properties of membranes. *Annu. Rev. Biophys. Biophys. Chem.* **18,** 113–136.

McLaughlin, S., and Aderem, A. (1995). The myristoyl-electrostatic switch: A modulator of reversible protein–membrane interactions. *Trends Biochem. Sci.* **20,** 272–276.

Maduke, M., and Roise, D. (1993). Import of a mitochondrial presequence into protein-free phospholipid vesicles. *Science* **260,** 364–367.

Maloy, W. L., and Kari, U. P. (1995). Structure-activity studies on magainins and other host defense peptides. *Biopolymers* **37,** 105–122.

Matsuzaki, K. (1997). Molecular action mechanisms and membrane recognition of membrane-acting antimicrobial peptides. *Yakugaku Zasshi* (*J. Pharm. Soc. Jpn.*) **117,** 253–264.

Matsuzaki, K., Murase, O., Tokuda, H., Funakoshi, S., Fujii, N., and Miyajima, K. (1994). Orientational and aggregational states of magainin 2 in phospholipid bilayers. *Biochemistry* **33,**3342–3349.

Moll, T. S., and Thompson, T. E. (1994). Semisynthetic proteins: Model systems for the study of the insertion of hydrophobic peptides into preformed lipid bilayers. *Biochemistry* **33,** 15469–15482.

Morein, S., Andersson, A., Rilfors, L., and Lindblom, G. (1996). Wild type *Escherichia coli* cells regulate the membrane lipid composition in a "window" between gel and non-lamellar structures. *J. Biol. Chem.* **271,** 6801–6809.

Mosior, M., and McLaughlin, S. A. (1992a). Binding of basic peptides to acidic lipids in membranes: Effects of inserting alanine(s) between the basic residues. *Biochemistry* **31,** 1767–1773.

Mosior, M., and McLaughlin, S. A. (1992b). "Electrostatics and Dimensionality Can Produce Apparent Cooperativity when Protein Kinase C and Its Substrates Bind to Acidic Lipids in Membranes." Ellis Horwood, Chester, England.

Murray, D., Hermida-Matsumoto, L., Buser, C. A., Tsang, J., Sigal, C. T., Ben-Tal, Honig, N. B., Resh, M. D., and McLaughlin, S. (1998). Electrostatics and the membrane association of src: Theory and experiment. *Biochemistry* **37,** 2145–2159.

North, C. L., Barranger-Mathys, M., and Cafiso, D. S. (1995). Membrane orientation of the N-terminal segment of alamethicin determined by solid-state ^{15}N NMR. *Biophys. J.* **69,** 2392–2397.

Oblatt-Montal, M., Buhler, L. K., Iwamoto, T., Tomich, J. M., and Montal, M. (1993). Synthetic peptides and four-helix bundle proteins as model systems for the pore-forming structure of channel proteins. I. Transmembrane segment M2 of the nicotinic cholinergic receptor channel is a key pore-lining structure. *J. Biol. Chem.* **268,** 14601–14607.

Osterberg, F., Rilfors, L., Wieslander, A., Lindblom, G., and Gruner, S. M. (1995). Lipid extracts from membranes of *Acholeplasma laidlawii* A grown with different fatty acids have a nearly constant spontaneous curvature. *Biochim. Biophys. Acta* **1257,** 18–24.

Parente, R. A., Nadasdi, L., Subbarao, N. K., and Szoka, F. C., Jr. (1990). Association of a pH-sensitive peptide with membrane vesicles: Role of amino acid sequence. *Biochemistry* **29,** 8713–8719.

Pawlak, M., Meseth, U., Dhanapal, B., Mutter, M., and Vogel, H. (1994). Template-assembled melittin: Structural and functional characterization of a designed, synthetic channel-forming protein. *Protein Sci.* **3,** 1788–1805.

Peitzsch, R. M., and McLaughlin, S. M. (1993). Binding of acylated peptides and fatty acids to phospholipid vesicles: Pertinence to myristoyated proteins. *Biochemistry* **32,** 10426–10443.

Polozov, I. V., Polozova, A. I., Mishra, V. K., Anantharamaiah, G. M., Segrest, J. P., and Epand, R. M. (1998). Studies of kinetics and equilibrium membrane binding of class A and class L model amphipathic peptides. *Biochim. Biophys. Acta* **1368,** 343–354.

Qin, Z., and Cafiso, D. S. (1996). Membrane structure of the PKC and calmodulin binding domain of MARCKS determined by site-directed spin-labeling. *Biochemistry* **35,** 2917–2925.

Reddy, G. L., Iwamoto, T., Tomich, J. M., and Montal, M. (1993). Synthetic peptides and four-helix bundle proteins as model systems for the pore-forming structure of channel proteins. II. Transmembrane segment M2 of the brain glycine receptor is a plausible candidate for the pore-lining structure. *J. Biol. Chem.* **268,** 14608–14615.

Ren, J., Lew, S., Wang, Z., and London, E. (1997). Transmembrane orientation of hydrophobic alpha-helices is regulated both by the relationship of helix length to bilayer thickness and by the cholesterol concentration. *Biochemistry* **36,** 10213–10220.

Rosen, A., Keenan, K. F., Thelen, M., Nairn, A. C., and Aderem, A. (1990). Activation of protein kinase C results in the displacement of its myristoylated, alanine-rich substrate from punctate structures in macrophage filopodia. *J. Exp. Med.* **172,** 1211–1215.

Roux, B., and Karplus, M. (1994). Molecular dynamics simulations of the gramicidin channel. *Annu. Rev. Biophys. Biomol. Struct.* **23,** 731–761.

Roux, M., Neumann, J.-M., Bloom, M. and Devaux, P. F. (1988). ^{2}H and ^{31}P NMR study of pentalysine interaction with headgroup deuterated phosphatidylcholine and phosphatidylserine. *Eur. Biophys. J.* **16,** 267–273.

Sansom, M. S. P. (1993). Alamethicin and related peptaibols—model ion channels. *Eur. Biophys. J.* **22,** 105–124.

Schwarz, G., and Beschiaschvili, G. (1989). Thermodynamic and kinetic studies on the association of melittin with a phospholipid bilayer. *Biochim. Biophys. Acta* **979,** 82–90.

Schwarz, G., Gerke, H., Rizzo, V., and Stankowski, S. (1987). Incorporation kinetics in a membrane, studied with the pore-forming peptide alamethicin. *Biophys. J.* **52,** 685–692.

Seelig, J., and Ganz, P. (1991). Nonclassical hydrophobic effect in membrane binding equilibria. *Biochemistry* **30,** 9354–9359.

Sharp, K. A., Nicholls, A., Friedman, R., and Honig, B. (1991). Extracting hydrophobic free energies from experimental data: Relationship to protein folding and theoretical models. *Biochemistry* **30,** 9686–9697.

Shirley, B. A., Stanssens, P., Hahn, U., and Pace, C. N. (1992). Contribution of hydrogen bonding to the conformational stability of ribonuclease T1. *Biochemistry* **31,** 725–732.

Silverman, L., and Resh, M. D. (1992). Lysine residues form an intergral component of a novel NH_2-terminal membrane targeting motif for myristoylated pp60v-src. *J. Cell Biol.* **119,** 415–425.

Simon, S. A., and McIntosh, T. J. (1989). Magnitude of the solvation pressure depends on dipole potential. *Proc. Natl. Acad. Sci. U.S.A.* **86,** 9263–9267.

Skehel, J. J., Bizebard, T., Bullough, P. A., Hughson, F. M., Knossow, M., Steinhauer, D. A., Wharton, S. A., and Wiley, D. C. (1995). Membrane fusion by influenza hemagglutinin. *Cold Spring Harbor Symp. Quant. Biol.* **60,** 573–580.

Slater, S. J., Kelly, M. B., Yeager, M. D., Larkin, J., Ho, C. and Stubbs, C. D. (1996). Polyunsaturation in cell membranes and lipid bilayers and its effects on membrane proteins. *Lipids* **31,** S189–S192.

Tamm, L. K. (1991). Membrane insertion and lateral mobility of synthetic amphiphilic signal peptides in lipid model membranes. *Biochim. Biophys. Acta* **1071,** 123–148.

Tanford, C. (1978). The hydrophobic effect and the organization of living matter. *Science* **200,** 1012–1018.

Taniguchi, H., and Manenti, S. (1993). Interaction of myristoylated alanine-rich protein kinase C substrate (MARCKS) with membrane phospholipids. *J. Biol. Chem.* **268,** 9960–9963.

Thorgeirsson, T. E., Yu, Y. G., and Shin, Y. K. (1995). A limiting law for the electrostatics of the binding of polypeptides to phospholipid bilayers. *Biochemistry* **34,** 5518–5522.

Thorgeirsson, T. E., Russell, C. J., King, D. S., and Shin, Y. K. (1996). Direct determination of the membrane affinities of individual amino acids. *Biochemistry* **35,** 1803–1809.

Towler, D. A., Gordon, J. I., Adams, S. P., and Glaser, L. (1988). The biology and enzymology or eukaryotic protein acylation. *Ann. Rev. Biochem.* **57,** 69–99.

van Klompenburg, W., Nilsson, I., von Heijne, G., and de Kruijff, B. (1997). Anionic phospholipids are determinants of membrane protein topology. *EMBO J.* **16,** 4261–4266.

Veatach, W. R., Fossel, E. T., and Blout, E. R. (1974). The conformation of gramicidin A. *Biochemistry* **13,** 5249–5256.

Victor, K., and Cafiso, D. S. (1998). Structure and position of the N-terminal membrane-binding domain of pp60src at the membrane interface. *Biochemistry* **37,** 3402–3410.

Vogel, H., and Jahnig, F. (1986). The structure of melittin in membranes. *Biophys. J.* **50,** 573–582.

Vogt, T. C., Killian, J. A., and de Kruijff, B. (1994). The influence of acylation on the lipid structure modulating properties of the transmembrane polypeptide gramicidin. *Biochim. Biophys. Acta* **1193,** 55–61.

von Heijne, G. (1992). Membrane protein structure prediction. Hydrophobicity analysis and the positive-inside rule. *J. Mol. Biol.* **225,** 487–494.

von Heijne, G. (1995). Membrane protein assembly: rules of the game. *BioEssays* **17,** 25–30.

von Heijne, G., and Gavel, Y. (1988). Topogenic signals in integral membrane proteins. *Eur. J. Biochem.* **174,** 671–678.

Wade, D., Andreu, D., Mitchell, S. A., Silveira, A. M., Boman, A., Boman, H. G., and Merifield, R. B. (1992). Antibacterial peptides designed as analogs or hybrids of cecropins and melittin. *Int. J. Pept. Protein. Res.* **40,** 429–436.

Wallace, B. A. (1996). The roles of tryptophan residues in the structure, function, and folding of the gramicidin transmembrane ion channel. *Adv. Exp. Med. Biol.* **398,** 607–614.

Wang, Y., and Weiner, H. (1994). Evaluation of electrostatic and hydrophobic effects on the interaction of mitochondrial signal sequences with phospholipid bilayers. *Biochemistry* **33,** 12860–12867.

Wattenbarger, M. R., Chan, H. S., Evans, D. R., Bloomfield, V. A., and Dill, K. A. (1990). Surface-induced enhancement of internal structure in polymers and proteins. *J. Chem. Phys.* **93,** 8343–8351.

Wertz, S. L., Savino, Y. and Cafiso, D. S. (1996). Solution and membrane bound structure of a peptide derived from the PKC substrate domain of neuromodulin. *Biochemistry* **35,** 11104–11112.

Westerhoff, H. V., Zasloff, M., Rosner, J. L., Hendler, R. W., De Waal, A., Vaz Gomes, A., Jongsma, P. M., Riethorst, A., and Juretic, D. (1995). Functional synergism of the magainins PGLa and magainin-2 in *Escherichia coli,* tumor cells and liposomes. *Eur. J. Biochem.* **228,** 257–264.

White, S. H., and Wimley, W. C. (1994). Peptides in lipid bilayers: Structural and thermodynamic basis for partitioning and folding. *Curr. Opin. Struct. Biol.* **4,** 79–86.

Wimley, W. C., and White, S. H. (1992). Partitioning of tryptophan side-chain analogs between water and cyclohexane. *Biochemistry* **31,** 12813–12818.

Wimley, W. C., and White, S. H. (1993). Membrane partitioning: Distinguishing bilayer effects from the hydrophobic effect. *Biochemistry* **32,** 6307–6312.

Wimley, W. C., and White, S. H. (1996). Experimentally determined hydrophobicity scale for proteins at membrane interfaces. *Nat. Struct. Biol.* **2,** 842–848.

Wimley, W. C., Creamer, T. P., and White, S. W. (1996). Solvation energies of amino acid side chains and backbone in a family of host-guest pentapeptides. *Biochemistry* **35,** 5109–5124.

Winiski, A. P., McLaughlin, A. C., McDaniel, R. V., Eisenberg, M., and McLaughlin, S. A. (1986). An experimental test of the discreteness-of-charge effect in positive and negative lipid bilayers. *Biochemistry* **25,** 8206–8214.

Wooley, G. A., and Wallace, B. A. (1992). Model ion channels: Gramicidin and alamethicin. *J. Membr. Biol.* **129,** 109–136.

Wyman, T. B., Nicol, F., Zelphati, O., Scaria, P. V., Plank, C., and Szoka, F. C., Jr. (1997). Design, synthesis, and characterization of a cationic peptide that binds to nucleic acids and permeabilizes bilayers. *Biochemistry* **36,** 3008–3017.

Yang, L., and Glaser, M. (1995). Membrane domains containing phosphatidylserine and substrate can be important for the activation of protein kinase C. *Biochemistry* **34,** 1500–1506.

You, S., Peng, S., Lien, L., Breed, J., Sansom, M. S., and Woolley, G. A. (1996). Engineering stabilized ion channels: Covalent dimers of alamethicin. *Biochemistry* **35,** 6225–6232.

Yu, Y. G., Thorgeirsson, T. E., and Shin, Y.-K. (1994). Topology of an amphipathic mitochondrial signal sequence in the membrane-inserted state. A spin labeling study. *Biochemistry* **33,** 14221–14226.

Zhang, Y. P., Lewis, R. N., Hodges, R. S., and McElhaney, R. N. (1992). Interaction of a peptide model of a hydrophobic transmembrane alpha-helical segment of a membrane protein with phosphatidylcholine bilayers: Differential scanning calorimetric and FTIR spectroscopic studies. *Biochemistry* **31,** 11579–11588.

Zhang, Y. P., Lewis, R. N., Hodges, R. S., and McElhaney, R. N. (1995). Interaction of a peptide model of a hydrophobic transmembrane alpha-helical segment of a membrane protein with phosphatidylethanolamine bilayers: Differential scanning calorimetric and Fourier transform infrared spectroscopic studies. *Biophys. J.* **68,** 847–857.

Zheng, N., and Gierasch, L. M. (1996). Signal sequences: The same yet different. *Cell.* (*Cambridge, Mass.*) **86,** 849–852.

CHAPTER 8

Lateral Diffusion of Lipids and Proteins

Michael J. Saxton
Institute of Theoretical Dynamics, University of California, Davis, California 95616

I. INTRODUCTION

A century ago, Overton began to examine one biologically essential function of the membrane, to serve as a barrier that allows selective trans-

Current Topics in Membranes, Volume 48

1063-5823/99 $30.00

port across the membrane. In the 1970s, biologists began to examine another essential function of the membrane to allow diffusion and reaction within the plane of the membrane.

This chapter focuses on the factors that affect the lateral diffusion coefficient *D,* the effects of heterogeneity on lateral diffusion, and two tools developed in physics to characterize heterogeneous systems, percolation theory and anomalous diffusion. The heterogeneity of cell membranes is an active area of research, and many excellent reviews are available with viewpoints ranging from computational statistical mechanics to detailed biochemical studies of receptors (Abney and Scalettar, 1995; Bergelson *et al.,* 1995; Edidin, 1992a, 1993, 1997; Holowka and Baird, 1996; Jain and White, 1977; Mouritsen and Jorgensen, 1997; Neubig, 1994; Tocanne *et al.,* 1994a; Vaz and Almeida, 1993; Welti and Glaser, 1994; Wolf, 1994, 1995). Many general reviews on membrane dynamics are available (Abney and Scalettar, 1993; Almeida and Vaz, 1995; Damjanovich *et al.,* 1994; Edidin, 1992b, 1994a,b, 1996; Jacobson *et al.,* 1995; Kusumi and Sako, 1996; Lipowsky and Sackmann, 1995; Sheets *et al.,* 1995; Sheetz, 1993; Tocanne *et al.,* 1994b; Zhang *et al.,* 1993).

A. The Fluid Mosaic Model and Its Modifications

According to the fluid mosaic model (Singer, 1971, 1974; Singer and Nicolson, 1972), the membrane is a lipid bilayer, forming a two-dimensional oriented viscous solution of integral proteins in a fluid lipid matrix. Integral proteins span the membrane; peripheral proteins do not cross the membrane but are bound noncovalently to it. Lateral diffusion in the plane of the membrane and rotational diffusion perpendicular to the membrane are allowed, but displacements perpendicular to the membrane and rotations toward the plane of the membrane are highly restricted. It is instructive to read the original papers to see how much of the original model is now textbook material, but some modifications in the model are now required to include other structural elements and to allow heterogeneity.

The most important structural element lacking in the original model is the cytoskeleton, which can immobilize proteins directly or indirectly, or hinder diffusion by the formation of corrals near the membrane surface. Our understanding of the membrane skeleton (Bennett and Gilligan, 1993; Morrow *et al.,* 1997) and membrane–cytoskeletal interactions (Janmey, 1995; Luna and Hitt, 1992) has increased considerably since the discussion by Singer (1974). Other membrane structures have been characterized, such as coated pits (Schmid, 1997), caveolae (Parton, 1996; Parton and Simons, 1995), and related detergent-insoluble, glycolipid-enriched domains (DIGs) (Ahmed *et al.,* 1997; Brown, 1998; Brown and London, 1997; Harder and Simons, 1997; Okamoto *et al.,* 1998). The original model emphasized integral and peripheral proteins, although it recognized the possibility of monotopic proteins, which are inserted into the bilayer but do not span the membrane. Few monotopic proteins have been identified; two examples are prostaglandin H synthase (Picot and Garavito, 1994; Picot *et al.,* 1994) and caveolin (Parton, 1996). In addition, there are glycosylphosphatidylinositol (GPI)-linked proteins (Englund, 1993; Nosjean *et al.,* 1997) and acylated proteins (Johnson *et al.,* 1994; Mumby, 1997; Zhang and Casey, 1996; Zurzolo and Rodriguez-Boulan, 1994).

The fluid mosaic model must also be modified to allow heterogeneity over a wide range of time and length scales. In particular, lipid domains may be present. Diffusion may be obstructed by proteins or gel-phase lipids, and membrane components frequently show non-Brownian motion. Metabolic turnover may contribute to heterogeneity (Gheber and Edidin, 1998). Recent experiments show the possibility of ordered superlattice domains in fluid-phase lipids. These structures, perhaps transient, are believed to result from long-range elastic repulsion among bulky, rigid species such as pyrene or sterols (Cheng *et al.,* 1997; Chong, 1994; Liu *et al.,* 1997; Parasassi *et al.,* 1995; Sugar *et al.,* 1994). As recognized in the original model, the distribution of phospholipids across the membrane may be asymmetric; the asymmetry is maintained by metabolic energy (Devaux, 1992).

B. The Physiological Importance of Lateral Diffusion

The lateral diffusion of membrane components is essential to a variety of physiological processes in cells (Axelrod, 1983; McCloskey and Poo, 1984). Mobile receptors are common. The first steps in the action of insulin, epidermal growth factor, and nerve growth factor are the binding of the hormone to its receptor and the cross-linking of the receptors (Schlessinger, 1980, 1993–1994; Heldin, 1995). Similarly, histamine release by mast cells is triggered by the aggregation of Fc receptors for IgE (Metzger and Kinet, 1988; Holowka and Baird, 1996). The importance of lateral diffusion in adhesion was demonstrated for the adherence of vesicles to cells (McCloskey and Poo, 1986a) and cells to supported lipid bilayers (Chan *et al.,* 1991; Dustin *et al.,* 1996). Lateral diffusion increases bond formation and allows adhesion molecules to accumulate at the contact region. Diffusion-coupled reactions are also essential. One of the first steps in visual transduction is the activation of ~500 molecules of G-protein by a single activated rhodopsin molecule (Stryer, 1986; Liebman *et al.,* 1987; Lamb, 1996). Electron transfer in mitochondria (Lenaz, 1988; Hackenbrock *et al.,* 1986; Chazotte and Hackenbrock, 1989) and chloroplasts (Lavergne and Joliot, 1991) requires diffusion of mobile redox carriers.

C. Experimental Techniques

The time and length scales of measurements of lateral motion are of crucial importance, particularly in an inhomogeneous system, where different methods of measurement see different inhomogeneities. This chapter focuses on diffusion at long time scales; for a lucid discussion of measurements at shorter time scales, see Bloom and Thewalt (1994, 1995), and for more details on nuclear magnetic resonance (NMR) length scales, see Karakatsanis and Bayerl (1996). The distance dependence of the diffusion coefficient is discussed elsewhere (Saxton, 1989a, 1994a; Vaz and Almeida, 1991). Shin *et al.,* (1991) described an elegant electron spin resonance (ESR) technique permitting measurement of short-range and long-range diffusion coefficients in the same experiment on the same sample, thereby removing one of the potential artifacts (Saxton, 1989a) in comparison of the two diffusion coefficients.

Space does not permit a review of experimental techniques here; see the general reviews by Edidin (1992b) and Tocanne *et al.,* (1994b). Fluoresence photobleaching recovery (FPR) is reviewed by Edidin (1994b), Jovin and Vaz (1989), Petersen *et al.,* (1986), and Wolf (1989). Single-particle tracking (SPT) is discussed briefly in Section V,A and is reviewed elsewhere (Saxton and Jacobson, 1997). NMR measurements are reviewed by Bloom and Thewalt (1994, 1995). Most NMR measurements are short range; long-range measurements are made by pulsed-gradient spin echo techniques (Lindblom and Orädd, 1994) or by using bilayers supported on glass microspheres (Hetzer *et al.,* 1998 and references cited therein). ESR techniques are reviewed by Freed (1994). Fluorescence quenching and excimer formation are short-range measurements in which the range is set by the distance a particle can diffuse during the fluorescence lifetime; the use of labels with long lifetimes extends the range significantly (Davenport, 1997; Terpetschnig *et al.,* 1997). Colocalization experiments at the resolution of light microscopy have been essential in work on the association of receptors with membrane domains (Holowka and Baird, 1996; Stauffer and Meyer, 1997), and various forms of high-resolution optical microscopy are being developed to characterize association at resolutions below the wavelength of light. Imaging fluorescence correlation spectroscopy can be used to measure diffusion, colocalization, cluster density, and cluster size (Petersen *et al.,* 1993; Wiseman *et al.,* 1997; Huang and Thompson, 1996). Fluorescence resonance energy transfer (FRET) (Matko and Edidin, 1997; Wu and Brand, 1994) and imaging FRET (Berland *et al.,* 1996; Kenworthy and Edidin, 1998; Koppel *et al.,* 1994; Schwille *et al.,* 1997) measure colocalization at high spatial resolution, as will SPT measurements in which two fluorescent labels are used simultaneously. FRET has been demonstrated between individual fluorophore molecules (Schütz *et al.,* 1997a). Laser tweezer measurements are reviewed by Svoboda and Block (1994) and Sheetz (1998). The barrier-free path (BFP) is discussed by Edidin *et al.,* (1991, 1994) and by Kusumi *et al.,* (1998). Finally, the width of the distribution of experimental and Monte Carlo diffusion coefficients may be used as a measure of the inhomogeneity of the system; the scatter in D ought to be viewed as signal as well as noise (Saxton, 1997).

II. THE KEY QUESTION

The key question in lateral diffusion in cell membranes is why diffusion coefficients for integral proteins in the plasma membrane are one or two orders of magnitude lower than diffusion coefficients for integral proteins in an artificial bilayer. (See, for example, Edidin, 1994b, 1996; Webb *et al.,* 1981; Zhang *et al.,* 1993.)

For lipids in the fluid phase of artificial bilayers, typical diffusion coefficients are 1–10 μm^2/sec. For lipids in plasma membranes, D is lower, in the range of 0.05–2 μm^2/sec. In the gel phase, diffusion is along defects and grain boundaries, and $D \leq 0.01$ μm^2/sec. In FPR experiments on pure fluid-phase lipids, the mobile fraction is usually close to one.

In some systems the diffusion coefficients of integral membrane proteins are close to the value predicted by the Saffman–Delbrück equation. These systems include artificial bilayers; the rod outer segment; spherocytic red blood cells, in which the membrane skeleton is missing due to mutation (Sheetz *et al.,* 1980); and blebs, in which the membrane appears to be detached from cytoskeletal attachments (Tank *et al.,* 1982; Webb *et al.,* 1981). However, the interaction of membrane and cytoskeleton is not always the controlling factor; removal of the glycosylation sites of mouse class I major histocompatibility complex (MHC) antigen L^d increased D to a value near the Saffman–Delbrück value (Wier and Edidin, 1988). Values of D for proteins in artificial bilayers are in the range of 1–5 μm^2/sec and for the mentioned other cases 0.2–1 μm^2/sec.

For most integral proteins in the plasma membrane, D is in the range of 0.01–0.2 μm^2/sec and the mobile fraction is 0.25 to 0.75. In a standard FPR experiment, the immobile fraction corresponds to $D < 0.0001$ μm^2/sec. The fastest-moving GPI-linked proteins have diffusion coefficients close to the values for lipids in plasma membranes, but the mobile fraction may be around 50%. Here the immobile fraction must be maintained with no direct cytoskeletal interactions.

Extensive tables of numerical data are given in various reviews. For lipid data, see Lee and Jacobson (1994), and for GPI-linked and transmembrane proteins, see Clegg and Vaz (1985) and Edidin (1992b, 1994b).

III. THEORETICAL DEVELOPMENTS

The main qualitative modification required for the Singer–Nicolson model is the recognition of heterogeneity of membrane structure and the resulting heterogeneity of motion. The quantitative tools developed in physics to characterize heterogeneity are percolation theory and anomalous diffusion (Bouchaud and Georges, 1990, Bunde and Havlin, 1991, 1994; Haus and Kehr, 1987; Havlin and Ben-Avraham, 1987; Hughes 1995, 1996; Shlesinger, 1988; Stauffer and Aharony, 1992; Weiss, 1994). The book by Dewey (1997) is an excellent introduction to percolation theory, anomalous diffusion, and fractals. This section discusses the physical principles; Section V discusses applications to membranes.

A. Percolation Theory

As Singer and Nicolson (1972) stated, "None of the evidence so far obtained for the bilayer form permits us to say whether the bilayer is *continuous* or *interrupted.*" The connectivity of obstructed systems is systematically treated in percolation theory. At low obstacle concentrations, the obstacles form islands in a continuous conducting phase, and at high obstacle concentrations, the conducting phase forms lakes in a continuous obstructed phase. The percolation transition is the geometric phase transition between these states. The concentration at which the transition occurs is called thc percolation threshold. The threshold is given here in terms of the area fraction of obstacles; in the physics literature it is more often given in terms of the area fraction of the conducting phase. Early applications of percolation to membranes are summarized by Vaz (1992).

Percolation thresholds are considered for the continuum and various lattices. Attention is restricted to site percolation, in which nodes of a lattice are blocked randomly; in bond percolation, the bonds connecting the nodes are cut randomly. A percolation threshold exists only when the obstacles are immobile. For mobile obstacles there is instead a glass transition, at which the diffusion coefficient of the mobile particles is decreased sharply by crowding. Below the glass transition, motion is hydrodynamic; above the transition, motion is controlled by thermally activated jump processes (Barrat and Klein, 1991; Fehr and Löwen, 1995). For a two-dimensional Lennard–Jones fluid at a dimensionless temperature of 0.50, the glass transition occurs at a dimensionless density of 0.83 (Ranganathan, 1994). As the concentration of immobile particles is increased, the system reaches the percolation threshold, whereas the system reaches the glass transition as the concentration of mobile particles are increased.

Some numerical values of the site percolation threshold in two dimensions are given in Table I. The threshold increases as the coordination number z of the lattice increases. For randomly centered, overlapping circles, the continuum percolation threshold is at an obstacle area fraction of C = 0.6764 (Lee and Torquato, 1990), and for hard disks, the threshold is C = 0.82 (Berryman, 1983). Intermediate cases are treated in the cherry pit model, in which the obstacles have impenetrable hard cores surrounded by shells that may overlap (Lee and Torquato, 1990). In the original work, the overlap was simply used as a device to generate complex obstacle shapes from simple elements (Torquato, 1991). This geometry, however, may be interpreted as an impenetrable obstacle surrounded by a region inaccessible

TABLE I

Two-Dimensional Percolation Thresholds[a]

Effect of lattice for random point obstacles

Lattice	z	C_P
Honeycomb[b]	3	0.303
Square[b]	4	0.407
Triangular[b]	6	0.500
Overlapping circles on continuum[c]		0.676
Impenetrable circles on continuum[d]		0.82

Effect of obstacle shape for triangular lattice[e]

Obstacle	C_P	
Hexagon	0.585	compact
Point	0.500	
Linear trimer	0.461	extended

[a] In terms of obstructed area fraction.
[b] From Stauffer and Aharony (1992).
[c] From Lee and Torquato (1990).
[d] From Berryman (1983).
[e] From Saxton (1993a).

to a tracer of finite radius (Kim and Torquato, 1992) or as an impenetrable protein surrounded by penetrable layers of boundary lipid.

The threshold also depends on obstacle shape, decreasing as the obstacle becomes more extended, as shown in Table I for the triangular lattice. For the continuum, Thorpe and collaborators calculated thresholds for identical overlapping ellipses with random centers and orientations and, obtained an empirical equation for the threshold as a function of the aspect ratio of the ellipses (Xia and Thorpe, 1988; Garboczi *et al.,* 1991). As the ellipses become more elongated, the percolation threshold goes to zero. This equation has been used to estimate the aspect ratio of obstacles from observed percolation thresholds in FPR measurements on lipid mixtures in the two-phase region (Almeida and Vaz, 1995). The aspect ratio provides a useful qualitative description, although it should not be taken too literally in a system of unknown and possibly complex geometry. Thresholds for some random fractal aggregates are given by Saxton (1993a).

So far we have assumed randomly positioned obstacles, but interactions among the obstacles may have a major effect on the threshold. A potential that tends to align obstacles linearly lowers the threshold. A repulsive interaction among obstacles raises the threshold (for a good example, see Dewey, 1997, pp. 229–231). For interacting obstacles, the situation is complex because the values of the threshold depend on the temperature and the distance at which particles are defined to be in the same cluster (Bug *et al.,* 1985; Cooper *et al.,* 1989; Heyes and Melrose, 1989).

We have also assumed a point diffusing particle, but in reality the tracer has a nonzero radius so the diffusion coefficient and the percolation threshold depend on the excluded area, not the obstacle area. For a diffusing species at infinite dilution in a pure lipid, the dependence of D on tracer size is given by the free volume model or the Saffman–Delbrück model (Section IV,A). Obstruction leads to an additional contribution to the size dependence. The effects can be large; for random point obstacles on a triangular lattice, as the tracer size is increased from a point to a hexagon of unit radius to a hexagon of radius two, the threshold decreases from 0.50 to 0.12 to 0.048 (Saxton, 1993a). The effect is also discussed by Kim and Torquato (1992) for the continuum in three dimensions and by Slater and Treurniet (1997) for square obstacles on the square lattice. For fractal obstacles there is little dependence on tracer size because fractal obstacles are self-similar, so tracers see on average the same hindrances to diffusion whatever the tracer size (Saxton, 1993a).

The percolation threshold is defined as the obstacle concentration for the percolation transition in the limit of an infinite system. To find a percolation threshold, one generates random configurations of obstacles at a given area fraction in a box of size L and tests whether there is a percolating cluster.

From this one finds the cumulative probability that a percolating cluster is present, as a function of the area fraction of obstacles. As shown in Fig. 1, this percolation probability grows sharper as the system size L increases, as in any phase transition. The width of the transition region is proportional to $1/L^{3/4}$. Ther percolation threshold is the limiting value as $L \rightarrow \infty$; the intersection of the curves for different L is one estimate of the threshold. Technical details and references to the physics literature are given elsewhere (Saxton, 1993a).

The sharpening of the percolation probability as L increases could affect experimental results because different experimental techniques have different length scales. A percolation threshold in NMR (Dolainsky *et al.*, 1997) may differ from one in FPR, and the sharpening may be detectable in a series of FPR experiments in which the bleach beam radius is varied. Consider the FPR mobile fraction as a function of obstacle concentration. Well below the threshold, the mobile fraction is one. Well above the threshold, the mobile fraction is zero, except for finite domains on the edge of the bleach spot. Near the threshold, some of the tracers are trapped in finite domains and do not contribute to the recovery unless they are at the edge of the bleach spot. The percolation probability grows sharper as the system size increases, and in the limit of an infinite system the total area of finite domains becomes negligible compared to the total area of the percolating cluster of conducting sites below the threshold or the total area of obstructed sites above the threshold. In the Monte Carlo calculation

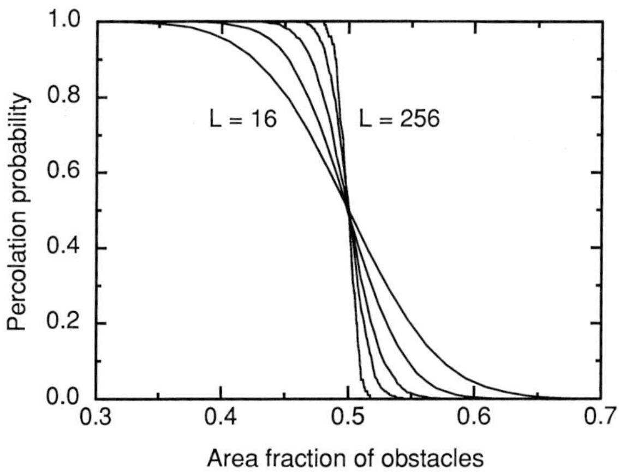

FIGURE 1 Cumulative probability that a percolating cluster is present, as a function of the area fraction of obstacles for random point obstacles on the triangular lattice with system sizes L = 16, 32, 64, 128, and 256. The percolation threshold is exactly 0.5. These are Monte Carlo curves (Saxton, 1993a) but individual data points are not shown.

leading to Fig. 1, the system size is the prescribed lattice size L and the obstacles are lattice points. In FPR, the system size is the radius over which recovery occurs, say factor of 3 larger than the bleach beam radius. For protein obstacles, the obstacle size is the protein size; for gel-phase lipid obstacles, it is the minimum size of a gel-phase domain stable over the time scale of the experiment.

B. The Diffusion Coefficient in Obstructed Systems

The next step is to consider the effect of obstruction on the diffusion coefficient (Scalettar and Abney, 1991). These results are based largely on Monte Carlo calculations for a lattice or the continuum. For speed and convenience of programming, many Monte Carlo calculations of obstructed diffusion are done on lattices. In either case, a tracer carries out a random walk in the presence of obstacles at a prescribed concentration, the mean-square displacement $\langle r^2 \rangle$ is found as a function of time, and the long-range diffusion coefficient is found as the limiting slope at large times. The mean-square displacement is averaged over many random walks with a particular obstacle configuration, and many obstacle configurations with the prescribed obstacle concentration. The obstacle concentration is given as the area fraction C.

Several approximations are made in these calculations: perturbation of lipid by obstacles is neglected, a hard-core repulsion is often assumed, and hydrodynamic interactions must be taken into account separately. These approximations affect both lattice and continuum calculations and will be discussed in detail in Section IV. Two additional approximations are required when a lattice model is used.

First, the lattice model says nothing about dynamics on time scales less than the time required for a particle to diffuse one lattice constant. For a lipid spacing of 0.8 nm and a diffusion coefficient of 5 μm^2/sec, the equation $\langle r^2 \rangle = 4\ Dt$ implies that the time to diffuse one lattice constant is 32 nsec. Faster processes (Bloom and Thewalt, 1994), such as trans-gauche isomerization in the fatty acid chains (0.1 nsec) and diffusional rotation of the lipid (1 nsec), are averaged over. Such processes are included implicitly in the diffusion coefficient D_0 at zero obstacle concentration.

Second, at short distances there may be errors because the lattice is discrete. This does not seem to be a major limitation for long-range diffusion coefficients. For a system of mobile particles, a continuum model based on Brownian dynamics (Scalettar and Abney, 1991) gave diffusion coefficients in good agreement with Monte Carlo results on a lattice (Pink, 1985, Saxton, 1987). However, the problem is particularly noticeable in histograms of parameters used to characterize short SPT trajectories and there a continuum calculation is more appropriate (Saxton, 1994b).

Obstructed diffusion problems can be classified according to the ratio γ of the jump rate of the tracer to the jump rate of the obstacle, which can be taken to be the ratio of the diffusion coefficients of the two species in the limit of low obstacle concentrations (Saxton, 1987). If $\gamma = 1$, all particles are mobile with the same diffusion coefficient, as for a single species of mobile protein in a bilayer. This is self-diffusion or tracer diffusion. Monte Carlo values of D as a function of the concentration of mobile particles for point and hexagonal particles on the triangular lattice are given by Pink (1985) and by Saxton (1987), who gives an empirical formula for D. If $1 < \gamma < \infty$, there are two mobile species with different mobility, as in the case of a lipid or small protein obstructed by a larger mobile protein. A simple formula for $D(C, \gamma)$ was given by Van Beijeren and Kutner (1985), and this formula agrees well with Monte Carlo results for the square lattice, except in the percolation limit (Saxton, 1987).

The case $\gamma \rightarrow \infty$ is the percolation limit, in which obstacles are immobile. Immobile obstacles have a much greater effect than mobile obstacles. The key factor determining D is the presence of a percolation threshold. In fact, a crude first approximation to the diffusion coefficient (normalized to one at zero obstacle concentration) is simply $D^* = 1 - C/C_P$ (Saxton, 1994a), although for the continuum it is significantly more accurate to use the equation of Garboczi *et al.* (1991) discussed later in this section. For a square lattice, the formula $D^* = 1 - (\pi - 1)C - 0.85571C^2$ gives the first three terms of the series expansion exactly (Nieuwenhuizen *et al.*, 1986).

Consider the diffusion of a point particle as immobile obstacles are added at random. Initially, diffusion is unhindered. As obstacles are added, diffusion is increasingly hindered, until the percolation threshold is reached. Above the threshold, long-range diffusion is blocked but local diffusion in finite domains is possible. As the obstacle concentration increases further, the size of the finite domains is reduced. Near the threshold, diffusion is hindered because the particle encounters dead ends, bottlenecks, and backbends on all length scales. Near C_P, the diffusion coefficient is $D \sim |C - C_P|^{\mu}$, where C is the obstacle concentration, C_P is the percolation threshold, and the conductivity exponent $\mu \simeq 1.3$. In two dimensions, the behavior of the diffusion coefficient near the threshold is the same for lattice and continuum cases (Feng *et al.*, 1987; Halperin *et al.*, 1985). This agreement is a surprising result. For an obstructed lattice, the narrowest path between two obstacles is a chain of single bonds, but for random circular obstacles on the continuum, there is a continuous distribution of neck widths between pairs of obstacles.

Compact obstacles such as hexagons are less effective obstacles than the same area fraction of random point obstacles, and random point obstacles are less effective than extended or fractal obstacles (Eisinger *et al.*, 1986; Garboczi *et al.*, 1991; Saxton, 1987, 1992, 1993a; Xia and Thorpe, 1988). Random fractal aggregates become more effective obstacles as the fractal dimension decreases, i.e., as their structure becomes more extended (Schram *et al.*, 1996).

Monte Carlo results are available for various obstacle geometries. For lattice diffusion, Eisinger *et al.* (1986) developed the milling crowd model to examine the effects of obstruction on excimer formation and found D for point tracers with hexagonal obstacles on the triangular lattice. Saxton (1993a) examined the effects of hexagonal obstacles. Schram *et al.* (1994) used FPR simulations to find the effect of immobile circular obstacles on point tracers and obtained an empirical formula for D as a function of obstacle concentration and radius. Saxton (1992) and Schram *et al.* (1996) examined the effects of random fractal aggregates. Saxton (1996) found diffusion coefficients for a variety of obstruction/binding and pure binding models. Saxton (1990a) examined the effect of mixtures of mobile and immobile obstacles as the mixture was varied at constant area fraction from completely mobile obstacles to completely immobile. An improved algorithm for lattice diffusion was developed by Slater and Guo (1996). For the continuum case, Torquato and Kim (1989) and Tobochnik (1990) presented similar improved algorithms and obtained D for circular overlapping obstacles. Garboczi *et al.* (1991) examined the effect of overlapping ellipses with random centers and random orientation and presented an empirical equation for D as a function of the obstacle concentration and the aspect ratio of the ellipses. This equation includes the proper power-law dependence of D on C near the percolation threshold and supersedes the empirical equation given by Saxton (1982). [Note that the equation of Garboczi *et al.* (1991) is given in terms of an area density of conducting phase x, which may be calculated from the area fraction of overlapping elliptical obstacles using the equation $x = [-4(1 + \rho)^2/(\pi\rho)]\ln(1 - C)$, where $\rho \leq 1$ is the ratio of the minor to major axes of the ellipse. This equation can be derived from results of Xia and Thorpe (1988).]

If the tracer can penetrate the obstacles, percolation still occurs but its effects are much smaller. This case includes gel-phase lipids that do not exclude the tracer completely and liquid–liquid lateral phase separation. One approximation used here is the Bruggeman–Landauer formula (Saxton, 1982), but a better approach is to use rigorous bounds on D. If the diffusion coefficients in the two phases differ by a factor of 10 or less, the upper and lower bounds are close and provide excellent approximations for D (Torquato, 1991, 1994). Almeida *et al.* (1992a) applied these bounds to lateral phase separation of the liquid-ordered and liquid-disordered phases in mixtures of dimyristoylphosphatidylcholine and cholesterol.

These bounds describe the effect of immobile obstacles in D for both penetrable and impenetrable obstacles (Torquato, 1991, 1994). They are based on variational principles and use methods of the statistical mechanics of fluids to describe the structure of the mixture. As more information is incorporated, the bounds become narrower. The first piece of information is simply the area fraction. The bounds are narrowed if it is known that the system is isotropic and can be narrowed further using microstructural information available for perfectly penetrable, partially penetrable, and impenetrable disks. The structural characterization in these bounds does not include the long-range correlations that occur at the percolation threshold, so the bounds are less useful when a percolation threshold is involved.

C. Anomalous and Normal Diffusion

In normal diffusion, the mean-square displacement is proportional to time, $\langle r^2 \rangle = 4Dt$, but in anomalous diffusion this relation does not hold. This chapter considers only anomalous subdiffusion, in which diffusion is hindered and $\langle r^2 \rangle \propto t^\alpha$ with the anomalous diffusion exponent $\alpha < 1$. The diffusion coefficient therefore decreases with time according to $D \propto t^{\alpha-1}$. In general, anomalous diffusion is the result of a deviation from the central limit theorem caused by a pathologically broad distribution of jump times or jump distances, or long-range correlations in diffusive motion (Bouchaud and Georges, 1990; Scher *et al.*, 1991).

Consider first the case of pure obstruction (Saxton, 1994a). At an obstacle concentration $C = 0$, diffusion is normal and $\alpha = 1$. For obstacle concentrations below the percolation threshold, diffusion is anomalous at times much shorter than the crossover time and normal at times much longer than the crossover time. As the obstacle concentration approaches the percolation threshold, diffusion becomes more anomalous for longer times: α decreases and the crossover time increases. At the percolation threshold, α reaches its limiting value in two dimensions of 0.696 (Havlin and Bunde, 1991) and the crossover time becomes infinite. Diffusion is then anomalous at all times and there is no crossover because the percolation cluster is self-similar. That is, the percolation cluster is statistically the same at all magnifications and, to the diffusing species, it appears to be the same on all time and length scales. Note that for pure obstruction, the anomalous diffusion exponent is between 0.696 and 1. An observed exponent significantly smaller than 0.696 implies a mechanism other than pure obstruction.

Next, consider the case of binding to immobile species (Bouchaud and Georges, 1990; Saxton, 1996). This case is complex, but one simple picture of hindered diffusion in the cell membrane is the obstruction/binding model (Saxton, 1996), similar to the post model of Zhang *et al.* (1993). Here point obstacles, representing immobile membrane proteins, bind tracers in nearest-neighbor sites. In this model, the type of diffusion is very sensitive to the initial conditions. Diffusion is highly anomalous if the tracer is initially at a random site. The system is then in a nonequilibrium state because the probability that a tracer starts at a site is independent of the depth of the potential well at that site. However, if the initial condition for the tracer is the thermal equilibrium distribution, binding reduces D but does not make diffusion any more anomalous. In fact, as Haus and Kehr (1987) show, a large class of plausible binding models cannot produce anomalous diffusion, provided the tracer is at thermal equilibrium (Saxton, 1996). As a result, observations of anomalous subdiffusion may help to test whether the membrane is in an equilibrium or a nonequilibrium state. The sensitivity to initial conditions implies a potential artifact in FPR experiments. If photobleaching changes the interaction of the tracer with the obstacles significantly, the initial condition is a nonequilibrium state.

Anomalous subdiffusion is also produced by other models, including the continuous-time random walk (CTRW) model used by Nagle (1992) in his work on the effects of anomalous subdiffusion on FPR. Here anomalous subdiffusion results from long-time tails in the waiting time distribution for a jump from a lattice site. The escape time is generated randomly at each time step, so there is temporal disorder but no spatial disorder. Diffusion is highly anomalous at all times, with no crossover to normal. Nagle (1992) simulated an FPR curve assuming a CTRW, fit the curve using the standard equation for FPR with normal diffusion (Axelrod *et al.,* 1976), and showed that the resulting values of D and the mobile fraction were strongly dependent on the time range fit.

D. Models of Cell Membrane Dynamics

Kusumi and Sako (1996) emphasize cytoskeletal corrals as a hindrance to diffusion and view the membrane as a series of potential energy barriers and wells in which the dominant barriers are a fairly regular lattice of corral walls. The anomalous subdiffusion model (Feder *et al.,* 1996) pictures the membrane as "a random array of continuously changing potential energy traps" resulting from interactions with all membrane components and varying in time and space. The range of binding energies is so broad that there is no average residence time; this assumption leads to a time-dependent diffusion coefficient (Bouchaud and Georges, 1990). The diffusion coefficient is assumed to be $D(t) \sim t^{\alpha-1}$, so the mean-square displacement $\langle r^2(t)\rangle \sim t^{\alpha}$ and diffusion is anomalous over all time scales, as in the CTRW model of Nagle (1992). Saxton and Jacobson (1997) attempted to construct an anomalous subdiffusion model out of biological parts to see whether the ideas of Kusumi and Sako (1996) and Feder *et al.* (1996) can be reconciled. Jacobson and collaborators (Jacobson *et al.,* 1995; Sheets *et al.,* 1995; Zhang *et al.,* 1993) presented models of hindered diffusion related directly to SPT experiments.

E. Structural Models

Molecular dynamics and the Pink 10-state model are important approaches to membrane modeling. Their strength is the wealth of structural detail they provide, although they are less directly applicable to long-range lateral diffusion.

Molecular dynamics uses Newtonian mechanics to calculate motion at the atomic level, assuming a potential energy function for atom–atom interactions (Jakobsson, 1997; Merz, 1997; Merz and Roux, 1996; Shen *et al.,* 1997). The system size is necessarily small, at most 100 lipids per monolayer and a few thousand water molecules, and the time scale is necessarily short, at most 1–10 nsec. The accessible time range could describe fluorescence quenching for some fluorophores but not long-range lateral diffusion. As the field develops, the emphasis may well be on short-range protein–lipid interactions, and it will be useful for work on long-range lateral diffusion to examine the extent of the perturbation of lipids near protein surfaces.

The Pink 10-state model provides a very useful level of detail, intermediate between molecular dynamics and the hard-disk obstacles of Section III,B (Mouritsen, 1990; Mouritsen *et al.,* 1996; Pink, 1990). In Monte Carlo calculations using the 10-state model, each lipid hydrocarbon chain is at a point on a triangular lattice. Chains may be in the gel state, the fluid state, or any of eight low-energy disordered states. In its original form the model was static, with the hydrocarbon chains stationary and restricted to lattice sites. An extension of the model added translational degrees of freedom on the lattice in order to account for the entropy of mixing in lateral phase separation. A novel variant uses random triangulation to allow continuous but restricted translational motion (Nielsen *et al.,* 1996). Lipid–protein interactions and hydrocarbon chain dynamics are much simpler in the 10-state model than in molecular dynamics. However, the 10-state model makes it practical to use a large enough system to model phase transitions, while still providing much molecular information and permitting detailed comparison with experiment. To characterize phase transitions, sophisticated techniques from computational statistical mechanics are used. An important result well described in terms of this model is hydrophobic matching (Mouritsen and Bloom, 1993). If hydrophobic lipid–lipid and lipid–integral protein interactions are included, the hydrophobic length of the boundary lipid must match that of the protein, leading to extension or contraction of boundary lipid acyl chains and to a statistical separation of lipids by chain length, i.e., to lipid domains surrounding proteins.

IV. FACTORS CONTROLLING LATERAL DIFFUSION

Several factors affect the lateral diffusion coefficient: lipid viscosity, obstructions, hydrodynamic interactions, binding to immobile species, and association among mobile species. The diffusion coefficient can be written as a product of terms

$$D(C) = D_0\, D^*_{\mathrm{obst}}(C)\, D^*_{\mathrm{hydro}}(C),$$

where C is the area fraction of obstacles D_0 is the diffusion coefficient at zero obstacle concentration, $D^*_{\mathrm{obst}}\,(C)$ accounts for the direct effect of obstruction, and $D^*_{\mathrm{hydro}}(C)$ accounts for hydrodynamic interactions. All the $D^*(C)$ are normalized to 1 at $C = 0$. For a detailed discussion of this approximation, see Bussell *et al.* (1995a) and Dodd *et al.* (1995). For lipid diffusion, an additional factor $D^*_{\mathrm{fv}}\,(C)$ may be included to account for altered free volume near the obstacles. The effect of binding to immobile species is to multiply D by the fraction of time the particle is freely diffusing (Koppel, 1981). Association among mobile species may affect every factor.

A. Hydrodynamic versus Free Volume Control

The diffusion coefficient at zero obstacle concentration is given by a hydrodynamic model or a free volume model. The assumptions of the models and comparison with the experiment were reviewed extensively by Almeida and Vaz (1995), Clegg and Vaz (1995), and Scalettar and Abney (1991).

Saffman and Delbrück (1975) presented a hydrodynamic model for translational and rotational diffusion in a two-dimensional continuum fluid, including interaction with the aqueous phase. The translational diffusion coefficient is

$$D = (kT/4\pi\eta h)\,[\ln(\eta h/\eta_{\mathrm{aq}}R) - \gamma_{\mathrm{E}}],$$

where η is the viscosity of the lipid phase, $\eta_{aq} \ll \eta$ is the viscosity of the aqueous phase, h is the height in the membrane of the diffusing species, R is its radius, k is the Boltzmann constant, T is temperature, and $\gamma_E = 0.5772 \ldots$ is Euler's constant. Typical values are $\eta = 1$ poise, $\eta_{aq} = 0.01$ poise, $h = 5$ nm, and $R = 2$ nm (Clegg and Vaz, 1985). The key points are the weak, logarithmic dependence on particle radius and the strong dependence on particle height. Further developments of this model are given by Evans and Sackmann (1988), Hughes *et al.* (1982), and Wiegel (1979).

According to the free volume model, diffusion is limited by the occurrence of a large enough fluctuation in free volume adjacent to the mobile species. In the form used by Almeida *et al.* (1992a).

$$D \propto \exp\left(-a_0/[a(T) - a_0]\right) \exp(-E_a/kT).$$

Here a_0 is the area of the molecule at close packing and $a(T)$ is the average free area per molecule, obtained by combining dilatometric measurements of volume and deuterium NMR measurements of thickness. The activation energy E_a is a fitted parameter. The effect of tracer radius is much greater in the free volume model than in the Saffman–Delbrück model.

Diffusion of lipid-sized species follows the free volume model, and diffusion of large proteins follows the Saffman–Delbrück equation (Clegg and Vaz, 1985). The main evidence supporting the free volume model for lipid diffusion is that the diffusion coefficient of lipids is independent of chain length (Balcom and Petersen, 1993; Vaz *et al.*, 1985) in agreement with the free volume model and in strong disagreement with the hydrodynamic model.

One obvious question is where the dividing line between the two models is? Clegg and Vaz (1985) estimated it to be an area of 1 nm^2. For a structurally dissimilar series of lipophilic probes, the dependence of D on molecular weight could be fit by a free volume expression below ~400 Da and a hydrodynamic expression above ~500 Da (Johnson *et al.*, 1996). Work from Petersen's laboratory defined the dividing line by measuring diffusion coefficients and activation energies for a series of macrocyclic polyamide amphiphiles in DMPC. The structural similarity of the probes revealed a clear break at 0.8 to 1 nm^2 (Liu *et al.*, 1997).

A second obvious question is, on which side of the dividing line does a protein with a single transmembrane helix lie? Single transmembrane helices span the transition region. A small transmembrane α-helix, polyalanine, has an area of 0.9 nm^2 (Furois-Corbin and Pullman, 1986). The channel form of gramicidin, a β(L, D)-helix, has an area of 2.5 nm^2 (Woolf and Roux, 1996). Diffusion measurements are in progress for a series of complexes of transmembrane helices with areas between 0.3 and 1.8 nm^2 (Lee *et al.*, 1998).

For large proteins, diffusion follows the Saffman–Delbrück equation. Vaz and collaborators compared diffusion coefficients of integral proteins of various sizes, including monomers, dimers, and oligomers of the acetylcholine receptor. The differences were small and consistent with appropriate modifications of the Saffman–Delbrück equation, as expected for a monomer with five transmembrane helices (Criado *et al.,* 1982; Vaz and Criado, 1985). Dimerization of the high-affinity IgE receptor reduces D by a factor of 1.6; higher oligomers have the same D but are mostly immobile. This receptor has seven transmembrane helices, and dimerization has the expected effect, but the immobilization of higher oligomers must involve biological interactions as well as hydrodynamic ones (Menon *et al.,* 1986).

B. Perturbation of Lipid-Free Volume by Obstacles

Obstacles, either membrane proteins or gel-phase lipid, may affect lateral diffusion indirectly by perturbing lipid dynamics. The effect of density fluctuations in the pure lipid is included implicitly in D_0, but the effect of obstacles on density fluctuations and ordering in the fluid-phase lipid must be treated separately. For lipid diffusion, obstacles may reduce the free volume in the lipid phase, and the diffusion coefficient may decrease significantly (Almeida *et al.,* 1992b; Minton, 1989). For protein diffusion, the effect of obstacles on viscosity is included in the hydrodynamic interaction.

A pure obstruction model gives too high a value of D, and several authors have tried to account for the discrepancy by including boundary lipid in the obstacle area (Almeida *et al.,* 1992b; Donaldson, 1989; Schram *et al.,* 1994). This approach is supported by experimental results of Blackwell and Whitmarsh (1990), who measured the effect of protein obstacles on lateral diffusion by fluorescence quenching of pyrene by plastoquinone. The integral proteins decreased collisional quenching significantly, but the fluorophore was unlikely to have collided with an obstacle during its fluorescence lifetime of 100 nsec. This result suggests that the interaction was lipid mediated. Almeida *et al.* (1992b) explained these results in terms of a model of lipid diffusion obstructed by gel-phase lipid, in which the free volume near the obstacles was reduced to account for the ordering of fluid-phase lipid by the gel-phase domains. The existence of such an effect and the values of the boundary layer thickness used are plausible, but caution is appropriate. First, the model of Almeida *et al.* (1992b) assumes ordering of boundary lipid, as is appropriate for gel-phase lipid obstacles. However, proteins may order or disorder boundary lipid (Shen *et al.,* 1997; Woolf and Roux, 1994), as determined in part by hydrophobic matching (Section III,E). Second, it is assumed implicitly that obstruction by protein plus boundary lipid is the only hindrance to diffusion, and the boundary layer thickness is chosen to account for the entire discrepancy between theory and experiment. Such boundary layer thicknesses must be viewed as upper limits until all other interactions are taken into account, particularly hydrodynamic interactions in the case of proteins obstructed by proteins. In the case of diffusion of lipids obstructed by gel-phase lipids or proteins, this approach is more justified. It would be useful to be able to include in the model an independent measure of the free volume near an obstacle.

C. Obstruction

The effect of obstacles was discussed in detail in Section III. Obstruction may occur in the transmembrane, extracellular, and cytoplasmic regions. In the first two regions, the effect is predominantly direct obstruction, but in the cytoplasmic region, the obstruction of proteins by the membrane skeleton is presumably dominant. As a first approximation, obstruction in the transmembrane region may be taken to be a pure hard-disk interaction, but in the extracellular region the obstacles may be more flexible and may interact by screened Coulomb forces. The area fraction of obstacles is in general different in the three regions (Section V,F).

D. Hydrodynamic Interactions

The hydrodynamic interaction among particles is a significant hindrance to lateral diffusion, as shown in an important series of theoretical papers by Hammer, Koch, and collaborators. For mobile obstacles, they derived the equivalent of the Saffman–Delbrück equation for a pair of interacting particles (Bussell *et al.,* 1992, 1995a). For immobile obstacles, they assumed that the mobile protein was diffusing through a fixed bed of obstacles and used a two-dimensional version of the Brinkman model for hydrodynamic interactions (Bussell *et al.,* 1994, 1995b). They also obtained numerical solutions of the hydrodynamic equations, in good agreement with the analytical results (Dodd *et al.,* 1995).

One of their main results is that for immobile obstacles, the hydrodynamic interaction is long range, so a small concentration of immobile obstacles can have a significant effect on lateral diffusion (Bussell *et al.,* 1995b). An area fraction of 2% immobile obstacles reduces D threefold; obstruction by the same area fraction would lower D by only a few percent.

Their other major result is that immobile obstacles have a much greater hydrodynamic effect than mobile obstacles, just as in the case of obstruction. The physical reason for the large effect is that immobile particles bound to the cytoskeleton can exert a net force on the lipid solute. The biological implication, as Bussell *et al.* (1995b) point out, is that lateral diffusion in the plasma membrane is hindered much more than lateral diffusion in organelle membranes, artificial vesicles, and blebs. The high diffusion coefficients of proteins in blebs have been attributed to the lack of cytoskeletal interactions (Tank *et al.,* 1982; Webb *et al.,* 1981), tacitly assumed to be binding or obstruction. When the membrane is detached from the cytoskeleton, however, there may be no immobilized proteins in the bleb, removing a significant hindrance to diffusion (Bussell *et al.,* 1995b).

For mobile obstacles, the effects of obstruction and hydrodynamic interaction are of similar magnitude. The combined effect can be represented as a product of a short-range diffusion coefficient reduced by hydrodynamic interactions and a long-range diffusion coefficient reduced by obstruction, and the combined effect agrees reasonably well with the limited experimental data available for bacteriorhodopsin (see Section IV,F) and mitochondrial proteins (Bussell *et al.,* 1995a; Dodd *et al.,* 1995).

For immobile obstacles, the theoretical predictions were compared with several sets of experiments (Bussell *et al.,* 1995b; Dodd *et al.,* 1995). For example, for the anion transport protein band 3, D is 56-fold lower in normal erythrocytes than in a spherocyte lacking a membrane skeleton (Sheetz *et al.,* 1980). The combined effects of obstruction and hydrodynamic interaction predict a 7- to 13-fold decrease; the remaining factor is presumably the result of interactions with the membrane skeleton. It would be disturbing if obstruction and hydrodynamic interaction accounted for the entire decrease in D, given the experimental results showing the effect of perturbation of the membrane skeleton on lateral diffusion (see Section V,E). Bussell *et al.* (1995b) found good agreement with experiment for anchorage modulation data from lymphocytes. Both the magnitude of D and its dependence on area fraction were correct, although the agreement would be more convincing if the data included lower area fractions of obstacles and if the actual area fraction of immobile obstacles were known. An unambiguous experimental test of the hydrodynamic interaction in a well-defined model system would be of considerable significance.

The experiments of Thomas *et al.* (1992) can be interpreted in terms of the hydrodynamic model. An electric field was applied to cells to concentrate membrane proteins at one pole, and then D for the IgE receptor was measured by pattern FPR. Despite a large difference in protein concentration at the two poles, there was no change in D. In the interpretation of Bussell *et al.* (1995b), the immobile proteins are not moved by the electric field, so the dominant hydrodynamic interactions are unchanged. The mobile protein concentrations are changed, but the obstruction effect is small because the obstacles are mobile (see Section IV,F). When the membrane was separated from the cytoskeleton by osmotic perturbation, D increased fivefold and became dependent on protein concentration. In the interpretation of Bussell *et al.* (1995b), obstruction and hydrodynamic interaction with immobile obstacles are now removed, and one sees the weaker dependence of D on the concentration of mobile obstacles.

Experimental tests of the dependence on particle size would be of considerable interest. Hammer and colleagues require the proteins to have multiple transmembrane helices so that the lipid solvent molecules are much smaller than the protein solute, and hydrodynamics can be applied legitimately. But, one might ask, in a fluid as complex as a bilayer, is the effective solvent molecule a lipid or a hydrocarbon chain or a single methylene group? It would be useful to determine experimentally how the hydrodynamic interaction actually varies as a function of obstacle and tracer size to see whether there is a clear transition from free-volume control to hydrodynamic control as found by Liu *et al.* (1997) for unobstructed diffusion.

E. Binding

So far this chapter has considered hard-core repulsion, but lateral diffusion may be affected significantly by attractive and long-range repulsive forces (Abney *et al.,* 1989; Abney and Scalettar, 1993; Donaldson, 1989; Pink *et al.,* 1986). Interactions leading to binding can be nonspecific, such as electrostatic, van der Waals, or lipid-mediated forces (Abney and Scalettar, 1993; Sintes and Baumgärtner, 1997), or highly specific, as in the case of neurofascin, where phosphorylation of a particular tyrosine abolishes ankyrin binding and increases the FPR mobile fraction (Garver *et al.,* 1997). At a given area fraction of particles, a pure hard-core repulsion has the smallest effect on diffusion. A long-range repulsive potential decreases D by increasing the effective radius of the particles, and an attractive potential decreases D because the mobile particles may form clusters or adhere to obstacles (Abney *et al.,* 1989). Binding or association must be involved in the indirect interactions that control the diffusion of some species, such as the immobile fraction of GPI-linked proteins. In some cases the biological effects of association far outweigh the purely physical effects. Cross-linking may trigger a chain of signaling events that involves interactions with the cytoskeleton, coated pits, caveolae, or DIGs.

Lateral diffusion may be hindered by transient binding to immobile species. At long times, the effect is to reduce *D* by a factor equal to the fraction of time the particle is freely diffusing (Koppel, 1981). Binding is favored in a two-dimensional system compared with a three-dimensional system (see Section V,B). Experimental evidence for binding comes from FPR and SPT. If exchange is fast on the FPR time scale, binding decreases *D;* if it is slow, it reduces the mobile fraction (Elson and Reidler, 1979). In SPT one can test a trajectory systematically for binding, asking whether a particle stays in one pixel longer than a freely diffusing particle would (Saxton, 1993b).

Association among mobile species hinders lateral diffusion. Association has little effect on the diffusion coefficient of a large protein at infinite dilution because by the Saffman–Delbrück equation, *D* depends only logarithmically on the protein radius. Association increases the effect of obstacles in the transmembrane and extracellular regions because the excluded area is increased. It would be particularly interesting if association increased the size of a diffusing species enough to change the mechanism controlling diffusion from nonhydrodynamic to hydrodynamic (see Sections IV,A, B, and D). Association is expected to increase the hindrance to diffusion by the membrane skeleton because the diffusing species requires a larger fluctuation in the corral wall to cross the barrier (see Edidin *et al.,* 1994 and Section V,E).

F. What Controls Lateral Diffusion?

We return to the basic question: why is protein diffusion in the plasma membrane 10- to 100-fold lower than protein diffusion in artificial bilayers, organelle membranes, or blebs? The simplest hypothesis is that obstruction in the transmembrane region is responsible for the reduction in *D.* Monte Carlo results indicate that obstruction alone is insufficient (see, e.g., Blackwell and Whitmarsh, 1990; Pink, 1985; Pink *et al.,* 1986; Saxton, 1987, 1990a). The next hypothesis is that obstruction and hydrodynamic interactions together are responsible (Bussell *et al.,* 1995a,b; Dodd *et al.,* 1995). The combined effects are closer to the experimental results, but the predicted value of *D* is still higher than is observed in the plasma membrane. This disagreement with experiment does not disprove either mechanism; it merely indicates that several hindrances are operating simultaneously.

For an integral protein in a synthetic lipid bilayer, the hindrances to diffusion are obstruction, hydrodynamic interaction, and perturbation of lipid viscosity. The various theoretical curves are shown in Figs. 2a–2c, and the combined theoretical curves are compared with experimental data for bacteriorhodopsin (BR) in Fig. 2d, updating and extending those of Bussell *et al.* (1995a). The agreement is good but not yet quantitative, and theory predicts correctly that the normalized D for BR is higher than that for lipid.

The effect of obstruction alone is shown in Fig. 2a. Curve PT shows the case of percolation for random point obstacles on the triangular lattice (Saxton, 1994a) for which the percolation threshold is 0.5. Curve PC shows the corresponding curve for overlapping circular obstacles in the continuum (Garboczi *et al.*, 1991, Eq. 15), for which $C_P = 0.676$. Curve M shows tracer diffusion, as in the case of BR obstructed by mobile BR, for the triangular lattice (Saxton, 1987, Eq. 12). The curve goes to zero above $C = 0.7$, although in the Monte Carlo calculations, no attempt was made to locate the glass transition precisely. Curve MM shows the effect of slow mobile obstacles on a faster mobile species, as in the case of lipid obstructed by mobile BR (Van Beijeren and Kutner, 1985; Saxton, 1987, Eqs. 4–6).

The effect of obstacles on D through the reduction of free volume is shown in Fig. 2b, from the model of Almeida *et al.* (1992b, Eq. A12). It is assumed here that BR orders nearby lipids as a domain of gel-phase lipid does, a major assumption. Here D^*_{fv} depends on the ratio R/ξ of the particle radius to the coherence length of the ordered lipid. The effect of hydrodynamic interactions alone is shown in Fig. 2c. Curve M shows the effect of mobile particles (Bussell *et al.*, 1995a, Eq. 2) and curve I shows the effect of immobile particles (Bussell *et al.*, 1995b, Eq. 3).

The combined curves are constructed as products of diffusion coefficients as follows. For BR obstructed by mobile BR, obstruction is from tracer diffusion (curve M, Fig. 2a), and the hydrodynamic interaction is for the mobile case (curve M, Fig. 2c). For lipid obstructed by mobile BR, viscosity is from the free volume model (Fig. 2b) with R/ξ equal to 1.05, obstruction is by mobile obstacles with a different jump rate (curve MM, Fig. 2a), and no hydrodynamic interactions are included because the lipid is the solvent. For a small protein obstructed by hypothetical immobile BR, obstruction is from the percolation model (curve PC, Fig. 2a), and hydrodynamic interactions are for the immobile case (curve I, Fig. 2c). In calculating the effect of obstruction here, the tracer size is assumed to be negligibly small. If the tracer were the size of BR, diffusion would be hindered more strongly because D^*_{obst} is controlled by the excluded area fraction, not the area fraction of obstacles.

For proteins in the plasma membrane, diffusion may be hindered further by interactions with the membrane skeleton and obstruction in the extracellular region. These interactions are more biologically specific, varying with

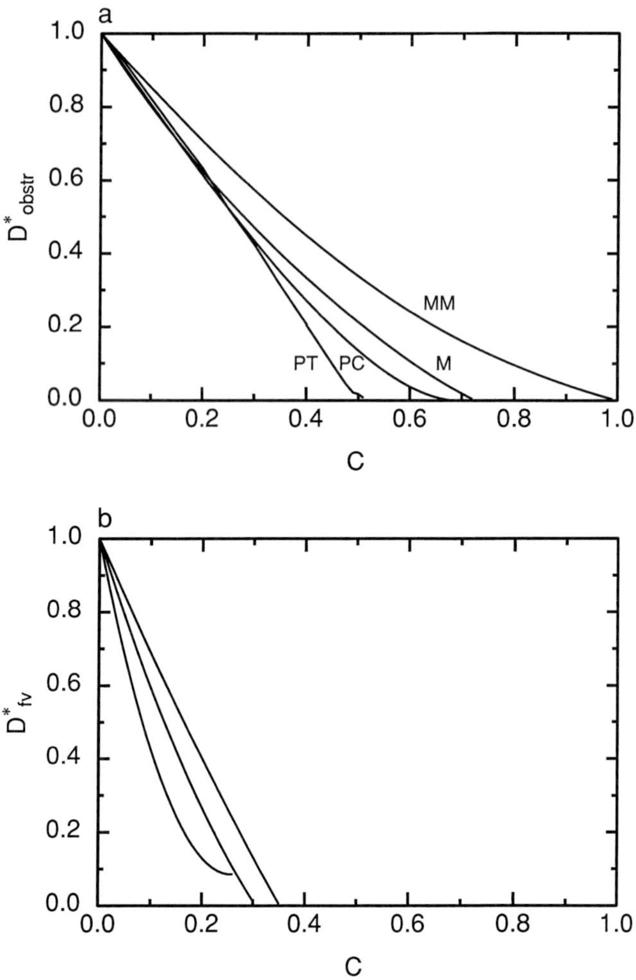

FIGURE 2 Theoretical and experimental diffusion coefficients D as a function of obstacle area fraction C. All experimental points and theoretical curves are normalized by the values at $C = 0$. (a) Effect of obstruction. Curve PT, immobile point obstacles on the triangular lattice. Curve PC, immobile overlapping circular obstacles on the continuum. Curve T, tracer diffusion. Curve MM, effect of slow mobile obstacles on a faster mobile species, with a ratio of jump rates of $\gamma = 2.1$. (b) Effect of obstacles on D through the free volume. Curves for $R/\xi = 1.0$, 1.5, and 2.0 (from left to right). Empirical values of the ratio are around 1.05 for proteins with multiple transmembrane helices (Almeida *et al.*, 1992a). (c) Effect of hydrodynamic interactions. Curve M, mobile particles. Curve I, immobile particles, normalized by the Saffman–Delbrück D for $\eta h/\eta_{aq}R = 250$ (Section IV, A). (d) Comparison of theoretical curves with FPR data for bacteriorhodopsin (BR) and fluorescent lipid analogs in the presence of BR obstacles. Small circles, data of Schram *et al.* (1994) for NBD-PC in the presence in BR in egg PC at 20°C. Large circles, data of Peters and Cherry (1982) for diO in the presence of BR in DMPC at 32°C. Crosses, data of Peters and Cherry (1982) for BR in the presence of BR in DMPC at 32°C. Lines, combined theoretical curves for BR obstructed by mobile BR (curve BR), lipid obstructed by BR (Lipid), and a small protein obstructed by hypothetical immobile BR (Imm BR).

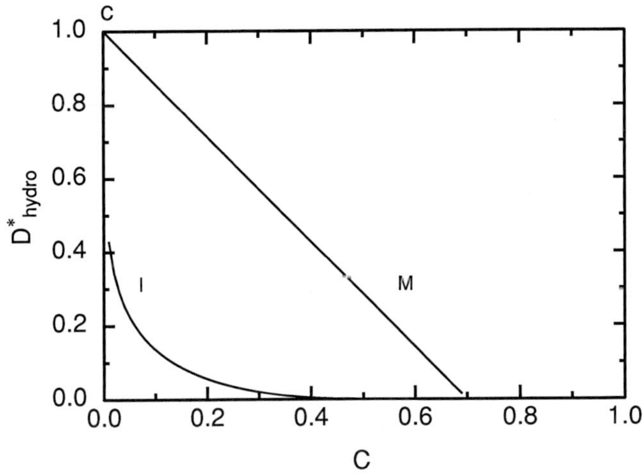

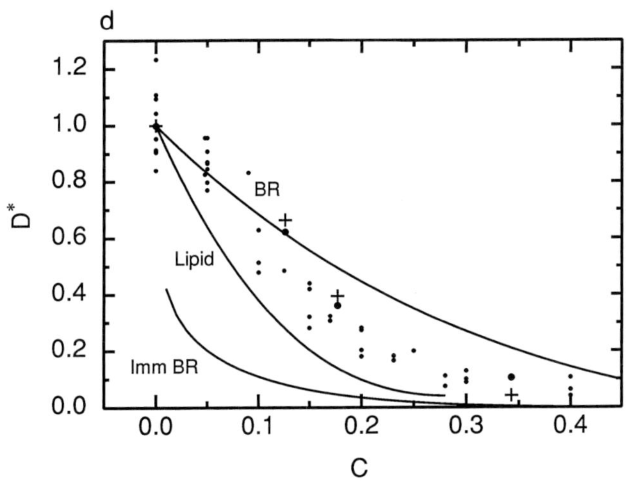

FIGURE 2 (*Continued*)

protein and cell type. For detailed discussions of these mechanisms, see Edidin (1994a, 1996) and Zhang *et al.* (1993).

In some cases, interaction with the membrane skeleton appears to be the controlling factor. Experiments on the effect of perturbations of the erythrocyte membrane skeleton on lateral diffusion were essential in developing our views of diffusion in membranes. The early photobleaching experiment of Peters *et al.* showed that band 3 was highly immobile. A key experiment (Sheetz *et al.,* 1980) compared diffusion of band 3 in normal erythrocytes with diffusion in spherocytes lacking a membrane skeleton and showed that $D(\text{erythrocyte}) = \frac{1}{56} D(\text{spherocyte})$. Many experiments on erythrocytes showed that treatments that partially dissociate the membrane skeleton increase the diffusion coefficient of band 3 but treatments that cross-link the membrane skeleton decrease D (Sheetz, 1983; Tsuji and Ohnishi, 1986; Golan, 1989). Similar results were obtained for diffusion of Na,K-ATPase in renal epithelial cells (Paller, 1994).

Barrier-free path measurements on class I MHC antigen in mouse hepatoma cells are consistent with control of lateral motion by cytoplasmic structures. Here proteins labeled with colloidal gold were moved across the cell surface with laser tweezers until they encountered a barrier strong enough to force them out of the optical trap. At 34°C, the BFP for the transmembrane-linked form was 3.5 μm but the BFP for a homologous GPI-linked form was 8.5 μm (Edidin *et al.*, 1991). Similarly, for class I MHC mutants truncated in the cytoplasmic region, proteins with 0 and 4 amino acids in the cytoplasmic region had a BFP similar to the homologous GPI-linked protein, but a mutant with 7 cytoplasmic amino acids had a smaller BFP, similar to that of the native protein with 31 cytoplasmic amino acids (Edidin *et al.*, 1994). Complicating the picture, however, are FPR measurements in L cells showing similar diffusion coefficients for the forms with 4, 7, and 31 cytoplasmic amino acids (Edidin and Zuniga, 1984). The explanation may be the difference in cell type or the greater aggregation of protein in the BFP experiments as a result of the gold label.

In other cases, obstruction in the extracellular region appears to be the dominant factor, and the cytoplasmic and transmembrane regions have little effect. For monkey COS cells, truncation of 533 out of the 542 amino acids in the cytoplasmic region of the EGF receptor had no effect on D or the mobile fraction (Livneh *et al.*, 1986). For the vesicular stomatitis virus G glycoprotein, the diffusion coefficient for a chimeric GPI-linked form was only a factor of 1.6 higher than the native transmembrane form (Zhang *et al.*, 1991). Similarly, in 3T3 cells, five isoforms of the neural cell adhesion molecule showed similar diffusion coefficients and mobile fractions, even though the size of the cytoplasmic region was varied and the membrane anchorages included GPI-linked and transmembrane isoforms (Jacobson *et al.*, 1997). SPT measurements comparing the low-density lipoprotein (LDL) receptor with a mutant lacking the cytoplasmic tail showed similar motion, including anomalous subdiffusion, in the two forms (Ghosh, 1991). In all these experiments, if the cytoskeleton has any effect it must be indirect or involve a region very close to the membrane.

Labeling may contribute to obstruction in the extracellular region. For the GPI-linked protein Thy-1 in lymphoma cells, D decreased threefold as the label was changed from rhodamine to Fab fragment to $F(ab)_2$ to IgG

(Zhang *et al.,* 1992). In FPR experiments on the high-affinity IgE receptor in rat basophilic leukemia cells, the mobile fraction was 79% for an IgE label and 18% for the much larger dioctadecylindocarbocyanine (diI)–LDL–IgE label. The difference was attributed to increased drag in the extracellular matrix (Feder *et al.,* 1996). Lee *et al.* (1993) have estimated the viscosity of the pericellular matrix from SPT measurements on lipids labeled with colloidal gold. In keratocytes the SPT diffusion coefficient increased twofold on degradation of the pericellular matrix with heparinase.

Measurements of the effect of the transmembrane region on lateral diffusion were summarized in Section IV,A. Goncalves *et al.* (1993) found that a mutation of a single amino acid in the transmembrane region of the insulin receptor increased the diffusion coefficient by a factor of 2.8. The mutation changed the kinked structure of the wild type to an α-helix. Diffusion measurements were done only in cells, so the effect may have been direct or indirect.

To test mechanisms controlling diffusion, Kucik *et al.* (1998) used SPT to measure the lateral diffusion of concanavalin A-coated beads with diameters of 40, 190, and 550 nm on fish keratocytes. The diffusion coefficient varied by at most a factor of two among the bead sizes. Kucik *et al.* (1998) argued that this result implies viscosity control, not control by transient binding or "microcorrals" below the resolution of SPT. For transient binding, if the probability that one protein binds to one site is P, the probability that N proteins bind to N sites is $\sim P^N$, so the diffusion coefficient decreases very rapidly with bead size. A similar argument applies to trapping by microcorrals. The observed weak dependence on bead size for these very large clusters is more consistent with the size dependence predicted by the Saffman–Delbrück equation.

Over two decades after the publication of the fluid mosaic model and the development of the fluorescence photobleaching recovery technique, we still do not fully understand the hindrances to diffusion. If a single mechanism were dominant, the problem would have been solved long ago. The root of the difficulty is that there are several mechanisms hindering diffusion, the mechanisms operate in parallel, and the importance of the mechanisms varies with the cell type and the mobile species. Furthermore, labeling may perturb the system directly or by cross-linking receptors and inducing interactions with structural elements. In addition, the inhomogeneity of the membrane is important. Domains exist over a wide range of length and time scales, and different types of measurements with different time scales see different processes. Finally, the composition of the membrane is complex. A fundamental quantity in modeling hindered diffusion is the area fraction of obstacles, and this is not well known.

V. APPLICATIONS TO MEMBRANES

This section discusses some applications of anomalous subdiffusion and percolation to membranes. The experimental demonstration of the applicability of these physical concepts to membranes is only beginning.

A. Single-Particle Tracking

Single-particle tracking makes it possible to observe the motion of an individual particle on the cell surface (Saxton and Jacobson, 1997). Proteins or lipids—individual molecules or small clusters—are labeled with a fluorophore or colloidal gold, and computer-enhanced video microscopy is used to observe the motion. The time resolution is typically tens of milliseconds, as in standard FPR. The spatial resolution is typically tens of nanometers, approximately 1.5 to 2 orders of magnitude greater than in FPR. This increase in spatial resolution helps in understanding the reasons for the FPR immobile fraction; Wilson *et al.* (1996) observed different modes of motion of the class II MHC antigen in fibroblasts, even though the molecules moved so slowly that they would be classed as immobile in FPR. The power of the technique is illustrated by elegant SPT experiments measuring the motion of individual rhodamine-labeled lipid analogs in planar-supported bilayers (Schmidt *et al.*, 1995, 1996; Schütz *et al.*, 1997b).

The major result of SPT so far is the demonstration that particles can display several modes of motion: normal diffusion, anomalous subdiffusion, confinement, directed motion, and immobility (for brief reviews, see Jacobson *et al.*, 1995; Sheets *et al.*, 1995). Inhomogeneity of the membrane is suggested by the occurrence of anomalous subdiffusion and by the observed transitions between modes of motion (Ghosh, 1991; Sheets *et al.*, 1997; Simson *et al.*, 1998). The standard analysis of FPR data (Axelrod *et al.*, 1976) assumes that all particles are either freely diffusing or immobile, but the motion as seen in SPT is more complex. Diffusion coefficients from SPT and FPR experiments have not yet been reconciled; the effect of the larger label used in SPT appears to be important, and the occurrence of anomalous subdiffusion and confined motion may affect FPR data analysis (Saxton and Jacobson, 1997). Table II gives some published examples of classification of modes of motion. These classifications are not directly comparable because different proteins, cell types, and methods of classification were used. However, all find only a minority of particles to be freely diffusing or immobile. All the methods of analysis need to be characterized more rigorously to establish the probability that the motion observed is in fact from the mode identified and not a random fluctuation in another

TABLE II
Modes of Motion from SPT

E-cadherin in epithelial cells[a]		
Simple diffusion	11%	
Stationary	22%	
Restricted	30%	
Directed	37%	
IgE receptor in rat basophilic leukemia cells[b]		
Normal diffusion	12%	
Immobile	25%	
Constrained diffusion	59%	
Directed	4%	
Thy-1 in fibroblasts[c]		
Fast	29%	(normal diffusion)
Stationary	10%	
Slow	24%	(anomalous subdiffusion)
Confined	37%	(anomalous subdiffusion; confinement)
Directed	0%	
NCAM-180 in 3T3 cells[d]		
Mobile	50%	
Stationary	15%	
Slow	14%	(anomalous subdiffusion)
Corralled	21%	(anomalous subdiffusion; confinement)
Directed	0%	

[a] From Kusumi *et al.* (1998).
[b] From Slattery (1995).
[c] From Sheets *et al.* (1997).
[d] From Simson *et al.* (1998).

mode. Once this is accomplished, it will be possible to examine transitions among modes of motion and ask whether it is likely that the cause is, say, a random fluctuation, entry into a domain, or a biological modulation event such as a change in phosphorylation state.

B. Kinetics

What are the implications for reaction kinetics of the fluid mosaic model as modified to allow for heterogeneity? The simple reduction of dimensionality from three to two dimensions has several important consequences, and anomalous subdiffusion would also have a major effect.

Adam and Delbrück (1968) showed that a three-dimensional diffusional search of a ligand for a receptor might be speeded if the ligand is first adsorbed nonspecifically to a membrane and then diffuses in two dimensions along the membrane surface to the receptor. Whether there is in fact an enhancement depends on details of the system, particularly the ratio of the three- and two-dimensional diffusion coefficients. [See the discussions by McCloskey and Poo (1986b) and Wiegel and DeLisi (1982), who argue that enhancement is almost never important in cells; Axelrod and Wang (1994); and Wang *et al.* (1992).] Reduction of dimensionality has been applied to several specific examples, such as the myristoyl-electrostatic switch (McLaughlin and Aderem, 1995), prothrombin (Giesen *et al.*, 1991a,b), and low-affinity growth factor receptors (Schlessinger *et al.*, 1995).

The reason for the enhancement is that diffusion in two dimensions is recurrent but diffusion in three dimensions is transient. That is, in two dimensions a diffusing particle tends to stay in its original vicinity and revisits its starting place with probability one, although the average return time is infinite. In three dimensions, however, there is a significant probability (e.g., 0.341 for the simple cubic lattice) that a diffusing particle will never return to its starting point (de Gennes, 1982; Kopelman, 1988, 1989; Hughes, 1995). Another way of looking at this (Dewey, 1997; Pfeifer, 1987) is that for a random walk of t steps on a d-dimensional lattice, the number of sites actually visited is t; these sites are not necessarily distinct. The mean-square displacement is $r^2 \propto t$, so in time t the diffusing particle could potentially visit any site in the volume $r^d \propto t^{d/2}$. The fraction F of the accessible sites that are actually visited is then $F \propto t/t^{d/2}$. At large times, $F \rightarrow 1$ in two dimensions, but $F \rightarrow 0$ in three dimensions. Some have pictured the enhancement as a concentration effect. For a spherical cell of volume V containing one spherical organelle of radius R, bound reactants are concentrated from a volume V to a volume $4\pi R^2 \Delta R$, where ΔR is the thickness of the protein layer on the organelle. A more appropriate value of ΔR would include the fluctuations in height of the protein with respect to the membrane and the fluctuations in the membrane perpendicular to the membrane plane. Whatever the choice of ΔR, the limitation of this picture is that ΔR is introduced arbitrarily and the argument of Adam and Delbrück (1968) can be carried out without it.

A second consequence of the reduction in dimensionality is that association is favored (Grasberger *et al.*, 1986). Binding is more favored in the membrane than in bulk solution because on going from three to two dimensions, one translational degree of freedom and two rotational degrees of freedom are almost completely lost, so that much of the entropy cost of association is prepaid. Grasberger *et al.* (1986) estimated that these factors may increase the likelihood of dimerization by a factor of 10^6. Brandts and Jacobson (1983) and Shiozawa *et al.* (1989) proposed a clustering mechanism of transmembrane signaling based on the idea of prepayment of entropy cost.

A third consequence is that diffusional mixing of reactants is slowed. In three dimensions, diffusion is fast enough that there is a steady-state solution to the diffusion equation for diffusion to a sink with the boundary condition of a prescribed reactant concentration at infinity. However, two-dimensional diffusion is so slow that no such steady-state solution exists and the boundary condition must be imposed at a finite distance. As a result of slow diffusional mixing in two dimensions, kinetics may be dominated by concentration fluctuations instead of the mean concentration. Depletion zones may form around reactants and reduce the reaction rates. In a theoretical paper on the kinetics of signal transduction, Shea *et al.* (1997) demonstrated that collision coupling and receptor cross-linking may be slowed as a result of depletion zones in two dimensions. If anomalous subdiffusion occurs, diffusional mixing is even less efficient and depletion zones are even more important.

Consider a simple reaction $A + B \rightarrow$ products. In an unobstructed three-dimensional system, the rate law is classical: rate $\propto D[A][B]$. In an unobstructed two-dimensional system, the rate coefficient is time dependent, going to zero as $1/\ln t$ (Torney and McConnell, 1983). On a two-dimensional percolation cluster, if A and B are added at random in equal concentrations, the order of the reaction is 2.55, and segregation of reactants occurs (Newhouse and Kopelman, 1988). In general, reaction rates in low-dimensional, fractal, and finite systems may show nonclassical behavior, such as time-dependent rate coefficients, anomalously high reaction orders, and self-ordering of reactants. This nonclassical behavior is sensitive to the initial conditions, the size of the system, the source terms, whether the steady-state or transient case is considered, and whether the concentrations of A and B are strictly equal or statistically equal. Classical rate laws assume well-mixed systems, but diffusion in low dimensions does not necessarily mix the reactants well enough for the classical laws to apply (Argyrakis and Kopelman, 1987; Dewey, 1997; Kopelman, 1988, 1989; Kopelman and Koo, 1991; Lindenberg *et al.,* 1989, 1990; Pfeifer, 1987; Shi and Kopelman, 1992).

This nonclassical behavior may affect the interpretation of experiments on processes such as signal transduction. A reaction is said to be chemically controlled if the reaction time is much greater than the diffusion time. If anomalous subdiffusion occurs, however, the value of the diffusion time is increased and the diffusion coefficient depends on the distance over which diffusion occurs and therefore on the initial distribution of reactants and the details of replenishment of reactants.

Under certain conditions, inefficient mixing due to anomalous subdiffusion on a percolation cluster may actually increase reaction rates (Saxton, 1998). Suppose a reaction has a very low probability of occurrence per collision. The obstructions slow the initial encounter of the reactants, but they also slow the separation of the reactants after an unsuccessful collision. At low enough reaction probabilities, the second factor predominates, and the rate is larger on a percolation cluster than on an unobstructed lattice. Carrying out a reaction on a percolation cluster slows the fast reactions considerably and speeds the slow ones somewhat, so that a wide range of reaction rates in an unobstructed system is compressed into a narrower range on a percolation cluster.

There is not yet evidence for low-dimensional kinetic effects in cell membranes, except for the adenylate cyclase experiments discussed in Section V,D and the chloroplast model discussed in Section V,C, but if anomalous subdiffusion occurs in a domain, reaction rates there are necessarily affected. Such effects may lie buried in effective diffusion coefficients and rate constants, just as the complexity of motion observed by SPT (Table II) has been absorbed into values of D and the mobile fraction. Also, diffusion may be normal in much of the membrane and is anomalous only in certain domains. Savageau (1995) considered the modifications in the Michaelis–Menten equation required for fractal kinetics.

C. Percolation Effects on Kinetics

Even if diffusion is normal, confinement may affect reactions by changing activity coefficients (Zimmerman and Minton, 1993) or by disconnecting the system. Confinement of reacting species in disconnected regions can affect both kinetics and equilibrium (Melo *et al.,* 1992; Thompson *et al.,* 1995; Vaz, 1994, 1996). In a disconnected system with low concentrations of reactants, the number of particles able to find reaction partners is reduced. The resulting reaction rates and equilibrium concentrations may be much different from those in an unconfined system. For this effect to be significant, the number of reactant molecules in a compartment must be around 10 or fewer, a plausible number for receptors (Thompson *et al.,* 1995).

The structure producing confinement could be gel-phase lipid, the membrane skeleton, or a fluid–fluid lateral phase separation, although the latter case would place stringent requirements on the partition coefficients of the reactants. If a system is poised near the percolation threshold, the extent of a reaction could be changed by small perturbations or adjusted by small, metabolically driven changes. The parameter causing the threshold crossing could be, for example, lipid composition, pH, ionic strength, a change in lateral pressure due to osmotic swelling, or, in the case of membrane skeleton, the phosphorylation state (Thompson *et al.,* 1995; Vaz, 1994, 1996).

Lavergne and Joliot (1991) proposed a percolation model of electron transfer from photosystem II to plastoquinone in chloroplasts. Plastoquinone is trapped in small regions of the membrane, so there is a fast equilibrium within these regions at a time scale set by the diffusion time of plastoquinone within the regions. However, the obstacles can move on longer time scales, allowing slow equilibration among regions. So measurements of reaction kinetics on a short time scale see a quasi-equilibrium; measurements on a long time scale see the global equilibrium; and measurements of redox potential see the global equilibrium because exogenous redox mediators in the aqueous phase have access to all regions. This model is sensitive to the protein concentrations assumed (Drepper *et al.*, 1993), and the transition is more like a glass transition than a percolation transition, but it is a specific, testable example of percolation effects on a reaction.

As the chloroplast model indicates, the kinetics can be affected significantly even in the absence of a strict percolation threshold. Temporary disconnection on a time scale greater than the reaction time is sufficient. In some cells the membrane skeleton might provide enough disconnection; SPT results of Sako and Kusumi (1994) for various integral proteins showed a dwell time around 25 sec in regions of area 0.1 to 1.0 μm^2. For this to work, of course, the reactants would have to be in the same compartment to begin with or the cell would have to remove a barrier keeping them apart. Peters (1988) criticized the mobile receptor hypothesis on these grounds, pointing out that in the erythrocyte the concentration of required proteins was so low and the membrane corrals were so small that it was extremely unlikely for the required proteins to be in the same corral. Diffusion across corral boundaries was too slow to account for the observed rates, Peters argued, so the diffusion step must take place in the aqueous phase. Edidin (1994a,b) reversed this argument to say that for these signaling reactions, short-range diffusion within a corral of dimensions of hundreds of nanometers is more important than long-range diffusion over micrometer distances as measured in FPR. By the same argument, short-range anomalous subdiffusion may be more important than long-range normal diffusion. Edidin's argument requires some mechanism of partitioning or loose association to have the required proteins in the same corral.

D. Percolation Experiments

An experimental demonstration of percolation in synthetic lipid bilayers was carried out by Vaz, Thompson, and collaborators (Almeida and Vaz, 1995; Vaz, 1992). FPR measurements were made on multibilayers of binary lipid mixtures in the region of lateral phase separation, using a fluorescent probe excluded from the gel phase. The area fraction of gel-phase obstacles was obtained from the phase diagrams. As the temperature was changed at constant composition, sharp changes were observed in the diffusion coefficient and the mobile fraction. The plot of mobile fraction as a function of area fraction of gel phase shows an inflection point, and the inflection point marks the percolation transition. Observed changes in D near the threshold are complicated by fast recovery in finite domains (Almeida *et al.,* 1992b; Schram *et al.,* 1994; Vaz *et al.,* 1989) and diffusion along boundary defects in the gel phase.

When the locus of percolation transitions from FPR was plotted on the lipid-phase diagrams, the position of the percolation line in the two-phase region varied considerably among lipid mixtures. In the nearly ideal DMPC/DPPC mixture, the transition was at a mass fraction of 0.5, corresponding to a gel-phase area fraction of 0.375. For a DMPC/DSPC mixture, however, the area fraction of gel phase was only 0.16, which suggests very elongated gel-phase domains disconnecting the fluid phase. In part of the phase diagram for a mixture of di$C_{17:0}$PC and $C_{22:0}C_{22:0}$PC, the fluid phase remained continuous even at a mass fraction of 90% gel phase, implying a network of fluid phase surrounding large domains of gel-phase lipid, a structure observed by Hwang *et al.* (1995) by fluorescence near-field scanning optical microscopy (NSOM) of monolayers of lipid mixtures. Almeida *et al.* (1993) extended this work to a ternary mixture with liquid–liquid-phase separation, DMPC, DSPC, and cholesterol, and interpreted the results quantitatively in terms of a free volume model. Coelho *et al.* (1997) modeled FPR recovery using overlapping random elliptical obstructions, calculated the mobile fraction as a function of area fraction of obstacles, and calculated the shape of the recovery curve.

Several sets of experiments have been done on other model systems, including membranes containing the integral proteins bacteriorhodospin or pOmp A (Piknová *et al.,* 1996, 1997; Sankaram *et al.,* 1994). The reactions studied were quenching of fluorophore-labeled lipid by spin-labeled lipid and spin–spin interactions in ESR. FPR experiments indicated that the addition of bacteriorhodopsin led to more centrosymmetric gel domains, as indicated by the shift of the percolation threshold from 0.28 to 0.43 (Schram and Thompson, 1997). On the basis of FPR measurements on a series of binary lipid mixtures Schram *et al.* (1996) argued that gel-phase domains became more effective obstacles as the ideality of the mixture increased.

NMR measurements of percolation (Dolainsky *et al.,* 1997) eliminate potential artifacts from fluorescent or spin labels and allow variation of the length scale of the measurement, although over lengths shorter than in

FPR. These experiments are likely to provide important information once the calculations of obstructed diffusion are refined sufficiently to permit quantitative analysis of the data. The dependence of the percolation probability on the length scale of the measurement (see Section III,A) is of critical importance in comparing NMR and FPR experiments on percolation. The size distribution of finite fluid domains near the percolation threshold must also be considered.

The interpretation of all these experiments is complicated by the fact that a change in temperature at constant composition could change the size of the gel domains at a constant number of domains, the number of domains at constant size, or both (Vaz and Almeida, 1993).

Experiments on adenylate cyclase activity in reticulocyte membranes support a percolation model (Sobolev *et al.,* 1988; Zakharova *et al.,* 1995). The activity was measured as a function of the area fraction of fluid-phase lipid as the fraction was varied by phospholipase A treatment or by polylysine binding. The fraction of fluid-phase lipid was measured by ESR using 2,2,6,6-tetramethylpiperidinyl-l-oxyl. The activity and the diffusion coefficient of band 3 showed a threshold and the expected power-law dependence on the distance from the percolation threshold, although the exponent for D was closer to the value for the backbone than for a percolation cluster. (The backbone of a percolation cluster is the percolation cluster with the dead ends removed.)

E. Membrane Skeleton

The diffusion of some proteins in some cells is controlled primarily by interactions with the membrane skeleton (see Section IV,F). To review the structure of the erythrocyte membrane skeleton briefly, spectrin tetramers form the bonds of a network and actin oligomers form the nodes. Approximately six tetramers are bound to each node. The network is attached to the cytoplasmic side of the membrane by connections to membrane components; spectrin is linked to band 3 by ankyrin and to glycophorin by band 4.1. Similar structures have been found in other types of cells (Bennett and Gilligan, 1993; Morrow *et al.,* 1997).

The corral model of the membrane skeleton was first portrayed as literal corral walls confining cattle (integral proteins) but allowing poultry (lipids) to cross (Sheetz, 1983). This picture led to the membrane skeleton fence model, in which the cytoplasmic surface of the plasma membrane is divided into compartments by barriers. Membrane proteins can cross the barriers as a result of transient events: conformational fluctuations in the barriers and membrane, dissociation of the structural proteins forming the barriers, or dissociation of part of the network from the membrane (Kusumi *et al.*, 1998; Kusumi and Sako, 1996; Sheets *et al.*, 1995; Tsuji and Ohnishi, 1986).

Long-range diffusion in the fence model can be described in terms of bond percolation, where a bond represents a diffusional path from one corral to an adjacent one (Saxton, 1989b, 1990b,c). This approach makes it possible to examine the effect of imperfections in the membrane skeleton on lateral diffusion and mechanical stability. The percolation model predicts that the erythrocyte loses its resistance to mechanical stress if less than 35% of the normal amount of spectrin is present, a value roughly consistent with the threshold observed in patients with hereditary hemolytic anemia. Furthermore, the model predicts that for a static network, lateral diffusion and mechanical reinforcement are incompatible. If the membrane skeleton is connected enough to provide mechanical reinforcement, it is connected enough to block long-range lateral diffusion. This condition is relaxed for fluctuating barriers. Diffusion can then take place above the percolation threshold, but at a rate set by the fluctuations of the fence.

An elegant paper by Boal (1994) used Monte Carlo simulations and scaling arguments to examine a triangular network of polymer chains in which each chain was attached at its midpoint to an impenetrable membrane. The model includes only steric interactions and the entropy of the chains, and predicts the geometry and elasticity of the network. Boal (1994) showed that the essential features of the model were the steric interactions within a chain, the steric interactions among chains, and the attachment to the membrane. The polymer network, percolation, and fence models are not really competing models but developments of a common theme. Conformational fluctuations in the polymer network would provide gate openings, and defects in the network would show percolation.

In the novel technique of fluorescence-imaged microdeformation, erythrocyte ghosts are partially aspirated into a micropipette, and concentrations of labeled membrane and membrane skeleton components are measured as a function of position by fluorescence microscopy (Discher *et al.*, 1994; Discher and Mohandas, 1996; Knowles *et al.*, 1997). The results are analyzed in terms of elasticity theory. The importance of this technique is that it measures and correlates local molecular concentrations of several species under well-defined conditions of mechanical stress obtained from the measured geometry, the aspiration pressure, and the elastic properties of the erythrocyte. The mobile and immobile fractions of labeled species can be measured.

SPT measurements showed confinement within corrals for several integral proteins in several mammalian cell lines. On average the proteins remained in regions of area 0.1 to 1.0 μm^2 for approximately 25 sec (Kusumi and Sako, 1996; Kusumi *et al.,* 1998), far longer than a freely diffusing particle would (Saxton, 1995). The long-range diffusion coefficient is determined by the rate of movement among corrals. Experiments using laser tweezers have been used to estimate corral size and energy barriers and to distinguish corrals from tethers (Kusumi *et al.,* 1998; Sako and Kusumi, 1994, 1995). Preliminary SPT results on erythrocytes have been presented (Tomishige and Kusumi, 1998).

F. Area Fractions of Obstacles

The area fraction of obstacles is the key quantity in evaluating the effect of hindrances to lateral diffusion. Some estimates of area fractions in cell membranes are tabulated elsewhere (Saxton, 1989a), but the area fractions are not well known. Furthermore, many integral proteins have large extracellular regions but only a single transmembrane helix, so the area fractions in the two regions are often much different. Obstruction is likely to be controlled by the area fraction in the extracellular region, whereas hydrodynamic interactions are controlled by the area fraction in the transmembrane region. Beyond the technical question of area fractions, some workers have considered the general question of what the plasma membrane looks like on the molecular scale (Barclay *et al.,* 1997; Jentoft, 1990; Viitala and Järnefelt, 1985; and the remarkable book by Goodsell, 1993). The importance of this work is that it provides the image underlying our thinking about the cell membrane.

One way to find the area fraction is to count the proteins, add up their areas, and divide by the total membrane area. Each step is problematic, as shown in a careful attempt to estimate the area fraction for leukocytes (Barclay *et al.,* 1997). For T cells, the area fraction in the extracellular region is 60% if the proteins are assumed to lie flat on the membrane. From the compilation of protein abundances and areas of Barclay *et al.* (1997), one finds that the area fraction in the extracellular region is only 10% if the proteins are fully extended away from the membrane and one finds that the area of the transmembrane helices is less than 1%. Barclay *et al.* (1997) considered only leukocyte-specific antigens, so one must also add to the transmembrane area fraction the contribution from transport proteins. Finally, these estimates must be reduced by a factor of 3 or 4 to account for the area of microvilli, leading Barclay *et al.* (1997) to a final estimated area fraction of 20% for the extracellular region.

Two experimental approaches show promise. Fluorescence-imaged microdeformation was described in Section V,E. Ryan *et al.* (1988) measured the area fraction of membrane proteins by applying an electric field to cells to move membrane proteins to one pole, mostly by electroosmosis. The concentration distribution of a labeled protein was found by digital imaging microscopy, and the nonideality of the distribution was attributed to steric exclusion. This method yields the fraction f_0 of accessible surface occupied by mobile particles; f_0 was found to be independent of the mobile protein labeled. The accessible surface is defined as the area not occupied by immobile obstacles. The method also gives the immobile fraction, but this depends on the protein labeled, just as in FPR. For rat basophilic leukemia cells, $f_0 = 0.45$, and the total area fraction occupied by mobile or immobile proteins was estimated to be about 75%. The problem of microvilli remains.

A related question is the existence, size, and area fraction of gel-phase domains. In artificial bilayers, the area fraction of gel phase can be determined from the phase diagram, but the estimates of domain size appear to be limited by the range of the technique used and the need to use a tractable model of a geometrically complex phase. Experimental results are summarized by Almeida and Vaz (1995) and Coelho *et al.* (1997).

In the plasma membrane, there is no direct evidence for gel domains except in specialized membranes such as myelin (Stoffel and Bosio, 1997), stratum corneum (Bouwstra *et al.,* 1997), and sperm (Wolf, 1994, 1995). In studies on 11 different cell types, Parasassi *et al.* (1993) found no evidence for gel-phase lipid on the basis of measurements of Laurdan fluorescence. A general argument against the presence of gel-phase lipid (McElhaney, 1984) is that if gel-phase and fluid-phase lipid coexist, the permeability of the membrane may increase significantly, leading to leakage (Clerc and Thompson, 1995). Evidence supporting the existence of gel-phase domains in cell membranes is reviewed by Welti and Glaser (1994).

VI. SUMMARY

The major qualitative modification required in the Singer–Nicolson model is to allow for heterogeneity. Percolation theory and anomalous diffusion are standard tools used in physics to characterize heterogeneity. Heterogeneity in cell biology may be more heterogeneous than heterogeneity in physics. In physics, one typically assumes a single mechanism that gives a self-similar structure. For example, random point obstacles at the percolation threshold give a self-similar percolation cluster, or in the CTRW model, a single probability distribution of waiting times gives an infinite range of values, producing anomalous subdiffusion on all time scales. It is plausible, however, that in the plasma membrane there are multiple mechanisms operating on different time and space scales (Saxton and Jacobson, 1997). For example, at short range, obstruction and binding may control lateral diffusion; at intermediate range, diffusion may be controlled by the membrane skeleton; and at long times, metabolism or biological modulation may affect diffusion. In general, lipid domains can range from small transient fluctuations lasting milliseconds or less to permanent features of the cell surface such as the polarized domains of epithelial (Devarajan and Morrow, 1996) and sperm (Wolf, 1994, 1995) cells.

One interesting area is the view of the membrane as a random fluctuating potential energy surface. Diffusion on such an energy surface has been examined in the physics literature (Bouchaud and Georges, 1990; Isichenko, 1992). This type of model may seem remote from cell biology, but some leading researchers are beginning to view the membrane in such terms. Kusumi and Sako (1996) picture the membrane as a series of energy mountains and valleys of various heights, with the corral walls of the membrane skeleton as the dominant feature. Webb's laboratory is beginning to measure the energy landscape, using a scanning force microscope based on an optical trap to make continuous measurements of the local forces on a membrane protein (Switz *et al.,* 1996; Stout and Webb, 1998). The instrument can measure piconewton forces on a time scale of 1 msec with 10-nm resolution (Ghislain *et al.,* 1994) so direct contact with experiments may be possible.

An important unresolved question is the extent to which the plasma membrane is in a nonequilibrium state. Edidin (1992a) says that it is not in equilibrium, quoting the statement of V. A. Parsegian that "there is nothing so dead as a system in equilibrium." Vesicle traffic and metabolically driven membrane flows lead to a nonequilibrium state (Edidin, 1996, 1997). Gheber and Edidin (1998) modeled membrane turnover in a membrane with barriers and found that a combination of barriers and vesicle traffic can produce patchiness of the membrane. The anomalous subdiffusion model is based on the assumption of a nonequilibrium state (Feder *et al.,* 1996). In Webb's laboratory, Ghosh (1991) showed a change from anomalous to normal diffusion for the LDL receptor on treatment with azide and deoxyglucose, but Slattery (1995) found no such effect for a different cell and receptor. Monte Carlo results show that diffusion can be much more anomalous in a nonequilibrium system than in an equilibrium one (Saxton, 1996).

Acknowledgments

This work was supported in part by NIH Grant GM38133 and the Office of Research, University of California, Davis. I thank P. F. F. Almeida, T. G. Dewey, and R. N. McElhaney for helpful discussions, D. F. Kucik for a preprint, and A. Lopez for experimental data in Fig. 2.

References

Abney, J. R., and Scalettar, B. A. (1993). Molecular crowding and protein organization in biological membranes. *In* "Thermodynamics of Membrane Receptors and Channels" (M. B. Jackson, ed.), pp. 183–226, CRC Press, Boca Raton, Florida.

Abney, J. R., and Scalettar, B. A. (1995). Fluctuations and membrane heterogeneity. *Biophys. Chem.* **57,** 27–36.

Abney, J. R., Scalettar, B. A., and Owicki, J. C. (1989). Self diffusion of interacting membrane proteins. *Biophys. J.* **55,** 817–833.

Adam, G., Delbrück, M. (1968). Reduction of dimensionality in biological diffusion processes. *In* "Structural Chemistry and Molecular Biology" (A. Rich and N. Davidson, eds.), pp. 198–213. Freeman, San Francisco.

Ahmed, S. N., Brown, D. A., and London, E. (1997). On the origin of sphingolipid/cholesterol-rich detergent-insoluble cell membranes: Physiological concentrations of cholesterol and sphingolipid induce formation of a detergent-insoluble, liquid-ordered lipid phase in model membranes. *Biochemistry* **36,** 10944–10953.

Almeida, P. F. F., and Vaz, W. L. C. (1995). Lateral diffusion in membranes. *In* "Structure and Dynamics of Membranes. From Cells to Vesicles" (R. Lipowsky and E. Sackmann, eds.), Vol. 1A, pp. 305–357. Elsevier, Amsterdam.

Almeida, P. F. F., Vaz, W. L. C., and Thompson, T. E. (1992a). Lateral diffusion in the liquid phases of dimyristoylphosphatidylcholine/cholesterol lipid bilayers: A free volume analysis. *Biochemistry* **31,** 6739–6747.

Almeida, P. F. F., Vaz, W. L. C., and Thompson, T. E. (1992b). Lateral diffusion and percolation in two-phase, two-component lipid bilayers. Topology of the solid-phase domains in-plane and across the lipid bilayer. *Biochemistry* **31,** 7198–7210.

Almeida, P. F. F., Vaz, W. L. C., and Thompson, T. E. (1993). Percolation and diffusion in three-component lipid bilayers: Effect of cholesterol on an equimolar mixture of two phosphatidylcholines. *Biophys. J.* **64,** 399–412.

Argyrakis, P., and Kopelman, R. (1987). Self-stirred vs. well-stirred reaction kinetics. *J. Phys. Chem.* **91,** 2699–2701.

Axelrod, D. (1983). Lateral motion of membrane proteins and biological function. *J. Membr. Biol.* **75,** 1–10.

Axelrod, D., and Wang, M. D. (1994). Reduction-of-dimensionality kinetics at reaction-limited cell surface receptors. *Biophys. J.* **66,** 588–600.

Axelrod, D., Koppel, D. E., Schlessinger, J., Elson, E., and Webb, W. W. (1976). Mobility measurement by analysis of fluorescence photobleaching recovery kinetics. *Biophys. J.* **16,** 1055–1069.

Balcom, B. J., and Petersen, N. O. (1993). Lateral diffusion in model membranes is independent of the size of the hydrophobic region of molecules. *Biophys. J.* **65,** 630–637.

Barclay, A. N., Brown, M. H., Law, S. K. A., McKnight, A. J., Tomlinson, M. G., and van der Merwe, P. A. (1997). "The Leucocyte Antigen FactsBook," 2nd Ed., pp. 118–123. Academic Press, San Diego.

Barrat, J.-L., and Klein, M. L. (1991). Molecular dynamics simulations of supercooled liquids near the glass transition. *Annu. Rev. Phys. Chem.* **42,** 23–53.

Bennett, V., and Gilligan, D. M., (1993). The spectrin-based membrane skeleton and micron-scale organization of the plasma membrane. *Annu. Rev. Cell Biol.* **9,** 27–66.

Bergelson, L. D., Gawrisch, K., Ferretti, J. A., and Blumenthal, R., eds. (1995). Domain organization in biological membranes. *Mol. Membr. Biol.* **12,** 1–162.

Berland, K. M., So, P. T. C., Chen, Y., Mantulin, W. W., and Gratton, E. (1996). Scanning two-photon fluctuation correlation spectroscopy: Particle counting measurements for detection of molecular aggregation. *Biophys. J.* **71,** 410–420.

Berryman, J. G. (1983). Random close packing of hard spheres and disks. *Phys. Rev. A* **27,** 1053–1061.

Blackwell, M. F., and Whitmarsh, J. (1990). Effect of integral membrane proteins on the lateral mobility of plastoquinone in phosphatidylcholine proteoliposomes. *Biophys. J.* **58,** 1259–1271.

Bloom, M., and Thewalt, J. (1994). Spectroscopic determination of lipid dynamics in membranes. *Chem. Phys. Lipids* **73,** 27–38.

Bloom, M., and Thewalt, J. L. (1995). Time and distance scales of membrane domain organization. *Mol. Membr. Biol.* **12,** 9–13.

Boal, D. H. (1994). Computer simulation of a model network for the erythrocyte cytoskeleton. *Biophys. J.* **67,** 521–529.

Bouchaud, J.-P., and Georges, A. (1990). Anomalous diffusion in disordered media: Statistical mechanisms, models and physical applications. *Phys. Rep.* **195,** 127–293.

Bouwstra, J. A., Thewalt, J. Gooris, G. S., and Kitson, N. (1997). A model membrane approach to the epidermal permeability barrier: An X-ray diffraction study. *Biochemistry* **36,** 7717–7725.

Brandts, J. F., and Jacobson, B. S. (1983). A general mechanism for transmembrane signalling based on clustering of receptors. *Surv. Synth. Pathol. Res.* **2,** 107–114.

Brown, D. A., and London, E. (1997). Structure of detergent-resistant membrane domains: Does phase separation occur in biological membranes? *Biochem. Biophys. Res. Commun.* **240,** 1–7.

Brown, R. E. (1998). Sphingolipid organization in biomembranes: What physical studies of model membranes reveal. *J. Cell Sci.* **111,** 1–9.

Bug, A. L. R., Safran, S. A., Grest, G. S., and Webman, I. (1985). Do interactions raise or lower a percolation threshold? *Phys. Rev. Lett.* **55,** 1896–1899.

Bunde, A., and Havlin, S. eds. (1991). "Fractals and Disordered Systems." Springer-Verlag, Berlin.

Bunde, A., and Havlin, S., eds. (1994). "Fractals in Science." Springer-Verlag, Berlin.

Bussell, S. J., Koch, D. L., and Hammer, D. A. (1992). The resistivity and mobility functions for a model system of two equal-sized proteins in a lipid bilayer. *J. Fluid Mech.* **243,** 679–697.

Bussell, S. J., Hammer, D. A., and Koch, D. L. (1994). The effect of hydrodynamic interactions on the tracer and gradient diffusion of integral membrane proteins in lipid bilayers. *J. Fluid Mech.* **258,** 167–190.

Bussell, S. J., Koch, D. L., and Hammer, D. A. (1995a). Effect of hydrodynamic interactions on the diffusion of integral membrane proteins: Tracer diffusion in organelle and reconstituted membranes. *Biophys. J.* **68,** 1828–1835.

Bussell, S. J., Koch, D. L., and Hammer, D. A. (1995b). Effect of hydrodynamic interactions on the diffusion of integral membrane proteins: Diffusion in plasma membranes. *Biophys. J.* **68,** 1836–1849.

Chan, P.-Y., Lawrence, M. B., Dustin, M. L., Ferguson, L. M., Golan, D. E., and Springer, T. A. (1991). Influence of receptor lateral mobility on adhesion strengthening between membranes containing LFA-3 and CD2. *J. Cell Biol.* **115,** 245–255.

Chazotte, B., and Hackenbrock, C. R. (1989). Lateral diffusion as a rate-limiting step in ubiquinone-mediated mitochondrial electron transport. *J. Biol. Chem.* **264,** 4978–4985.

Cheng, K. H., Ruonala, M., Virtanen, J., and Somerharju, P. (1997). Evidence for superlattice arrangements in fluid phosphatidylcholine/phosphatidylethanolamine bilayers. *Biophys. J.* **73,** 1967–1976.

Chong, P. L.-G. (1994). Evidence for regular distribution of sterols in liquid crystalline phosphatidylcholine bilayers. *Proc. Natl. Acad. Sci. U.S.A.* **91,** 10069–10073.

Clegg, R. M., and Vaz, W. L. C. (1985). Translational diffusion of proteins and lipids in artificial lipid bilayer membranes. A comparison of experiment with theory. *In* "Progress in Protein–Lipid Interactions" (A. Watts and J. J. H. H. M. De Pont, eds.), pp. 173–229. Elsevier, Amsterdam.

Clerc, S. G., and Thompson, T. E. (1995). Permeability of dimyristoyl phosphatidylcholine/dipalmitoyl phosphatidylcholine bilayer membranes with coexisting gel and liquid-crystalline phases. *Biophys. J.* **68,** 2333–2341.

Coelho, F. P., Vaz, W. L. C., and Melo, E. (1997). Phase topology and percolation in two-component lipid bilayers: A Monte Carlo approach. *Biophys. J.* **72,** 1501–1511.

Cooper, N., Tedder, A., Heyes, D. M., and Melrose, J. R. (1989). Percolation cluster statistics of 2D Lennard-Jones phases. *J. Phys. Condensed Matter* **1,** 6217–6230.

Criado, M., Vaz, W. L. C., Barrantes, F. J., and Jovin, T. M. (1982). Translational diffusion of acetylcholine receptor (monomeric and dimeric forms) of *Torpedo marmorata* reconstituted into phospholipid bilayers studied by fluorescence recovery after photobleaching. *Biochemistry* **21,** 5750–5755.

Damjanovich, S., Szöllősi, J., Trón, L., and Edidin, M. (1994). "Mobility and Proximity in Biological Membranes." CRC Press, Boca Raton, Florida.

Davenport, L. (1997). Fluorescence probes for studying membrane heterogeneity. *In* "Methods in Enzymology" (L. Brand and M. L. Johnson, eds.), Vol. 278, pp. 487–512. Academic Press, San Diego.

de Gennes, P. G. (1982). Kinetics of diffusion-controlled processes in dense polymer systems. I. Nonentangled regimes. *J. Chem. Phys.* **76,** 3316–3321.

Devarajan, P., and Morrow, J. S. (1996). The spectrin cytoskeleton and organization of polarized epithelial cell membranes. *Curr. Top. Membr.* **43,** 97–128.

Devaux, P. F. (1992). Protein involvement in transmembrane lipid asymmetry. *Annu. Rev. Biophys. Biomol. Struct.* **21,** 417–439.

Dewey, T. G. (1997). "Fractals in Molecular Biophysics." Oxford Univ. Press, Oxford.

Discher, D. E., and Mohandas, N. (1996). Kinematics of red cell aspiration by fluorescence-imaged microdeformation. *Biophys. J.* **71,** 1680–1694.

Discher, D. E., Mohandas, N., and Evans, E. A. (1994). Molecular maps of red cell deformation: Hidden elasticity and *in situ* connectivity. *Science* **266,** 1032–1035.

Dodd, T. L., Hammer, D. A., Sangani, A. S., and Koch, D. L. (1995). Numerical simulations of the effect of hydrodynamic interactions on diffusivities of integral membrane proteins. *J. Fluid Mech.* **293,** 147–180.

Dolainsky, C., Karakatsanis, P., and Bayerl, T. M. (1997). Lipid domains as obstacles for lateral diffusion in supported bilayers probed at different time and length scales by two-dimensional exchange and field gradient solid state NMR. *Phys. Rev. E* **55,** 4512–4521.

Donaldson, P. J. (1989). "Modulation of Lateral Diffusion on Lipid Bilayer Membranes." Ph.D. Thesis, Cornell Univ. Ithaca, New York.

Drepper, F. Carlberg, I., Andersson, B., and Haehnel, W. (1993). Lateral diffusion of an integral membrane protein: Monte Carlo analysis of the migration of phosphorylated light-harvesting complex II in the thylakoid membrane. *Biochemistry* **32,** 11915–11922.

Dustin, M. L., Ferguson, L. M., Chan, P.-Y., Springer, T. A., and Golan, D. E. (1996). Visualization of CD2 interaction with LFA-3 and determination of the two-dimensional dissociation constant for adhesion receptors in a contact area. *J. Cell Biol.* **132,** 465–474.

Edidin, M. (1992a). The variety of cell surface membrane domains. *Comments Mol. Cell. Biophys.* **8,** 73–82.

Edidin, M. (1992b). Translational diffusion of membrane proteins. *In* "The Structure of Biological Membranes" (P. Yeagle, ed.), pp. 539–572. CRC Press, Boca Raton, Florida.

Edidin, M. (1993). Patches and fences: Probing for plasma membrane domains. *J. Cell Sci. Suppl.* **17,** 165–169.

Edidin, M. (1994a). Lateral mobility of membrane proteins—a journey from heterokaryons to laser tweezers. *In* "The Legacy of Cell Fusion" (S. Gordon, ed.), pp. 101–114. Oxford Univ. Press, Oxford.

Edidin, M. (1994b). Fluorescence photobleaching and recovery, FPR, in the analysis of membrane structure and dynamics. *In* "Mobility and Proximity in Biological Membranes" (S. Damjanovich, J. Szöllősi, L. Trón, and M. Edidin eds.), pp. 109–135. CRC Press, Boca Raton, Florida.

Edidin, M. (1996). Getting there is only half the fun. *Curr. Top. Membr.* **43,** 1–13.

Edidin, M. (1997). Lipid microdomains in cell surface membranes. *Curr. Opin. Struct. Biol.* **7,** 528–532.

Edidin, M., and Zuniga, M. (1984). Lateral diffusion of wild-type and mutant L^d antigens in L cells. *J. Cell Biol.* **99,** 2333–2335.

Edidin, M., Kuo, S. C., and Sheetz, M. P. (1991). Lateral movements of membrane glycoproteins restricted by dynamic cytoplasmic barriers. *Science* **254,** 1379–1382.

Edidin, M., Zúñiga, M. C., and Sheetz, M. P. (1994). Truncation mutants define and locate cytoplasmic barriers to lateral mobility of membrane glycoproteins. *Proc. Natl. Acad. Sci. U.S.A.* **91,** 3378–3382.

Eisinger, J., Flores, J., and Petersen, W. P. (1986). A milling crowd model for local and long-range obstructed lateral diffusion. Mobility of excimeric probes in the membrane of intact erythrocytes. *Biophys. J.* **49,** 987–1001.

Elson, E. L., and Reidler, J. A. (1979). Analysis of cell surface interactions by measurements of lateral mobility. *J. Supramol. Struct.* **12,** 481–489.

Englund, P. T. (1993). The structure and biosynthesis of glycosyl phosphatidylinositol protein anchors. *Annu. Rev. Biochem.* **62,** 121–138.

Evans, E., and Sackmann, E. (1988). Translational and rotational drag coefficients for a disk moving in a liquid membrane associated with a rigid substrate. *J. Fluid Mech.* **194,** 553–561.

Feder, T. J., Brust-Mascher, I., Slattery, J. P., Baird, B., and Webb, W. W. (1996). Constrained diffusion or immobile fraction on cell surfaces: A new interpretation. *Biophys. J.* **70,** 2767–2773.

Fehr, T., and Löwen, H. (1995). Glass transition in confined geometry. *Phys. Rev. E* **52,** 4016–4025.

Feng, S. Halperin, B. I., and Sen, P. N. (1987). Transport properties of continuum systems near the percolation threshold. *Phys. Rev. B* **35,** 197–214.

Freed, J. H. (1994). Field gradient ESR and molecular diffusion in model membranes. *Annu. Rev. Biophys. Biomol. Struct.* **23,** 1–25.

Furois-Corbin, S., and Pullman, A. (1986). Theoretical study of the packing of α-helices by energy minimization: Effect of the length of the helices on the packing energy and on the optimal configuration of a pair. *Chem. Phys. Lett.* **123,** 305–310.

Garboczi, E. J., Thorpe, M. F., DeVries, M. S., and Day, A. R. (1991). Universal conductivity curve for a plane containing random holes. *Phys. Rev. A* **43,** 6473–6482.

Garver, T. D., Ren, Q., Tuvia, S., and Bennett, V. (1997). Tyrosine phosphorylation at a site highly conserved in the L1 family of cell adhesion molecules abolishes ankyrin binding and increases lateral mobility of neurofascin. *J. Cell Biol.* **137,** 703–714.

Gheber, L. A., and Edidin, M. (1998). Vesicle traffic and barriers to lateral mobility create microdomains in the plasma membrane—a simulation. *Biophys. J.* **74,** A223 (abstract).

Ghislain, L. P., Switz, N. A., and Webb, W. W. (1994). Measurement of piconewton forces using a simple optical force microscope. *Biophys. J.* **66,** A278 (abstract).

Ghosh, R. N. (1991). "Mobility and Clustering of Individual Low-Density Lipoprotein Receptor Molecules on the Surface of Human Skin Fibroblasts." Ph.D. Thesis. Cornell Univ., Ithaca, New York.

Giesen, P. L. A., Willems, G. M., and Hermens, W. T. (1991a). Production of thrombin by the prothrombinase complex is regulated by membrane-mediated transport of prothrombin. *J. Biol. Chem.* **266,** 1379–1382.

Giesen, P. L. A., Willems, G. M., Hemker, H. C., and Hermens, W. T. (1991b). Membrane-mediated assembly of the prothrombinase complex. *J. Biol. Chem.* **266,** 18720–18725.

Golan, D. E. (1989). Red blood cell membrane protein and lipid diffusion. *In* "Red Blood Cell Membranes: Structure, Function, Clinical Implications" (P. Agre and J. C. Parker, eds.), pp. 367–400. Dekker, New York.

Goncalves, E., Yamada, K., Thatte, H. S., Backer, J. M., Golan, D. E., Kahn, C. R., and Shoelson, S. E. (1993). Optimizing transmembrane domain helicity accelerates insulin receptor internalization and lateral mobility. *Proc. Natl. Acad. Sci. U.S.A.* **90,** 5762–5766.

Goodsell, D. S. (1993). "The Machinery of Life." Springer-Verlag, New York.

Grasberger, B., Minton, A. P., DeLisi, C., and Metzger, H. (1986). Interaction between proteins localized in membranes. *Proc. Natl. Acad. Sci. U.S.A.* **83,** 6258–6262.

Hackenbrock, C. R., Chazotte, B., and Gupte, S. S. (1986). The random collision model and a critical assessment of diffusion and collision in mitochondrial electron transport. *J. Bioenerg. Biomembr.* **18,** 331–368.

Halperin, B. I., Feng, S., and Sen, P. N. (1985). Differences between lattice and continuum percolation transport exponents. *Phys. Rev. Lett.* **54,** 2391–2394.

Harder, T., and Simons, K. (1997). Caveolae, DIGs, and the dynamics of sphingolipid-cholesterol microdomains. *Curr. Opin. Cell Biol.* **9,** 534–542.

Haus, J. W., and Kehr, K. W. (1987). Diffusion in regular and disordered lattices. *Phys. Rep.* **150,** 263–406.

Havlin, S., and Ben-Avraham, D. (1987). Diffusion in disordered media. *Adv. Phys.* **36,** 695–798.

Havlin, S., and Bunde, A. (1991). Percolation II. *In* "Fractals and Disordered Systems" (A. Bunde and S. Havlin, eds.), pp. 96–149. Springer-Verlag, Berlin.

Heldin, C. H. (1995). Dimerization of cell surface receptors in signal transduction. *Cell (Cambridge, Mass.)* **80,** 213–223.

Hetzer, M., Heinz, S., Grage, S., and Bayerl, T. M. (1998). Asymmetric molecular friction in supported phospholipid bilayers revealed by NMR measurements of lipid diffusion. *Langmuir* **14,** 982–984.

Heyes, D. M., and Melrose, J. R. (1989). Continuum percolation of 2D Lennard-Jones and square-well phases. *Mol. Phys.* **68,** 359–379.

Holowka, D., and Baird, B. (1996). Antigen-mediated IgE receptor aggregation and signaling: A window on cell surface structure and dynamics. *Annu. Rev. Biophys. Biomol. Struct.* **25,** 79–112.

Huang, Z., and Thompson, N. L. (1996). Imaging fluorescence correlation spectroscopy: Nonuniform IgE distributions on planar membranes. *Biophys. J.* **70,** 2001–2007.

Hughes, B. D. (1995). "Random Walks and Random Environments. Volume 1: Random Walks." Oxford Univ. Press (Clarendon), Oxford.

Hughes, B. D. (1996). "Random Walks and Random Environments. Volume 2: Random Environments." Oxford. Univ. Press (Clarendon), Oxford.

Hughes, B. D., Pailthorpe, B. A., White, L. R., and Sawyer, W. H. (1982). Extraction of membrane microviscosity from translational and rotational diffusion coefficients. *Biophys. J.* **37,** 673–676.

Hwang, J., Tamm, L. K., Böhm, C., Ramalingam, T. S., Betzig, E., and Edidin, M. (1995). Nanoscale complexity of phospholipid monolayers investigated by near-field scanning optical microscopy. *Science* **270,** 610–614.

Isichenko, M. B. (1992). Percolation, statistical topography, and transport in random media. *Rev. Mod. Phys.* **64,** 961–1043.

Jacobson, K., Sheets, E. D., and Simson, R. (1995). Revisiting the fluid mosaic model of membranes. *Science* **268,** 1441–1442.

Jacobson, K. A., Moore, S. E., Yang, B., Doherty, P., Gordon, G. W., and Walsh, F. S., (1997). Cellular determinants of the lateral mobility of neural cell adhesion molecules. *Biochim. Biophys. Acta* **1330,** 138–144.

Jain, M. K., and White III, H. B. (1977). Long-range order in biomembranes. *Adv. Lipid Res.* **15,** 1–60.

Jakobsson, E. (1997). Computer simulation studies of biological membranes: Progress, promise and pitfalls. *Trends Biochem. Sci.* **22,** 339–344.

Janmey, P. (1995). Cell membranes and the cytoskeleton. *In* "Structure and Dynamics of Membranes. Generic and Specific Interactions" (R. Lipowsky and E. Sackmann, eds.), Vol. 1B, pp. 805–849. Elsevier, Amsterdam

Jentoft, N. (1990). Why are proteins O-glycosylated? *Trends Biochem. Sci.* **15,** 291–294.

Johnson, D. R., Bhatnagar, R. S., Knoll, L. J., and Gordon, J. I. (1994). Genetic and biochemical studies of protein N-myristoylation. *Annu. Rev. Biochem.* **63,** 869–914.

Johnson, M. E., Berk, D. A., Blankschtein, D., Golan, D. E., Jain, R. K., and Langer, R. S. (1996). Lateral diffusion of small compounds in human stratum corneum and model lipid bilayer systems. *Biophys. J.* **71,** 2656–2668.

Jovin, T. M., and Vaz, W. L. C. (1989). Rotational and translational diffusion in membranes measured by fluorescence and phosphorescence methods. *In* "Methods in Enzymology" (S. Fleischer and B. Fleischer, eds.), Vol. 172, pp. 471–513. Academic Press. San Diego.

Karakatsanis, P., and Bayerl, T. M. (1996). Diffusion measurements in oriented phospholipid bilayers by ^{1}H-NMR in a static fringe field gradient. *Phys. Rev. E* **54,** 1785–1790.

Kenworthy, A. K., and Edidin, M. (1998). Imaging FRET detects clustering of ganglioside GM_1 molecules with one another, but not with a GPI-anchored protein, 5′NT, on the apical surface of MDCK cells. *Biophys. J.* **74,** A372 (abstract).

Kim, I. C., and Torquato, S. (1992). Diffusion of finite-sized Brownian particles in porous media. *J. Chem. Phys.* **96,** 1498–1503.

Knowles, D. W., Tilley, L., Mohandas, N., and Chasis, J. A. (1997). Erythrocyte membrane vesiculation: Model for the molecular mechanism of protein sorting. *Proc. Natl. Acad. Sci. U.S.A.* **94,** 12969–12974.

Kopelman, R. (1988). Fractal reaction kinetics. *Science* **241,** 1620–1626.

Kopelman, R. (1989). Diffusion-controlled reaction kinetics. *In* "The Fractal Approach to Heterogeneous Chemistry" (D. Avnir, ed.), pp. 295–309. Wiley, Chichester.

Kopelman, R., and Koo, Y.-E. (1991). Reaction kinetics in restricted spaces. *Isr. J. Chem.* **31,** 147–157.

Koppel, D. E. (1981). Association dynamics and lateral transport in biological membranes. *J. Supramol. Struct. Cell Biochem.* **17,** 61–67.

Koppel, D. E., Morgan, F., Cowan, A. E., and Carson, J. H. (1994). Scanning concentration correlation spectroscopy using the confocal laser microscope. *Biophys. J.* **66,** 502–507.

Kucik, D. F., Elson, E. L., and Sheetz, M. P. (1999). Weak dependence of mobility of membrane protein aggregates on aggregate size supports a viscous model of retardation of diffusion. *Biophys. J.* **76,** 314–322.

Kusumi, A., and Sako, Y. (1996). Cell surface organization by the membrane skeleton. *Curr. Opin. Cell Biol.* **8,** 566–574.

Kusumi, A., Sako, Y., and Yamamoto, M. (1993). *Biophys. J.* **65,** 2021–2040.

Kusumi, A., Sako, Y., Fujiwara, T., and Tomishige, M. (1998). Application of laser tweezers to studies of the fences and tethers of the membrane skeleton that regulate the movements of plasma membrane proteins. *Methods Cell Biol.* **55,** 173–194.

Lamb, T. D. (1996). Gain and kinetics of activation in the G-protein cascade of phototransduction. *Proc. Natl. Acad. Sci. U.S.A.* **93,** 566–570.

Lavergne, J., and Joliot, P. (1991). Restricted diffusion in photosynthetic membranes. *Trends Biochem. Sci.* **16,** 129–134.

Lee, C., Lui, C., Srivastava, M., and Petersen, N. O. (1998). The dynamics of lateral diffusion: From small molecules to aggregates. *Biophys. J.* **74,** A222 (abstract).

Lee, G. M., and Jacobson, K. (1994). Lateral mobility of lipids in membranes. *Curr. Top. Membr.* **40,** 111–142.

Lee, G. M., Zhang, F., Ishihara, A., McNeil, C. L., and Jacobson, K. A. (1993). Unconfined lateral diffusion and an estimate of pericellular matrix viscosity revealed by measuring the mobility of gold-tagged lipids. *J. Cell Biol.* **120,** 25–35.

Lee, S. B., and Torquato, S. (1990). Monte Carlo study of correlated continuum percolation: Universality and percolation thresholds. *Phys. Rev. A* **41,** 5338–5344.

Lenaz, G. (1988). Role of mobility of redox components in the inner mitochondrial membrane. *J. Membr. Biol.* **104,** 193–209.

Liebman, P. A., Parker, K. R., and Dratz, E. A. (1987). The molecular mechanism of visual excitation and its relation to the structure and composition of the rod outer segment. *Annu. Rev. Physiol.* **49,** 765–791.

Lindblom, G., and Orädd, G. (1994). NMR studies of translational diffusion in lyotropic liquid crystals and lipid membranes. *Prog. Nucl. Magn. Reson. Spectrosc.* **26,** 483–515.

Lindenberg, K., West, B. J., and Kopelman, R. (1989). Diffusion-limited $A + B \rightarrow 0$ reaction: Spontaneous segregation. *In* "Noise and Chaos in Nonlinear Dynamical Systems" (F. Moss, L. Lugiato, and W. Schleich, eds.), pp. 142–171. Cambridge Univ. Press. Cambridge.

Lindenberg, K., West, B. J., and Kopelman, R. (1990). Diffusion-limited $A + B \rightarrow 0$ reaction: correlated initial condition. *Phys. Rev. A* **42,** 890–894.

Lipowsky, R., and Sackmann, E., eds. (1995). "Structure and Dynamics of Membranes. From Cells to Vesicles." Elsevier, Amsterdam.

Liu, C., Paprica, A., and Petersen, N. O. (1997). Effects of size of macrocyclic polyamides on their rate of diffusion in model membranes. *Biophys. J.* **73,** 2580–2587.

Livneh, E., Benveniste, M., Prywes, R., Felder, S., Kam, Z., and Schlessinger, J. (1986). Large deletions in the cytoplasmic kinase domain of the epidermal growth factor receptor do not affect its lateral mobility. *J. Cell Biol.* **103,** 327–331.

Luna, E. J., and Hitt, A. L. (1992). Cytoskeleton–plasma membrane interactions. *Science* **258,** 955–964.

McCloskey, M., and Poo, M.-m. (1984). Protein diffusion in cell membranes: Some biological implications. *Int. Rev. Cytol.* **87,** 19–81.

McCloskey, M. A., and Poo, M.-m. (1986a). Contact-induced redistribution of specific membrane components: Local accumulation and development of adhesion. *J. Cell Biol.* **102,** 2185–2196.

McCloskey, M. A., and Poo, M.-m. (1986b). Rates of membrane-associated reactions: Reduction of dimensionality revisited. *J. Cell Biol.* **102,** 88–96.

McElhaney, R. N. (1984). The relationship between membrane lipid fluidity and phase state and the ability of bacteria and mycoplasmas to grow and survive at various temperatures. *In* "Membrane Fluidity" (M. Kates and L. A. Manson, eds.), pp. 249–278. Plenum, New York.

McLaughlin, S., and Aderem, A. (1995). The myristoyl-electrostatic switch: A modulator of reversible protein–membrane interactions. *Trends Biochem. Sci.* **20,** 272–276.

Matko, J., and Edidin, M. (1997). Energy transfer methods for detecting molecular clusters on cell surfaces. *In* "Methods in Enzymology" (L. Brand and M. L. Johnson, eds.), Vol. 278, pp. 444–462. Academic Press, San Diego.

Melo, E. C., Lourtie, I. M. G., Sankaram, M. B., Thompson, T. E., and Vaz, W. L. C. (1992). Effects of domain connection and disconnection on the yields of in-plane bimolecular reactions in membranes. *Biophys. J.* **63,** 1506–1512.

Menon, A. K., Holowka, D., Webb, W. W., and Baird, B. (1986). Cross-linking of receptor-bound IgE to aggregates larger than dimers leads to rapid immobilization. *J. Cell Biol.* **102,** 541–550.

Merz, K. M., Jr. (1997). Molecular dynamics simulations of lipid bilayers. *Curr. Opin. Struct. Biol.* **7,** 511–517.

Merz, K. M., Jr., and Roux, B. (1996). "Biological Membranes: A Molecular Perspective from Computation and Experiment." Birkhäuser, Boston.

Metzger, H., and Kinet, J.-P. (1988). How antibodies work: Focus on Fc receptors. *FASEB J.* **2,** 3–11.

Minton, A. P. (1989). Lateral diffusion of membrane proteins in protein-rich membranes. A simple hard particle model for concentration dependence of the two-dimensional diffusion coefficient. *Biophys. J.* **55,** 805–808.

Morrow, J. S., Rimm, D. L., Kennedy, S. P., Cianci, C. D., Sinard, J. H., and Weed, S. A. (1997). Of membrane stability and mosaics: The spectrin cytoskeleton. *In* "Handbook of Physiology. Section 14: Cell Physiology" (J. F. Hoffman and J. D. Jamieson, eds.), pp. 485–540. Oxford Univ. Press, New York.

Mouritsen, O. G. (1990). Computer simulation of cooperative phenomena in lipid membranes. *In* "Molecular Description of Biological Membranes by Computer Aided Conformational Analysis" (R. Brasseur, ed.), Vol. 1, pp. 3–83. CRC Press, Boca Raton, Florida.

Mouritsen, O. G., and Bloom, M. (1993). Models of lipid–protein interactions in membranes. *Annu. Rev. Biophys. Biomol. Struct.* **22,** 145–171.

Mouritsen, O. G., Jorgensen, K. (1997). Small-scale lipid-membrane structure: Simulation versus experiment. *Curr. Opin. Struct. Biol.* **7,** 518–527.

Mouritsen, O. G., Sperotto, M. M., Risbo, J., Zhang, Z., and Zuckermann, M. J. (1996). Computational approach to lipid–protein interactions in membranes. *In* "Advances in Computational Biology" (H. O. Villar, ed.), Vol. 2, pp. 15–64. JAI Press, Greenwich, Connecticut.

Mumby, S. M. (1997). Reversible palmitoylation of signaling proteins. *Curr. Opin. Cell Biol.* **9,** 148–154.

Nagle, J. F. (1992). Long tail kinetics in biophysics? *Biophys. J.* **63,** 366–370.

Neubig, R. R. (1994). Membrane organization in G-protein mechanisms. *FASEB J.* **8,** 939–946.

Newhouse, J. S., and Kopelman, R. (1988). Steady-state chemical kinetics on surface clusters and islands: Segregation of reactants. *J. Phys. Chem.* **92,** 1538–1541.

Nielsen, M., Miao, L., Ipsen, J. H., Mouritsen, O. G., and Zuckermann, M. J. (1996). Random-lattice models and simulation algorithms for the phase equilibria in two-dimensional condensed systems of particles with coupled internal and translational degrees of freedom. *Phys. Rev. E* **54,** 6889–6905.

Nieuwenhuizen, T. M., van Velthoven, P. F. J., and Ernst, M. H. (1986). Diffusion and long-time tails in a two-dimensional site-percolation model. *Phys. Rev. Lett.* **57,** 2477–2480.

Nosjean, O., Briolay, A., and Roux, B. (1997). Mammalian GPI proteins: Sorting, membrane residence and functions. *Biochim. Biophys. Acta* **1331,** 153–186.

Okamoto, T., Schlegel, A., Scherer, P. E., and Lisanti, M. P. (1998). Caveolins, a family of scaffolding proteins for organizing "preassembled signaling complexes" at the plasma membrane. *J. Biol. Chem.* **273,** 5419–5422.

Paller, M. S. (1994). Lateral mobility of Na,K-ATPase and membrane lipids in renal cells. Importance of cytoskeletal integrity. *J. Membr. Biol.* **142,** 127–135.

Parasassi, T., Loiero, M., Raimondi, M., Ravagnan, G., and Gratton, E. (1993). Absence of lipid gel-phase domains in seven mammalian cell lines and in four primary cell types. *Biochim. Biophys. Acta* **1153,** 143–154.

Parasassi, T., Giusti, A. M., Raimondi, M., and Gratton, E. (1995). Abrupt modifications of phospholipid bilayer properties at critical cholesterol concentrations. *Biophys. J.* **68,** 1895–1902.

Parton, R. G. (1996). Caveolae and caveolins. *Curr. Opin. Cell Biol.* **8,** 542–548.

Parton, R. G., and Simons, K. (1995). Digging into caveolae. *Science* **269,** 1398–1399.

Peters, R. (1988). Lateral mobility of proteins and lipids in the red cell membrane and the activation of adenylate cyclase by β-adrenergic receptors. *FEBS Lett.* **234,** 1–7.

Peters, R., and Cherry, R. J. (1982). Lateral and rotational diffusion of bacteriorhodopsin in lipid bilayers: Experimental test of the Saffman–Delbrück equations. *Proc. Natl. Acad. Sci. U.S.A.* **79,** 4317–4321.

Peters, R., Peters, J., Tews, K. H., and Bähr, W. (1974). A microfluorimetric study of translational diffusion in erythrocyte membranes. *Biochim. Biophys. Acta* **367,** 282–294.

Petersen, N. O., Felder, S., and Elson, E. L. (1986). Measurement of lateral diffusion by fluorescence photobleaching recovery. *In* "Handbook of Experimental Immunology" (D. M. Weir, L. A. Herzenberg, C. Blackwell, and L. A. Herzenberg, eds.), 4th Ed., Vol. 1, pp. 24.1–24.23. Blackwell Scientific, Oxford.

Petersen, N. O., Höddelius, P. L., Wiseman, P. W. Seger, O., and Magnusson, K.-E. (1993). Quantitation of membrane receptor distributions by image correlation spectroscopy: Concept and application. *Biophys. J.* **65,** 1135–1146.

Pfeifer, P. (1987). Characterization of surface irregularity. *In* "Preparative Chemistry Using Supported Reagents" (P. Laszlo, ed.), pp. 13–33. Academic Press, San Diego.

Picot, D., and Garavito, R. M. (1994). Prostaglandin H synthase: Implications for membrane structure. *FEBS Lett.* **346,** 21–25.

Picot, D., Loll, P. J., and Garavito, R. M. (1994). The X-ray crystal structure of the membrane protein prostaglandin H_2 synthase-1. *Nature* (*London*) **367,** 243–249.

Piknová, B., Marsh, D., and Thompson, T. E. (1996). Fluorescence-quenching study of percolation and compartmentalization in two-phase lipid bilayers. *Biophys. J.* **71,** 892–897.

Piknová, B., Marsh, D., and Thompson, T. E. (1997). Fluorescence quenching and electron spin resonance study of percolation in a two-phase lipid bilayer containing bacteriorhodopsin. *Biophys. J.* **72,** 2660–2668.

Pink, D. A. (1985). Protein lateral movement in lipid bilayers: Simulation studies of its dependence upon protein concentration. *Biochim. Biophys. Acta* **818,** 200–204.

Pink, D. A. (1990). Computer simulation of biological membranes. *In* "Molecular Description of Biological Membranes by Computer Aided Conformational Analysis" (R. Brasseur, ed.), Vol. 1, pp. 151–170. CRC Press, Boca Raton, Florida.

Pink, D. A., Laidlaw, D. J., and Chisholm, D. M. (1986). Protein lateral movement in lipid bilayers. Monte Carlo simulation studies of its dependence upon attractive protein–protein interactions. *Biochim. Biophys. Acta* **863,** 9–17.

Ranganathan, S. (1994). Localization and glass transition in two-dimensional liquids. *J. Phys. Condensed Matter* **6,** 1299–1308.

Ryan, T. A., Myers, J., Holowka, D., Baird, B., and Webb, W. W. (1988). Molecular crowding on the cell surface. *Science* **239,** 61–64.

Saffman, P. G., and Delbrück, M. (1975). Brownian motion in biological membranes. *Proc. Natl. Acad. Sci. U.S.A.* **72,** 3111–3113.

Sako, Y., and Kusumi, A. (1994). Compartmentalized structure of the plasma membrane for receptor movements as revealed by a nanometer-level motion analysis. *J. Cell Biol.* **125,** 1251–1264.

Sako, Y., and Kusumi, A. (1995). Barriers for lateral diffusion of transferrin receptor in the plasma membrane as characterized by receptor dragging by laser tweezers: Fence versus tether. *J. Cell Biol.* **129,** 1559–1574.

Sankaram, M. B., Marsh, D., Gierasch, L. M., and Thompson, T. E. (1994). Reorganization of lipid domain structure in membranes by a transmembrane peptide: An ESR spin label study on the effect of the *Escherichia coli* outer membrane protein A signal peptide on the fluid lipid domain connectivity in binary mixtures of dimyristoyl phosphatidylcholine and distearoyl phosphatidylcholine. *Biophys. J.* **66,** 1959–1968.

Savageau, M. A. (1995). Michaelis–Menten mechanism reconsidered: Implications of fractal kinetics. *J. Theor. Biol.* **176,** 115–124.

Saxton, M. J. (1982). Lateral diffusion in an archipelago. Effects of impermeable patches on diffusion in a cell membrane. *Biophys. J.* **39,** 165–173.

Saxton, M. J. (1987). Lateral diffusion in an archipelago. The effect of mobile obstacles. *Biophys. J.* **52,** 989–997.

Saxton, M. J. (1989a). Lateral diffusion in an archipelago. Distance dependence of the diffusion coefficient. *Biophys. J.* **56,** 615–622.

Saxton, M. J. (1989b). The spectrin network as a barrier to lateral diffusion in erythrocytes. A percolation analysis. *Biophys. J.* **55,** 21–28.

Saxton, M. J. (1990a). Lateral diffusion in a mixture of mobile and immobile particles. A Monte Carlo study. *Biophys. J.* **58,** 1303–1306.

Saxton, M. J. (1990b). The membrane skeleton of erythrocytes. A percolation model. *Biophys. J.* **57,** 1167–1177.

Saxton, M. J. (1990c). The membrane skeleton of erythrocytes: Models of its effect on lateral diffusion. *Int. J. Biochem.* **22,** 801–809.

Saxton, M. J. (1992). Lateral diffusion and aggregation. A Monte Carlo study. *Biophys. J.* **61,** 119–128.

Saxton, M. J. (1993a). Lateral diffusion in an archipelago. Dependence on tracer size. *Biophys. J.* **64,** 1053–1062.

Saxton, M. J. (1993b). Lateral diffusion in an archipelago. Single-particle diffusion. *Biophys. J.* **64,** 1766–1780.

Saxton, M. J. (1994a). Anomalous diffusion due to obstacles: A Monte Carlo study. *Biophys. J.* **66,** 394–401.

Saxton, M. J. (1994b). Single-particle tracking: Models of directed transport. *Biophys. J.* **67,** 2110–2119.

Saxton, M. J. (1995). Single-particle tracking: Effects of corrals. *Biophys. J.* **69,** 389–398.

Saxton, M. J. (1996). Anomalous diffusion due to binding. A Monte Carlo study. *Biophys. J.* **70,** 1250–1262.

Saxton, M. J. (1997). Single-particle tracking: The distribution of diffusion coefficients. *Biophys. J.* **72,** 1744–1753.

Saxton, M. J. (1998). Lateral diffusion in an archipelago: Chemically limited reactions. *Biophys. J.* **74,** A223 (abstract).

Saxton, M. J., and Jacobson, K. (1997). Single-particle tracking: Applications to membrane dynamics. *Annu. Rev. Biophys. Biomol. Struct.* **26,** 373–399.

Scalettar, B. A., and Abney, J. R. (1991). Molecular crowding and protein diffusion in biological membranes. *Comments Mol. Cell Biophys.* **7,** 79–107.

Scher, H., Shlesinger, M. F., and Bendler, J. T. (1991). Time-scale invariance in transport and relaxation. *Phys. Today* **44**(No. 1), 26–34.

Schlessinger, J. (1980). The mechanism and role of hormone-induced clustering of membrane receptors. *Trends Biochem. Sci.* **5,** 210–214.

Schlessinger, J. (1993–1994). Cellular signaling by receptor tyrosine kinases. *Harvey Lect.* **89,** 105–123.

Schlessinger, J., Lax, I., and Lemmon, M. (1995). Regulation of growth factor activation by proteoglycans: What is the role of the low affinity receptors? *Cell (Cambridge, Mass.)* **83,** 357–360.

Schmid, S. L. (1997). Clathrin-coated vesicle formation and protein sorting: An integrated process. *Annu. Rev. Biochem.* **66,** 511–548.

Schmidt, T., Schütz, G. J., Baumgartner, W., Gruber, H. J., and Schindler, H. (1995). Characterization of photophysics and mobility of single molecules in a fluid lipid membrane. *J. Phys. Chem.* **99,** 17662–17668.

Schmidt, T., Schütz, G. J., Baumgartner, W., Gruber, H. J., and Schindler, H. (1996). Imaging of single molecule diffusion. *Proc. Natl. Acad. Sci. U.S.A.* **93,** 2926–2929.

Schram, V., and Thompson, T. E. (1997). Influence of the intrinsic membrane protein bacteriorhodopsin on gel-phase domain topology in two-component phase-separated bilayers. *Biophys. J.* **72,** 2217–2225.

Schram, V., Tocanne, J.-F., and Lopez, A. (1994). Influence of obstacles on lipid lateral diffusion: Computer simulation of FRAP experiments and application to proteoliposomes and biomembranes. *Eur. Biophys. J.* **23,** 337–348.

Schram, V., Lin, H.-N., and Thompson, T. E. (1996). Topology of gel-phase domains and lipid mixing properties in phase-separated two-component phosphatidylcholine bilayers. *Biophys. J.* **71,** 1811–1822.

Schütz, G. J., Schindler, H., and Schmidt, T. (1997a). Details of lipid motion in biomembranes studied by single molecule microscopy. *Biophys. J.* **72,** A24 (abstract).

Schütz, G. J., Schindler, H., and Schmidt, T. (1997b). Single-molecule microscopy on model membranes reveals anomalous diffusion. *Biophys. J.* **73,** 1073–1080.

Schwille, P., Meyer-Almes, F.-J., and Rigler, R. (1997). Dual-color fluorescence cross-correlation spectroscopy for multicomponent diffusional analysis in solution. *Biophys. J.* **72,** 1878–1886.

Shea, L. D., Omann, G. M., and Linderman, J. J. (1997). Calculation of diffusion-limited kinetics for the reactions in collision coupling and receptor cross-linking. *Biophys. J.* **73,** 2949–2959.

Sheets, E. D., Simson, R., and Jacobson, K. (1995). New insights into membrane dynamics from the analysis of cell surface interactions by physical methods. *Curr. Opin. Cell Biol.* **7,** 707–714.

Sheets, E. D., Lee, G. M., Simson, R., and Jacobson, K. (1997). Transient confinement of a glycosylphosphatidylinositol-anchored protein in the plasma membrane. *Biochemistry* **36,** 12449–12458.

Sheetz, M. P. (1983). Membrane skeletal dynamics: Role in modulation of red cell deformability, mobility of transmembrane proteins, and shape. *Semin. Hematol.* **20,** 175–188.

Sheetz, M. P. (1993). Glycoprotein motility and dynamic domains in fluid plasma membranes. *Annu. Rev. Biophys. Biomol. Struct.* **22,** 417–431.

Sheetz, M. P., ed. (1998). Laser Tweezers in Cell Biology. *Methods Cell Biol.* Vol. 55.

Sheetz, M. P., Schindler, M., and Koppel, D. E. (1980). Lateral mobility of integral membrane proteins is increased in spherocytic erythrocytes. *Nature (London)* **285,** 510–512.

Shen, L., Bassolino, D., and Stouch, T. (1997). Transmembrane helix structure, dynamics, and interactions: Multi-nanosecond molecular dynamics simulations. *Biophys. J.* **73,** 3–20.

Shi, Z.-Y., and Kopelman, R. (1992). Nonclassical kinetics and reaction probability for bimolecular reactions in low-dimensional media. *J. Phys. Chem.* **96,** 6858–6861.

Shin, Y.-K., Ewert, U., Budil, D. E., and Freed, J. H. (1991). Microscopic versus macroscopic diffusion in model membranes by electron spin resonance spectral-spatial imaging. *Biophys. J.* **59,** 950–957.

Shiozawa, J. A., Brandts, J. F., and Jacobson, B. S. (1989). Binding of plasma membrane glycoproteins to the cytoskeleton during patching and capping is consistent with an entropy-enhancement model. *Biochim. Biophys. Acta* **980,** 361–366.

Shlesinger, M. F. (1988). Fractal time in condensed matter. *Annu. Rev. Phys. Chem.* **39,** 269–290.

Simson, R., Yang, B., Moore, S. E., Doherty, P., Walsh, F. S., and Jacobson, K. A. (1998). Structural mosaicism on the submicron scale in the plasma membrane. *Biophys. J.* **74,** 297–308.

Singer, S. J. (1971). The molecular organization of biological membranes. *In* "Structure and Function of Biological Membranes" (L. I. Rothfield, ed.), pp. 145–222. Academic Press, New York.

Singer, S. J. (1974). The molecular organization of membranes. *Annu. Rev. Biochem.* **43,** 805–833.

Singer, S. J., and Nicolson, G. L. (1972). The fluid mosaic model of the structure of cell membranes. *Science* **175,** 720–731.

Sintes, T., and Baumgärtner, A. (1997). Protein attraction in membranes induced by lipid fluctuations. *Biophys. J.* **73,** 2251–2259.

Slater, G. W., and Guo, H. L. (1996). An exactly solvable Ogston model of gel electrophoresis: I. The role of the symmetry and randomness of the gel structure. *Electrophoresis* (*Weinheim*) **17,** 977–988.

Slater, G. W., and Treurniet, J. R. (1997). Exactly solvable Ogston model of gel electrophoresis. III. Percolation and sieving through two-dimensional gels. *J. Chromatogr. A* **772,** 39–48.

Slattery, J. P. (1995). "Lateral Mobility of $Fc_{\epsilon}RI$ on Rat Basophilic Leukemia Cells as Measured by Single Particle Tracking Using a Novel Bright Fluorescent Probe." Ph.D. Thesis. Cornell Univ., Ithaca, New York.

Sobolev, A. S., Kazarov, A. R., and Rosenkranz, A. A. (1988). Application of percolation theory principles to the analysis of interaction of adenylate cyclase complex proteins in cell membranes. *Mol. Cell. Biochem.* **81,** 19–28.

Stauffer, D., and Aharony, A. (1992). "Introduction to Percolation Theory," 2nd Ed. Taylor & Francis, London.

Stauffer, T. P., and Meyer, T. (1997). Compartmentalized IgE receptor-mediated signal transduction in living cells. *J. Cell Biol.* **139,** 1447–1454.

Stoffel, W., and Bosio, A. (1997). Myelin glycolipids and their functions. *Curr. Opin. Neurobiol.* **7,** 654–661.

Stout, A. L., and Webb, W. W. (1998). Optical force microscopy. *Methods Cell Biol.* **55,** 99–116.

Stryer, L. (1986). Cyclic GMP cascade of vision. *Annu. Rev. Neurosci.* **9,** 87–119.

Sugar, I. P., Tang, D., and Chong, P. L.-G. (1994). Monte Carlo simulation of lateral distribution of molecules in a two-component lipid membrane. Effect of long-range repulsive interactions. *J. Phys. Chem.* **98,** 7201–7210.

Svoboda, K., and Block, S. M. (1994). Biological applications of optical forces. *Annu. Rev. Biophys. Biomol. Struct.* **23,** 247–285.

Switz, N. A., Mertz, J., and Webb, W. W. (1996). A feedback-modified optical trap for probing local viscosity and examining diffusive behavior on cell membranes. *Biophys. J.* **70,** A334 (abstract).

Tank, D. W., Wu, E.-S., and Webb, W. W. (1982). Enhanced molecular diffusibility in muscle membrane blebs: Release of lateral constraints. *J. Cell Biol.* **92,** 207–212.

Terpetschnig, E., Szmacinski, H., and Lakowicz, J. R. (1997). Long-lifetime metal–ligand complexes as probes in biophysics and clinical chemistry. *In* "Methods in Enzymology" (L. Brand and M. L. Johnson, eds.), Vol. 278, pp. 295–321. Academic Press, San Diego.

Thomas, J. L., Feder, T. J., and Webb, W. W. (1992). Effects of protein concentration on IgE receptor mobility in rat basophilic leukemia cell plasma membranes. *Biophys. J.* **61,** 1402–1412.

Thompson, T. E., Sankaram, M. B., Biltonen, R. L., Marsh, D., and Vaz, W. L. C. (1995). Effects of domain structure on in-plane reactions and interactions. *Mol. Membr. Biol.* **12,** 157–162.

Tobochnik, J. (1990). Efficient random walk algorithm for computing conductivity in continuum percolation systems. *Comput. Phys.* **4,** 181–184.

Tocanne, J.-F., Cézanne, L., Lopez, A., Piknova, B., Schram, V., Tournier, J.-F., and Welby, M. (1994a). Lipid domains and lipid/protein interactions in biological membranes. *Chem. Phys. Lipids* **73,** 139–158.

Tocanne, J.-F., Dupou-Cézanne, L., and Lopez, A. (1994b). Lateral diffusion of lipids in model and natural membranes. *Prog. Lipid Res.* **33,** 203–237.

Tomishige, M., and Kusumi, A. (1998). Regulation of lateral movements of erythrocyte band 3 by membrane skeleton as studied by single particle tracking and optical tweezers. *Biophys. J.* **74,** A223 (abstract).

Torney, D. C., and McConnell, H. M. (1983). Diffusion-limited reaction rate theory for two-dimensional systems. *Proc. R. Soc. London A* **387,** 147–170.

Torquato, S. (1991). Random heterogeneous media: Microstructure and improved bounds on effective properties. *Appl. Mech. Rev.* **44,** 37–76.

Torquato, S. (1994). Unified methodology to quantify the morphology and properties of inhomogeneous media. *Physica A (Amsterdam)* **207,** 79–91.

Torquato, S., and Kim, I. C. (1989). Efficient simulation technique to compute effective properties of heterogeneous media. *Appl. Phys. Lett.* **55,** 1847–1849.

Tsuji, A., and Ohnishi, S. (1986). Restriction of the lateral motion of band 3 in the erythrocyte membrane by the cytoskeletal network: Dependence on spectrin association state. *Biochemistry* **25,** 6133–6139.

van Beijeren, H., and Kutner, R. (1985). Mean square displacement of a tracer particle in a hard-core lattice gas. *Phys. Rev. Lett.* **55,** 238–241.

Vaz, W. L. C. (1992). Translational diffusion in phase-separated lipid bilayer membranes. *Comments Mol. Cell. Biophys.* **8,** 17–36.

Vaz, W. L. C. (1994). Diffusion and chemical reactions in phase-separated membranes. *Biophys. Chem.* **50,** 139–145.

Vaz, W. L. C. (1996). Consequences of phase separations in membranes. *In* "Handbook of Nonmedical Applications of Liposomes" (Y. Barenholz and D. D. Lasic, eds.), Vol. 2, pp. 51–60. CRC Press, Boca Raton, Florida.

Vaz, W. L. C., and Almeida, P. F. (1991). Microscopic versus macroscopic diffusion in one-component fluid phase lipid bilayer membranes. *Biophys. J.* **60,** 1553–1554.

Vaz, W. L. C., and Almeida, P. F. F. (1993). Phase topology and percolation in multiphase lipid bilayers: Is the biological membrane a domain mosaic? *Curr. Opin. Struct. Biol.* **3,** 482–488.

Vaz, W. L. C., and Criado, M. (1985). A comparison of the translational diffusion of a monomer and an oligomer of the acetylcholine receptor protein reconstituted into soybean lipid bilayers. *Biochim. Biophys. Acta* **819,** 18–22.

Vaz, W. L. C., Clegg, R. M., and Hallmann, D. (1985). Translational diffusion of lipids in liquid crystalline phase phosphatidylcholine multibilayers. A comparison of experiment with theory. *Biochemistry* **24,** 781–786.

Vaz, W. L. C., Melo, E. C. C., and Thompson, T. E. (1989). Translational diffusion and fluid domain connectivity in a two-component, two-phase phospholipid bilayer. *Biophys. J.* **56,** 869–876.

Viitala, J., and Järnefelt, J. (1985). The red cell surface revisited. *Trends Biochem. Sci.* **10,** 392–395.

Wang, D., Gou, S.-Y., and Axelrod, D. (1992). Reaction rate enhancement by surface diffusion of adsorbates. *Biophys. Chem.* **43,** 117–137.

Webb, W. W., Barak, L. S., Tank, D. W., and Wu, E.-S. (1981). Molecular mobility on the cell surface. *Biochem. Soc. Symp.* **46,** 191–205.

Weiss, G. H. (1994). "Aspects and Applications of the Random Walk." North-Holland, Amsterdam.

Welti, R., and Glaser, M. (1994). Lipid domains in model and biological membranes. *Chem. Phys. Lipids* **73,** 121–137.

Wiegel, F. W. (1979). Hydrodynamics of a permeable patch in the fluid membrane. *J. Theor. Biol.* **77,** 189–193.

Wiegel, F. W., and DeLisi, C. (1982). Evaluation of reaction rate enhancement by reduction in dimensionality. *Am. J. Physiol.* **243,** R475–R479.

Wier, M., and Edidin, M. (1988). Constraint of the translational diffusion of a membrane glycoprotein by its external domains. *Science* **242,** 412–414.

Wilson, K. M., Morrison, I. E. G., Smith, P. R., Fernandez, N., and Cherry, R. J. (1996). Single particle tracking of cell-surface HLA-DR molecules using R-phycoerythrin labeled monoclonal antibodies and fluorescence digital imaging. *J. Cell Sci.* **109,** 2101–2109.

Wiseman, P. W., Höddelius, P., Petersen, N. O., and Magnusson, K.-E. (1997). Aggregation of PDGF-β receptors in human skin fibroblasts: Characterization by image correlation spectroscopy (ICS). *FEBS Lett.* **401,** 43–48.

Wolf, D. E. (1989). Designing, building, and using a fluorescence recovery after photobleaching instrument. *Methods Cell Biol.* **30,** 271–306.

Wolf, D. E. (1994). Microheterogeneity in biological membranes. *Curr. Top. Membr.* **40,** 143–165.

Wolf, D. E. (1995). Lipid domains in sperm plasma membranes. *Mol. Membr. Biol.* **12,** 101–104.

Woolf, T. B., and Roux, B. (1994). Molecular dynamics simulation of the gramicidin channel in a phospholipid bilayer. *Proc. Natl. Acad. Sci. U.S.A.* **91,** 11631–11635.

Woolf, T. B., and Roux, B. (1996). Structure, energetics, and dynamics of lipid–protein interactions: A molecular dynamics study of the gramicidin A channel in a DMPC bilayer. *Proteins: Struct, Funct. Genet.* **24,** 92–114.

Wu, P., and Brand, L. (1994). Resonance energy transfer: Methods and application. *Anal. Biochem.* **218,** 1–13.

Xia, W., and Thorpe, M. F. (1988). Percolation properties of random ellipses. *Phys. Rev. A* **38,** 2650–2656.

Zakharova, O. M., Rosenkranz, A. A., and Sobolev, A. S. (1995). Modification of fluid lipid and mobile protein fractions of reticulocyte plasma membranes affects agonist-stimulated adenylate cyclase. Application of the percolation theory. *Biochim. Biophys. Acta* **1236,** 177–184.

Zhang, F. L., and Casey, P. J. (1996). Protein prenylation: Molecular mechanisms and functional consequences. *Annu. Rev. Biochem.* **65,** 241–269.

Zhang, F., Crise, B., Su, B., Hou, Y., Rose, J. K., Bothwell, A., and Jacobson, K. (1991). Lateral diffusion of membrane-spanning and glycosylphosphatidylinositol-linked proteins: Toward establishing rules governing the lateral mobility of membrane proteins. *J. Cell Biol.* **115,** 75–84.

Zhang, F., Schmidt, W. G., Hou, Y., Williams, A. F., and Jacobson, K. (1992). Spontaneous incorporation of the glycosyl-phosphatidylinositol-linked protein Thy-1 into cell membranes. *Proc. Natl. Acad. Sci. U.S.A.* **89,** 5231–5235.

Zhang, F., Lee, G. M., and Jacobson, K. (1993). Protein lateral mobility as a reflection of membrane microstructure. *BioEssays* **15,** 579–588.

Zimmerman, S. B., and Minton, A. P. (1993). Macromolecular crowding: Biochemical, biophysical, and physiological consequences. *Annu. Rev. Biophys. Biomol. Struct.* **22,** 27–65.

Zurzolo, C., and Rodriguez-Boulan, E. (1994). Lipid-tagged proteins. *Curr. Top. Membr.* **40,** 295–318.

CHAPTER 9

A Short History of Ion Channels and Signal Propagation

Clay M. Armstrong
Department of Physiology, University of Pennsylvania, Philadelphia, Pennsylvania 19104

I. INTRODUCTION

The investigation of electrical activity in cells is surely one of the most fascinating stories in biology. Dependent on quantities that cannot be seen and that were difficult to measure until this century, electrical properties in living tissues must have seemed almost mystical. Important chapters in the story, however, were written by scientists equipped with an excellent understanding of the physics of their time and apparently not at all in awe

1063-5823/99 $30.00

of the subject. In 1872, Hermann had an accurate understanding of the requirements for electrical transmission in nerve fibers, based on the cable equations. Nernst's (1888) studies of diffusion potentials gave the crucial insight that differences in ion concentrations could give rise to voltages. Development of a voltage required not only a concentration gradient, but that one of the ions of a salt be more diffusible than the other. For example, if Cl^- can diffuse faster than Na^+, Cl^- will outrun Na^+, making the leading edge of the concentration wave negative. Bernstein then made a brilliant proposal, that a resting cell membrane prohibits the diffusion of all ions except K^+, thus creating an extreme form of a Nernst diffusion potential. Further, according to Bernstein (1902), voltage changes could be generated by postulating a sudden loss of the membrane's selective K^+ permeability. At about the same time, Overton (1902) pointed to the essential role of Na^+ in maintaining the excitability of muscle fibers. The prescience of these studies and ideas is all the more remarkable on remembering that cell theory was not well established and that the bilipid membrane was proposed more than 20 years later.

II. IONIC CURRENTS IN AXONS

The next major step awaited the improvement of tools for electrical measurements, the rediscovery of the squid giant axon (Young, 1936), and the attention of a trio of brilliant minds, Hodgkin, Huxley, and Katz. In pioneering experiments on squid axons, Cole and Curtis (1939) discovered a sudden drop in the membrane resistance of the axon as an impulse traveled by, in good accord with Bernstein's (1902) hypothesis. It was soon found that the voltage went positive, much beyond the value expected for a loss of selective permeability (Hodgkin and Huxley, 1939). The sodium theory some years later explained this discrepancy: a sudden rise in Na^+ permeability carried the voltage almost to the sodium equilibrium potential. The importance of Na^+, first pointed out by Overton (1902), could now be understood.

A. Amplification of the Signal

At this point the origin of biological voltages was clear. Further, the propagation of signals in telephone cables was well understood and formed a good analogy for nerve transmission (Fig. 1). A square step of current introduced in New York gradually loses its sharpness and gets smaller as it travels toward London as a result of imperfections in the insulator. First,

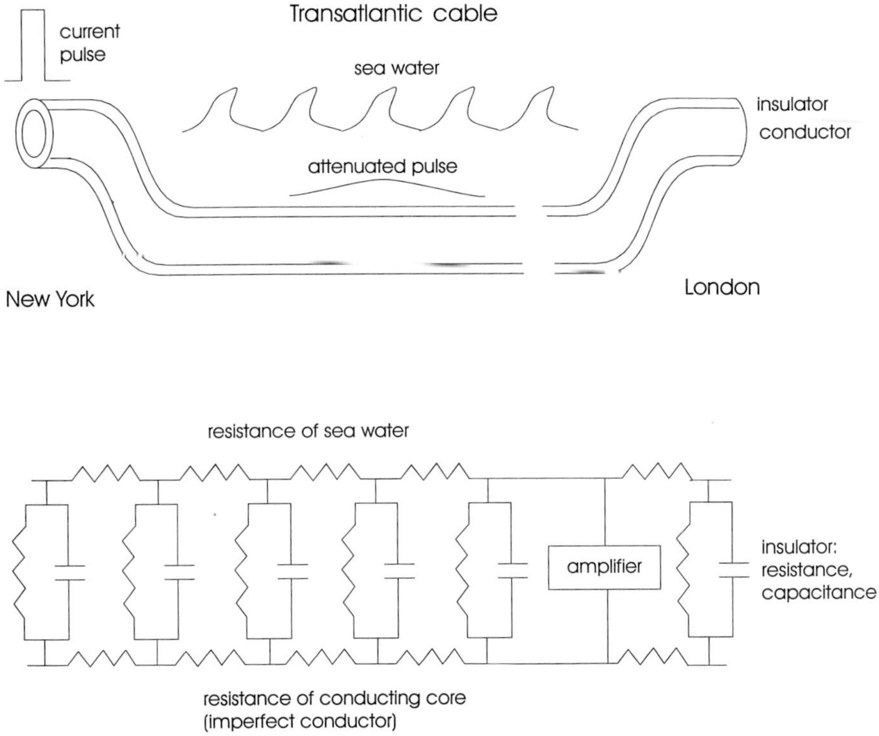

FIGURE 1 Cable analogy for impulse conduction in an axon. (Top) Degradation of a current pulse as it travels through the cable. The loss is due to finite resistance of the cable's insulation and unavoidable parasitic capacitance. An amplifier must be introduced periodically to reinforce the signal.

the resistance of the insulator is finite rather than infinite. Second, there is, unavoidably, parasitic capacitance in every foot of the cable, formed by the separation of two conductors (the conducting core of the cable and sea water) by the insulator. Resistive leak makes the signal smaller, and the necessity of charging the parasitic capacitance slows it down. As a result, the signal must be boosted by an amplifier at many points along the way to keep its speed and amplitude (Fig. 1). The nerve had been visualized as a cable by Hermann (1872), but what formed the amplifiers?

Hodgkin and Huxley (1939) identified the amplifiers with a series of brilliant experiments that must have been guided by a divine hand. They were aided by two essential tools that had come out of Cole's laboratory. The first was the 'space clamp,' a wire inserted along the axis of a squid giant axon that shorted out the resistance of the cytoplasm (Marmont,

1949). Using Fig. 1 (bottom) as an appropriate model for an axon, the resistance of sea water is negligible, so a wire in the interior reduces the core resistance to the point that the capacitors and resistors of the insulator are in parallel. The result is (simplifying a bit) that the transmembrane voltage is the same everywhere throughout the axon: it behaves as one giant patch. The second tool was the voltage clamp, devised originally by Marmont as a current clamp operating on the principal of feedback. Cole (1949) converted this clamp into an early voltage clamp. Finally, Hodgkin *et al.* (1952) turned the voltage clamp idea into a precise and useful tool.

B. Insights from Voltage Clamping

With these tools and a great deal of insight, they explained the action potential as resulting from the interaction of two currents: I_{Na} carried by in-rushing Na^+ ions to depolarize and I_K carried by out-moving K^+ ions to repolarize. Depolarization spreads downstream from an excited patch of membrane by means of axon cable properties, and the resulting depolarization triggers an increase of Na permeability. A delayed consequence of the depolarization is an increase in K permeability, which restores V_m to its resting level. These ideas were perfectly in line with the behavior of a transatlantic telephone cable as analyzed by Kelvin (1855). The axon, with its conducting core, is the cable, and the Na conductance serves as the amplifier. A crucial point is that the Na permeability, like an amplifier, is stimulated to activity by a voltage change, spreading from the active patch upstream.

The Na^+ and K^+ conductances under voltage clamp are shown in Fig. 2. Following a step depolarization from −60 to 0 mV, g_{Na} "activates" or rises in amplitude and then spontaneously decays or "inactivates" if the depolarization is maintained. One could say that the permability pathway simply closes again, but Hodgkin and Huxley (1952) found that it can "close" in two quite different ways. If the voltage is returned to −60 mV before inactivation has occurred appreciably, conductance decays quickly or "deactivates." Deactivation is simply the reverse of activation. Crucially, after deactivating, the conductance can be activated again almost immediately by a second depolarization. The decay is much slower if the step is maintained, and the slow decay is called 'inactivation' (Fig. 2). With great foresight, Hodgkin and Huxley (1952) described the conductances as resulting from the interplay of four factors, as summarized in Fig. 2 (bottom). The first two factors are "gating" factors and account for the time course of the conductance, its rise and fall. m^3 rises quickly from zero on depolarization and stays high during depolarization. It accounts for the rapid activation of I_{Na}. *h,* the inactivation gating factor, starts high (all pathways available)

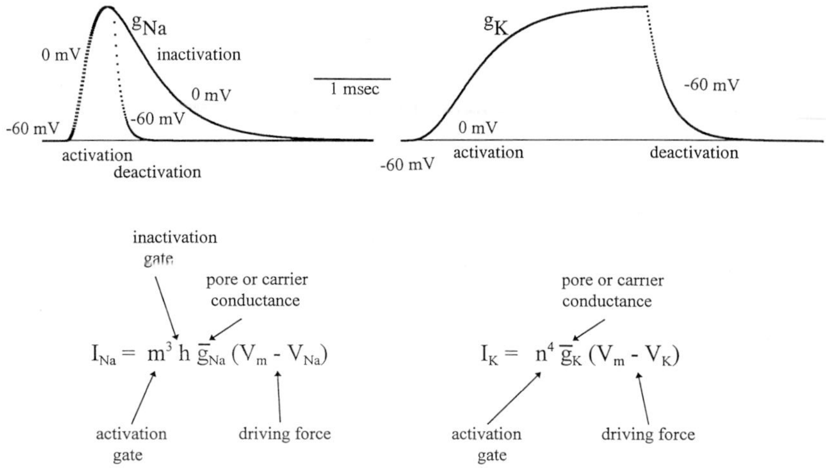

FIGURE 2 Conductances of axon membrane according to Hodgkin and Huxley (1952). g_{Na} the Na^+ conductance, activates with a sigmoid time course on depolarizing the membrane from −60 to 0 mV. If depolarization is maintained, the current inactivates within a few milliseconds, but if the membrane voltage is returned to −60 mV, g_{Na} deactivates rapidly. g_K has simpler behavior: it activates relatively slowly and deactivates on repolarization. The equations are Hodgkin and Huxley's description of this behavior.

and decays slowly, bringing the conductance down with it. The next factor is g_{Na}, the product of the total number of pathways by the conductance of each. The use of a conductance to describe the current–voltage relation of each activated pathway, i.e., current is proportional to the driving force, was a conscious simplification, reasonably accurate only in normal ionic conditions. The final term is the driving force, how far V_m is from the "equilibrium" voltage where current through the channel is zero. This term can be thought of as expressing the selectivity of the permeability path. V_{Na} implies that it is perfectly selective for Na^+ , which was amended by later experiments showing the imperfect selectivity of the channel.

g_K is shown in Fig. 2C. It rises more slowly than g_{Na} and does not inactivate. The equation for g_K resembles the one for g_{Na}, but has only a single gating factor, n^4. g_K is slower because of the relatively slow kinetics of n^4. A crucial feature of Na and K channel gating is that activation has a sigmoid time course and deactivation is exponential. Years later this accurate observation was seen to reflect the subunit structure of the channel.

III. CARRIERS, PORES, GATES, AND SELECTIVITY BEFORE CLONING

Amplifiers in the axonal transmission cable, consist of two apparently separate permeability pathways, each with its characteristic equilibrium potential and gating factors. At the conclusion of Hodgkin and Huxley's work, there were several major questions. (1) Are there two separate ion pathways? One possibility was that sodium inactivation was not simply shutting of the pathway, but conversion into a K^+ selective pathway (Mullins, 1959). (2) What is the nature of the conducting pathway(s)? (3) Are the pathways made of lipid, carbohydrate, or protein? How do the pathways select between Na and K? Is ion binding important? Are steric factors involved? 4) What is the machinery described by the "gating" factors? Hodgkin and Huxley (1952) proposed that each elemental path is controlled by several charged particles that move in response to changes of the membrane field, but the physical mechanism was unclear. (5) How are activation and inactivation of the Na pathway related?

A. Are There Two Separate Conducting Pathways?

These questions kept a generation of membrane biophysicists quite busy. Perhaps the most pressing early question was whether there are two ion pathways or a single one that is convertible from initial Na^+ selectivity to K^+ selectivity. The discovery of the highly selective tetrodotoxin (TTX; Narahashi *et al.,* 1964; Hille, 1971) helped, but did not completely answer the question. It blocked I_{Na} without altering I_K, but it could be, and was, argued that it simply prevented conversion to the K^+ permeable form of the pathway. Similarly the relatively selective I_K blocker TEA^+ could simply be said to prevent conversion into the K^+ selective form. Answers to these questions tended to come in bunches. Mullins, who insisted that the distinctness of the pathways must be demonstrated rigorously, professed himself satisfied after viewing the results of perfusion with pronase internally. This mixture of proteolytic enzymes was used initially to remove axoplasm and thus facilitate internal perfusion. It was seen to alter the ionic currents (Rojas and Atwater, 1967), an effect that was eventually analyzed as removal of the ability of the Na^+ pathway to inactivate (Armstrong *et al.,* 1973). The K^+ pathway remained intact, with unaltered properties, as could be demonstrated by adding TTX. In these circumstances it could not be argued that an Na pathway was being converted to a K pathway: the Na^+ and K^+ paths across the membrane were clearly separate.

B. Carriers or Pores?

What is the nature of the conducting pathway? Hodgkin and Katz (1949) thought it most likely that Na^+ ions were conveyed across the membrane in nonionic form by means of a lipid soluble carrier. However, the provocative tracer experiments of Hodgkin and Keynes (1955) suggested a long pore that contained two ions at a time that could not pass each other. The carrier idea received a strong boost from the discovery of valinomycin (Harold and Baarda, 1967; Pressman *et al.,* 1967), a small molecule, almost certainly a carrier, that conveyed K^+ selectivity on a membrane to which it was added. A major difference in the theoretically anticipated properties of carrier vs pore was that the latter should conduct ions at a much higher rate. Two early sets of experiments suggested that the actual conduction rate was much too fast for a carrier mechanism. Counts of Na pathway density based on TTX labeling to estimate conductance, arriving at a high conductance, close to the diffusion maximum (Hille, 1970). In K channels, the rate of TEA block of K pathways could be measured and was found to be very high, again suggesting a pore mechanism (Armstrong, 1966). The final proof for pores rather than carriers were the single channel measurements made possible by development of the patch clamp (Hamill *et al.,* 1981). These measurements showed that the rate of ion movement was very high, compatible only with a pore, and that the channels were either on or off, as expected from the Hodgkin and Huxley formulation, translated into terms of pores (Sigworth and Neher, 1980; Zagotta *et al.,* 1988).

C. Pore Selectivity

Further study of block by TEA and its derivatives gave a picture of the gross architecture of a K channel that is still valid. TEA has a diameter of about 8 Å, which is close to the size of a K ion with a single hydration shell. A model for block by TEA is shown in Fig. 3. The channel portrayed has a wide inner mouth, about 8 Å in diameter, and a narrower portion about the size of a dehydrated K, 2.66 Å in diameter. K can move through the narrow part of the pore by shedding its hydration waters, which are replaced by interaction with oxygens in the pore wall. TEA can enter the inner mouth of the channel because of its resemblance to a hydrated K. Further progress through the pore is impossible because TEA cannot shed its covalently linked ethyl side chains. The diameter of the narrow part of the pore, now called the selectivity filter, was determined from the size of the cations that could penetrate (Hille, 1973). Rb (2.94 Å) was reasonably permeant, whereas Cs (3.34 Å) was impermeant, suggesting that the narrow pore was approximately 3 Å in diameter. Ions above this size were excluded by sieving, but how could the impermeance of Na (1.9 Å) be explained? A reasonable postulate was that the permeating ion must be a close fit to

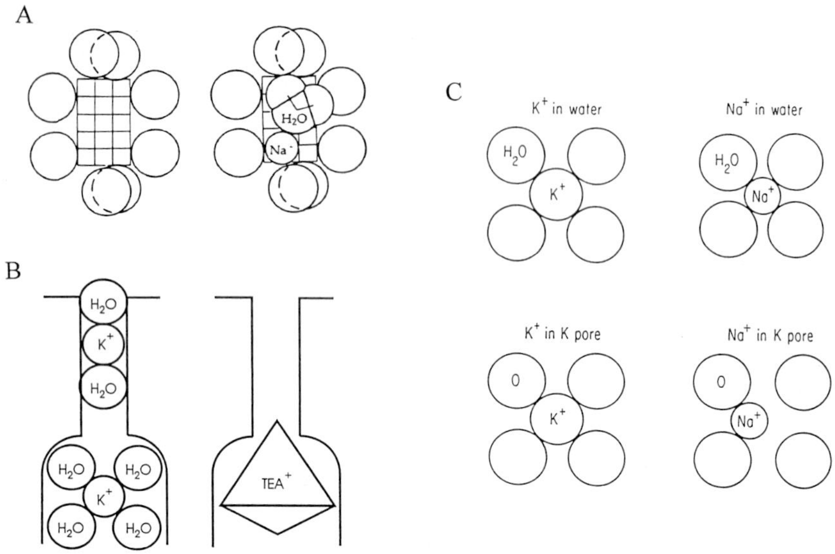

FIGURE 3 Ideas about selectivity. (A) The sodium channel admits any ion that, when hydrogen bonded, will fit through a 3 × 5-Å aperture (from Hille, 1971, 1972). (B) Experiments with the K channel blocker TEA^+, which is about the size of a hydrated K^+, suggested that it blocked because it could not follow K^+, which can dehydrate into the narrow part of the pore. (C) The K pore excludes Na^+ ions because they are too small to complex with the pore walls.

the pore, an idea introduced to the membrane field by Mullins (1959). A specific model is shown in Fig. 3B, which shows K and Na ions in water, surrounded by the oxygens of water molecules; and in the narrow part of a K pore. The pore is postulated to have rigidly fixed diameter that makes K a good fit. The K ion then has similar energy in water and in the pore. The rigidity of the pore wall in theory prevents it from closing in around the Na ion. Na is thus not well complexed in the pore, with the consequence that its energy is higher than in water. Thus Na ions face a large energy barrier when they enter the narrow part of the pore. The precise size of the energy barrier depends on the rigidity of the pore wall, but an idea of the magnitude can be obtained from calculating the electrostatic energy required to pull a single water molecule 1 Å away from close contact with a Na^+ ion. This energy is 8.6 kcal/mole, corresponding to a barrier that one in two million ions would have the thermal energy to surmount. This suggests that the wall has some flexibility, as the observed K/Na permeability ratio is much lower.

In elegantly simple experiments, Hille (1971, 1972) measured the size limits for organic and inorganic cations permeating the Na channel. He found that any cation that would fit through a 3 × 5-Å aperture could permeate reasonably well. Some cations with van der Waal's dimensions larger than this limit could nonetheless permeate if they effectively shrunk their size by forming hydrogen bonds, suggesting interaction with carbonyl oxygens in the pore wall. A Na ion can fit through a 3 × 5-Å aperture with one water molecule abreast. The Na pore shows a preference for Na over K of only about 10:1. This preference could be explained by saying that the larger K ion cannot fit easily with one water molecule abreast. Hille (1968) also pointed out the likelihood that a permeating Na ion interacted with a charged carboxyl group in the pore. This postulated carboxyl also provided a convenient explanation for the pore-blocking effects of protons.

D. Gates and Their Location in the Pore

Pores not only select and conduct ions, but also activate and deactivate on a millisecond time scale. Does this mean they assembled from several pieces during activation? Did the pore swell in diameter on opening and contract when closing or was there a gate that swung in to close a pore that remained otherwise unchanged? An answer to these questions was suggested by experiments with TEA and derivatives. It was noted that TEA entered the pore at a measurable rate after the pore was activated to the conducting state (Armstrong, 1966). This occurred only with TEA applied from the inside and suggested that a receptor became available only when the pore began to conduct. Was the receptor created on activation or preexisting but protected? Experiments with TEA derivatives, e.g., 3-phenylpropyltriethylammonium ion (ϕ-C3), gave what seemed a credible answer: the channel could close and trap the TEA derivative inside, showing that the receptor remains intact after the channel closed (Armstrong, 1971). The top part of Fig. 4 shows the time course of the normal, maintained K^+ conductance and its apparent "inactivation" when ~100 μM ϕ-C3 is present in the axoplasm. As shown in the bottom part of Fig. 4, the channels have uncertain occupancy when fully closed and are not occupied by ϕ-C3. After the channel is opened by depolarization, it conducts normally for a few milliseconds and is then blocked by ϕ-C3. The blocker hinders but does not entirely prevent closing of the gate on repolarization, resulting in a state where ϕ-C3 can be "trapped" in the channel for many seconds. This question has been examined with much better precision by Holmgren *et al.* (1997), using *Shaker* K channels. They found that wild-type *Shaker* channels could not close when occupied by C10 (decyltriethylammonium ion). However, changing an isoleucine in the sixth transmembrane crossing (S6) to a cysteine produced K channels that could close and trap C10 inside.

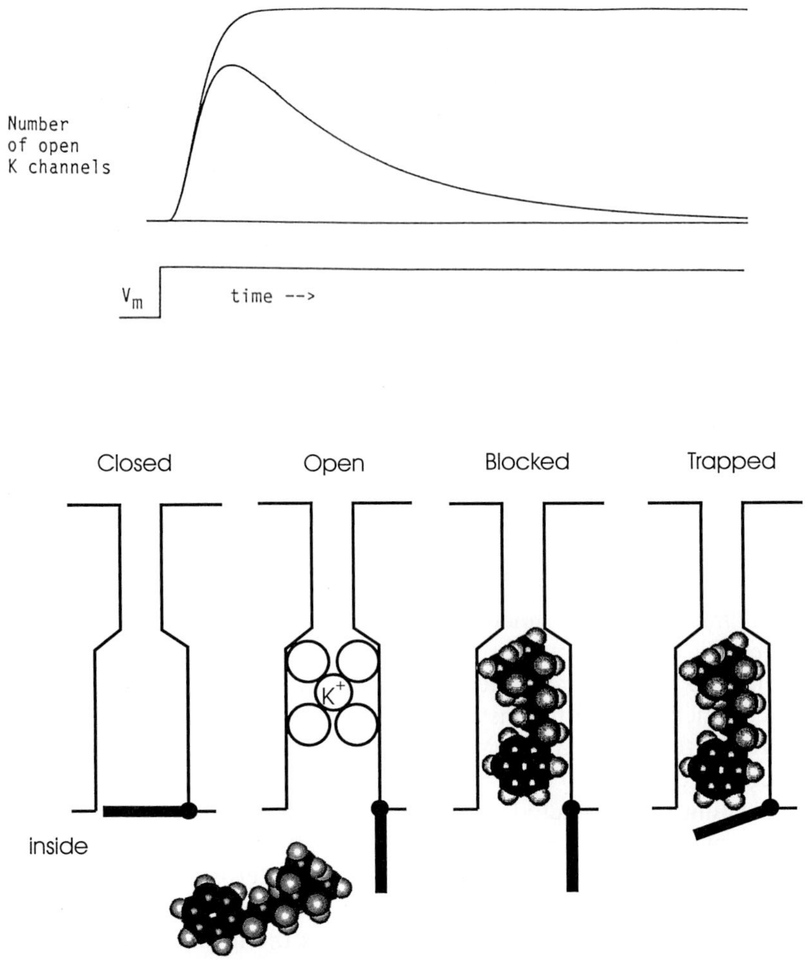

FIGURE 4 Block of K channels by a TEA$^+$ derivative. 3-Phenylpropyltriethylammonium ion (Φ-C3) enters the channel after it opens and has a dwell time in the channel that is long enough to cause a semblance of inactivation. Φ-C3 hinders but does not prevent closing of the gate (trapped state). These effects show that the gate is inside.

It is tempting to think that the mutation simply increased the volume available within the closed channel, making room for C10. The presence of C10 in the pore increased the energy of the closed state only slightly (probably less than in squid), again showing that the receptor remains intact when the gate closes.

Experiments with TEA derivatives also gave a good indication of the location of the gate. C10 and others block only when applied inside and enter the channel only when fully activated (Armstrong, 1971; Choi *et al.*, 1993). Evidence for the latter conclusion is shown in Fig. 4. The top trace shows the control current and the bottom trace shows current in the presence of C10. The two superimpose initially, showing that none of the channels are blocked at rest and that the channel gate must open before C10 can enter. This shows that the gate must be at the inner end of the channel, between the axoplasm and the C10 receptor.

C10 and ϕ-C3 were useful in these studies because they have much higher affinity for the channel than TEA. Up to a certain point, at about 10–12 methylenes, affinity increases with each added methylene, suggesting that the hydrophobic tail of the molecule interacts with hydrophobic elements of the C10–TEA receptor.

The temporal pattern of the currents observed with ϕ-C3 and other derivatives is strikingly similar to that of inactivation in Na channels. This obviously suggested that the mechanism of Na inactivation might be similar to that of block by ϕ-C3. Because the inactivation particle of an Na channel cannot be washed away by internal perfusion, it was clear that it must be attached to the channel structure. This idea got strong support from experiments with internal pronase, mentioned earlier. These showed that internal proteolysis could destroy fast inactivation of Na channels without affecting activation (Fig. 5). The idea that an internally located inactivation particle was clipped off by pronase seemed irresistible. The half-time of decay of the current that was still inactivated after a short time in pronase made it clear that a single hit was sufficient to destroy inactivation: the half-time did not change as the enzyme acted.

It seemed plausible that the inactivation particle might have a receptor analogous to the TEA–C10 receptor in K channels and that the receptor might be available only when the channel is open. Here, however, there was a great problem as Hodgkin and Huxley had described the h gating factor as completely independent of activation m^3. In the ϕ-C3-like model, the gates were not independent, as the activation gate had to be open before ϕ-C3 or the inactivation particle could block the channel. A convincing resolution of this question had to wait for measurement of gating current.

E. Gating Current

Hodgkin and Huxley clearly realized that the voltage sensitivity of Na and K conductances could only mean that the conductances are controlled by charges that move within the membrane in response to changes of the membrane's electric field. Early attempts failed to demonstrate such currents, but they were eventually found (Armstrong and Bezanilla, 1973; Keynes and Rojas, 1974) and everyone sighed with relief that what *had to be* in fact *was.*

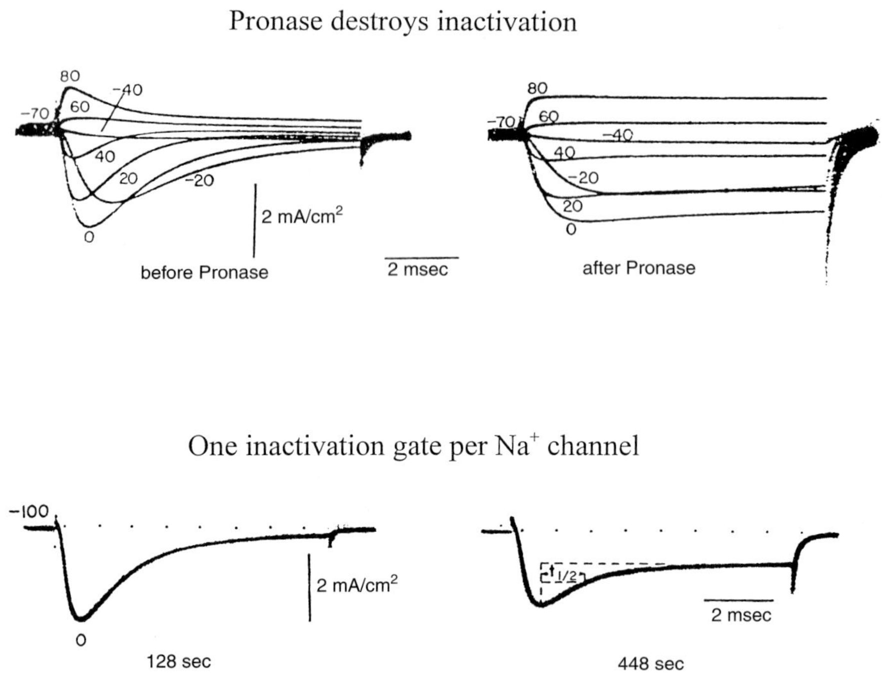

FIGURE 5 (Top) Destroying the inactivation gate proteolytically. Internal perfusion of an axon with pronase destroys inactivation without affecting activation. (Bottom) The half-time of inactivation does not change as pronase acts, showing that a single hit by the enzyme is sufficient.

The gating current is shown in Fig. 6 during the activation of a channel on depolarization, on repolarization after a short pulse with little time for inactivation, and after a long one that allows complete inactivation. Current is outward during activation, reflecting the outward movement of what are now known to be S4 segments. After a short pulse with little inactivation, all of the charge moves back promptly as the channels deactivate. In the Hodgkin and Huxley scheme, the gating current on repolarization should be almost independent of pulse duration. In fact, I_g after a long pulse decays substantially faster, and total charge movement during the first millisecond after repolarization is reduced to about one-third, an effect called charge immobilization (Bezanilla and Armstrong, 1977; Armstrong and Bezanilla, 1977). This made it obvious that activation and inactivation are not independent (C. F. Aldrich *et al.*, 1983) and gave support to a C9-like model for Na inactivation, for the following reason. C9 was known from squid experiments to hinder but not prevent closing of the K channel activation gate, by a "foot in the

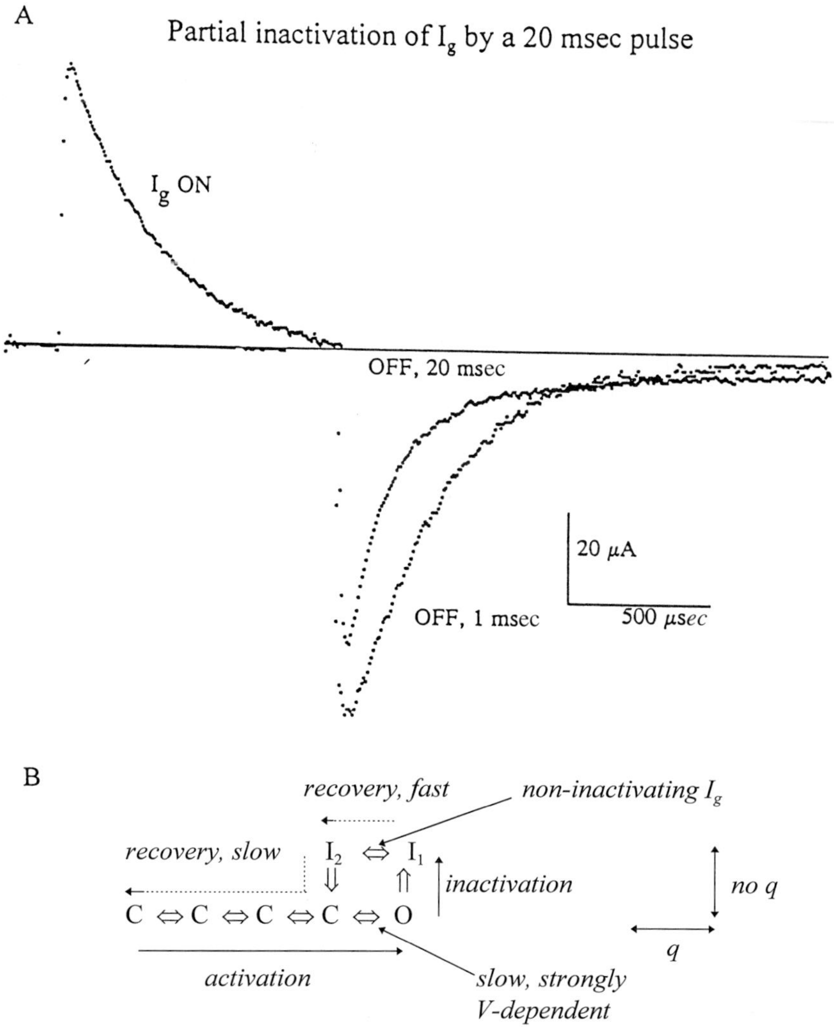

FIGURE 6 (A) Gating current of Na channels. Gating current is outward on depolarization and inward on repolarization, consistent with the movement of a passive "voltage sensor" in the membrane. After inactivation has occurred, total inward charge movement in the first few milliseconds is reduced, an effect called "immobilization" of charge. This phenomenon means that inactivation hinders gate closing, as Φ-C3 hinders closing by a K channel (Fig. 4), by a "foot in the door" effect. (B) A state diagram for an Na channel. There are four closed states (C), one open state (O), and two inactivated states. Channels are in the left-most closed state at rest. Activation has a sigmoid time course because the channels must pass through the closed states before reaching O. These horizontal steps involve charge movement and are voltage sensitive. Inactivation has no associated charge movement. On repolarization, inactivated channels move from I_1 to I_2, generating the component of gating current that does not inactivate. This provides a recovery path that bypasses the open state.

door" effect. The explanation of reduced I_g on repolarization of an inactivated Na channel was the same: the inactivating particle in the channel behaves as a foot in the door, preventing the flow of a portion of the gating charge.

The great advantage of gating current measurement is that it gives information about the closed states of the channel. A state diagram for the inactivation model derived in part from I_g measurements is shown in Fig. 6B (Armstrong and Gilly, 1979). There are four closed states, C4 through C1, a single open state, O, and two inactivated states, I1 and I2. At rest, all of the channels are assumed to be in state C4. After a large depolarization, the channels move to the right in the diagram and conductance rises with the population of state O. The activation gate opens in the single step from C1 to O, but all of the horizontal steps contribute to gating current. Once open, channels can inactivate to state I1, with a rate constant k, and as this happens the current decays away. On repolarization, the channels largely transfer from state I1 to I2, generating the fast decaying component of the gating current (OFF, 20ms) seen in Fig. 6. The I1 > I2 step has the useful function of preventing the reflux of channels through the open state during recovery from inactivation. Recovery follows the path shown by the dashed arrows, while activation follows the solid arrow.

IV. THE AGE OF CLONING

By 1980 a good outline of pore structure, selectivity, and gating behavior was in hand, provided by the techniques of membrane biophysics (reviewed in Armstrong, 1981). It was time for new information, which burst upon the channel world from the laboratory of Shosaku Numa. The accomplishments of this great scientist and his laboratory, as well as his insight into the arcane world of channels, are still astonishing. His cloning of the Na channel was followed rather promptly by cloning of the K channel in the Jan (Tempel *et al.,* 1987), Pongs (Pongs *et al.,* 1988), and Tanouye (Tanouye *et al.,* 1986) laboratories. It is probably fortunate that the Na channel was cloned first, as it was a single peptide with four similar but not identical domains (Fig. 7C). This immediately suggested that the four domains sur-

FIGURE 7 (A) Sequence and probable membrane folding topology of a subunit of the *Shaker* B K channel. The S4 transmembrane segment, the hallmark of voltage-gated channels, is enclosed in a box. Near the N terminus (left), 172 amino acids are omitted, whereas 177 amino acids are omitted at the C terminus. (B) The *Shaker* K channel is composed of four of the subunits from A around a central pore. (C) A sodium channel is made of a single peptide, containing four repeating domains. Each domain resembles a *Shaker* K channel subunit.

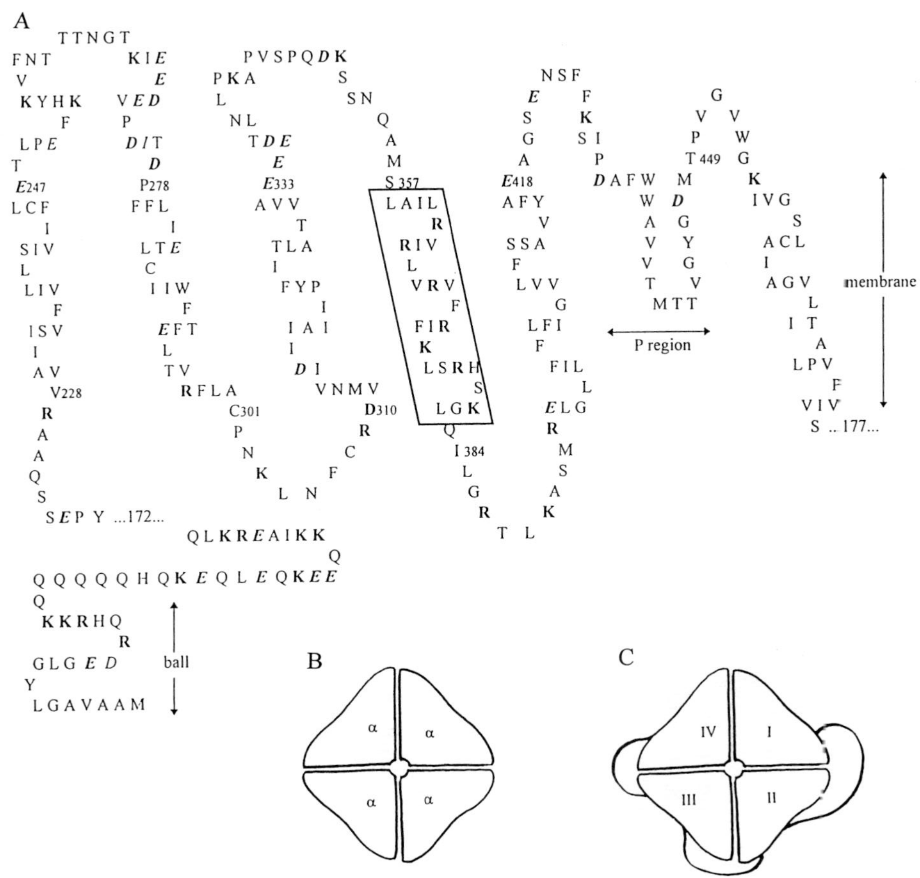
A
B
C
membrane
P region
ball
S 357
I 384
E418
T 449
E247
P278
E333
V228
C301
D310
SEPY ...172...
S .. 177...
α
I
II
III
IV

round a central pore. The K channel peptide is about one-fourth the size and resembles a single domain of the Na channel. It seemed likely that the channel was formed from four of these peptides (Fig. 7B), which was proved by MacKinnon (1991).

Within each domain or subunit is a structure ideally suited for voltage sensing: a region of about 20 amino acids with a positive residue (arginine or lysine) in every third position. After a brief sojourn in the cytoplasm, this region was modeled as an amphipathic membrane helix called S4 (Fig. 7A). It is the hallmark of voltage-sensitive channels. The probable transmembrane topology of a K channel subunit is given in Fig. 7A.

The subunit/domain structure in retrospect was first foreshadowed by the accurate observations of Hodgkin and Huxley (1952), who described the sigmoid activation time course of K and Na channels. The fourth power of the K channel gating factor, n^4, was immediately interpretable: a transition, almost certainly involving the outward movement of S4, had to occur in each of the four subunits and then the gate could open.

A. Mapping Functions to the Sequence: Where Is the Pore?

After these easy and rather obvious conclusions began the more rigorous assignment of function to various segments of the sequence. An early target of inquiry was the pore itself, through which ions moved. Guy and Seetharamulu (1986) placed the pore in the segment joining S5 and S6, partly in answer to a question mentioned earlier, to theoretically provide negative counter charges for the S4. Whatever the reason, this was a good guess, and the later experiments of MacKinnon, Miller, Yellen, and others on *Shaker* K channels soon localized the pore or P region to the S5–S6 linker with certainty (Fig. 7A; MacKinnon and Miller, 1989; Yellen *et al.*, 1991; Hartmann *et al.*, 1991). How was this region arranged and how did it provide the K channel with selectivity? Many clever experiments were performed to answer this question, in many cases using cysteine substitution and labeling with Karlin reagents (Akabas *et al.*, 1992), Ag, or Cd. A few of the many significant findings are given here. Ag labeling suggested that the first half of the P region was α helical (Lü and Miller, 1995). A "signature sequence," TXXTXGYG, was recognized that is highly conserved in known K channels (Heginbotham, *et al.*, 1994). This sequence contained a number of hydrophilic residues, whose side chains provided potential complexing sites for ions in transit through the pore. Mutation of these hydrophilic side chains to nonhydrophilic substitutes generally made no difference: the side chains seemed not to matter. This led Heginbotham *et al.* (1994) to consider the possibility that loop structures in the

P region turned backbone carbonyls of the peptide chain inward to form the pore lining.

MacKinnon at this stage decided that further progress required a new approach and began transforming himself into a crystallographer. Elucidation of the P region and the signature sequence made it easy to recognize K channels in the protein database, and K channels with only two transmembrane crossings (2TM) were soon discovered (Ho *et al.,* 1993; Kubo *et al.,* 1993). The two crossings were quickly recognized as likely homologues of S5 and S6 from the six transmembrane crossing (6TM) channels. Interestingly, the 2TM variety was not voltage gated: it was clear that nature had separated conduction from gating, just as Hodgkin and Huxley had done conceptually. Presumably the 2TM variety is more primitive, and the gating properties were added at some point in evolution. Discovery of K channels in bacteria, specifically the KcsA channel found in *Streptomyces lividans,* made it relatively easy to get large quantities of channel protein, just as MacKinnon was completing his transformation into a crystallographer. Within a very short time he and his group had crystallized and solved the structure of the KcsA channel.

The structure is shown in Fig. 8. Beginning with the backbone, the two transmembrane helices from each tetramer run almost in parallel and cross the membrane at a 25° angle to the normal. In the words of Doyle *et al.* (1998), each helix pair, forms one of the four support poles of an inverted

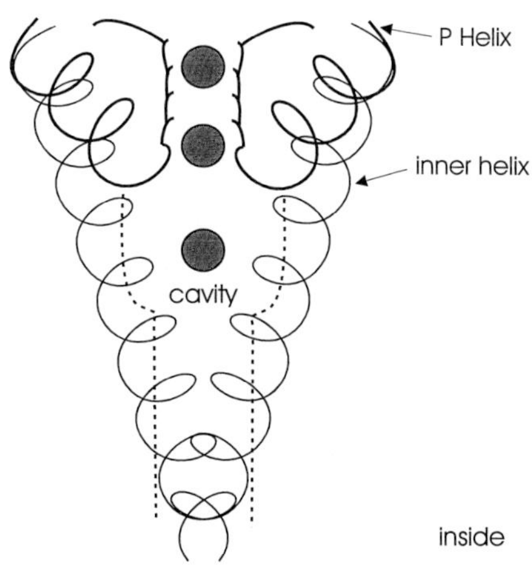

FIGURE 8 Excerpts from the structure of the KcsA K channel, after Doyle *et al.* (1998). See text for a description. Outer helices run almost in parallel with the inner helix and have been omitted for clarity. Two ions are shown in the selectivity filter and one in the "cavity." Dashed lines give the approximate dimensions of the cavity and the inner part of the pore.

teepee. The inner helix of each pair, corresponding to S6 in *Shaker,* lies close to the pore axis, and the outer helix provides added support. The four teepee poles are far apart at the outer surface of the membrane, but converge and cross near the inner surface. The volume contained is a large cone with base up at the outer surface, and a small cone with base down at the inner. The large outer cone is filled by the P region. The first nine residues of the segment linking the two helices (corresponding to the S5–S6 linker of *Shaker*) form a "turret" at the outer surface that is important in binding scorpion toxins. The first half of the P region follows. Here 11 residues form three turns of an α helix that runs at about a 45° angle toward the pore axis. The helix is followed by the "signature sequence," a strand that winds its way toward the outside (TVGYG), forming the selectivity filter. A point of major importance is that the backbone carbonyls of the selectivity filter are turned inward to line the pore: ions are complexed by the backbone carbonyls, not by the side chains, which had been shown not to be crucially important (Heginbotham *et al.,* 1994). After leaving the selectivity filter, the strand then turns out from the pore axis (DYPV). The following residues turn inward and form the inner helix, which runs through the membrane toward the inside. The side chains of the residues of the inner two-thirds of this helix (corresponding to S6 in *Shaker*) form the walls of the pore, internal to the selectivity filter.

A K ion permeating from the outside would first come in close contact with the pore at residues Y and G at the mouth of the selectivity filter. At this point the water molecules surrounding the ion would be replaced by the backbone carbonyls of G and Y. The ion would be close on the heels of a preceding K ion at the other end of the selectivity filter, but separated from it by a water molecule. After moving to the inner site of the selectivity filter (residues G,V,T), the ion would reacquire its hydration shell and enter the "cavity," a 10-Å-diameter water-filled cavern. On leaving the cavity it would race through a long, inhospitable tube lined with hydrophobic residues, narrow (about 6 Å) and forbidding. Ion-binding sites within the pore were identified by transferring crystals from the normal high K medium to Rb and by measuring the electron density difference when Rb replaced the lighter K ion. Two ions were found in the selectivity filter, a third in the cavity, and none in the inner tube. An ion in the cavity feels the stabilizing influence of the dipole moment of the P-helix dipole, which is cavity negative.

This structure accounts pleasingly for physiological data. Beginning with selectivity, crystallography data make it clear that the selectivity filter is only wide enough for a dehydrated ion, although the individual carbonyls cannot be resolved. Surrounding the filter is a ring structure composed of tyrosine and tryptophan side chains that is thought to hold the lumen of the selectivity filter within a narrow range of diameters. This diameter is too small for a Cs to permeate comfortably (Cs blocks the pore, presumably by getting stuck in the filter) and too large to be a close fit for Na (see earlier discussion). The basis of the long pore effect is provided by the presence at any instant of several ions in the pore. The binding site for internally applied TEA is the cavity, and it is clear that TEA could not move on through the narrow selectivity filter. Binding for the hydrophobic tail of C10 or ϕ-C3 is provided by hydrophobic residues at the margin of the cavity. The external TEA site is not so readily understood. The tyrosine residues that determine external TEA sensitivity are too far apart to closely encircle a TEA unless there is a good deal of flexibility in this region. That such flexibility is present is suggested by the data of Liu *et al.* (1996).

There are very plausible theories as well for the structure of the selectivity filter of sodium and calcium channels (Lipkind and Fozzard, 1994; Ellinor *et al.,* 1995). In both cases the models are quite different from the KcsA structure in that ions in transit are postulated to interact with side chains rather than backbone carbonyls. There is reason to think, particularly in the case of doubly charged Ca, that interaction with a strongly negative group is required, in part to help with dehydration of the ion. It will be of great interest to see if nature used a quite different mechanism for selecting Na and Ca, as these models suggest.

B. The S4 Segment as Voltage Sensor

As described earlier, the activation gate of the both Na and K channels is at the inner end of the channel. What is the gate and what makes it open and close in response to voltage changes? The obvious first guess was the S4 as voltage sensor, and it was a target for many experiments that sought to measure the effects on voltage sensitivity of neutralizing positive charges in this segment. This pursuit was complicated by two factors. It soon became apparent that channels expressed poorly if more than two of the arginines or lysines were replaced by neutral residues, making it hard to observe sizeable changes in the voltage sensitivity. Second, changing even neutral residues in this sensitive region caused large shifts in the midpoint of the activation curve, together with changes in the steepness. These results were not readily interpretable, and could only be catalogued.

Clear results followed the introduction of the thiol-labeling compounds introduced by Karlin and associates (Akabas *et al.,* 1992). Horn and coworkers examined the effects of substituting a cysteine residue for a native residue in the S4 (Yang *et al.,* 1996). Cysteine substitution in the S4 of domain 4 in many cases perturbs the kinetics of inactivation without greatly

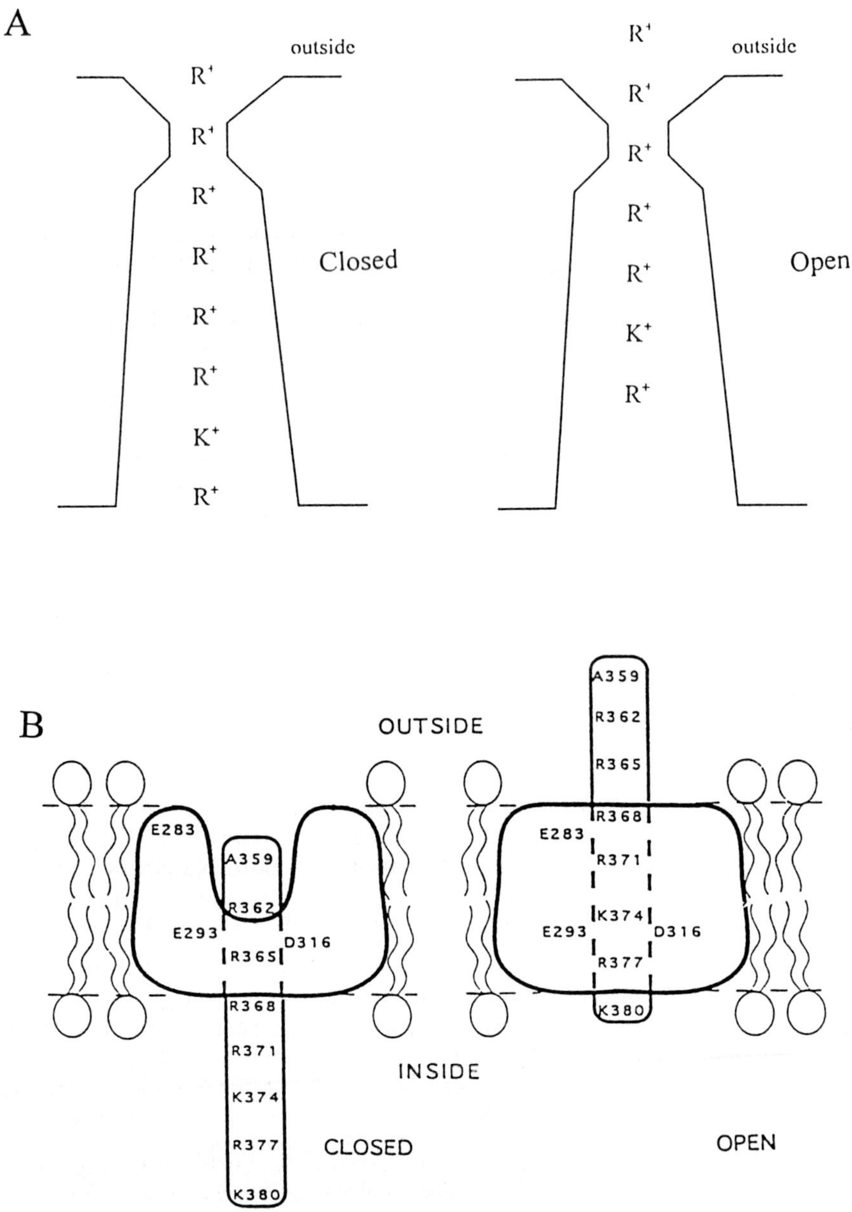

FIGURE 9 Summaries of the S4-labeling experiments on Na channels (A; Yang *et al.*, 1996) and K channels (B; Larsson *et al.*, 1996). (A). In a typical experiment, a cysteine was substituted for one of the charged residues shown in S4 domain IV (see Fig. 7C). The accessibility of the substituted cysteine to labeling with cysteine-reactive agents, e.g., MTSET (Akabas *et al.*, 1992), was then tested. Cysteines in the lower six charged residue positions can be labeled in the closed state, but only four in the open state. A cysteine substituted for

altering activation of the channel. Labeling the introduced cysteine then caused a detectable (but not necessarily interpretable) change in inactivation properties, and the rate of this change gave a measure of the labeling rate. As an example, some of the introduced residues were found to be labeled only by internally applied thiol reactors. Some could be labeled from inside in the closed state and from the outside in the open state. Overall the results gave clear evidence for the outward movement of S4 during activation. A surprise was that many positions near the inner end of S4 could be labeled when the channel was closed. This led to the interesting proposal that a hydrophilic region surrounds the inner part of S4, making it easily accessible to internally applied reagents. If so, this could help solve one of the long unanswered mysteries regarding the channel: the identity of negative countercharges for the positive residues of S4. Such countercharges are necessary in theory to lower the energy arginines and lysines. Horn's proposal raises the possibility that the counterions are in fact anions from the internal solution.

Similar results soon followed for the K channel (Larsson *et al.,* 1996). Results for both channels are summarized in Fig. 9. These results make two major points clear: that S4 is a transmembrane segment and that it moves outward during activation. Both studies provide much additional food for thought about details of the activation process.

C. Physical Identity of the Activation Gate

From the evidence just cited it seems clear that S4 segments serve as voltage-sensing elements of the channel and that they move outward as the channel opens. What happens next? The S4 segments must tug on the ~15 residue segment (*Shaker* B) linking S4 to S5. This segment is suggested on reasonable grounds to be α-helical, with predominantly hydrophobic residues on one face and hydrophilic at the other (McCormack *et al.,* 1989; Holmgren *et al.,* 1996). Exposure of some of the residues to MTSET labeling varies between the open and the closed state (Holmgren *et al.,* 1996). Also, mutations in this region strongly affect the affinity of the inactivation ball (see later) for the channel, e.g., the completeness of inactivation is increased

the second R^+ from the top can be labeled when the channel is open, but is inaccessible from both sides when it is closed. (B). Cysteine substitution and labeling were performed by similar methods in *Shaker,* yielding the labeling patterns shown. Cysteines at positions within the heavy lines were inaccessible to labeling from both sides. Altogether, experiments from both channels show that S4 is transmembrane and that it moves outward during activation.

by the mutation L385A and decreased by T388A (Isacoff *et al.,* 1991). Both residues lie on the predominantly hydrophobic face, which may be exposed on activation, forming a receptor for the inactivation ball. Do the four S4–S5 segments serve as a flap over the end of the channel or do they distort the inner end of the pore, formed by the S5 and S6 helices? A major surprise from data of MacKinnon and colleagues (Doyle *et al.,* 1998) is the length and narrowness of the tube internal to the TEA site. This tube is lined with hydrophobic side chains, many of which are valine and alanine. MacKinnon and colleagues raise the possibility that in voltage-gated channels, movement of the S4 units imparts via the S4–S5 segments a twist to the tube which collapses or opens it. The tube is sufficiently long that a pinch at its inner end would hardly affect the TEA site, as required by the data (Liu *et al.,* 1997). Residues in the innermost part of S6 (482–486) can be labeled by MTSET when the channel is either open or closed. Between 470 and 479, the residues that can be labeled show a strong preference for open state labeling. How does one account for the sharp boundary between always accessible and open-state accessible residues? Perhaps there is a pinch in the neighborhood of residue 480, which is near the narrowest part of the tube in the KcsA channel (Doyle *et al.,* 1998).

Although it is controlled by the S4 units in four subunits, the activation gate of the K channel behaves as a single gate. This is a necessary conclusion from the experiments of Hodgkin and Huxley, who said that all four of their *n* particles must move before the channel opens: the sigmoid kinetics of K channel opening are not compatible with partial opening of a channel after movement of less than all of the *n* particles. Single channel measurements are also compatible with a single opening step. Some models propose that each S4 segment moves in two steps. In the first step, the S4 units move one by one, but the second step is "concerted," meaning that the second phase of motion of all the S4 units is simultaneous (Zagotta *et al.,* 1994; Schoppa *et al.,* 1992) and accompanied by the event that opens the channel for conduction.

Steps in the activation of K channels can be studied conveniently using the nonconducting *Shaker* mutant W434F (Stefani *et al.,* 1994), which has gating currents but does not conduct ionic current. These currents presumably reflect movement of the S4 units, which go through the correct motions, but open the gate of a channel that has a stopper in it. Interpreting a bit, the S4 units move briskly outward during a small step and return briskly on return of V_m to rest; after a large depolarization, which opens the channels, the return of the S4 units to resting position is very much slower (Fig. 10A). There seems to be a qualitative difference between the early steps of activation and a late/last step that opens the gate. A similar conclusion has followed from examination of the mutation L2V (Schoppa *et al.,*

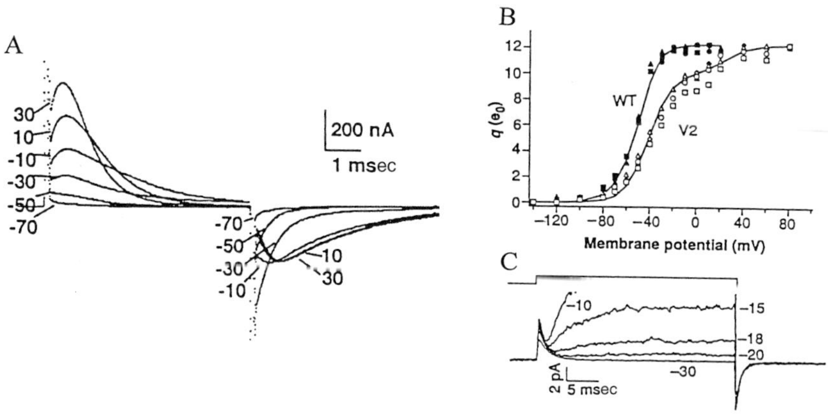

FIGURE 10 A concerted late step in channel activation. (A) Gating currents from the non-conducting *Shaker* mutant W434F (Stefani *et al.* 1994). Numerals indicate the voltage (mV) during the pulse that caused the gating current. The inward (downward) currents at pulse end were all recorded at ~80 mV, after pulses to the indicated voltage. After small depolarizing pulses to <~10 mV, the tails decay quickly and exponentially. Above this voltage, in the range where conduction normally occurs, the tails are much slower and have a rising phase. This shows that a transition near the end of the activation chain and close to the open state (see Fig. 6B) is slowly reversible. (B) Gating charge movement (q) of a wild-type *Shaker* K channel (WT) and from the mutant L2V (leucine substituted for valine near the inner end of S4). The mutation leaves gating current in the range of −100 to −20 mV almost unchanged, but shifts a component associated with the opening transition far to the right on the voltage axis. The result is shown in (C) At −20 mV, the gating current is nearly normal, but very little K^+ current is activated.

1992). In this case it seems that the early steps occur normally, but the voltage range for the late/last step that opens the channels is pushed far to the right (Fig. 10B). Where these clues will lead remains to be seen. It now seems possible that the first insights will come when a voltage-gated K channel is crystallized.

D. Inactivation of Na Channels

The Na channel inactivation gate lies in the segment linking domains 3 and 4 of the channel peptide. A single cut in this segment, performed by genetic engineering, slows inactivation drastically (Stühmer *et al.*, 1989), which is very reminiscent of the action of pronase. Mutational analysis has since shown that inactivation is very sensitive to a sequence IFM present in this linking segment (West *et al.*, 1992). Precisely what happens when inactivation occurs is not yet clear. Because of the linkage between activation and inactivation, the exact mechanism cannot be known without precise knowledge of the activation gate. Physiologically it is clear that a foot in the door mechanism is a simplification because the channel can inactivate to some degree before the channel conducts and because, during recovery from inactivation, the channel does not leak during withdrawal of the foot. The kinetic scheme shown in Fig. 6B provides for both of these properties formally, but does not provide telling insight regarding the linkage between S4 movement, the activation gate, and inactivation. Another problem that is handled at best formally by the kinetic diagram is the sensitivity of the inactivation recovery rate to V_m (Keynes, 1991). A formal solution is to draw a slanting arrow from state I_2 to C_3, but the physical meaning can only be imagined.

E. Inactivation of K Channels

Some types of K channels, e.g., *Shaker* B, inactivate with a rapid time course, whereas others do not inactivate appreciably. Inactivation in *Shaker* has been worked out in a series of elegant experiments by Aldrich and colleagues (Hoshi *et al.,* 1990). They unambiguously identified the inactivation region of these channels as the N-terminal of each of the four tetramers (Fig. 7A). The first 20 residues form a "ball" that enters and inactivates the channel when it is open. The ball cannot inactivate the channel unless it is open; conversely, the channel cannot close when the ball is in the channel (Demo and Yellen, 1991; Ruppersberg *et al.,* 1991). If the ball is removed by genetic engineering, the resultant channel does not inactivate. Reciprocally, a ball peptide consisting of the first 20 amino acids can, when added to the cytoplasm, cause an engineered noninactivating channel to inactivate. The next 63 amino acids form a "chain" linking the ball to the remainder of the channel. When the chain is shortened by deleting some residues, thus providing less conformational flexibility, inactivation speeds up. A single ball out of the four normally present is sufficient to inactivate the channel, but, in engineered channels, the rate of inactivation is proportional to the number of balls present. Similarly, if the balls are cut off with papain, the inactivation rate slows progressively as the enzyme acts (Gomez-Lagunas and Armstrong, 1995). This is quite different from the inactivation of sodium channels, where a single cut from pronase completely destroys inactivation for the affected channel.

V. SUMMARY

There has been great progress, in this century in the understanding of cellular electrical properties and signaling. This progress was based on a strong foundation that was provided by pioneers such as Overton. In conclusion, it is of interest to ask once again why nerve fibers developed the elaborate machinery of the Na channel rather than, as Bernstein suggested, simply devising a means for breaking down the K selectivity of the resting membrane. There are three points to think about. The first is that rapid propagation of an impulse requires lowering the membrane resistance, thus providing the large currents necessary for rapidly charging the membrane capacitance. An increase in the Na^+ conductance provides this. In addition, the electrical driving force is larger, thanks to the selectivity of the Na channel. This helps to drive current through the axon core and also helps increase propagation speed. Finally, energy efficiency is increased: during the upstroke of the action potential, the largest flux is carried by the ion that is effective in driving the voltage positive. Movement of K^+ during this phase would be counterproductive and would increase the load on the Na–K pump.

References

Akabas, M. H., Stauffer, D. A., Xu, M., and Karlin, A. (1992). Acetylcholine receptor channel structure probed in cysteine-substitution mutants. *Science* **258,** 307–310.

Aldrich, R. W., Corey, D. P., and Stevens, C. F. (1983). A reinterpretation of mammalian sodium channel gating based on single channel recording. *Nature* (*London*) **306,** 436–441.

Armstrong, C. M. (1966). Time course of TEA^+-induced anomalous rectification in squid giant axons. *J. Gen. Physiol.* **50,** 491–503.

Armstrong, C. M. (1971). Interaction of tetraethylammonium ion derivaties with the potassium channels of giant axons. *J. Gen. Physiol.* **58,** 413–437.

Armstrong, C. M. (1981). Sodium channels and gating currents. *Physiol. Rev.* **61**(3), 644–683.

Armstrong, C. M., and Bezanilla, F. (1973). Currents related to movement of the gating particles of the sodium channels. *Nature* (*London*) **242,** 459–461.

Armstrong, C. M., and Bezanilla, F. (1977). Inactivation of the sodium channel. II. Gating current experiments. *J. Gen. Physiol.* **70**(5), 567–590.

Armstrong, C. M., and Gilly, W. F. (1979). Fast and slow steps in the activation of sodium channels. *J. Gen. Physiol.* **74,** 691–711.

Armstrong, C. M., Bezanilla, F., and Rojas, E. (1973). Destruction of sodium conductance inactivation in squid axons perfused with pronase. *J. Gen. Physiol.* **62,** 375–391.

Bernstein, J. (1902). Untersuchungen zur Thermodynamik der bioelektrischen Ströme. Erster Theil. *Pfluegers Arch.* **92,** 521–562.

Bezanilla, F., and Armstrong, C. M. (1977). Inactivation of the sodium channel. I. Sodium current experiments. *J. Gen. Physiol.* **70,** 549–566.

Choi, K. L., Mossman, C., Aube, J., and Yellen, G. (1993). The internal quaternary ammonium receptor site of Shaker potassium channels. *Neuron* **10,** 533–541.

Cole, K. S. (1949). Dynamic electrical characteristics of the squid axon membrane. *Arch. Sci. Physiol.* **3,** 253–258.

Cole, K. S., and Curtis, H. J. (1939). Electrical impedance of the squid giant axon during activity. *J. Gen. Physiol.* **22,** 649–670.

Demo, S. D., and Yellen, G. (1991). The inactivation gate of the Shaker K+ Channel behaves like an open-blocker. *Neuron* **7**(5), 743–753.

Doyle, D. A., Cabral, J. M., Pfuetzner, R. A., Kuo, A. Gulbis, J. M., Cohen, S. L., Chait, B. T., and MacKinnon, R. (1998). The structure of the potassium channel: Molecular basis of K^+ conduction and selectivity. *Science* **280,** 69–77.

Ellinor, P. T., Yang, J., Sather, W. A., Zhang, J. F., and Tsien, R. W. (1995). Ca^{2+} channel selectivity at a single locus for high-affinity Ca^{2+} interactions. *Neuron* **15,** 1121–1132.

Gomez-Lagunas, F., and Armstrong, C. M. (1995). Inactivation in Shaker B K+ channels: A test for the number of inactivating particles on each channel. *Biophys. J.* **68,** 89–95.

Guy, H. R., and Seetharamulu, P. (1986). Molecular model of the action potential sodium channel. *Proc. Natl. Acad. Sci. U.S.A.* **83,** 508–512.

Hamill, O. P., Marty, A., Neher, E., Sakmann, B., and Sigworth, F. J. (1981). Improved patch-clamp techniques for high-resolution current recording from cells and cell-free membrane patches. *Pfluegers Arch.* (Eur. *J. Physiol.*) **391,** 85–100.

Harold, F. M., and Baarda, J. R. (1967). *J. Bacteriol.* **94,** 53.

Hartmann, H. A., Kirsch, G. E., Drewe, J. A., Taglialatela, M., Joho, R. H., and Brown, A. M. (1991). Exchange of conduction pathways between two related K^+ channels. *Science* **215,** 942–945.

Heginbotham, L., Lu, Z., Abramson, T., and MacKinnon, R. (1994). Mutations in the K^+ channel signature sequence. *Biophys. J.* **66,** 1061–1067.

Hermann, L. (1872). "Grundriss der Physiologie," 4th Ed., Quoted in Hermann, L. (1899), Zur Theorie der Erregungsleitung und der elektrischen Erregung. *Pfluegers Arch.* **75,** 574–590.

Hille, B. (1968). Charges and potentials at the nerve surface, Divalents and pH. *J. Gen. Physiol.* **51,** 221–236.

Hille, B. (1970). Ionic channels in nerve membrane. *Prog. Biophys. Mol. Biol.* **21,** 1–32.

Hille, B. (1971). The permeability of the sodium channel to organic cations in myelinated nerve. *J. Gen. Physiol.* **58,** 599–619.

Hille, B. (1972). The permeability of the sodium channel to metal cations in myelinated nerve. *J. Gen. Physiol.* **59,** 637–658.

Hille, B. (1973). Potassium channels in myelinated nerve. Selective permeability to small cations. *J. Gen. Physiol.* **61,** 669–686.

Ho, K., Nichols, C. G., Lederer, W. J., Lytton, J., Vassilev, P. M., Kanazirska, M. V., and Hebert, S. C. (1993). Cloning and expression of an inwardly rectifying ATP-regulated potassium channel. *Nature* (*London*) **362,** 31–38.

Hodgkin, A. L., and Huxley, A. F. (1939). Action potentials recorded from inside a nerve fibre. *Nature* (*London*) **144,** 710–711.

Hodgkin, A. L., and Huxley, A. F. (1952). A quantitative description of membrane current and its application to conduction and excitation in nerve. *J. Physiol.* (*London*) **117,** 500–544.

Hodgkin, A. L., and Katz, B. (1949). The effect of sodium ions on the electrical activity of the giant axon of the squid. *J. Physiol.* (*London*) **108,** 37–77.

Hodgkin, A. L., and Keynes, R. D. (1955). The potassium permeability of a giant nerve fiber. *J. Physiol.* (*London*) **128,** 61–88.

Hodgkin, A. L., Huxley, A. F., and Katz, B. (1952). Measurement of current-voltage relations in the membrane of the giant axon of *Loligo. J. Physiol.* (*London*) **116,** 424–448.

Holmgren, M., Jurman, M. E., and Yellen, G. (1996). N-type inactivation and the S4–S5 region of the Shaker K^+ channel. *J. Gen. Physiol.* **108,** 195–206.

Holmgren, M., Smith, P. L., and Yellen, G. (1997). Trapping of organic blockers by closing of voltage-dependent K+ channels, evidence for a trap door mechanism of activation gating. *J. Gen. Physiol.* **109,** 527–535.

Hoshi, T., Zagotta, W. N., and Aldrich, R. W. (1990). Biophys. and molecular mechanisms of *Shaker* potassium channel inactivation. *Science* **250,** 533–538.

Isacoff, E. Y., Jan, Y. N., and Jan, L. Y. (1991). Putative receptor for the cytoplasmic inactivation gate in the *Shaker* K^+ channel. *Nature* (*London*) **353,** 86–90.

Kelvin, Lord (William Thompson) (1855). On the theory of the electric telegraph. *Proc. R. Soc.* **7,** 382–399.

Keynes, R. D. (1991). On the voltage dependence of inactivation in the sodium channel of the squid giant axon. *Proc. R. Soc. London B* **249,** 107–112.

Keynes, R. D., and Rojas, E. (1974). Kinetics and steady-state properties of the charged system controlling sodium conductance in the squid giant axon. *J. Physiol.* (*London*) **239,** 393–434.

Kubo, Y., Baldwin, T. J., Jan, Y. N., and Jan, L. Y. (1993). Primary structure and functional expression of a mouse inward rectifier potassium channel. *Nature* (*London*) **362,** 127–233.

Larsson, H. P., Baker, O. S., Dhillon, D. S., and Isacoff, E. Y. (1996). Transmembrane movement of the Shaker K^+ channel S4. *Neuron* **16,** 387–397.

Lipkind, G. M., and Fozzard, H. A. (1994). A structural model of the tetrodotoxin and saxitoxin binding site of the Na^+ channel. *Biophys. J.* **66,** 1–13.

Liu, Y., Jurman, M. E., and Yellen, G. (1996). Dynamic rearrangement of the outer mouth of a K^+ channel during gating. *Neuron* **16,** 859–867.

Liu, Y., Holmgren, M., Jurman, M. E., and Yellen, G. (1997). Gated access to the pore of a voltage-dependent K^+ channel. *Neuron* **19,** 175–184.

Lü, Q., and Miller, C. (1995). Silver as a probe of pore-forming residues in a potassium channel. *Science* **268,** 304–307.

McCormack, K., Campanelli, J. T., Ramaswami, M., Mathew, M. K., and Tanouye, M. (1989). Leucine-zipper update. *Nature* (*London*) **340,** 103–104.

MacKinnon, R. (1991). Determination of the subunit stoichiometry of a voltage-activated potassium channel. *Nature* (*London*) **350,** 232–235.

MacKinnon, R., and Miller, C. (1989). Mutant potassium channels with altered binding of charybdotoxin, a pore-blocking peptide inhibitor. *Science* **245,** 1382–1385.

MacKinnon, R., Aldrich, R. W., and Lee, A. W. (1993). Functional stoichiometry of Shaker potassium channel inactivation. *Science* **262,** 757–759.

Marmont, G. (1949). Studies of the axon membrane. I. A new method. *J. Cell. Comp. Physiol.* **34,** 351–382.

Mueller, P., and Rudin, P. O. (1967). Development of K^+-Na^+ discrimination in experimental bimolecular lipid membranes by macrocyclic antibiotics. *Biochem. Biophys. Res. Commun.* **26,** 398.

Mullins, L. J. (1959). An analysis of conductance changes in squid axon. *J. Gen. Physiol.* **42,** 1013–1035.

Narahashi, T., Moore, J. W., and Scott, W. R. (1964). Tetrodotoxin blockage of sodium conductance increase in lobster giant axons. *J. Gen. Physiol.* **47,** 965–974.

Nernst, W. (1888). Zur Kinetik der in Lösung befindlichen Körper: Theorie der Diffusion. *Z. Phys. Chem.,* 613–637.

Noda, M., Shimizu, T., Takai, T., Kayano, T., Ikeda, T., Takashi, H., Nakayama, H. Kanaoka, Y., Minamino, N., Kangawa, K., Matsuo, H., Raftery, M. A., Hirose, T., Inayama, S., Hayashida, H., Miyata, T., and Numa, S. (1984). Primary structure of *Electrophorous electricus* sodium channel deduced for cDNA sequence. *Nature* (*London*) **312,** 121–127.

Overton, E. (1902). Beiträge zur allgemeinen MuskelundNervenphysiologie. II. Über die Untentbehrlichkeit von Natrium-(oder Lithium) Ionen für den Contractsionact des Muskels. *Pfluegers Arch.* **92,** 346–386.

Pongs, O., Kecskemethy, N., Muller, R., Krah–Jentgens, I., Baumann, A., Kiltz, H. H., Canal, I., Llamazares, S., and Ferrus, A. (1988). Shaker encodes a family of putative potassium channel proteins in the nervous system of *Drosophila. EMBO J.* **7,** 1087–1096.

Pressman, B. K., Harris, E. J., Jagger, W. S., and Johnson, J. H. (1967). *Proc. Natl. Acad. Sci. U.S.A.* **58,** 1949.

Rojas, E., and Atwater, I. (1967). Blocking of potassium currents by pronase in perfused giant axons. *Nature (London)* **215,** 850.

Ruppersberg, J. P., Frank, R., Pongs, O. and Stocker, M. (1991). Cloned neuronal $I_K(A)$ channels reopen during recovery from inactivation. *Nature (London)* **353,** 657–660.

Schoppa, N. E., McCormack, K., Tanouye, M. A., and Sigworth, F. J. (1992). The size of gating charge in wild-type and mutant Shaker potassium channels. *Science* **255,** 1712–1715.

Sigworth, F., and Neher, E. (1980). Single Na^+ channel currents observed in cultured rat muscle cells. *Nature (London)* **287,** 447–449.

Stefani, E., Toro, L., Perozo, E., and Bezanilla, F. (1994). Gating of Shaker K^+ channels, I. Ionic and gating currents. *Biophys. J.* **66,** 996–1010.

Stühmer, W., Conti, F., Suzuki, H., Wang, X. D., Noda, M., Yahagi, N., Kubo, H., and Numa, S. (1989). Structural parts involved in activation and inactivation of the sodium channel. *Nature (London)* **339,** 597–603.

Tanouye, M. A., Kamb, C. A., Iverson, L. E., Salkoff, L. (1986). Genetics and molecular biology of ionic channels in *Drosophila. Ann. Rev. of Neurosci.* **9,** 255–276.

Tempel, B. L., Papazian, D. M., Schwarz, T. L., Jan, Y. N., and Jan, L. Y. (1987). Sequence of a probable potassium channel component encoded at Shaker locus of *Drosophila. Science* **237,** 770–775.

West, J. W., Patton, D. E., Scheuer, T., Wang, Y., Goldin, A. L., and Catterall, W. A. (1992). A cluster of hydrophobic amino acid residues required for fast Na^+-channel inactivation. *Proc. Natl. Acad. Sci. U.S.A.* **89,** 10910–10914.

Yang, N., George, A. L., and Horn, R. (1996). Molecular basis of charge movement in voltage-gated sodium channels. *Neuron* **16,** 113–122.

Yellen, G., Jurman, M. E., Abramson, T., and MacKinnon, R. (1991). Mutations affecting internal TEA blockade identify the probable pore-forming region of a K^+ channel. *Science* **215,** 939–942.

Yellen, G. D., Sodickson, D., Chen, T.-Y., and Jurman, M. E. (1994). An engineered cysteine in the external mouth of a K^+ channel allows inactivation to be modulated by metal binding. *Biophys. J.* **66,** 1068–1075.

Young, J. Z. (1936). Structure of nerve fibers and synapses in some invertebrates. *Cold Spring Harbor Symp. Quant. Biol.* **4,** 1–6.

Zagotta, W. N., Brainard, M. S., and Aldrich, R. W. (1988). Single-channel analysis of four distinct classes of potassium channels in *Drosophila* muscle. *J. Neurosci.* **8,** 4765–4779.

Zagotta, W. N., Hoshi, T., and Aldrich, R. W. (1994). *Shaker* potassium channel gating III, evaluation of kinetic models for activation. *J. Gen. Physiol.* **103,** 321–362.

CHAPTER 10

Lipid Membrane and Ligand-Gated Ion Channels in General Anesthetic Action

Warren S. Sandberg*, Brian W. Dubois*, and Keith W. Miller*†
*Department of Anesthesia and Critical Care, Massachusetts General Hospital, and
†Department of Biological Chemistry and Molecular Pharmacology, Harvard Medical School, Boston, Massachusetts 02114

I. INTRODUCTION

General anesthesia can be produced by a diverse group of structurally unrelated chemical compounds. Ernest Overton, simultaneously with Hans Meyer, discovered in 1899 that the anesthetic potency of a compound *in vivo* could be correlated with its lipid solubility (oil–water partitioning)

1063-5823/99 $30.00

(Fig. 1). This correlation has been termed the Meyer–Overton rule and has provided the context for almost a century of research into the mechanisms of general anesthesia (Overton, 1991—this reference is to an English translation of the original monograph published in 1901).

The first part of the chapter begins by reviewing the lipid-based theories of anesthesia. Lipid–membrane theories provide an attempt to explain the Meyer–Overton rule based on the ability of an anesthetic to dissolve in, and alter, the physical properties of lipid membranes. The theories account for the structural diversity of compounds capable of producing anesthesia and for our inability to find a chemical antagonist of the anesthetic state. The theories have floundered, however, because bulk membrane physical properties are changed very little at clinically relevant anesthetic concentrations. Consequently, research has shifted to focusing on lipid microenvironments, such as at the lipid–protein interface, and on direct anesthetic–protein interactions.

Subsequently, this chapter presents evidence that anesthetics interact directly with proteins. At issue is whether a single hydrophobic site on a presumptive target protein can account for the Meyer–Overton rule. The issue *in vivo* is undoubtedly complicated by the fact that more than one target protein is likely to be involved in producing the anesthetic state. This chapter provides evidence that anesthetics interact directly with members of a superfamily of ligand-gated ion channels. The ion channels likely play a

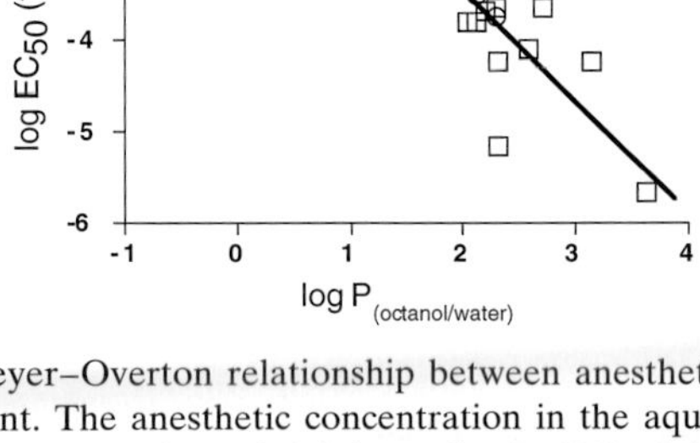

FIGURE 1 The Meyer–Overton relationship between anesthetic potency and octanol/water partition coefficient. The anesthetic concentration in the aqueous phase surrounding amphibious animals that causes a loss of righting reflex in 50% of the tested animals (EC_{50}) correlates with the octanol/water partition coefficient; data are from Firestone *et al.* (1986). The anesthetics tested include barbituates, alkanols, and volatile agents and the animals are tadpoles (□), frogs (◇), and newts (○).

role in producing the anesthetic state and are the subject of ongoing and future research into anesthetic mechanisms.

II. THE MEYER-OVERTON HYPOTHESIS AND THE EVOLUTION OF LIPID-BASED THEORIES OF ANESTHESIA

A. *The Membrane as a Site of Anesthetic Action*

Overton's central observation, that the potency of an anesthetic could be correlated with its solubility in oil, led to the logical presumption that anesthetics acted on cellular lipids. As cellular architecture became known, and with the discovery of excitable membranes, the lipid bilayer became the obvious first place to look for the site of anesthetic action. Several properties of the lipid bilayer made it an attractive potential site of anesthetic action. First, the hydrocarbon interior provided a lipid environment as predicted by the Meyer–Overton rule. Second, the nonrigid structure of the lipid bilayer appeared accommodating to the many diverse structural compounds capable of producing the anesthetic state (Fig. 2).

During Overton's era, protein structure was viewed as rigid and, as introduced by Emil Fischer in 1890, permitted interaction with a specific substrate in a "lock-and key" manner. In proposing a single target for all anesthetics, it was conceptually simpler to propose the lipid bilayer as the target than to suppose that each anesthetic had a unique protein target. It was proposed in 1875 (Bernard, 1875) that all general anesthetics worked through a common target, and this proposal became known as the "unitary hypothesis" of anesthesia. Reconciling the inherent substrate specificities of proteins with the structural diversity of anesthetics is a problem being revisited as proteins have come to be regarded as the more likely targets of general anesthetics.

B. *Membrane Perturbation Theories and Their Limitations*

As anesthetics dissolve in membranes, the physical properties of the membrane are expected to be altered. Identifying and assessing the significance of these alterations led to the formulation of several hypotheses of how anesthetics acted.

The observation that the effects of general anesthetics could be reversed by pressure (Johnson and Flagler, 1950; Lever *et al.,* 1971) provided an important test for these theories. Another test was provided by the observation that as one ascended a homologous series of hydrophobic compounds,

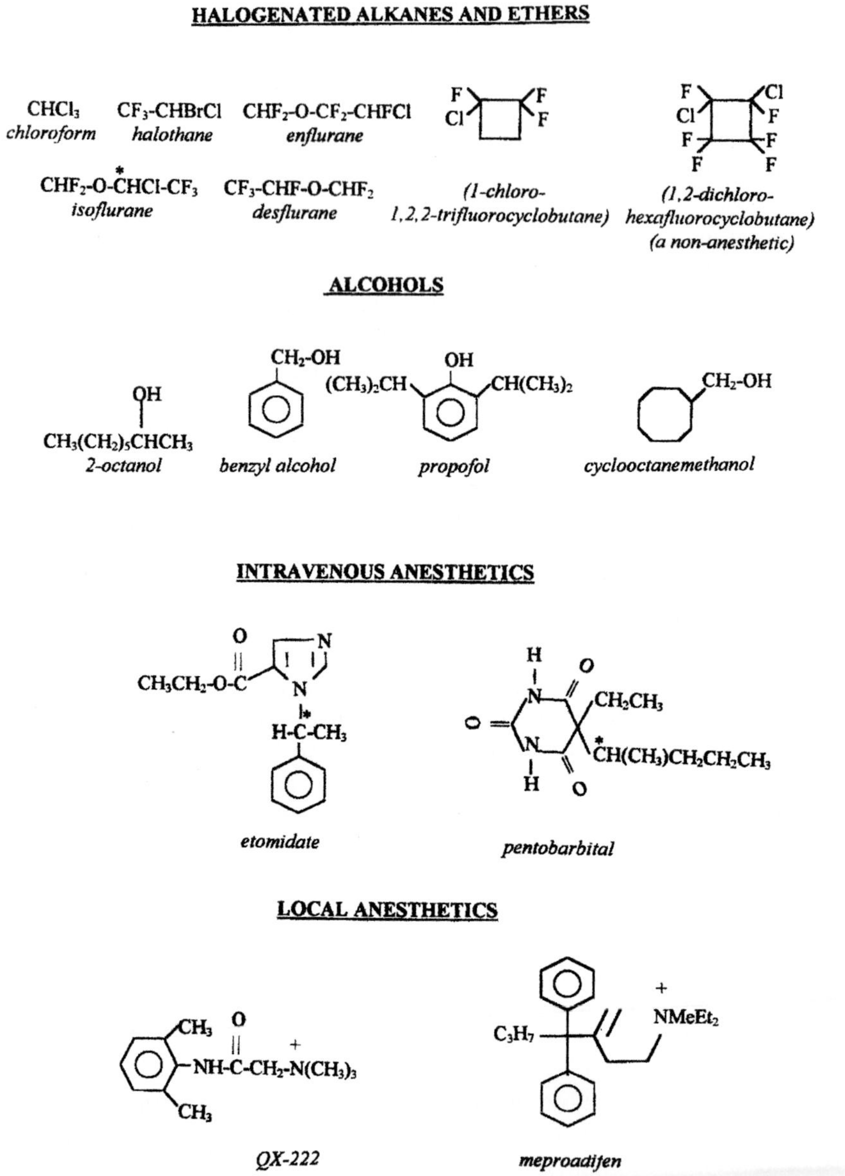

FIGURE 2 Structures of selected anesthetics discussed in this chapter. Many of the anesthetics have chiral carbons, those mentioned in the text are indicated with asterisks.

the largest members, predicted to be potent anesthetics by the Meyer–Overton rule, were in fact nonanesthetic in animals (i.e., the "cutoff" effect).

1. The Critical Volume Hypothesis

This hypothesis stated that as anesthetics dissolve in the membrane they increase the total volume of the system until some critical threshold is reached, beyond which anesthesia ensues. Membrane volume (Kita *et al.,* 1981) expands by a few tenths of a percent at clinical concentrations of anesthetics, and such effects can be reversed by pressure (Miller *et al.,* 1973). Expansion is anisotropic, anesthetics have small or no demonstrable effects on membrane thickness (Franks and Lieb, 1978; Reyes and Latorre, 1979; Turner and Oldfield, 1979), which results in their area expanding more than their volume (Ueda *et al.,* 1974; Bull *et al.,* 1982).

The critical volume hypothesis fails to explain the *cutoff effect* which is observed with *n*-alcohols (Pringle *et al.,* 1981), cycloalkanemethanols (Raines *et al.,* 1993), perfluoroalkanes (Liu *et al.,* 1994), and possibly *n*-alkanes (Liu *et al.,* 1993). Long chain alcohols, which do not cause anesthesia, such as tetradecanol and hexadecanol, are nonetheless freely soluble in the bilayer (Franks and Lieb, 1986; Miller *et al.,* 1989) and indeed cause membrane expansion (Bull *et al.,* 1982).

The temperature dependence of anesthetic potency in frogs follows the temperature dependence of each agent's partition coefficient (Meyer, 1901), an observation also made by Overton (1991, p. 180). Temperature is a complex variable that is sensed by the whole physiological system; low temperature can cause insensibility in the absence of anesthetic. Furthermore, in a membrane, both protein and lipid experience a temperature change whereas only the lipid is perturbed by the anesthetic. Nonetheless, the simple prediction that an increase in volume caused by temperature should cause anesthesia is not met and this failure is compounded by the small size of the required temperature change. Membrane area (Seeman, 1969) and volume (Melchior *et al.,* 1980) increase, whereas thickness decreases (Rand and Pangborn, 1973), all by about 0.1% per degree C temperature increase. Theories of anesthetic action based solely on changes in membrane dimensions cannot account for these observations.

2. Membrane Disordering and Phase Transition Theories

If the dimensional changes in membranes were small, what about the changes in other physical properties? Anesthetics were observed to increase membrane fluidity (Metcalfe *et al.,* 1968), leading to the hypothesis that changes in lipid fluidity affected membrane proteins and produced anesthesia. Similarly, it was proposed that nerve membrane lipids might undergo a phase transition based on the observation that most anesthetics depress the gel→liquid crystal phase transition temperature for pure lipid systems (Kamaya *et al.,* 1979; Mountcastle *et al.,* 1978). Both theories accounted for pressure reversal, quantitatively in the case of the fluidity model but rather too sensitively in the case of the phase transition model (Boggs *et al.,* 1976), when compared to the pressures of 150 atm or more required to reverse anesthesia in animals (Johnson and Flagler, 1950; Lever *et al.,* 1971). The fluidization theory also accounted for the cutoff phenomenon because small alcohols were found to decrease membrane order measured on the 12th acyl carbon (favoring anesthesia), whereas larger alcohols actually enhanced membrane order (Miller *et al.,* 1989). To explain cutoff on the basis of changing membrane order, long chain alcohols should be anesthetic antagonists (not the case). However, subsequent studies showed that when probed at the 5th acyl carbon, short chain alcohols still disordered but long chain ones now failed to change order in any direction (L. L. Firestone and K. W. Miller, unpublished data). Thus, the fluidity model only correctly accounts for cutoff when shallow regions of the bilayer are probed (region 2 in Fig. 3). This region is also the focus of much recent work (see below).

Both theories ultimately fail because the effects produced by anesthetics can be reproduced by increases in temperature of a few tenths of a degree C (Lieb *et al.,* 1982; Pang *et al.,* 1980; Bradley and Richards, 1984; Eger, *et al.,* 1965; Steffey and Eger, 1974).

3. Theories Based on Cation Permeability Changes

At clinically relevant concentrations, anesthetics raise liposomal membrane permeability to cations (Bangham *et al.,* 1965; Johnson and Bangham, 1969; Johnson *et al.,* 1973). These effects are about 10 times larger than the structural perturbations considered earlier and can be reversed by pressure in a manner that closely parallels *in vivo* pressure reversal (Johnson *et al.,* 1973). Using solvent-free black membranes, anesthetics such as benzyl alcohol and chloroform were shown to increase the permeability to organic ions by as much as 40% (Reyes and Latorre, 1979). All these processes are thought to be rate limited by processes at the lipid–water interface; a theme that is echoed in more recent work (see later). Similar results were obtained when cation permeability mediated by simple ionophores, such as the carrier valinomycin and the channel gramicidin, was measured (Johnson *et al.,* 1973), showing that the lipid perturbation could be sensed by these model "proteins."

These observations were developed into a theory suggesting that anesthetics worked by collapsing pH gradients across cell membranes (Bangham and Mason, 1980) Catecholamines are trapped and stored in synaptic vesicles by virtue of being protonated (and hence, charged) at the normally low intravesicular pH. Anesthetics were proposed to increase membrane permeability to protons, leading to the collapse of this pH gradient, in turn allowing the stored catecholamines to leak out, thus depleting the nerve of neurotransmitter. This theory has not been borne out in practice, largely because the effects were shown to be small in subsequent measurements (Akeson and Deamer, 1989) and because not all anesthetics change membrane permeability to ions (Raines and Cafiso, 1989)

C. Current Trends in Lipid Theories

Weaknesses in the lipid theories just reviewed led many researchers to turn to protein-based theories, which will be reviewed next. Despite this, interest in lipid theories has remained strong, perhaps because of their inherent elegance and the seductive power of the Meyer–Overton rule. In any case, anesthetics rapidly achieve high concentrations (~25 mM) within the membrane due to their lipophilicity, and lipids may yet play a major role in the mechanism of anesthesia, either as the route by which anesthetics reach their final targets or as critical participants in a multicomponent complex. Recent discoveries and ideas about anesthetic–bilayer interactions are reviewed next.

1. Physical Chemical Studies of Anesthetic–Lipid Interaction

a. Locations of Anesthetics within the Bilayer. The location of anesthetics within the lipid bilayer may hold important clues about how they function and where they exert their effects. General anesthetics vary tremendously in their polarity and in the arrangements of their polar atoms. Similarly, the environment offered by the bilayer ranges from polar to almost completely nonpolar (Fig. 3). For the purpose of this discussion, the bilayer can be divided into three regions: (1) the head groups facing the aqueous phase and invariably bearing charges (e.g., the zwitterionic phosphatidylcholine head group); (2) the shallow part of the hydrocarbon layer, including the glycerol backbone and, usually, ester carbonyls, having moderate polarity; and (3) the central portions consisting of the hydrocarbon tails which are almost completely nonpolar. As expected, alcohols spend much of their time in the bilayer, presumably with their hydroxyl groups in the more polar region (Colley and Metcalfe, 1972; Pope *et al.*, 1984) whereas alkanes are found predominantly in the central hydrocarbon chain region (Colley and Metcalfe, 1972; Jacobs and White, 1984; Pope *et al.*, 1984).

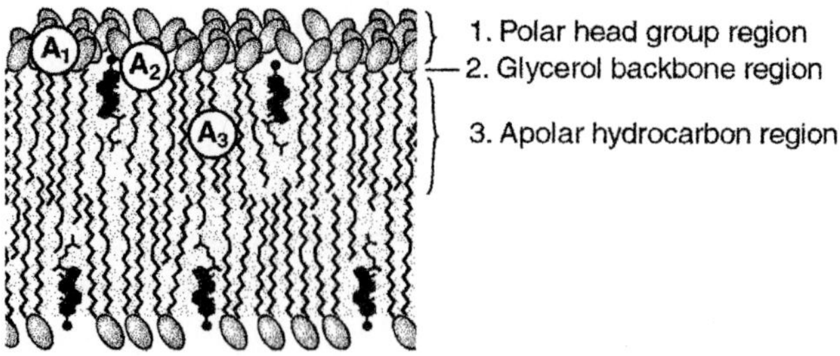

FIGURE 3 Anesthetics experience three types of interactions depending on their position within the lipid bilayer. A hypothetical anesthetic is shown near the phospholipid-charged head groups (A_1), near the moderately polar glyerol backbone (A_2), and near the nonpolar hydrocarbon tails (acyl chains) (A_3). Region 2 is implicated most strongly in general anesthetic action (see Sections II,B and II,C).

Nuclear magnetic resonance (NMR) studies have revealed that the halogenated volatile anesthetics, halothane, isoflurane, and enflurane, are indeed located preferentially in the shallower portion of the bilayer nearer the glycerol backbone, with some in the interior and virtually none in the head group region (Baber *et al.,* 1995). These studies have been extended by comparing anesthetics with structurally similar compounds that are nonanesthetic, even at concentrations well in excess of those predicted to be anesthetizing based on their lipid solubility (Koblin *et al.,* 1994). Nonanesthetic compounds all share the property of being almost completely nonpolar and reside primarily in the deeper layers of the bilayer (North and Cafiso, 1997; Tang *et al.,* 1997) (see also Chapter 3 in this volume).

b. Alterations in Membrane Dipole Moment. The slight preference of anesthetics for the shallower part of the lipid bilayer (region 2 in Fig. 3) has been invoked to explain the interesting observation that anesthetics reduce the magnitude of the membrane dipole potential (Qin *et al.,* 1995). Membranes have a large internal potential, perhaps due to a dipole layer at the membrane–solution interface arising from the net orientation of water or lipid carbonyl atoms (Cafiso, 1995). Thus, the hydrocarbon tails of the phospholipids are positive with respect to the shallower, interfacial region. Because the membrane has two such dipole layers with opposite orientation, there is no net transmembrane dipole potential. However, a transmembrane protein will experience a quadrupolar interaction with the bilayer. Anesthetics dissolved in the membrane apparently lower the dipole potential in each leaflet. The change in dipole potential was estimated to be about 10 mV at physiological concentrations (Qin *et al.,* 1995), a change that could alter the conformational equilibria of voltage-sensitive membrane proteins.

2. Anesthetic-Induced Changes in Membrane Mechanical Properties

a. Lateral Pressure within the Bilayer. Lipid bilayers in water have a large interfacial free energy concentrated over the very narrow membrane thickness, resulting in large lateral pressures within the membrane (Cantor, 1997). This pressure is not uniform throughout the membrane, varying systematically from the membrane–water interface to deeper layers in the membrane. Small compounds that dissolve into the membrane with a preference for the membrane–water interface (said to be "interfacially active") are predicted by modeling studies to cause large increases in the lateral pressure in shallower layers of the membrane, with concomitant decreases in the lateral pressure in deeper layers (Cantor, 1997).

Changing the lateral pressure profile within monolayers, either by adding anesthetics (e.g., halothane, isoflurane) or by physical compression, influences the activity of membrane proteins. For example, manipulation of the lateral pressure profile at first enhanced and then abolished the hydrolysis of dipalmitoylphosphatidylcholine by membrane phospholipase C (Goodman *et al.,* 1996). By affecting the lateral pressure profile, anesthetics (which distribute preferentially to the interface) could conceivably alter ion channel open (or closed) probabilities without interacting directly with the protein (Cantor, 1997). This hypothesis has been extended to explain why short chain alcohols are anesthetics, while long chain alcohols are not, assuming that the polar hydroxyl group is located preferentially at the interface (Fig. 4). Short chain *n*-alcohols are predicted to cause large increases in the lateral pressure at the interface, relative to deeper levels, whereas long chain *n*-alcohols have similar effects at all levels (Cantor, 1997).

b. Curvature Stress within the Bilayer. Membrane phospholipids are subject to a variety of attractive and repulsive forces when assembled as a *monolayer,* and these forces are modified in bilayers (Gruner and Shyamsunder, 1991). In monolayers, the forces between molecules include van der Waals attractions and steric repulsions in the hydrocarbon portion, with added polar and electrostatic interactions occurring in the glycerol head group regions. The net attractive force within these regions is usually not equal in magnitude, and the monolayer curves spontaneously (Gruner and Shyamsunder, 1991), a tendency that is a function of phospholipid head group size and charge as well as the composition of the hydrocarbon tails. Upon forming a bilayer (which is locally a planar structure) this curvature must be flattened (Fig. 5), requiring the input of energy that is

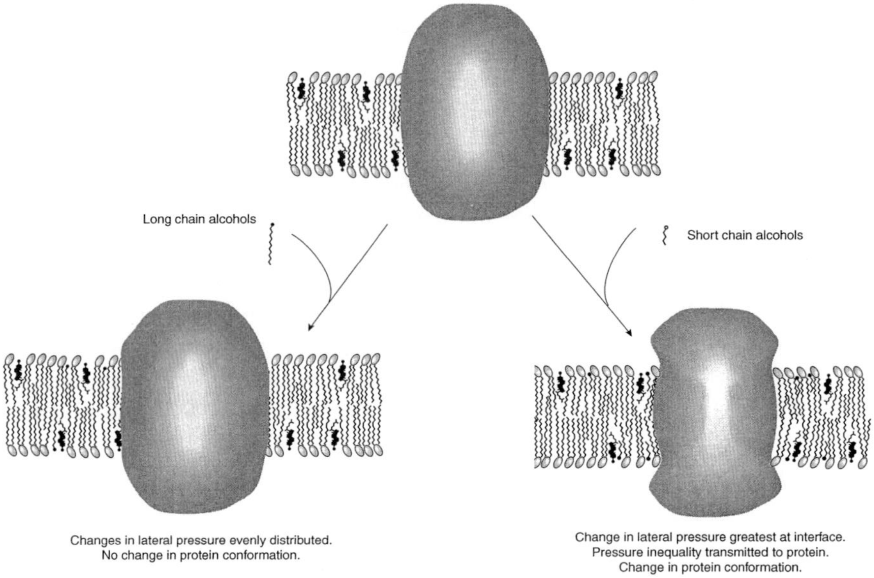

FIGURE 4 The effect of alkanols on the lateral pressure profile within the bilayer. In the absence of alkanol, the lipid bilayer exerts a uniform lateral pressure against the membrane protein. Long chain alcohols preserve the uniform lateral pressure when dissolved in the membrane. Short chain alcohols, however, produce changes in the lateral pressure profile that may change protein conformational equilibrium. Based on Cantor (1997).

stored in the form of an elastic curvature energy (Gruner and Shyamsunder, 1991).

The elastic curvature energy is related to but distinct from the lateral pressure arising from interfacial tension, but it modifies membrane protein activity and can, in turn, be modified by anesthetics. For example, the Ca^{2+}-pumping efficiency of the Ca^{2+}/ATPase is greater in membranes possessing high elastic curvature energy than in membranes containing low elastic curvature energy. This behavior is independent of the chemical nature of the lipids in the membrane, but instead is a function of their tendency to curve in monolayers (and hence, of their elastic curvature stress in a planar bilayer) (Navarro *et al.,* 1984). Studies of the effect of membrane curvature on the gramicidin A channel assembly show that curvature stress can also influence protein conformation (Lundbaek *et al.,* 1997). Short chain *n*-alcohols stabilize the bilayer, whereas longer alcohols raise the elastic curvature energy within bilayers; this effect is countered by pressure (Gruner and Shyamsunder, 1991)

Curvature stress and lateral pressure are both likely to increase with temperature (Gruner and Shyamsunder, 1991). Hence, both proposed

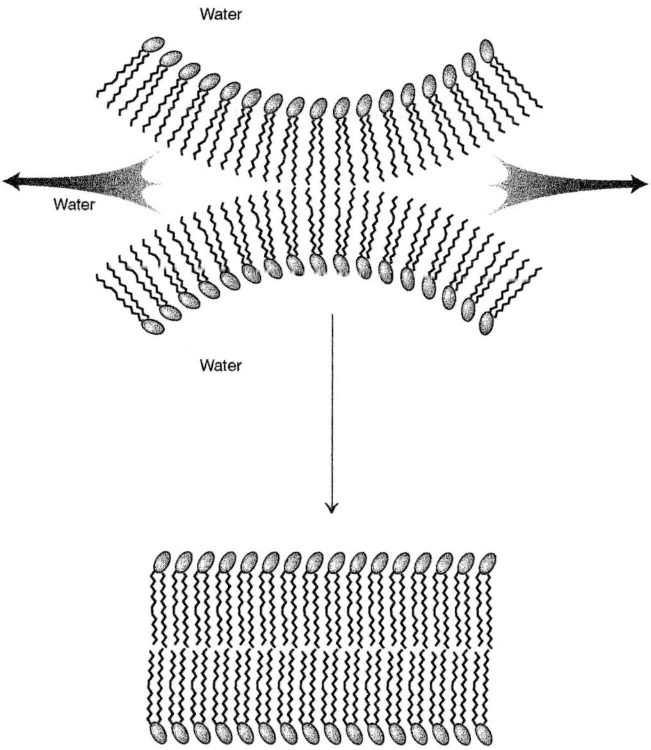

FIGURE 5 Depiction of curvature stress within the bilayer. Depending on the relative cross-sectional areas of their head groups and acyl chains, different phospholipids display different tendencies to curve when forming a monolayer. A phospholipid whose head group has a smaller cross-sectional area than its acyl chains is shown. This curvature is necessarily flattened during bilayer formation in water in order to minimize the contact between hydrocarbon and water, giving rise to a curvature stress. Curvature stress can be transmitted to integral membrane proteins and may be modified by anesthetics dissolving in the bilayer. Based on Gruner and Shyamsunder (1991).

mechanisms predict that the effects of anesthetics should be mimicked by changes in temperature, but the magnitude of these predictions has not been examined experimentally.

3. Molecular Dynamics Simulations of Membrane–Anesthetic Interactions

Molecular dynamic simulations modeling the interactions among anesthetics, the membrane, surrounding water, and the solvent–membrane interface reveal a picture of anesthetic–membrane interactions consistent with the recent discoveries described earlier. This topic is covered in detail in Chapter 3 of this volume. Briefly, the nonanesthetics as a class are almost completely nonpolar, whereas full anesthetics are amphiphillic. Simulations of the two classes of compounds in water–membrane systems predict that amphiphillic compounds achieve high concentrations at the membrane–water interface, whereas the completely nonpolar nonanesthetics do not. This prediction has been confirmed by NMR studies (North and Cafiso, 1997; Tang *et al.,* 1997). The resulting hypothesis is that anesthetics act at or must pass through an amphiphillic site and that the nonanesthetics, lacking any polar character, do not satisfy the requirement for some polar interaction in the site where anesthetics act.

D. Summary

Taken together, these results indicate that a role for the lipid bilayer in the mechanism of general anesthesia cannot be written off entirely. As a relatively nonordered bulk material, the membrane provides a seductively simple and unitary explanation for the remarkable lack of selectivity seen for most general anesthetics, while accounting for the behavior of hydrophobic nonanesthetics, cutoff and pressure reversal. Anesthetics are present in the membrane at high concentrations under clinical conditions and exert major influences on membrane behavior, even though their structural effects are small. There is a convergence of data that general anesthetics exert their major effect on the bilayer at the outer end of the acyl chains and the glycerol backbone region (region 2; Fig. 3). Many of the putative protein targets of anesthetics are integral membrane proteins, whose function is dependent to some degree on the surrounding lipid bilayer, and it seems likely that the mechanism of general anesthesia will involve both components. However, an experimental demonstration that the small perturbations that general anesthetics induce in bilayers can indeed modulate membrane protein function is still awaited. Meanwhile, a case for general anesthetics acting at protein sites has accumulated.

III. THE RISE OF PROTEIN-BASED THEORIES OF ANESTHETIC ACTION

A. Observations of Soluble Proteins Suggesting a Protein Site of Action for Anesthetics

Few proteins have anesthetic sites and the first to be studied had sites that excluded most anesthetics. However, following up on an old observation that anesthetics inhibited the chemiluminescence of certain bacteria (Harvey, 1915), a body of data now shows that the firefly, but interestingly not the bacterial, luciferase is inhibited by a wide range of inhalation agents with potencies that correlate as well with *in vivo* anesthetic potencies as does lipid solubility (Franks and Lieb, 1984), establishing the possibility that proteins could possess ligand-binding sites permissive enough to be consistent with the Meyer–Overton rule. Nevertheless, the model does not account for the action of intravenous anesthetics, and the inhibitory effect of anesthetics could not be reversed by pressure (Moss *et al.,* 1991), which is at odds with results in whole animals. This suggests that general anesthetic action at protein sites is incompatible with the unitary hypothesis.

A ligand-binding site on a protein has defined dimensions, beyond which exogenous ligands are sterically precluded from effectively binding. This could provide a natural explanation based on steric hindrance for the cutoff, emulating the success of the lipid theories in this regard. A particularly instructive example is provided by the inhibitory action of alcohols on firefly luciferase. These agents compete for the luciferin-binding pocket. Their IC_{50} increases linearly with chain length up to dodecanol and then remains approximately constant until hexadecanol. Above this the alcohols are too insoluble to achieve an IC_{50} concentration and no inhibition is observed. The plausible explanation put forward is that only the first 12 carbons fit in the substrate-binding site, with additional carbons remaining in the aqueous phase and contributing no binding energy (Franks and Lieb, 1984). As a model for anesthesia this particular site is inexact because the *in vivo* cutoff at dodecanol is approached smoothly (Alifimoff *et al.,* 1989).

There are also sometimes striking potency differences between stereoisomers of some, but not all, anesthetics (see Fig. 2 for examples of enantiomeric pairs). For example, no potency differences were found between secondary alcohol enantiomers for the production of anesthesia in tadpoles (Alifimoff *et al.,* 1987). Different investigators have observed modest (Lysko *et al.,* 1994) or no (Eger *et al.,* 1997a) differences in the anesthetic potency of isoflurane enantiomers in rats as well. The most striking example of stereoselectivity is (+)-etomidate, which in tadpoles is 15 times as potent as (−)-etomidate (Ashton and Wauquier, 1985; Tomlin *et al.,* 1998). Such a large difference implicates a protein site, although interaction with chiral lipids such as cholesterol has not been ruled out experimentally.

B. Ligand-Gated Ion Channels as Potential Targets of General Anesthetics

1. The Ligand-Gated Channel Superfamily

The ligand-gated ion channel superfamily includes the nicotinic acetylcholine receptor (nAcChoR) and the γ-aminobutyric acid (GABA), glycine and 5-HT_3 receptors. Each is thought to have four transmembrane α helices in each of five subunits (Fig. 6), and all are sensitive to anesthetics, although the degree of sensitivity depends on the receptor type. Anesthetics have numerous effects on members of this channel family, but the most is known about their effects on the nAcChoR and the GABA receptor.

2. Studies of the nAcChoR: Anesthetics Can Both Stabilize and Inhibit the Open Channel

a. Characteristics of the nAcChoR. The muscle type nicotinic acetylcholine receptor was readily accessible to electrophysiologists long before the present era of convenient expression systems. Consequently, its gating kinetics are well understood and many studies with anesthetics have been performed. It also has the unique advantage over all other members of the superfamily of being relatively abundant and available at high specific activity from the electric organ of the *Torpedo*. Because of this, numerous biochemical and structural studies have been performed on the nAcChoR,

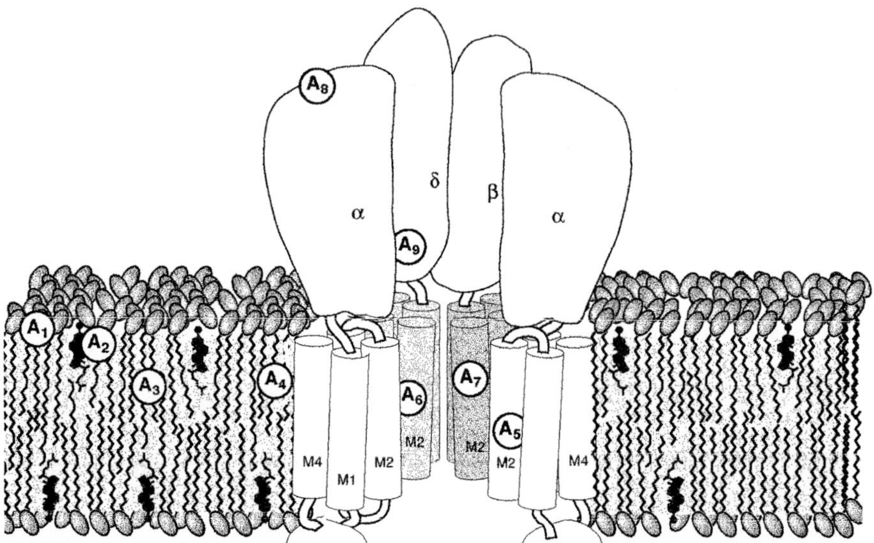

FIGURE 6 Depiction of a ligand-gated ion channel from the four transmembrane helix superfamily, which includes the $GABA_A$ and nicotinic acetylcholine receptors. One of the five subunits is removed to reveal the structure. Possible anesthetic sites are shown at hypothetical sites in the lipid–protein interface (A_4), between transmembrane α helices (A_5), between subunits in the transmembrane region (A_6), near the pore–facing surface of the M2 helix (A_7), at the extracelluar surface (A_8), and between subunits in the extracellular region (A_9).

and it serves as a paradigm for understanding the other ligand-gated receptors. The receptor has five subunits arranged around a central pore lined by the second of four transmembrane α helices (one from each subunit) (Unwin, 1995). The distance from the agonist site to the central pore is roughly 30 Å, and the mechanism of channel gating is thought to be highly allosteric. Consistent with this supposition, mutations and chemical perturbation directed at nearly any of the main structural features, including the lipid bilayer, affect gating, and large structural changes are seen between resting and open states (Unwin, 1995). The nAcChoR undergoes a series of rapid conformational shifts when an agonist is added to the resting (closed) channel. Resting channels open rapidly on agonist binding and then proceed quickly to a fast desensitized state. This is followed by a distinct slow desensitized state and then regeneration of the resting receptor after the agonist dissociates (Lingle *et al.,* 1992).

b. The nAcChoR Responds to a Variety of Anesthetics at Clinical Concentrations. In the muscle and *Torpedo* type nAcChoRs, acetylcholine-gated activity is inhibited noncompetitively by a wide range of inhalation and intravenous general anesthetics at clinical concentrations. To a first approximation this action is consistent with the hypothesis that anesthetics interact with the channel when it opens, but the most detailed studies reveal considerable complexity. In addition, some small agents shift the agonist concentration–response curve for opening to the left either with or without inhibition (Dilger and Vidal, 1994; Wood *et al.,* 1991). Finally, all inhalation agents enhance slow desensitization. These several actions undoubtedly have distinct underlying mechanisms. In some cases, outlined next, evidence for general anesthetic sites is good, but in others the underlying mechanisms are far from clear. A study of barbiturates emphasizes the complexity (Roth *et al.,* 1989).

i. A Site Where Large Alcohols Inhibit the Open State. Electrophysiological studies suggest that alcohols from propanol to dodecanol inhibit the open state of the nAcChoR, causing brief closings of the channel, resulting either from brief pore-plugging events or from alcohols stabilizing a transient closed state accessible from the open state (Dilger and Vidal, 1994). Rapid kinetic studies of nAcChoR-rich vesicles loaded with radioactive cations have provided indirect evidence for a discrete inhibitory site. In these experiments, agonist and alcohol are added simultaneously and in less than 1 msec to the receptor. The fast on rate of the agonist, the high open probability, and the brief period over which the cation flux through the open channel is monitored ensure that the alcohols interact only with the open channel. Under these conditions, the Hill coefficient for inhibition by alcohols is close to one, suggesting that a single site or multiple, noninteracting sites with equal affinities mediate inhibition. Additionally, inhibitory activity is constrained sterically. Alcohols with molecular volumes above 340 Å^3 or below 100 Å^3 fail to inhibit the receptor (Raines *et al.,* 1993; Wood *et al.,* 1991, 1993). A kinetic strategy was developed to provide more convincing evidence for a general anesthetic site on the nAcChoR. For two distinct inhibitors acting simultaneously on a protein, four actions may occur:

1. Both bind to the same site with steric competition.
2. Both bind to separate, allosterically coupled binding sites, so that binding of one site affects the affinity of the other.
3. Both bind to separate, noninteracting sites.
4. Both bind to one very large inhibitory site at which no interactions between bound ligands will be observed.

When inhibition of the open nAcChoR by octanol was studied in the presence of a series of fixed concentrations of heptanol, the apparent K_I for octanol increased with increasing heptanol concentration. The K_I for heptanol changed similarly with respect to increasing octanol concentration. This result excludes any model in which octanol and heptanol do not interact (3 and 4 in the list), because noninteracting models predict that each alcohol's K_I should be independent of the concentration of the other.

Distinguishing between models 1 and 2 (direct competition vs negative allosterism) is experimentally difficult, particularly if the allosteric interaction is strong. Strict competition between inhibitors predicts a linear increase in the apparent K_I of the first inhibitor as the concentration of the second inhibitor increases. An allosteric mechanism predicts a curvilinear (diminishing) increase in K_I under these conditions, but this will only be observed if a sufficient concentration range of inhibitor is tested. The interaction between octanol and heptanol in the inhibition of the nAcChoR is consistent with a strict competition mechanism, but a strong negative allosteric interaction cannot be excluded.

II. A SITE WHERE SMALL ALCOHOLS ENHANCE CHANNEL OPENING. In contrast to the inhibitory effects of large alcohols, the open state of the nAcChoR is stabilized by alcohols up to butanol (Wood *et al.,* 1993; Wu *et al.,* 1993). Stabilization of the open state could arise from small alcohols binding to the octanol site but failing to occlude the channel (Bradley *et al.,* 1984). However, the K_I for octanol is unaffected, even by very high concentrations of ethanol, excluding mechanisms whereby large and small alcohols interact (1 and 2 in the list). Separate, noninteracting sites (mechanism 3) are excluded because open channel stability increases linearly with anesthetic concentration. Thus mechanism 4, i.e., one large site to which alcohols bind without interacting, seems most likely as the mechanism for stabilizing the open state. This is in contrast to inhibition of the open state by larger alcohols, where interaction seems to occur. Such a large site might involve the interfaces between subunits or the lipid–protein interface, both of which are likely to be large, apolar, and important in modulating the relative conformational stability of the open state. This, in turn, emphasizes the importance of lipid protein interactions, which modulate both the activity and the anesthetic sensitivity of the nAcChoRs (Rankin *et al.,* 1994) and the GABA receptors (Bennett and Simmonds, 1996).

The discussion until now has focused on two discrete sites for alcohol binding: one where small alcohols stabilize the open channel without interactions between ligands and one where large alcohols inhibit channel opening, with an interaction between competing ligands. However, the nAcChoR also possesses further sites where local anesthetics bind. For example, octanol and the local anesthetic procaine inhibit the nAcChoR at separate but allosterically coupled sites; when one site is occupied, the affinity of the other inhibitor site decreased sixfold (Wood *et al.,* 1995). Similarly, the duration of open channel block by the quaternary local anesthetic QX-222 was increased by either butanol or diethyl ether, implying a positive heterotrophic interaction between separate sites (Dilger and Vidal, 1994).

III. DIRECT DETECTION OF A BARBITURATE SITE ON THE RESTING STATE. One would not expect to be able to detect the binding of an anesthetic to a membrane protein because such binding would be lost amidst a high background of anesthetic dissolved in the membrane's lipid. However, in acetylcholine receptor-rich postsynaptic membranes isolated from the electroplaques of *Torpedo,* there are only a few hundred lipid molecules per receptor (this is about 10,000-fold less lipid per receptor than is found in brain). This advantageous situation has been exploited to directly demonstrate that [^{14}C]pentobarbital and [^{14}C]amobarbital bind in a saturable, displaceable, and stereoselective manner to an allosteric site on the resting state of the nAcChoR. To date, this is the only unequivocal general anesthetic site to be detected on the ligand-gated ion channel superfamily. It is not shared by alcohols (Dodson *et al.,* 1987; Miller *et al.,* 1982).

c. Proposed Locations of the Inhibitory Sites on the nAcChoR. Having established the presence of at least three sites where anesthetics interact with the nAcChoR, one for local anesthetics and two for alcohols, the remaining question is where are these sites located on the protein?

Photolabeling experiments with charged inhibitors identified amino acid residues on the pore-facing side of each of the five channel-lining M2 helices

(Fig. 7). Subsequent site-directed mutagenesis experiments confirmed that this was the general location for QX-222 binding. The site was at the second and third turn of the helix from the cytoplasmic end of M2 (the 6′ and 10′ positions) (Changeux *et al.*, 1992; Leonard *et al.*, 1991).

Because the local anesthetic site and the alcohol inhibitory site interact, it seemed likely that the alcohol site was in the general area of the local anesthetic site. Additionally, the kinetics of inhibition of the open nAcChoR by octanol were consistent with a bimolecular reaction, in which octanol plugged the channel. Finally, the extracellular end of the channel (above the 10′ position) contains an extended region of hydrophobic amino acids absent in the GABA receptor. The fact that the GABA receptor is not inhibited by alcohols suggested that this region might be important for alcohol inhibition of the nAcChoR. Accordingly, mutagenesis efforts were directed at these residues.

The serine and threonine residues lining the channel at the 10′ site in the different subunits, as well as residues at the start of the long hydrophobic stretch, were mutated successively to more hydrophobic residues such as valine and isoleucine (Forman *et al.*, 1995). As the hydrophobicity of the substituting residues increased, general anesthetics inhibited the channel at successively lower concentrations (Forman *et al.*, 1995). This result suggests that general anesthetics inhibit the receptor by interacting with this region of the channel. Additionally, mutations deeper in the channel, such as phenylalanine replacing serine at the 6′ position, had a much smaller effect on inhibitory potency than mutations at the 10′ position.

3. Studies on the GABA Receptor

a. Characteristics of the GABA Receptor. The GABA receptor is a family of related receptors within the ligand-gated ion channel superfamily. Binding of GABA to the receptor induces a chloride current that hyperpolarizes the postsynaptic membrane. Thus, GABAergic neurons in the central nervous system (CNS) are inhibitory and their tonic or enhanced activation could produce anesthesia via a generalized inhibition of neuronal firing. At least three major subtypes are known [suffixed A, B, and C ($GABA_C$ is also known as $GABA_{Rho}$)], but most work currently focuses on the $GABA_A$ receptor. The $GABA_A$ receptor is distributed widely throughout the CNS, where it exists as a heteropentamer (Nayeem and Green, 1994) of distinct polypeptides ($\alpha_{1\text{-}6}$, $\beta_{1\text{-}3}$, $\gamma_{1\text{-}3}$, δ, ε). These subunits are grouped on the basis of amino acid sequence, and the subunit composition of naturally occurring GABA receptors varies throughout the CNS. Although the subunit stoichiometry and composition of the GABA receptor in particular regions of the CNS are often unknown, functional GABA homo- and heterooligomeric receptors have been created *in vitro* by expression in

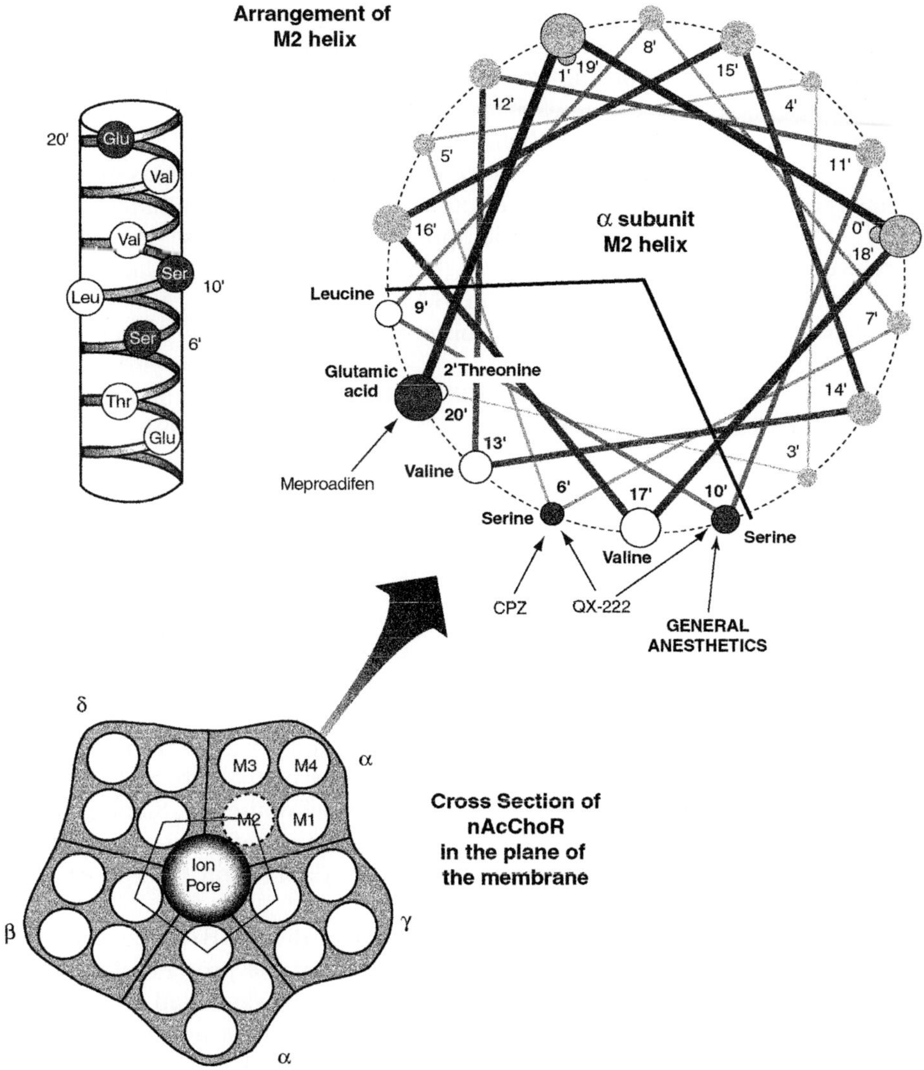

FIGURE 7 Schematic depiction of the α subunit of the acetylcholine recpetor. The pore-facing surface of the M2 helix of the α subunit contains amino acid residues (darker fill) that are implicated in anesthetic binding either by photolabeling or by site-directed mutagenesis (see text). CPZ, chlorpromazine.

mammalian cell lines or *Xenopus* oocytes, and the effects of different subunits and mutations on general anesthetic action have been studied by electrophysiologic recording from the host cells.

b. All General Anesthetics Enhance the Response to GABA and Some Activate the Receptor Directly. The GABA receptor in its various subtypes has received much attention as a potential target for general anesthetics. At low concentrations, anesthetics potentiate the response to GABA, referred to as an "indirect" action. High concentrations of anesthetics can also induce chloride currents in GABA receptor preparations in the absence of GABA. This latter response is termed a "direct" effect (Sigel *et al.,* 1990).

Alcohols (Dildy-Mayfield *et al.,* 1996; Nakahiro *et al.,* 1991) and anesthetics (Harrison *et al.,* 1993; Parker *et al.,* 1986; Yeh *et al.,* 1991) enhance GABA receptor activity at clinical concentrations. In the case of volatile agents, this enhancement of GABA receptor activity is due to their ability to prolong channel open time in the presence of GABA (Yeh *et al.,* 1991).

c. Mutants of the GABA Receptor Delineate Sites Sensitive to Anesthetics. Molecular biological experiments have shown that anesthetic efficacy and potency are modulated by GABA receptor subunit composition. Additionally, responses of a given receptor subtype to volatile anesthetics differ from responses of structurally distinct intravenous drugs (Harris *et al.,* 1995). This implies different loci for the actions of these two classes of anesthetics. Similarly, the direct and indirect actions of the anesthetic etomidate on the GABA receptor can be separated by mutation, implying that the two effects arise from different loci within the protein (Moody *et al.,* 1997).

It has been shown that anesthetic sensitivity of the GABA receptor can be modulated by changes in very small regions of the protein (Belelli *et al.,* 1997; Mihic *et al.,* 1997). One set of experiments was based on fusions between α subunits from the $GABA_C$ receptor subtype and the glycine receptor. The $GABA_C$ receptor differs from other GABA receptors in that it is inhibited (rather than potentiated) by anesthetics (Johnston, 1996). The glycine receptor is potentiated by anesthetics. The goal was to find what portions of the glycine receptor α subunit, when substituted for the native $GABA_C$ receptor sequence, yielded a $GABA_C$ receptor *potentiated* by anesthetics (in this case, enflurane). Only 45 contiguous amino acids, comprising the second and part of the third transmembrane helices, were required to confer enflurane potentiation of the agonist effect on the GABA receptor (Mihic *et al.,* 1997). This site was further narrowed to a single amino acid. Substitution of Ser 267 in the second transmembrane helix of the glycine receptor by the Ile found in the $GABA_C$ receptor yielded a glycine receptor mutant inhibited by anesthetics (Mihic *et al.,* 1997). How a single substitution brings about such a change in behavior cannot be demonstrated from this study, but several mechanisms were suggested. For example, Ser 267 is predicted by the results of Cys-scanning/chemical modification experiments (Xu and Akabas, 1996) to be pointed away from the solvent accessible channel pore. Thus, it might face the membrane lipids, another transmembrane helix in the same subunit, or be part of a subunit interface involved in allosterism. Another critical residue (288) on the the third transmembrane helix suggested that the second possibility is the most likely. Consequently, Ser 267 could be part of a potentiating ligand (anesthetic)-binding site, whose functional properties critical to the site cannot be replaced by Ile. What remains unclear is how a single substitution can at once abolish one response to anesthetics (potentiation of agonist effect) and confer another (inhibition of agonist effect). However, until more detailed electrophysiological experiments have been undertaken, it remains possible that inhibition and enhancement both occur to some degree in all receptors. Thus, for example, a mutation that apparently confers more anesthetic enhancement might actually be acting by reducing inhibition.

Similar experiments with the β subunit show that sensitivity to etomidate can be altered by very small changes in the protein. Single amino acid substitutions Ser $\rightarrow$ Asn at β_1 position 290 render the normally insensitive β_1 subunit sensitive to etomidate. Conversely, a Ser $\rightarrow$ Asn substitution at the homologous position (residue 289) in the etomidate-sensitive β_3 subunit renders it insensitive (Belelli *et al.,* 1997).

4. Stereoisomeric Differences in Anesthetic Potency at Ligand-Gated Ion Channels

Many anesthetics are racemic mixtures, and the differential potencies of enantiomeric pairs have been interpreted as evidence that anesthetics act at proteins.

Proteins commonly display substrate stereoselectivity, although it is more difficult to conceive of how a disordered medium such as the lipid bilayer could do so. However, membrane phospholipids are chiral, and parameters such as membrane ordering are altered differentially by stereoisomers of the partial anesthetic tetrahydrocannabinol (Lawrence and Gill, 1975), demonstrating that the membrane can be sensitive to stereoisomerism.

Experiments with isoflurane on anesthetic-sensitive mulluscan ion channels showed that the (+)isomer was more potent than the (−)isomer (Franks and Lieb, 1991). In contrast, the two isomers had identical effects on membrane-melting temperatures, suggesting that they did not interact stereoselectively with the phospholipids (Franks and Lieb, 1991). It is currently unclear whether isoflurane stereoisomers have different potencies in whole animals. Some workers have found (+)-isoflurane to be about 50% more potent than (−)-isoflurane (Lysko *et al.,* 1994) whereas others find no difference (Eger *et al.,* 1997a). Only six animals per group were used in each of these studies. In contrast, a study with 200 tadpoles showed no differences (Firestone *et al.,* 1992). The case for the stereoselectivity of isoflurane as a general anesthetic is still inconclusive.

Stereoisomers of volatile anesthetics show modest potency differences on the GABA receptor. For example, (+)-isoflurane was about two-fold more effective than (−)-isoflurane in stimulating GABA receptor agonist binding and in retarding GABA receptor antagonist binding to mouse neuronal membranes (Harris *et al.,* 1994; Moody *et al.,* 1993). Similar modest differences for enhancing agonist binding were seen between enantiomers of barbiturates [(−)-pentobarbital was about 1.5 times as potent as (+)-pentobarbital] (Olsen *et al.,* 1986). Pentobarbital also directly induces chloride currents in the GABA receptor, with (−)-pentobarbital being several-fold more potent than (+)-pentobarbital (Akaike *et al.,* 1985; Huang and Barker, 1980).

Etomidate stereoisomers show dramatic differences in potency at the GABA receptor. For example, (+)-etomidate, but not (−)-etomidate potentiated GABAergic inhibitory firing in hippocampal slices *in vitro* (Ashton and Wauquier, 1985). More recently, (+)-etomidate was found to be 16-fold more potent than (−)-etomidate in producing anesthesia in tadpoles and was more potent at potentiating GABA-induced currents in cells expressing bovine $\alpha_1\beta_2\gamma_{2L}$ receptors (Tomlin *et al.,* 1998). The nAcChoR also shows stereoselectivity of anesthetic binding (Miller *et al.,* 1982; Roth *et al.,* 1989). For instance, (−)-pentobarbital binds the receptor, but with a fourfold lower affinity than (+)-pentobarbital (the potency differences are opposite those seen for pentobarbital–GABA receptor interactions). However, inhibition of agonist-induced cation flux showed no stereoselectivity (Roth *et al.,* 1989).

Exploration of the stereoselective effects of anesthetics on their potential targets is an active area of endeavor, with the potential to add much to our knowledge of the mechanism of action of anesthetics. Stereoselectivity of anesthesia is regarded as strong evidence in favor of a protein site of anesthetic action because proteins can most readily provide the structural basis for stereoselectivity. Demonstration of stereospecific anesthetic effects on ligand-gated ion channels is an important test of their plausible involvement in anesthesia *in vivo.*

5. Role of Lipid–Protein Interactions in Protein-Based Theories of Anesthesia

The interaction between membrane proteins and their surrounding lipids is an inescapable consequence of their transmembrane structure. Lipid–protein interactions can be divided conceptually into those arising from bulk properties of the membrane phospholipids (i.e., relatively nonspecific interactions) and those involving specific interactions between protein and lipid. Nonspecific interactions can be modulated by mechanical properties of the membrane, such as thickness (a manifestation of hydrocarbon chain length and order), asymmetry, lateral pressure within the membrane, and the tendency of a membrane to curve. Membrane charge, internal dipole potential, head group size, and hydrocarbon unsaturation are all properties that could affect protein function in relatively nonspecific ways, and these chemical properties give rise to the mechanical characteristics of the membrane.

Specific lipid–protein interactions imply the existence of lipid binding sites that clearly would be potential targets for anesthetics. Specific lipid interactions with transmembrane proteins can be classified as annular [meaning between the receptor and the annulus of immobilized lipids surrounding it (Antollini *et al.,* 1996; Marsh and Barrantes, 1978)] or nonannular (meaning at sites distinct from the protein–lipid interface). Annular-binding sites may be less structurally selective because they occur in the part of the protein exposed to bulk lipids, but this is not necessarily so, because membrane phospholipids are chiral and are capable of specific interactions.

Lipid–protein interaction studies require large quantities of protein, and in the ligand-gated ion channel superfamily, only the *Torpedo* nAcChoR is sufficiently abundant. Negatively charged lipids such as phosphatidylserine have a preference over phosphatidylcholine for the lipid annulus of the nAcChoR (Dreger *et al.,* 1997; Raines and Miller, 1993). However, cholesterol and fatty acid spin-labeled analogs bind much less effectively to the receptor after proteolytic removal of the extracellular regions. Under these conditions, the secondary structure of the remaining transmembrane portion of the protein appears to be preserved, and phospholipid interaction is preserved (although somewhat reduced). Keeping in mind that "shaving" the receptor almost certainly alters the remaining structure, this result suggests that cholesterol and fatty acids bind to the extracellular portion of the protein (i.e., in a nonannular position) (Dreger *et al.,* 1997).

Although nonannular lipid–protein interactions have been well established, their functional role is still being elucidated. In the case of nAcChoR, cholesterol is required in the membrane for activity (Fong and McNamee, 1986), but the site of the effect of cholesterol on the protein has not been established. Indirect evidence that cholesterol acts at nonannular sites on the nAcChoR comes from combining the results of two studies. In the first study (Jones and McNamee, 1988), quenching of the fluorescence of the receptor by brominated cholesterol could be attributed to cholesterol binding at nonannular sites. In the second study, rapid conformational changes during activation of the receptor were also found to require cholesterol (Rankin *et al.,* 1997) with a concentration dependence similar to that of dibromocholesterol quenching. This suggests that the functional effect of cholesterol may be at a nonannular-binding site. However, other neutral lipids (even structurally dissimilar compounds such as tocopherol) can be substituted for cholesterol and also support nAcChoR activation in reconstituted vesicles (Sunshine and McNamee, 1992). This observation requires that nonannular sites for cholesterol have fairly lax structural specificity or that cholesterol and neutral lipids exert their effects elsewhere. A competing hypothesis is that cholesterol, because it can flip from one membrane leaflet to the other, serves to relieve curvature stress induced in the membrane during receptor conformational cycling (Rankin *et al.,* 1997). However, this appears not to be the case, as cholesterol analogs bearing standing charges support nAcChoR activation, despite their inability to flip between leaflets (Addona *et al.,* 1998). In fact, cholesterol tethered to a phospholipid backbone by a short spacer also supports nAcChoR activation. Thus, the cholesterol site required for activation appears to be nonannular, but it must not be deep in the receptor because cholesterol analogs with bulky substituents and standing charges are functional (Addona *et al.,* 1998).

If nonannular binding proves to be functionally important for ligand-gated ion channel functioning, then the lipid-binding site clearly deserves scrutiny as a potential site for anesthetic effects. A nonannular lipid-binding site with a large, amphiphillic natural ligand, lax selectivity requirements, and a strong effect on protein function certainly meets the criteria for a plausible protein site of anesthetic binding, in that it would likely obey the Meyer–Overton rule.

C. Other Potential Sites of Action on Membrane-Associated Proteins

Although this chapter is concerned primarily with the roles of the membrane and transmembrane channels in the mechanism of anesthesia, no such discussion can be complete without the discussion of other possible sites of action for anesthetics. The critical role of the luciferin-binding site on luciferase in the development of protein theories has already been discussed. Some other membrane-associated proteins also provide an environment for anesthetic binding entirely consistent with the observation that potency is a function of lipid solubility.

1. Characteristics of Protein Kinase C (PKC) and Their Potential Relevance to Anesthesia

Upon activation, PKC phosphorylates cellular proteins, altering their function and further amplifying the original signal. The endogenous activator of PKC is diacylglycerol (DAG) liberated from membrane phospholipid by the hydrolysis of phosphatidylinositol. PKC is also activated strongly and persistently by phorbol esters. Both activators are large hydrophobic compounds with polar groups at one end conferring an amphipathic character. The ligand-binding site on PKC is a large hydrophobic cleft (Zhang *et al.,* 1995) with obvious possibilities as a potential binding site for anesthetics. Activated PKC is obligatorily associated with the membrane *in vivo,* and the composition of the membrane modulates this activity (Slater *et al.,* 1994). The effects of PKC on cell behavior are protean, but a few observations bear special attention with respect to the potential role of the enzyme in anesthesia. First, protein kinase C affects neurotransmitter release (Shapira *et al.,* 1987) and affects the properties of ion channels (West *et al.,* 1991), including the GABA receptor (Browning *et al.,* 1990). Additionally, mice lacking PKCγ have evidently normal responses to acute pain, but no longer develop neuropathic (i.e., prolonged and profound response to ordinarily nonpainful stimuli after an inciting injury) pain responses (Malmberg *et al.,* 1997). Furthermore, specific inhibition of PKC by staurosporine abolishes learning in mice (Jerusalinsky *et al.,* 1994). Finally, staurosporine potentiates the effects of inhalation anesthetics on tadpoles (Firestone *et al.,* 1993). Anesthetics and alcohols inhibit PKC (α isoform) *in vitro* by competing for the phorbol ester-binding site (Slater *et al.,* 1993, 1997). In a lipid-free assay system the inhibitory potency of alcohols and anesthetics is a linear function of their hydrophobicity, although about three-fold more halothane is required to achieve half-maximal inhibition than is required for anesthesia in humans (Slater *et al.,* 1993). In the presence of phospholipids, the effect of alcohols becomes more complex, with short chain alcohols acting as inhibitors of, while long chain alcohols enhanced, phorbol-induced activity (Slater *et al.,* 1997).

2. Cytochromes P450 as Surrogates of the Site of Action of General Anesthetics

The cytochrome P450 system is a large family of homologous heme proteins that metabolize drugs as well as hydrocarbon compounds from dietary and endogenous sources. This family of enzymes is characterized by a broad substrate specificity for each member and a large number of related subtypes with partially overlapping but distinct substrate preferences. Thus, P450 monooxygenases are a conceptually attractive model of a protein site of action for general anesthetics in that the multiplicity of binding sites found throughout the family could accommodate the diversity of general anesthetic structures (LaBella *et al.,* 1997; LaBella and Queen, 1993). P450 monooxygenases are inhibited competitively by alcohols, and inhibitory potency is correlated directly with lipophilicity (estimated by the number of carbon atoms in each alcohol, Fig. 8a) (LaBella *et al.,* 1997). Furthermore, K_I for P450 inhibition shows a 1 : 1 correlation with EC_{50} (the concentration of alcohol for general anesthesia in 50% of a group of animals) for tadpoles (Fig. 8b) (LaBella *et al.,* 1997), which has led to the speculation that anesthesia may result from generalized, reversible inhibition of heme proteins (with selective sparing of heme proteins required for cell respiration) and the consequent changes in the levels of their second messenger products (LaBella *et al.,* 1997; LaBella and Queen, 1993).

The examples just cited reveal that anesthetics rapidly gain access to and interact with many cellular proteins beyond the channels currently thought to be their likely sites of action. This multiplicity of proteins affected by anesthetics should serve as a warning that anesthetics may not exert their effects at a single, dominant locus. Each of the systems described is limited in that they all fail to account for some behaviors of anesthetics in animals and several require high concentrations of anesthetic before an effect is seen. Nevertheless, each serves as a surrogate of the as yet to be discovered

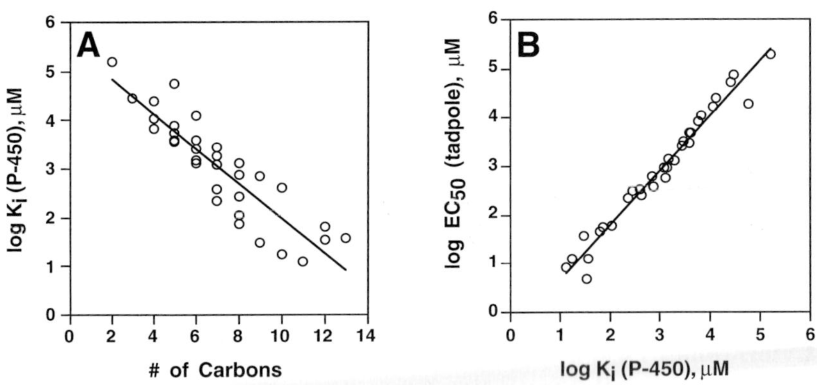

FIGURE 8 Correlation between the general anesthetic concentration of alkanols and those inhibiting cytochrome P450. (A) For each alkanol, the chain length is correlated with its binding constant, K_I, for cytochrome P450 inhibition. (B) Alkanol anesthetic potency in tadpoles is correlated with cytochrome P450 inhibition (data from LaBella *et al.,* 1997; LaBella and Queen, 1993).

physiological site(s) of action of general anesthetics that may be useful in advancing our knowledge of protein–anesthetic interactions.

IV. Future Directions

A. *Likely Collapse of Unitary Theories of Anesthesia*

1. How to Reconcile Meyer–Overton Behavior with a Unique Protein Binding Site?

General anesthetics encompass a huge range of dissimilar chemical structures, yet the Meyer–Overton rule (i.e., potency predicted almost purely by hydrophobicity) describes the behavior of the overwhelming majority of them. A seeming conundrum thus arises: membrane-based theories cannot explain how anesthesia occurs, but it is hard to imagine a protein site capable of accommodating such a diverse range of compounds.

Several characteristics of a potential single protein site of anesthesia can be predicted:

1. A moderately large and/or distensible binding site.
2. a. Lax structural specificity, implying primarily nonspecific attractive interactions and repulsive forces.
   b. Sufficient specificity to account for differences in potency among stereoisomers of intravenous general anesthetics.
3. Amphiphillicity of binding site to accommodate weakly polar anesthetic molecules.

A single site fulfilling predictions 1–3 could potentially accommodate all of the known anesthetics while excluding known nonanesthetics, but it is difficult to reconcile such a site with the relatively inflexible nature of protein structure (Richards, 1977). When considering large, multisubunit proteins, it is imaginable that binding between subunits or domains might provide a suitable site, but such interfaces are usually tightly packed and ordered, much the same as the protein interior (Chothia, 1976).

A conceptual alternative to the single-site model is to propose that there are multiple distinct but related binding sites on the same protein (or on a family of related proteins), all possessing to some degree the properties predicted in 1–3.

The immunoglobulins, P450 cytochromes, chemotaxis receptors, and olfactory receptors are all examples of proteins (or families of proteins) with broad substrate specificity. Invoking multiple sites on one (or multiple) protein(s) to account for the variety of compounds that produce anesthesia is a modification of the unified hypothesis of anesthesia. This modified hypothesis is unitary in the sense that all anesthetics are proposed to produce an identical end result (anesthesia) via more or less *similar* actions on one or more *protein* targets, mediated through multiple possible sites.

For example, even within the GABA receptor, different subunit compositions confer differing responses to anesthetics. Anesthetic efficacy and potency can be modulated by subunit composition, with responses to volatile anesthetics differing from responses to structurally distinct nonvolatile drugs (Harris *et al.,* 1995). This implies different loci for the actions of these two classes of anesthetics. Similarly, the direct and indirect actions of the anesthetic etomidate on the GABA receptor can be separated by mutation, implying that the two effects arise from different loci within the protein (Moody *et al.,* 1997). The effects of long and short chain alcohols on the nAcChoR (ion flux enhancement and channel blocking, respectively) appear to occur at distinct sites on that receptor as well (Wood *et al.,* 1991).

Given the multiplicity of effects on one example of ligand-gated channels and the number of such channels of this type found in the central nervous system that are sensitive to anesthetics *in vitro,* it is unlikely that only one receptor type is critical for the production of anesthesia. This argument can be extended to include the multiple classes of proteins beyond receptors found to be sensitive to anesthetics as well.

2. Demise of Unitary Theories of Anesthesia in Whole Animals

The discussion until now has focused on the molecular target of anesthetics, without addressing whether the clinical effects of anesthesia (a state characterized by unconsciousness and immobility) are all caused by interference with a single molecular target. The discovery of nonanesthetic (nonimmobilizing) hydrocarbons (Koblin *et al.,* 1994) has allowed new insights into whether these two effects occur by the same or different mechanisms in animals. Low (i.e., subanesthetic) concentrations of volatile anesthetics suppress learning in volunteers (Dwyer *et al.,* 1992), whereas higher concentrations produce all of the clinical manifestations of anesthesia. However, nonanesthetics [also termed nonimmobilizers (Eger *et al.,* 1997b)] can suppress learning in animals (at concentrations predicted to suppress learning by analogy to similar full anesthetic compounds) but are incapable of producing immobility (Kandel *et al.,* 1996). This result has been interpreted to mean that unconsciousness and immobility are two effects of anesthetics that occur at different sites, potentially by different mechanisms (Eger *et al.,* 1997b). Unconsciousness, seen as an extension of learning suppression, can be produced by completely apolar compounds, which is taken as evidence that the molecular target leading to unconsciousness is similarly apolar. In contrast, immobility can only be produced by amphiphillic compounds (the full anesthetics), which is interpreted to mean that the molecular target producing immobility has some polar character, implying that the two sites are distinct (Eger *et al.,* 1997b).

The notion that two distinct targets contribute to anesthesia in animals has been brought back to investigations of potential subcellular sites of anesthetic action, with interesting results. For example, structurally similar anesthetic and nonimmobilizing compounds have different effects on the kinetics of acetylcholine receptor desensitization, with the nonimmobilizers being much less effective (Raines, 1996). In experiments with the GABA receptor, nonimmobilizers fail to potentiate GABA induced chloride currents in *Xenopus* oocytes (Mihic *et al.,* 1994), whereas true anesthetics do. These same compounds had no membrane disordering effect, whereas true anesthetics did (Mihic *et al.,* 1994). Based on these results, immobility has been attributed to anesthetic action at the GABA receptor, whereas suppression of learning is attributed to an as yet unidentified site. Evidence suggests that, unlike desensitization, the open channel of the nAcChoR is inhibited by both anesthetic and nonimmobilizing compounds (Forman, 1998), suggesting an explanation for the observations described previously.

B. Potential Advances for Protein-Based Theories of General Anesthetic Action

1. Structural Information for Transmembrane Proteins

Because much attention has been focused on the interactions of anesthetics with ion channels, our knowledge of the functional effects of mutations and chemical modification of these interactions has outstripped our ability to relate these changes in behavior to changes in the protein structure. Structural studies of membrane proteins are notoriously difficult. They are large, requiring lipids to stabilize the folded state, which, in turn, makes it unlikely that they can be crystallized for diffraction analysis. Ion channels are multisubunit enzymes, which compounds the size problem, preventing elucidation of their solution structure via NMR, which is impossible because membrane fragments tumble too slowly to average nuclear anisotropy on the NMR time scale. Structural information available to date has been obtained from electron microscopy (at 9-Å resolution), chemical modification studies, and predictions based on sequence analysis. These data allow constraints to be placed on the locations of certain portions of the protein, but do not sustain a structure–function analysis of the type required to identify the modes of anesthetic action on a protein.

Structure function relationships are particularly difficult to deduce from static structures of allosteric proteins that can undergo large conformational shifts (Mattevi *et al.,* 1996). This certainly applies to ligand-gated ion channels, where ligand binding at an extracellular site tens of Ångstroms distant from the channel implies dramatic conformational shifts during normal cycling and where mutations are likely to affect conformational equilibria independently of their potential interactions with anesthetics.

Our understanding of anesthetic–protein interactions depends on a future improvement in imaging the proposed targets of action (i.e., receptors) and the current study of surrogate models (such as luciferase). Luciferase is itself a large enzyme, but it is soluble and binds a range of anesthetics. Structural studies of luciferase are in progress, with the objective of studying the critical anesthetic-binding site interactions at the atomic level, but the only conformation to be crystallized to date is the one with low affinity for general anesthetics. Developments in the area of membrane protein crystallography include a description of lipid cubic phases, in which a cubic lattice of phospholipids solubilizes, stabilizes, and promotes crystal formation of membrane proteins, feeding them into the matrix via lateral diffusion from a detergent micelle phase (Landau and Rosenbusch, 1996). In the prototypical system described, bacteriorhodopsin crystals were obtained and diffracted readily to 2-Å resolution.

2. Mutagenesis of Ligand-Gated Channels

Cloning of ligand-gated channels, followed by mutagenesis, expression, and characterization of the new receptors in suitable backgrounds, is a powerful tool that has already advanced our knowledge of anesthetic mechanisms while still in its infancy. It is reasonable to expect a torrent of new information about which positions in various ligand gated channels lead to new anesthetic responses when the wild-type residue is mutated. However, interpreting these new data without more structural information may prove frustrating.

Furthermore, the functional importance of the mutations characterized so far only *in vitro* may potentially be assessed in animals via transgenic knockouts. This possibility has already been alluded to for substitution mutants of the GABA receptor α subunit (Mihic *et al.*, 1997). Knockout mice in which the β_3 subunit has been deleted show complex changes in their response to some, but not all, anesthetics (Quinlan *et al.*, 1998). Mice deficient in the β_3 subunit were no different from wild-type mice in their response to obtundation by volatile anesthetics, but required higher anesthetic doses to blunt the response to painful stimuli. Sleep times in response to ethanol and pentobarbital were identical for wild-type and β_3-deleted mice, whereas β_3-deleted mice had shorter midazolam and etomidate sleep times than wild-type mice. Interpretation of these already confusing results is complicated by the fact that many of these knockout mice do not survive to adulthood, and those that do have severe craniofacial abnormalities

(Quinlan *et al.,* 1998). Although these studies are promising, it is unclear how useful it is to compare the behavior of mutant and wild-type mice in the face of such global physiological abnormalities. New techniques may overcome these difficulties, however.

Finally, protein kinase C knockout mice have demonstrated the role of that protein in modulating the response to pain (Malmberg *et al.,* 1997).

3. Rapid Photolabeling to Identify Residues Contacted by Inhibitors of the Open State of nAcChoR

Despite all that has been learned about anesthetic interactions with membranes and membrane proteins, we still lack the ability to rationally design new anesthetic drugs. To do so will require the identification of the "site" of anesthetic action and a detailed knowledge of the intermolecular interactions between anesthetics and their targets. Because diffraction and NMR structural data for membrane channels are limited, other techniques must be found to provide this information. Mutagenesis experiments that perturb channel function provide much useful information, but they do not establish the physical relationship between the mutated residue and the ligand. Photolabeling can identify the amino acids in the primary structure of the protein that interact with photoactivateable ligands.

Most of the photolabeling data available for the nAcChoR pertain to the resting or desensitized receptor (Changeux *et al.,* 1992; White and Cohen, 1992), whereas anesthetics interact significantly with the short-lived open channel state as well. The authors have embarked on a series of experiments designed to achieve time-resolved photolabeling of the nAcChoR. Photolabeling will be carried out under conditions identical to those used for kinetic analysis of receptor inhibition in rapid ion flux assays of receptor-enriched vesicles. Comparing the rate of photolabeling at a given site on the open channel with the rate of channel inhibition should allow the functional site to be located (Miller *et al.,* 1998).

Technical advances required to carry out these experiments have included the synthesis of radiolabeled alkanol analogs of the general formula CH_3–$(CH_3)_m$–CN_2–$(CH_2)_n$–OH, which can be photoactivated to form a covalent adduct, and the development of a rapid mixing, rapid-freeze clamp apparatus (Fig. 9). The latter consists of two pneumatically driven syringes, one containing receptor-loaded membranes and the other containing ligands. The two components are driven rapidly through a ball mixer (mixing time $<$500 msec) and then into a time delay loop of variable length to allow different interaction times. The mixture exits onto a spinning stainless-steel disc (that is constantly cooled by floating on a bath of liquid nitrogen) where it freezes in less than 1 msec. The frozen membrane/ligand complex can then be photoactivated for extended periods, enhancing the yield of

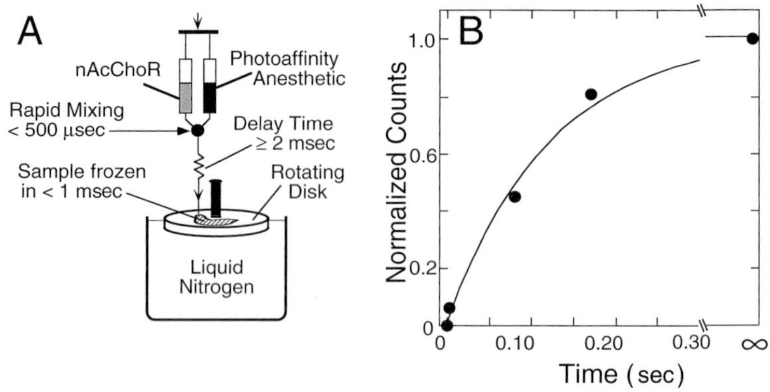

FIGURE 9 Time-resolved photolabeling of the acetylcholine receptor. (A) Illustration of the freeze-clamp apparatus. The contents of two pneumatically driven syringes (one containing native acetylcholine receptors and the other containing a labeled photolabile anesthetic) are pushed rapidly through a ball mixer (mixing time <500 msec) and into a variable length delay loop that allows various interaction times. The mixture is then ejected onto a rapidly rotating stainless-steel disk that has been precooled in liquid nitrogen. The freezing time has been determined to be less than 1 msec. The frozen mixture is then illuminated, collected, and subjected to SDS–PAGE for protein chemistry. (B) Illustration of time-resolved (freeze-clamped) labeling of the acetylcholine receptor (Miller *et al.,* 1998).

covalent adducts (Miller *et al.,* 1998). Protein fragmentation and peptide analysis techniques should ultimately allow the determination of the amino acids contacted by the photolabel and their position in the primary structure of any conformation of the receptor.

This new photolabeling technique may allow us to determine whether the photolabile general anesthetic approaches its channel inhibition site normal or parallel to the plane of the membrane (Fig. 10). If the approach is normal to the bilayer and the label gains access to the site by diffusing down the receptor channel, then residues deeper and deeper in the channel lining should be photolabeled as a function of time. Alternatively, a different set of residues will be identified if the label gains access to the site by diffusing through the membrane and past the more peripheral transmembrane helices before arriving at its target site.

V. Summary

Since Overton's monograph in 1899, much has been learned about the mechanism of anesthesia, but his seminal observation remains the touchstone for all research in the field. Any hypothesis about the mechanism of

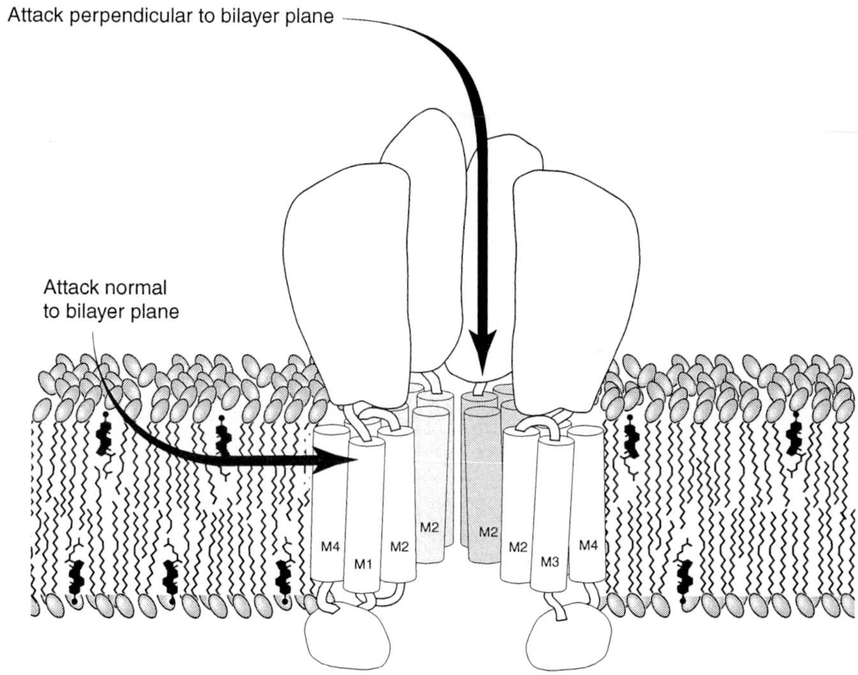

FIGURE 10 Time-resolved photolabeling of the acetylcholine receptor. If an anesthetic gains access to its binding site in the channel by diffusing down the aqueous lumen of the pore, then residues lining the channel should be labeled prior to the anesthetic arriving at its target site. Alternatively, if an anesthetic gains access to the same site through the membrane, then transmembrane helices must be labeled prior to the arrival of the anesthetic at its target site.

general anesthesia must account for the Meyer–Overton rule. Membrane-based theories account for this rule easily, but have had little success because the physical effects of anesthetics on membranes are so small that mechanisms by which they could affect protein function have yet to be established.

Proteins have specific binding sites for anesthetics, and their activities can be modified by anesthetic binding. However, most known general anesthetic-binding sites on proteins do not obey the Meyer–Overton rule. Efforts to reconcile a protein-based theory of anesthesia with the Meyer–Overton rule require abandoning or modifying unitary theories of general anesthetic action to posit multiple sites, and possibly multiple mechanisms, of action.

Ultimately, the quest to assign the site of action of anesthetics solely to the membrane or to proteins may be an oversimplification. It is easy to imagine a situation in which anesthetics initially dissolve in the membrane, then diffuse laterally to membrane protein targets. Evidence from work with the GABA receptor (Mihic *et al.,* 1997) suggests that anesthetic-binding sites occur between the protein–lipid interface and the pore of the channel. If the primary site of anesthetic action turns out to be at the lipid–protein interface, then the successful theory of anesthetic action will have to account for the behaviors of anesthetics in both materials.

Finally, it is appropriate to quote from the English translation of Overton's monograph (Overton, 1991):

> Non-specific narcotics . . . . can be transported into the lecithin and cholesterol-related components of cells and in this fashion alter their physical state . . . . in such a way that they themselves can no longer fulfill their normal functions within the cell, or they have a disturbing effect on the functions of other cell components.

References

Addona, G., Sanderman, H., Kloczewiak, M., Husain, S., and Miller, K. (1998). Where does cholesterol act during activation of the nicotinic acetylcholine receptor? *Biochim. Biophys. Acta* **1370,** 299–309.

Akaike, N., Hattori, K., Inomata, N., and Oomura, Y. (1985). γ-Aminobutyric-acid- and pentobarbitone-gated chloride currents in internally perfused frog sensory neurones. *J. Physiol.* (*London*) **360,** 367–386.

Akeson, M., and Deamer, D. (1989). Steady-state catecholamine distribution in chromaffin granule preparations: A test of the pump-leak hypothesis of general anesthesia. *Biochemistry* **28,** 5120–5127.

Alifimoff, J., Firestone, L., and Miller, K. (1987). Anesthetic potencies of secondary alcohol enantiomers. *Anesthesiology* **66,** 55–59.

Alifimoff, J., Firestone, L., and Miller, K. (1989). Anesthetic potencies of primary alkanols: Implications for the molecular dimensions of the anesthetic site. *Br. J. Pharmacol.* **96,** 9–16.

Antollini, S., Soto, M., Bonini de Romanelli, I., Gutierrez-Merino, C., Sotomayor, P. and Barrantes, F. (1996). Physical state of bulk and protein-associated lipid in nicotinic acetylcholine receptor-rich membrane studied by Laurdan generalized polarization and fluorescence energy transfer. *Biophys. J.* **70,** 1275–1284.

Ashton, D., and Wauquier, A. (1985). Modulation of a GABA-ergic inhibitory circuit in the *in vitro* hippocampus by etomidate isomers. *Anesth. Analg.* (*N.Y.*) **64,** 975–980.

Baber, J., Ellena, J., and Cafiso, D. (1995). Distribution of general anesthetics in phospholipid bilayers determined using ^{2}H NMR and ^{1}H-^{1}H NOE spectroscopy. *Biochemistry* **34,** 6533–6539.

Bangham, A., and Mason, W. (1980). Anesthetics may act by collapsing pH gradients. *Anesthesiology* **53,** 135–141.

Bangham, A., Standish, M., and Miller, N. (1965). Cation permeability of phospolipid model membranes: Effect of narcotics. *Nature* (*London*) **208,** 1295–1297.

Belelli, D., Lambert, J. J., Peters, J. A., Wafford, K., and Whiting, P.J. (1997). The interaction of the general anesthetic etomidate with the γ-aminobutyric acid type A receptor is influenced by a single amino acid. *Proc. Nat. Acad. Sci. U.S.A.* **94,** 11031–11036.

Bennett, P., and Simmonds, M. (1996). The influence of membrane cholesterol on the $GABA_A$ receptor. *Brit. J. Pharmacol.,* **117,** 87–92.

Bernard, C. (1875). "Lecons sur les Anesthetiques et sur l'Asphyxie." Bailliere, Paris.

Boggs, J., Yoong, T., and Hsia, J. (1976). Site and mechanism of anesthetic action: I. Effect of anesthetics and pressure on fluidity of spin-labeled lipid vesicles. *Mol. Pharmacol.* **12,** 127–135.

Bradley, D., and Richards, C. (1984). Temperature dependence of the action of nerve blocking agents and its relationship to membrane-buffer partition coefficients: Thermodynamic implications for the site of action of local anesthetics. *Br. J. Pharmacol.* **81,** 161–167.

Bradley, R., Sterz, R., and Peper, K. (1984). The effects of alcohols and diols at the nicotinic acetylcholine receptor of the neuromuscular junction. *Brain Res.* **295,** 101–112.

Browning, M., Bureau, M., Dudek, E., and Olsen, R. (1990). Protein kinase C and cAMPdependent protein kinase phosphorylate the β subunit of the purified γ-aminobutyric acid A receptor. *Proc. Nat. Acad. Sci. U.S.A.* **87,** 1315–1318.

Bull, M., Brailsford, J., and Bull, B. (1982). Erythrocyte membrane expansion due to the volatile anesthetics, the 1-alkanols, and benzyl alcohol. *Anesthesiology* **57,** 399–403.

Cafiso, D. (1995). Influence of charges and dipoles on macromolecular adsorption and permeability. *In* "Permeability and Stability of Lipid Bilayers." (E. Disalvo and S. Simon, eds.) pp. 179–195. CRC Press, Boca Raton, Florida.

Cantor, R. (1997). The lateral pressure profile in membranes: A physical mechanism of general anesthesia. *Biochemistry* **36,** 2339–2344.

Changeux, J.-P., Galzi, J.-L., Devillers-Thiery, A., and Bertrand, D. (1992). The functional architecture of the acetylcholine nicotinic receptor explored by affinity labelling and site-directed mutagenesis. *Q. Rev. Biophys.* **25,** 395–432.

Chothia, C. (1976). The nature of the accessible and buried surfaces in proteins. *J. Mol. Biol.* **105,** 1–14.

Colley, C., and Metcalfe, J. (1972). The localisation of small molecules in lipid bilayers. *FEBS Lett.* **24,** 241–246.

Dildy-Mayfield, J., Mihic, S., Liu, Y., Deitrich, R., and Harris, R. (1996). Actions of long chain alcohols on $GABA_A$ and glutamate receptors: Relation to *in vivo* effects. *Br. J. Pharmacol.* **118,** 378–384.

Dilger, J., and Vidal, A. (1994). Cooperative interactions between general anesthetics and QX-222 within the pore of the acetylcholine receptor ion channel. *Mol. Pharmacol.* **45,** 169–175.

Dodson, B., Braswell, L., and Miller, K. (1987). Barbiturates bind to an allosteric regulatory site on nicotinic acetylcholine receptor-rich membranes. *Mol. Pharmacol.* **32,** 119–126.

Dreger, M., Krauss, M., Herrmann, A., and Hucho, F. (1997). Interactions of the nicotinic acetylcholine receptor transmembrane segments with the lipid bilayer in native receptor-rich membranes. *Biochemistry* **36,** 839–847.

Dwyer, R., Bennett, H., Eger II, E., and Heilbron, D. (1992). Effects of isoflurane and nitrous oxide in subanesthetic concentrations on memory and responsiveness in volunteers. *Anesthesiology* **77,** 888–898.

Eger II, E., Saidman, L., and Brandstater, B. (1965). Temperature dependence of halothane and cyclopropane anesthesia in dogs: Correlation with some theories of anesthetic action. *Anesthesiology* **26,** 764–770.

Eger II, E. I., Koblin, D. D., Laster, M. J., Schurig, V., Juza, M., Ionescu, P., and Gong, D. (1997a). Minimum alveolar anesthetic concentration values for the enantiomers of isoflurane differ minimally. *Anesth. Analg.* (*N.Y.*) **85,** 188–92.

Eger II, E., Koblin, D., Harris, R., Kendig, J., Pohorille, A., Halsey, M., and Trudell, J. (1997b). Hypothesis: Inhaled anesthetics produce immobility and amnesia by different mechanisms at different sites. *Anesth. Analg.* (*N.Y.*) **84,** 915–918.

Firestone, L. L., Miller, J. C., and Miller, K. W. (1986). Tables of physical and pharmacological properties of anesthetics. *In* "Molecular and Cellular Mechanisms of Anesthetics" (S. H. Roth and K. W. Miller, (eds.), pp. 455–470. Plenum, New York.

Firestone, S., Ferguson, C., and Firestone, L. (1992). Isoflurane's optical isomers are equipotent in *Rana pipiens* tadpoles. *Anesthesiology* **77,** A758.

Firestone, S., Firestone, L., Ferguson, C., and Blanck, D. (1993). Staurosporine, a protein kinase C inhibitor, decreases the general anesthetic requirement in *Rana pipiens* tadpoles. *Anesth. Analg. (N.Y.)* **77,** 1026–1030.

Fong, T., and McNamee, M. (1986). Correlation between acetylcholine receptor function and structural properties of membranes. *Biochemistry* **25,** 830–840.

Forman, S. A. (1998). Direct interactions of anesthetics and nonanesthetics with the nicotinic acetylcholine receptor pore. *Toxicol. Lett.* **100,** 169–178.

Forman, S. A., Miller, K., and Yellen, G. (1995). A discrete site for general anesthetics on a postsynaptic receptor. *Mol. Pharmacol.* **48,** 547–581.

Franks, N., and Lieb, W. (1978). Where do general anesthetics act? *Nature (London)* **274,** 339–342.

Franks, N., and Lieb, W. (1984). Do general anaesthetics act by competitive binding to specific receptors? *Nature (London)* **310,** 599–601.

Franks, N., and Lieb, W. (1986). Partitioning of long-chain alcohols into lipid bilayers: Implications for mechanisms of general anesthesia. *Proc. Nat. Acad. Sci. U.S.A.* **83,** 5116–5120.

Franks, N., and Lieb, W. (1991). Stereospecific effects of inhalational general anesthetic optical isomers on nerve ion channels. *Science,* **254,** 427–430.

Goodman, D., Nemoto, E., Evans, R., and Winter, P. (1996). Anesthetics modulate phospholipase C hydrolysis of monolayer phospholipids by surface pressure. *Chem. Phys. of Lipids* **84,** 57–64.

Gruner, S., and Shyamsunder, E. (1991). Is the mechanism of general anesthesia related to lipid membrane spontaneous curvature? *Ann. N. Y. Acad. Sci.* **625,** 685–697.

Harris, B. D., Moody, E. J., Basile, A. S., and Skolnick, P. (1994). Volatile anesthetics bidirectionally and stereospecifically modulate ligand binding to GABA receptors. *Eur. J. Pharmacol.* **267,** 269–274.

Harris, B., Wong, G., Moody, E., and Skolnick, P. (1995). Different subunit requirements for volatile and nonvolatile anesthetics at γ-aminobutyric acid type A receptors. *Mol. Pharmacol.* **47,** 363–367.

Harrison, N., Kugler, J., Jones, M., Greenblatt, E., and Pritchett, D. (1993). Positive modulation of human γ-aminobutyric acid type A and glycine receptors by the inhalation anesthetic isoflurane. *Mol. Pharmacol.* **44,** 628–632.

Harvey, E. (1915). Effect of certain organic and inorganic substances upon light production by luminous bacteria. *Biol. Bull.* **29,** 308–311.

Huang, L.-Y. M., and Barker, J. L. (1980). Pentobarbital: Stereospecific actions of (+) and (−) isomers revealed on cultured mammalian neurons. *Science* **207,** 195–197.

Jacobs, R., and White, S. (1984). Behavior of hexane dissolved in dimyristoylphosphatidylcholine bilayers: An NMR and calorimetric study. *J. Am. Chem. Soc.* **106,** 915–920.

Jerusalinsky, D., Quillfeldt, J., Walz, R., Da Silva, R., Medina, J., and Izquierdo, I. (1994). Post-training intrahippocampal infusion of protein kiase C inhibitors causes amnesia in rats. *Behav. Neural Biol.* **61,** 107–109.

Johnson, F., and Flagler, E. (1950). Hydrostatic pressure reversal of narcosis in tadpoles. *Science* **112,** 91–92.

Johnson, S., and Bangham, A. (1969). The action of anesthetics on phospolipid membranes. *Biochim. Biophys. Acta* **193,** 92–104.

Johnson, S., Miller, K., and Bangham, A. (1973). The opposing effects of pressure and general anesthetics on the cation permeability of liposomes of varying lipid composition. *Biochim. Biophys. Acta* **307,** 42–57.

Johnston, G. (1996). $GABA_C$ receptors: Relatively simple transmitter gated ion channels? *Trends Pharmacol. Sci.* **17,** 319–323.

Jones, O., and McNamee, M. (1988). Annular and nonannular binding sites for cholesterol associated with the nicotinic acetylcholine receptor. *Biochemistry* **27,** 2364–2374.

Kamaya, H., Ueda, I., Moore, P., and Eyring, H. (1979). Antagonism between high pressure and anesthetics in the thermal phase-transition of dipalmitoyl phosphatidylcholine bilayer. *Biochim. Biophys. Acta* **550,** 131–137.

Kandel, L., Chortkoff, B., Sonner, J., Laster, M., and Eger II, E. (1996). Nonanesthetics can suppress learning. *Anesth. Analg.* (*N.Y.*) **82,** 321–326.

Kita, Y., Bennett, L., and Miller, K. (1981). The partial molar volumes of anesthetics in lipid bilayers. *Biochim. Biophys. Acta* **647,** 130–139.

Koblin, D., Chortkoff, B., Laster, M., Eger II, E., Halsey, M., and Ionescu, P. (1994). Polyhalogenated and perfluorinated compounds that disobey the Meyer–Overton hypothesis. *Anesth. Analg.* (*N.Y.*) **79,** 1043–1048.

LaBella, F., and Queen, G. (1993). General anesthetics inhibit cytochrome P450 monooxygenases and arachidonic acid metabolism. *Can. J. Physiol. Pharmacol.* **71,** 48–53.

LaBella, F., Chen, Q.-M., Stein, D., and Queen, G. (1997). The site of general anaesthesia and cytochrome P450 oxygenases: Similarities defined by straight chain and cyclic alcohols. *Br. J. Pharmacol.* **120,** 1158–1164.

Landau, E., and Rosenbusch, J. (1996). Lipidic cubic phases: A novel concept for the crystallization of membrane proteins. *Proc. Nat. Acad. Sci. U.S.A.* **93,** 14532–14535.

Lawrence, D., and Gill, E. (1975). The effects of Δ^1-tetrahydrocannabinol and other cannabinoids on spin-labeled liposomes and their relationship to mechanisms of general anesthesia. *Mol. Pharmacol.* **11,** 595–602.

Leonard, R., Charnet, P., Labarca, C., Vogelaar, N., Czyzyk, L., Gouin, A., Davidson, N., and Lester, H. (1991). Reverse pharmacology of the nicotinic acetylcholine receptor: Mapping the local anesthetic binding site. *Ann. N. Y. Acad. Sci.* **625,** 588–599.

Lever, M., Miller, K., Paton, W., and Smith, E. (1971). Pressure reversal of anaesthesia. *Nature* (*London*) **231,** 368–371.

Lieb, W., Kovalycsik, M., and Mendelsohn, R. (1982). Do clinical levels of general anaesthetics affect lipid bilayers? Evidence from Raman scattering. *Biochim. Biophys. Acta* **688,** 388–398.

Lingle, C. J., Maconochie, D., and Steinbach, J. H. (1992). Activation of skeletal muscle nicotinic acetylcholine receptors. [Review]. *J. Membr. Biol.* **126,** 195–217.

Liu, J., Laster, M., Taheri, S., Eger II, E., Koblin, D., and Halsey, M. (1993). Is there a cutoff in anesthetic potency for the normal alkanes? *Anesth. Analg.* (*N.Y.*) 12–18.

Liu, J., Laster, M., Koblin, D., Eger II, E., Halsey, M., Tahere, S., and Chortkof, B. (1994). A cutoff in potency exists in the perfluoroalkanes. *Anesth. Analg.* (*N.Y.*) **79,** 238–244.

Lundbaek, J., Maer, A., and Anderson, O. (1997). Lipid bilayer electrostatic energy, curvature stress, and the assembly of gramicidin channels. *Biochemistry* **36,** 5695–5701.

Lysko, G., Robinson, J., Casto, R., and Ferrone, R. (1994). The stereopecific effects of isoflurane isomers *in vivo*. *Eur. J. Pharmacol.* **263,** 25–29.

Malmberg, A., Chen, C., Tonegawa, S., and Basbaum, A. (1997). Preserved acute pain and reduced neuropatic pain in mice lacking PKCγ. *Science* **278,** 279–283.

Marsh, D., and Barrantes, F. (1978). Immobilized lipid in acetylcholine receptor-rich membranes from *Torpedo marmorata*. *Proc. Nat. Acad. Sci. U.S.A.* **75,** 4329–4333.

Mattevi, A., Rizzi, M., and Bolognesi, M. (1996). New structures of allosteric proteins revealing remarkable conformational changes. [Review] [35 refs.]. *Curr. Opin. Struct. Biol.* **6,** 824–829.

Melchior, D., Scavitto, F., and Steim, J. (1980). Dilatometry of diplamitoyllecithin-cholesterol bilayers. *Biochemistry* **19,** 4828–4834.

Metcalfe, J., Seeman, P., and Burgen, A. (1968). The proton relaxation of benzyl alcohol in erythrocyte membranes. *Mol. Pharmacol.* **4,** 87–95.

Meyer, H. H. (1901). Zur Theorie der Alkoholnarkose. Der Einfluss wechselnder Temperatur auf Wirkungsstärke und Theilungscoefficient der Narcotica. *Arch. Exp. Pathol. Pharmakol.* **42,** 338–346.

Mihic, S., McQuilkin, S., Eger II, E., Ionescu, P., and Harris, R. (1994). Potentiation of γ-aminobutyric acid type A receptor mediated chloride currents by novel halogenated compounds correlates with their abilities to induce general anesthesia. *Mol. Pharmacol.* **46,** 851–857.

Mihic, S., Ye, Q., Wick, M., Koltchine, V., Krasowski, M., Finn, S., Mascia, M., Valenzuela, C., Hanson, K., Greenblatt, E., Harris, R., and Harrison, N. (1997). Sites of alcohol and volatile anaesthetic action on $GABA_A$ and glycine receptors. *Nature, (London)* **389,** 385–389.

Miller, K., Paton, W., Smith, R., and Smith, E. (1973). The pressure reversal of general anesthesia and the critical volume hypothesis. *Mol. Pharmacol.* **9,** 131–143.

Miller, K., Sauter, J., and Braswell, L. (1982). A stereoselective pentobarbital binding site in cholinergic membranes from *Torpedo californica. Biochem. Biophys. Res. Commun.* **105,** 659–666.

Miller, K., Firestone, L., Alifimoff, J., and Streicher, P. (1989). Nonanesthetic alcohols dissolve in synaptic membranes without perturbing their lipids. *Proc. Nat. Acad. Sci. U.S.A* **86,** 1084–1087.

Miller, K., Addona, G., and Kloczewiak, M. (1998). Approaches to proving there are general anesthetic sites of action on ligand gated channels. *Toxicol. Lett.* **100,** 139–147.

Moody, E., Harris, B., and Skolnick, P. (1993). Stereospecific actions of the inhalation anesthetic isoflurane at the $GABA_A$ receptor complex. *Brain Res.* **615,** 101–106.

Moody, E., Knauer, C., Granja, R., Strakhova, M., and Skolnick, P. (1997). Distinct loci mediate the direct and indirect actions of the anesthetic etomidate at $GABA_A$ receptors. *J. Neurochem.* **69,** 1310–1313.

Moss, G., Curry, S., Franks, N., and Lieb, W. (1991). Mapping the polarity profiles of general anesthetic target sites using n-alkane-(α,ω)-diols. *Biochemistry* **30,** 10551–10557.

Mountcastle, D., Biltonen, R., and Halsey, M. (1978). Effect of anesthetics and pressure on the thermotropic behavior of multilamellar dipalmitoylphosphatidylcholine liposomes. *Proc. Nat. Acad. Sci. U.S.A.* **75,** 4906–4910.

Nakahiro, M., Arakawa, O., and Narahashi, T. (1991). Modulation of γ-aminobutyric acid receptor-channel complex by alcohols. *J. Pharmacol. Exp. Ther.* **259,** 235–240.

Navarro, J., Toivio-Kinnucan, M., and Racker, E. (1984). Effect of lipid composition on the calcium/adenosine 5′-triphosphate coupling ratio of the Ca^{+2}-ATPase of sarcoplasmic reticulum. *Biochemistry* **23,** 130–135.

Nayeem, N., and Green, T. P. (1994). Quaternary structure of the native $GABA_A$ receptor determined by eloctron microscopic image analysis. *J. Neurochem.* **62,** 815–818.

North, C., and Cafiso, D. S. (1997). Contrasting membrane localization and behavior of halogenated cyclobutanes that follow or violate the Meyer-Overton hypothesis of general anesthetic potency. *Biophys. J.* **72,** 1754–1761.

Olsen, R. W., Fischer, J. B., and Dunwiddie, T. V. (1986). Barbiturate enhancement of γ-aminobutyric acid receptor binding and function as a mechanism of anesthesia. *In* "Molecular and Cellular Mechanisms of Anesthetics" (S. H. Roth and K. W. Miller, eds.), pp. 165–178. Plenum, New York.

Overton, C. E. (1991). Studies of narcosis and a contribution to general pharmacology. *In* "Studies of Narcosis" (R. L. Lipnick, ed.), pp. 1–203. Chapman and Hall and the Wood Library-Museum of Anesthesiology, London.

Pang, K.-Y., Braswell, L., Chang, L., Sommer, T., and Miller, K. (1980). The perturbation of lipid bilayers by general anesthetics: A quantitative test of the disordered lipid hypothesis. *Mol. Pharmacol.* **18,** 84–90.

Parker, I., Gundersen, C., and Miledi, R. (1986). Actions of pentobarbital on rat brain receptors expressed in *Xenopus* oocytes. *J. Neurosci.* **6,** 2290–2297.

Pope, J., Walker, L., and Dubro, D. (1984). On the ordering of *N*-alkane and *N*-alcohol solutes in phospholipid bilayer model membrane systems. *Chem. Phys. Lipids* **35,** 259–277.

Pringle, M., Brown, K., and Miller, K. (1981). Can the lipid theories of anesthesia account for the cutoff in anesthetic potency in homologous series of alcohols? *Mol. Pharmacol.* **19,** 49–55.

Qin, Z., Szabo, G., and Cafiso, D. (1995). Anesthetics reduce the magnitude of the membrane dipole potential. Measurements in lipid vesicles using voltage sensitive spin probes. *Biochemistry* **34,** 5536–5543.

Quinlan, J., Homanics, G., and Firestone, L. (1998). Anesthesia sensitivity in mice that lack the $\beta 3$ subunit of the γ-aminobutyric acid type A receptor. *Anesthesiology* **88,** 775–780.

Raines, D. (1996). Anesthetic and nonanesthetic halogenated volatile compounds have dissimilar activities on nicotinic acetycholine receptor desensitization kinetics. *Anesthesiology* **84,** 663–671.

Raines, D. E., and Cafiso, D. S. (1989). The enhancement of proton/hydroxyl flow across lipid vesicles by inhalation anesthetics. *Anesthesiology* **70,** 57–63.

Raines, D., and Miller, K. (1993). The role of charge in lipid selectivity for the nicotinic acetylcholine receptor. *Biophys. J.* **64,** 632–641.

Raines, D. E., Korten, S. E., Hill, A. G., and Miller, K. W. (1993). Anesthetic cutoff in cycloalkanemethanols. A test of current theories. *Anesthesiology* **78,** 918–927.

Rand, R., and Pangborn, W. (1973). A structural transition in egg lecithin-cholesterol bilayers at 12°C. *Biochim. Biophys. Acta* **318,** 299–305.

Rankin, S., Raines, D., and Miller, K. (1994). Isoflurane's actions on the nicotinic acetylcholine receptor are dependent upon the composition of the lipid bilayer. *Anesthesiology* **81,** A884.

Rankin, S., Addona, G., Kloczewiak, M., Bugge, B., and Miller, K. (1997). The cholesterol dependence of activation and fast desensitization of the nicotinic acetylcholine receptor. *Biophys. J.* **73,** 2446–2455.

Reyes, J., and Latorre, R. (1979). Effect of the anesthetics benzyl alcohol and chloroform on bilayers made from monolayers. *Biophys. J.* **28,** 259–280.

Richards, F. M. (1977). Areas, volumes, packing, and protein structure. *Annu. Rev. Biophys. Bioeng.* **6,** 151–176.

Roth, S. H., Forman, S. A., Braswell, L. M., and Miller, K. W. (1989). Actions of pentobarbital enantiomers on nicotinic cholinergic receptors. *Mol. Pharmacol.* **36,** 874–880.

Seeman, P. (1969). Temperature dependence of erythrocyte membrane expansion by alcohol anesthetics. Possible support for the partition theory of anesthesia. *Biochim. Biophys. Acta* **183,** 520–529.

Shapira, R., Silberberg, S., Ginsburg, S., and Rahamimoff, R. (1987). Activation of protein kinase C augments evoked transmitter release. *Nature, (London)* **325,** 58–60.

Sigel, E., Baur, R., Trube, G., Mohler, H., and Malherbe, P. (1990). The effect of subunit composition of rat brain $GABA_A$ receptors and channel function. *Neuron* **5,** 703–711.

Slater, S., Cox, K., Lombardi, J., Ho, C., Kelly, M., Rubin, E., and Stubbs, C. (1993). Inhibition of protein kinase C by alcohols and anaesthetics. *Nature, (London)* **364,** 82–84.

Slater, S., Kelly, M., Taddeo, F., Ho, C., Rubin, E., and Stubbs, C. (1994). The modulation of protein kinase C activity by membrane lipid bilayer structure. *J. Biol. Chem.* **269,** 4866–4871.

Slater, S., Kelly, M., Larkin, J., Ho, C., Mazurek, A., Taddeo, F., Yeager, M., and Stubbs, C. (1997). Interaction of alcohols and anesthetics with protein kinase C α. *J. Biol. Chem.* **272,** 6167–6173.

Steffey, E., and Eger II, E. (1974). Hyperthermia and halothane MAC in the dog. *Anesthesiology* **41,** 392–396.

Sunshine, C., and McNamee, M. (1992). Lipid modulation of nicotinic acetylcholine receptor function: The role of neutral and negatively charged lipids. *Biochim. Biophys. Acta* **1108,** 240–246.

Tang, P., Yan, B., and Xu, Y. (1997). Different distribution of fluorinated anesthetics and nonanesthetics in model membrane: A 19F NMR study. *Biophys. J.* **72,** 1676–1682.

Tomlin, S., Jenkins, A., Lied, W., and Franks, N. (1998). Stereoselective effects of etomidate optical isomers on γ-aminobutyric acid type A receptors and animals. *Anesthesiology* **88,** 708–717.

Turner, G., and Oldfield, E. (1979). Effect of a local anesthetic on hydrocarbon chain order in membranes. *Nature (London)* **277,** 669–670.

Ueda, I., Shieh, D., and Eyring, H. (1974). Anesthetic interaction with a model cell membrane: Expansion, phase transition, and melting of the lecithin monolayer. *Anesthesiology* **41,** 217–225.

Unwin, N. (1995). Acetylcholine receptor channel imaged in the open state. *Nature (London)* **373,** 37–43.

West, J., Numann, R., Murphy, B., Scheuer, T., and Catterall, W. (1991). A phosphorylation site in the Na^+ channel required for modulation by protein kinase C. *Science* **254,** 866–868.

White, B., and Cohen, J. (1992). Agonist-induced changes in the structure of the acetylcholine receptor M2 regions revealed by photoincorporation of an uncharged nicotinic noncompetitive antagonist. *J. Biol. Chem.* **267,** 15770–15783.

Wood, S., Forman, S., and Miller, K. (1991). Short chain and long chain alkanols have different sites of action on nicotinic acetylcholine receptor channels from *Torpedo. Mol. Pharmacol.* **39,** 332–338.

Wood, S., Hill, W., and Miller, K. (1993). Cycloalkanemethanols discriminate between volume- and length-dependent loss of activity of alcohols at the *Torpedo* nicotinic acetylcholine receptor. *Mol. Pharmacol.* **44,** 1219–1226.

Wood, S., Tonner, P., deArmendi, A., Bugge, B., and Miller, K. (1995). Channel inhibition by alkanols occurs at a binding site on the nicotinic acetylcholine receptor. *Mol. Pharmacol.* **47,** 121–130.

Wu, G., Tonner, P., and Miller, K. (1993). Ethanol stabilizes the open channel state of the *Torpedo* nicotinic acetylcholine receptor. *Mol. Pharmacol.* **45,** 102–108.

Xu, M., and Akabas, M. (1996). Identification of channel lining residues in the M2 membrane-spanning segment of the $GABA_A$ receptor α_1 subunit. *J. Gen. Physiol.* **107,** 195–205.

Yeh, J., Quandt, F., Tanguy, J., Nakahiro, M., Narahashi, T., and Brunner, E. (1991). General anesthetic action on γ-aminobutyric acid-activated channels. *Ann. N. Y. Acad. of Sci.* **625,** 155–173.

Zhang, G., Kazanietz, M., Blumberg, P., and Hurley, J. (1995). Crystal structure of the cys2 activator-binding domain of protein kinase C gamma in complex with phorbol ester. *Cell (Cambridge, Mass.)* **81,** 917–924.

CHAPTER 11

Plasma Membrane-Localized Signal Transduction

Stephen M. Robbins*,† and Morley D. Hollenberg*,‡,§,||
*Cancer Biology, †Immunology, ‡Endocrine, and §Smooth Muscle Research Groups, Departments of *Oncology, ‡Pharmacology Therapeutics, †Biochemistry and Molecular Biology and ||Medicine, The University of Calgary, Faculty of Medicine, Calgary Canada, T2N 4N1

I. INTRODUCTION

Stemming from the work of Ehrlich(1908) and Langley (1906), the essential concept of a "receptor" that governs the selective and tissue-specific action of a variety of drugs, neurotransmitters, and hormones was defined clearly at the turn of the last century. The concept has weathered the test

1063-5823/99 $30.00

of time well, and one could argue that, since the studies of Ehrilich and Langley, only the fine details have been added by recent work that has delineated many of the receptor structures and signaling pathways that will be outlined in this chapter. For instance, to act on a target cell such as the adipocyte, insulin must first be bound in a highly specific manner by its receptor (Cuatrecasas, 1969; Freychet *et al.,* 1971). The binding of insulin to its receptor must then be translated across the cell membrane into an amplified cellular signal that ultimately leads to the well-described effects of insulin on its target tissues. It was, in fact, research on the action of insulin in the late 1950s (e.g., Zierler, 1960) that indicated clearly that the insulin receptor was probably situated in the plasma membrane rather than in the cytoplasm, and it was the imaginative use of insulin–agarose derivatives, both for adipocyte signaling experiments (Cuatrecasas, 1969) and for receptor isolation, using affinity chromatography (Cuatrecasas, 1972), that set the stage for the solubilization, characterization, isolation, and cloning of a wide variety of plasma membrane-localized receptors. For purposes of this chapter, we will focus only on receptors that are situated in the plasma membrane, and the reader is thus referred elsewhere (Katzenellenbogen *et al.,* 1996; Forman and Evans, 1995) for a discussion of receptors, such as those for steroid hormones, that are situated in the cytoplasm or nucleus.

Pharmacologically, this chapter, will restrict the term "receptor" to designate a macromolecule that performs two closely related, but quite distinct, functions. *First,* a receptor must bind its specific ligand with (commonly) high affinity (e.g., micromolar to picomolar $K_D$values) and stereospecific chemical specificity. *Second,* a receptor must respond to the binding of its specific ligand by a conformational change that ultimately results in an intracellular signal. It is this *dual* recognition-signaling property that distinguishes pharmacologic receptors from other membrane constituents, such as nutrient transporters or "tissue-targeting" proteins that may be involved, for example, only in cell localization, but not in transmembrane signaling. For "pharmacologic receptors," the signal-triggering conformational change can be caused by "agonists" but not by pure antagonists, which can bind to receptors with high affinity without activating the system. Such selective receptor antagonists have played key roles in defining receptor families, such as the β-adrenoreceptor system. Our focus in the sections that follow will be on the questions: (1) How are specific membrane receptors "activated" by their cognate ligands? (2) How can the activation of an individual receptor by a single ligand, such as insulin, lead to a complex constellation of cellular responses that may vary considerably from one cell type to another. (3) How might the membrane microenvironment in which the receptor resides play a role in the transmembrane signal transduction process? The focus of this chapter will be on signaling events that take place at or within the plasma membrane. Intracellular signaling pathways that regulate cytoplasmic enzymes directly or that result in nuclear transcriptional events will not be dealt with in any detail.

II. MEMBRANE RECEPTOR-MEDIATED SIGNALING: BIOCHEMICAL MECHANISMS

A. *Three Basic Paradigms for Membrane Receptor-Mediated Signaling*

For plasma membrane receptors, the general paradigms that lead to the initiation of a cellular signal can be seen to be few in number, as outlined in Fig. 1 and discussed in further detail elsewhere (Severson and Hollenberg, 1997): (A) the receptor may comprise a ligand-gated ion channel (R_I, Fig. 1) such as the nicotinic receptor for acetylcholine (receptor No. 1, Fig. 1;

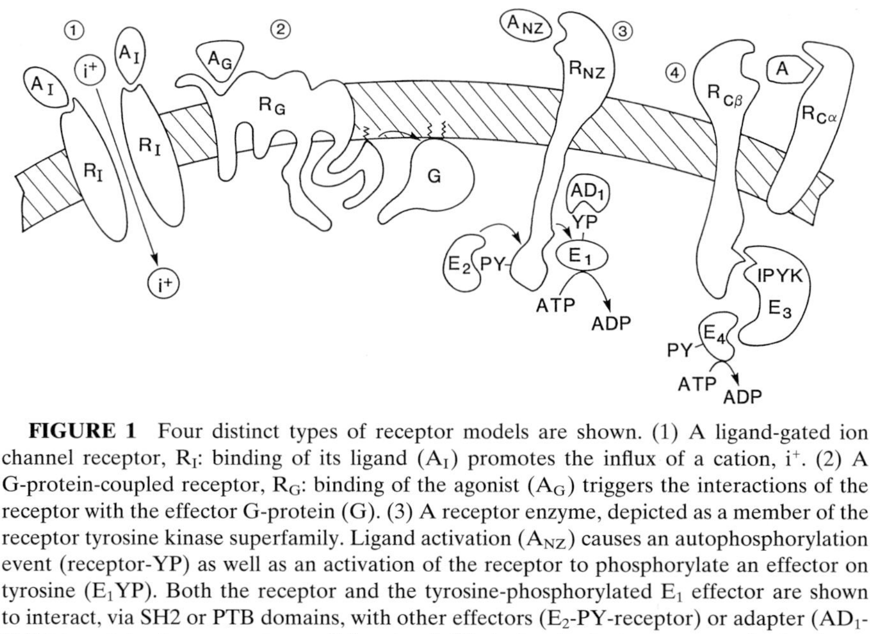

FIGURE 1 Four distinct types of receptor models are shown. (1) A ligand-gated ion channel receptor, R_I: binding of its ligand (A_I) promotes the influx of a cation, i^+. (2) A G-protein-coupled receptor, R_G: binding of the agonist (A_G) triggers the interactions of the receptor with the effector G-protein (G). (3) A receptor enzyme, depicted as a member of the receptor tyrosine kinase superfamily. Ligand activation (A_{NZ}) causes an autophosphorylation event (receptor-YP) as well as an activation of the receptor to phosphorylate an effector on tyrosine (E_1YP). Both the receptor and the tyrosine-phosphorylated E_1 effector are shown to interact, via SH2 or PTB domains, with other effectors (E_2-PY-receptor) or adapter (AD_1-PY-E_1) proteins to propagate a cellular signal. (4) A signal adapter receptor, depicted as a two-chain cytokine receptor [with a high-affinity-binding α subunit ($R_{C\alpha}$) and a cooperating signaling β subunit ($R_{C\beta}$)], that, via a cytoplasmic domain, interacts with and activates an intracellular tyrosine kinase effector (IPYK/E_3). The activated kinase tyrosine phosphorylates an effector (E_4) involved in signal propagation. None of the interacting species (receptor, effectors, G-proteins, etc.) are drawn to scale.

Conti-Tronconi *et al.,* 1994); (B) the receptor (R_{NZ}, Fig. 1) may be a ligand-regulated transmembrane enzyme, such as the tyrosine kinase receptors for platelet-derived growth factor (PDGF) or epidermal growth factor-urogastrone (EGF-URO) (receptor No. 3, Fig. 1: Schlessinger and Ullrich, 1992; Fantl *et al.,* 1993; Heldin, 1996); and (C) the receptor (R_G and R_C, Fig. 1) may not of itself possess either intrinsic enzymatic or transporter activity, but may upon being activated by its ligand interact specifically and directly with other membrane-localized signal adaptor molecules or effectors (E_1, E_2...Fig. 1). To date, this third type of paradigm is best represented by the family of G-protein-coupled receptors, like the ones for angiotensin II and adrenaline (receptor No. 2, Fig. 1). At the present time, well over 300 of these putative seven-transmembrane-spanning (so-called, serpentine) G-protein receptors have been cloned and sequenced (Strosberg, 1991; Savarese and Fraser, 1992; Guderman and Numberg 1995; Watson and Arkinstall, 1994). Because of the importance of G-protein-coupled receptors in regulating a wide variety of cell types, the signal transduction mechanisms for the "serpentine" receptors will be dealt with in some detail (see Section II,E). Apart from G-protein-coupled receptors, the third type of receptor paradigm may also be seen to apply to a variety of other receptor families that, like the seven-transmembrane "serpentine" receptors, do not possess intrinsic enzymatic activity. Paradigm C is also represented by the cytokine family of receptors (receptor No. 4, R_C, Fig. 1), such as the ones for growth hormone, interleukin 6 (IL-6), or granulocyte–macrophage colony-stimulating factor (GM-CSF). These receptors can be seen to act via the agonist-mediated recruitment of effectors (e.g., tyrosine kinases, such as the so-called JAK kinases, e.g., effector E_3, Fig. 1) into a signaling complex, wherein tyrosine phosphorylation of a variety of substrates plays a prominent role. In the case of the cytokine receptor family, a nonsignaling ligand-binding subunit ($R_{C\alpha}$, Fig. 1) has been found to interact with and modulate a signaling subunit ($R_{C\beta}$, Fig. 1); the ligand (A_C, Fig. 1) interacts with both subunits in the course of the signaling process. This recruitment, via the third receptor paradigm, of effector enzymes into a cell surface signaling complex can also be seen to play a role in the mechanism of action of a variety of other receptor systems, such as those receptor systems responsible for the activation of T and B lymphocytes (Wange and Samelson, 1996; Gold and DeFranco, 1994) and those receptors that are tethered to the cell surface via a glycosylphosphoinositide (GPI) linkage [e.g., ciliary neurotrophic factor (CNTF); Hirano *et al.,* 1997]. In essence, receptor paradigms B and C can both be seen as a variant of the "mobile" or "floating" receptor model of agonist action, as developed some time ago (Boeynaems and Dumont, 1997; De Häen, 1976; Cuatrecasas and Hollenberg, 1976). Because of the relevance of this conceptual framework for understanding receptor function, the model will be discussed briefly in the following section.

B. Receptor Mobility and Transmembrane Signaling

Two seminal sets of observations related to hormone action, along with independent measurements of the microaggregation and diffusion of fluorescently labeled receptors, have singled out receptor mobility and microaggregation as a key event for receptor-mediated signaling. It became evident early on in one set of studies of hormone action that a variety of specific hormone receptors (e.g., for adrenaline and glucagon) could activate independently what appeared to be a common membrane-associated adenylyl cyclase to yield cyclic AMP. Concurrently, in a second set of studies bearing on hormone action it was observed that insulin action could be mimicked by divalent but not by monovalent antireceptor antibodies (Kahn *et al.*, 1978, 1981), as well as by polyvalent plant lectins such as concanavalin A. At the same time, in the early 1970s, it became possible to monitor the rapid intermixing of cell surface proteins (Frye and Edidin, 1970) and to observe, with a fluorescence photobleaching recovery approach, the diffusion and microclustering of receptors such as the one for EGF-URO (Schlessinger *et al.*, 1978). Taken together, such data led to the birth of the "floating" or "mobile" receptor model of hormone action (dealt with at length elsewhere: De Häen, 1976; Boeynaems and Dumont, 1977; Cuatrecasas and Hollenberg, 1976), as depicted in Fig. 2. This model led to the hypothesis that receptor dimerization (or multimerization) was an essential event in the activation of growth factor receptors (Schlessinger and Ullrich,

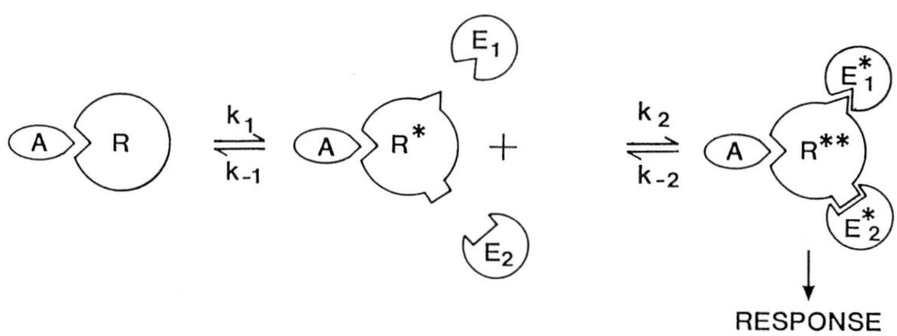

FIGURE 2 The mobile receptor model of hormone action. A single agonist (A) is shown to interact reversibly with an individual receptor, causing a conformational change that generates two distinct docking sites for distinct effector constituents. The binding of the two effectors (E_1 and E_2), which *need not* be mutually exclusive, as shown, then causes distinct cellular responses.

1992). Data implicating receptor dimerization as an essential process for signaling have also been obtained for the activation of G-protein-coupled receptors, akin to those present in ovarian cells and other tissues (Gregory *et al.,* 1982; Hazum and Keinan, 1985). The essential tenet of this model of receptor activation is that ligand binding initiates a number of membrane-localized receptor–effector interactions within the plane of the membrane. These interactions may promote not only the initial signaling event [usually a rapid ($<$ 1 sec) receptor microclustering process] but also the subsquent more delayed (seconds to tens of seconds) receptor patching and internatization process. Rapid signal intitiation is thought to be governed principally by the microaggregation/microclustering process; the much slower events of receptor patching and internalization (e.g., via the so-called "coated pits") are thought to be involved in either terminating the receptor signal or conveying the receptor physically to an intracellular site (e.g., the nucleus) for a "delayed" signal process (e.g., transcriptional regulation).

During the course of these mobile reactions that a ligand-occupied receptor may undergo, there is, in principle, no limit to the number of membrane effector moieties that might be activated, so as to generate multiple transmembrane signals from "single" ligand–receptor interactions (e.g., via effectors E_1 and E_2, Fig. 2). Based on this model, it would come as no surprise that a receptor like the one for angiotensin II could both activate a phospholipase and, via another pathway, inhibit adenylyl cyclase (see later). The receptor domains that may be involved in (1) ligand binding and (2) receptor–effector interactions (e.g., via so-called SH2 domains, see later) can each be thought of as related but independent sites on the receptor. Implications of this model for thinking about receptor function are at lease twofold: First, the model points to two types of domains on the receptor (ligand-binding domain and receptor–effector domain) that may be targets for drug development. Second, the model suggests that even in the absence of a receptor's specific ligand, receptor clustering (e.g., via antireceptor antibodies, polyvalent lectin-like molecules, or interactions with the extracellular matrix) could potentially modulate receptor function in a "hormone-like" fashion.

C. Signaling Modules and Receptor Function

For both G-protein-coupled receptors (Section II,E) and receptors that act via tyrosine kinase pathways (Section II,F) a number of prinicipal "modules" have been identified that play essential roles in the transmembrane signaling process. The involvement of signaling in these initial "adaptor" or "effector" modules can be seen as a variation of the mobile receptor paradigm. For agonists such as adrenaline that act via seven transmembrane "serpentine" receptors (receptor No. 2, Fig. 1), the discovery of the intermediate guanine nucleotide-dependent coupling protein (G-protein) responsible for the activation of adenylyl cyclase represented the key breakthrough (Rodbell, 1995; Gilman, 1995) in understanding hormone action. This insight foreshadowed the discovery of a host of guanine nucleotide-binding regulatory proteins that mediate the activation of a variety of effectors (adenylyl cyclase, phospholipases, ion channels) by multiple G-protein-coupled receptors, as outlined in Section II,E. In the case of the "serpentine" receptor family, the cardinal adaptor moiety (the G-protein) possesses intrinsic enzymatic activity (GTPase) that governs receptor–effector coupling. However, for those receptor signaling systems that employ tyrosine kinase activity (either intrinsic to the receptor per se or via enzymes that are either precoupled with or recruited to the receptor, e.g., IPYK, E_3 in Fig. 1), it has been discovered that a number of essential docking proteins, that of themselves are devoid of intrinsic enzymatic activity, play pivotal roles in the cell activation process. In the case of the receptor for EGF-URO, the docking proteins, GRB-2 (growth factor receptor-binding protein 2) and m-SOS (mammalian son-of-sevenless), have been found to act as essential docking intermediaries between the tyrosine-phosphorylated EGF-URO receptor (autophosphorylated in the presence of the ligand, EGF-URO) and the membrane-localized small G-protein, Ras (originally identified as a rat sarcoma oncogene). It is striking that the "serpentine" receptor systems and those receptors acting via tyrosine kinase pathways both employ an essential but distinct guanine nucleotide-regulated effector with intrinsic GTPase activity. In the case of G-protein-coupled receptor systems, the ligand-occupied receptor promotes the critical GDP/GTP exchange process; in the case of tyrosine kinase pathway-coupled receptor systems, the m-SOS moiety triggers the exchange of GTP/GDP on the effector, Ras (see Figure 3).

1. SH2 and SH3 Domains

Two essential protein motifs have been found to provide for the protein–protein interactions that link receptor-localized phosphotyrosine residues to the activation of effector moieties, such as Ras. The first critical motif, termed an SH2 domain (sarcoma homology domain 2), binds with high affinity and in an amino acid sequence context-specific manner to phosphotyrosine residues that may be found either in an autophosphorylated receptor domain (e.g., the one for EGF-URO or PDGF) or in a domain of an intermediate tyrosine-phosphorylated signal docking protein, such as the insulin receptor substrate-1 (IRS-1). IRS-1 becomes phosphorylated on tyrosine in the course of insulin action (e.g., effector E_1, Fig. 1) and acts

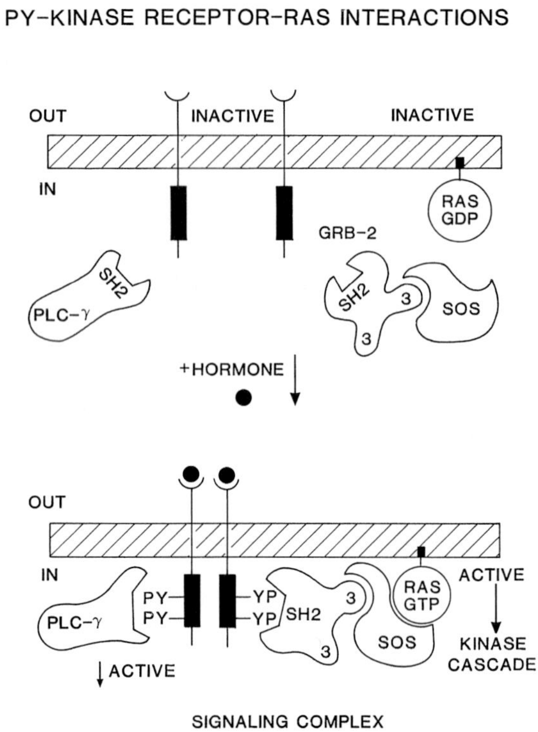

FIGURE 3 Tyrosine kinase receptor-mediated activation of Ras. The model shows the ligand (hormone)-dependent dimerization and activation of a growth factor receptor, with the appearance of autophosphorylated phosphotyrosine motifs on the cytoplasmic receptor domains. Interaction of receptor phosphotyrosines with SH2 domains of phospholipase C-γ (PLC-γ) results in enzyme activation. Interaction of the phosphotyrosine residues of the receptor with the SH2 domain of the adapter protein, GRB2, results, via the SH3 domain-mediated GRB2/SOS interaction, in the recruitment of SOS to the membrane, where its GDP/GTP exchange activity converts RAS to its GTP-bound active state. Activated RAS in turn kindles the Raf/MEK/MAP kinase cascade (adapted from McCormick, 1993).

as a docking site for other signal pathway adapter proteins (e.g., effector AD_1, Fig. 1: White and Yenush, 1998). The "SH2 domain" sequences were noticed originally as common elements in a variety of Src family tyrosine kinases; and it was pointed out that common "SH2" motifs were also found in a number of proteins recognized for their role in cell signaling (Pawson, 1995). Yet another common sequence in Src family tyrosine kinases, the so-called Src family homology domain 3 (SH3 domain), was also found as a common element in signaling-related constituents (Pawson, 1995). Importantly, the adapter molecule, GRB-2, that interacts with the Ras-targeted GDP/GTP exchange factor, m-SOS, and that via m-SOS couples phosphotyrosyl growth factor receptor-bound GRB-2 to the activation of Ras possesses both SH2 (binding to phosphotyrosyl receptor) and SH3 (binding to m-SOS) domains. A polyproline-rich motif on m-SOS has been found to participate in interacting with the SH3 domains of GRB-2 (Fig. 3). In keeping with the mobile receptor paradigm outlined earlier, the phosphotyrosyl residues in ligand-activated receptors, such as those for EGF-URO and PDGF (see Section II,F), are capable of interacting with SH2 domains on a variety of enzymatically active "effectors" (e.g., phospholipase C-γ) and on a number of noncatalytic docking/adapter proteins, such as GRB-2. Of importance in this regard are SH2 domains present in enzymes such as phospholipase C-γ and phosphatidylinositol 3′ kinase (PI3-kinase) (in the 85-kDa regulatory subunit), providing for the activation of these "effector" enzymes (e.g., effector E_2, Fig. 1) by interactions with phosphotyrosyl–PDGF and –EGF-URO receptors. The SH2 motifs also provide for important interactions between activated EGF-URO and PDGF receptors and c-Src itself (represented by E_2 in Fig. 1: Erpel and Courtneidge, 1995). Such phosphotyrosine–Src family kinase interactions, via SH2 domain adapter motifs, can also be seen to play essential roles in the activation of tissues by cytokines and in the activation of platelets by agents such as collagen. Thus, in accord with the mobile receptor model (Fig. 2), interactions between receptor phosphotyrosyl residues and adapter/effector SH2 domains can be seen as a branchpoint for multiple signaling events, mediated on the one hand via direct effector (e.g., phospholipase C-γ; PI3-kinase) activation, followed by the generation of signal mediator molecules (e.g., phosphoinositides and diacylglycerol, see Section III), and on the other hand mediated via an indirect phosphotyrosine-initiated adapter-coupled event (e.g., GRB-2/m-SOS) that results in the activation of Ras. In turn, activated Ras, via the recruitment and activation of the serine/threonine kinase, Raf-1, can set into play the downstream events catalyzed by the so-called "mitogen-activated" (MAP kinases) or "extracellular-regulated" (ERK kinases) family of signaling enzymes (Cobb *et al.,* 1996; Graves *et al.,* 1997; Su and Karin, 1996; Sugden and Clerk, 1997). As will be discussed in Section II,E, the activation of G-protein systems by "serpentine" receptors, such as the one for thrombin, can also lead to tyrosine kinase-mediated signaling events that presumably involve SH2 domain/effector interactions.

2. Common Themes

Taken together, there are a relatively limited number of common themes that govern the membrane receptor-mediated control of cellular function. The mobile receptor concept indicates how a single ligand/receptor interaction may be responsible for activating multiple cellular responses via distinct receptors. However, the extension of this model suggests that common modular domains (e.g., phosphotyrosyl motifs and SH2 domain interactions or common G-proteins) may play a role in the activation of common biochemical signaling pathways in an individual cell via multiple receptor systems. Thus, a variety of distinct cellular agonists, acting via independent receptor systems may, nonetheless, cause common end responses in cells. Finally, the fact that common protein sequence motifs may be involved in the response cells to a variety of stimuli provides a rational basis for the design of signal pathway-targeted drugs.

D. Ligand-Gated Ion Channels

In contrast with a number of ion channels such as the superfamily related to calcium and sodium transport (Chapters 9 and 10), the activity of a variety of hormones and neurotransmitters [e.g., acetylcholine, glycine, γ-aminobutyric acid (GABA)] depends on a ligand-specific regulation of ion conductance (receptor R_I in Figure 1). The nicotinic receptor for acetylcholine, which played a seminal role in Langley's development of the receptor concept, can be taken as a prototype for this category of receptor systems. Best characterized in work done with the nicotinic receptor isolated from the electric organ of electric eels and rays, this acetylcholine-regulated ion channel has been found to consist of a heterooligomeric complex (α_2, β, γ, δ) of glycoprotein subunits that form a central ion channel with common motifs in each subunit lining the channel. The agonist, acetylcholine, binds to the two αsubunits as well as interfaces between the α–γ and α–δ subunits (Stroud *et al.,* 1990; Conti-Tronconi *et al.,* 1994). The binding of acetylcholine (or other nicotinic agents) changes the conductance properties of the receptor channel, thereby causing depolarization and tissue activation. Extensive work has been done on the molecular pharmacology of the nicotinic receptor family, and comparable structural features characterize the excitatory neuronal cation channels gated by glutamate and serotonin ($5HT_3$ receptor) as well as the inhibitory neuronal chloride channels activated by GABA ($GABA_A$ receptor; Stephenson, 1995) and by glycine (Bechade *et al.,* 1994). All of these ligand-gated ion channels can be seen as forming a related superfamily (Table I).

E. G-Protein-Coupled Receptor Signaling

1. Guanine Nucleotide-Binding Proteins and Effector Regulation

Receptor-mediated G-protein activation represents perhaps one of the most commonly used mechanisms for generating intracellular signals (re-

TABLE I
Ligand-Gated Ion Channels

Agonist	Receptor subtype(s)	Ion transport	Effect
Acetylcholine	Cholinergic (nicototinic)	Na^+	Depolarization/excitation
Glutamate	NMDA	Ca^{2+}/Na^+	
	Kainate	Na^+	Depolarization/excitation
	AMPA (quisqualate)	Na^+	
GABA	$GABA_A$	Cl^-	Hyperpolarization/inhibition
Glycine	Glycine	Cl^-	Hyperpolarization/inhibition
Serotonin	$5HT_3$	Na^+, K^+	Depolarization

ceptor, R_G, Fig. 1). It was the recognition of the essential role of GTP in the regulation of membrane-associated adenylyl cyclase that led Rodbell and colleagues (1995) to propose the existence of an intermediary signal "transducer" that acted as a coupling device between the receptor and the cyclase. The subsequent identification of the heterotrimeric GTP-binding protein responsible for the stimulation of adenylyl cyclase (G_S) and the concurrent isolation of both the cyclase inhibitory protein (G_i) and the retinal cyclic AMP phosphodiesterase-regulating counterpart, transducin (G_T), set the stage for the cloning of multiple members of the G-protein superfamily (summarized by Simon *et al.,* 1991; Gilman, 1995). In the "resting state," the G-protein complex exists as a heterotrimer, composed of a GDP-bound α subunit complexed with a tightly associated $\beta\gamma$ dimer. The "catalytic" interaction between the ligand-occupied receptor and the oligomeric G-protein triggers the exchange of GTP for GDP on the α subunit, and it is this GTP/GDP exchange that leads to the dissociation of α_{GTP} from $\beta\gamma$, thereby promoting the interaction of α_{GTP} and $\beta\gamma$ with downstream effectors, such as adenylyl cyclases and phospholipases (see Fig. 5). In general, there are four subclasses of α subunits comprising 20 or more family members, including a number of mRNA splice product variants: (1) the G_S subgroups, known for their ability to activate adenylyl cyclase, (2) the G_i subgroup, originally identified as substrates for pertussis toxin (except for G_Z) and known to be involved in attenuating adenylyl cyclase and in modulating ion channel function, (3) the G_q subgroup involved in the activation of phospholipase C isoforms, and (4) the G_{12} subgroups, for which specific target effectors have yet to be identified unequivocally (Simon *et al.,* 1991). Multiple forms of the β (6 isoforms) and γ (11 isoforms) subunits are also known. The G-proteins are tethered at the inner face of the plasma membrane and can therefore be viewed as key inner membrane-localized "signal transducers."

A simplified overview of G-protein-coupled effector regulation is shown in Figure 4. Regulation of the production of a principal intracellular signal mediator, cyclic AMP, can be seen to be independently under both positive (e.g., in Fig. 4, ligand A_1 binding to the β adrenoreceptor, R_S, via G_S) and negative (e.g., in Fig. 4, ligand A_2 binding to the α_2 adrenoreceptor, R_i, via G_i) control. The specific interactions between the various receptors and the targeted G-proteins appear to be governed by complementary motifs on the receptors (e.g., intracellular loop 3) and on both the α and γ subunits of the G-protein heterotrimer (Taylor *et al.,* 1994; Bourne, 1997). Although a given receptor may display a preferential affinity for one class of G-proteins (e.g., most G_S-targeted receptors, such as the β-adrenoreceptor for adrenaline, do not interact appreciably with G_i or G_q), it is also the case that receptors such as the PAR_1 receptor for thrombin can, in keeping with the mobile receptor model, interact with two subclasses of G-proteins so as to be able not only to inhibit adenylyl cyclase (via G_i) but also to stimulate phosphoinositide turnover (via G_q). In Figure 4, receptors R_i and

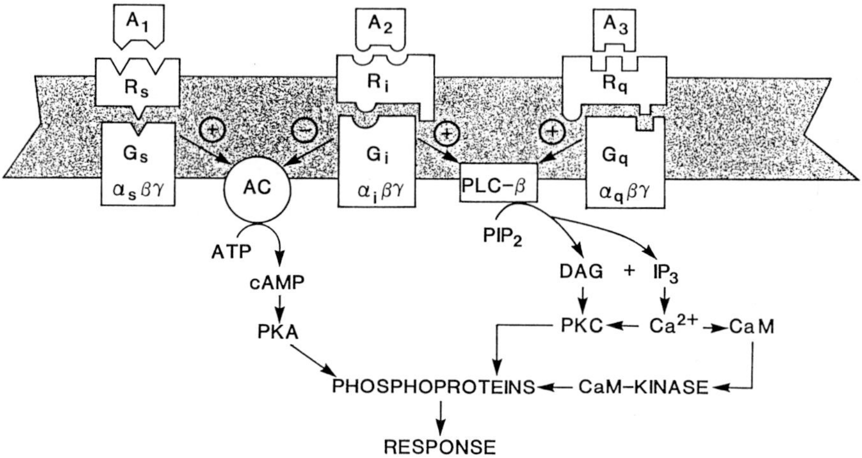

FIGURE 4 G-protein-coupled receptors. Three distinct receptors (R_s, R_i, R_q) are shown wherein the three selective agonists (A_1, A_2, A_3) promote receptor–G-protein interactions. Although R_s is shown to have a strict preference for G_s, R_i and R_q are able to interact and activate either G_i or G_q, depending on the cellular context. Dual control of adenylyl cyclase by G_s (positive) and G_i (negative) is shown as is the potential dual control (positive) of phospholipase C- (PLC-) by G_q and G_i. Signal amplification steps resulting from G-protein-coupled effector activation, involving the production of the "second messenger" cyclic AMP (cAMP), dicylglycerol (DAG), inositol 1,4,5-trisphosphate (IP_3), and an elevation of intracellular calcium, are also shown. These mediators are shown to activate cyclic AMP-dependent protein kinase (PKA), protein kinase C (PKC), and calcium/calmodulin (CaM)-dependent kinases.

R_q are depicted as interacting potentially with either G_q or G_i. The relative abundance of the interacting species (i.e., the receptor and the various G-proteins) and the protein–protein affinities would, by mass action, govern the predominance of the effect of an individual agonist in a particular setting, such as the platelet. The entire process of G-protein–effector activation is started by the "catalytic" receptor-mediated GTP/GDP exchange reaction. Once activated, the $\alpha_{GTP}/\beta\gamma$ form of the G-protein can go on to regulate a number of membrane-localized effectors, the best understood of which is adenylyl cyclase (Rodbell, 1995; Gilman, 1995; Levitzki *et al.,* 1993).

2. The G-Protein GTP/GDP Cycle, Posttranslational Modification Reactions, and Effector Modulation

As discusssed earlier, when the G-protein is activated by the receptor-catalyzed GTP/GDP exchange reaction, both the α and $\beta\gamma$ subunits can go on to regulate a number of effectors, including adenylyl cyclases, phospholipases, and ion channels. The time frame of these G-protein–effector interactions is governed by the rate of hydrolysis of GTP bound to the α subunit; this hydrolysis returns α_{GTP} to its "ground state," α_{GDP}, permitting a downregulation of effector activity and a reassociation of the α subunit with the $\beta\gamma$ dimer (Fig. 5). The domains that govern the mutual interactions

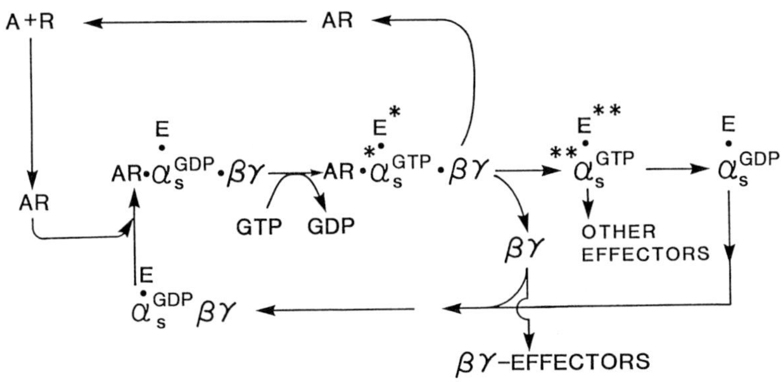

FIGURE 5 The G-protein GDP/GTP dissociation/association cycle. The binding of an agonist (A) to its receptor (R) is shown, in a catalytic manner, to trigger the exchange of GTP for GDP on the G-protein α subunit, depicted as the cyclase-activating subunit, α_s-GDP. The heterotrimeric protein is shown to be precoupled to an effector, E, as suggested by kinetic data of Levitzki *et al.* (1993) for adenylyl cyclase. Upon exchanging GTP for GDP, the activated α subunit (*αGTP) can stimulate its effector target (E*), which can proceed to a fully activated state (**αGTPE**) as the agonist receptor complex and $\beta\gamma$ subunit dissociate. The $\beta\gamma$ subunit can go on to regulate other effectors (not shown), while the hydrolsis of GTP bound to α_s returns the α_s subunit and the associated effector to a ground state that is capable of recombining with the free β γ subunit. This reassociation completes the GDP/GTP cycle.

between the G-proteins and both their target effectors and their receptors (Bourne, 1997) could, in theory, provide targets for the development of drugs. Although a popular model of G-protein function envisions the α subunit shuttling freely between the ligand-occupied receptor and the effector (e.g., adenylyl cyclase, Gilman, 1995), it it also possible that the G-protein and effector may be already precoupled in the membrane (Levitzki and Bar Sinai, 1991; Levitzki *et al.,* 1993). Based on kinetic considerations, this situation appears to be the case for the G_S/cyclase system (Levitzki and Bar Sinai, 1991), and it can be postulated that other "effectors" may also exist in the membrane, already "precoupled" with a "ground-state" G-protein. This situation might rationalize the very rapid time course of G-protein-coupled signaling, which would be limited only by a receptor diffusion process and not by a concurrent diffusion of the activated G-protein to the effector. Further, this possibility would provide for the cellular compartmentalization of G-protein–effector systems at discrete membrane loci such as the caveolae (see later).

The compartmentalization of G-proteins and their interactions with effectors may possibly be governed by a number of posttranslational covalent modifications of the G-protein subunits, of which the addition of lipid moieties may be particularly significant (Wedegaertner *et al.,* 1995). For instance, α-subunit palmitoylation (Linder *et al.,* 1993), with or without concurrent myristolation, may govern the interaction of the α subunits with effectors, as appears to be the case for the affinity of α_i subtypes for the $\beta\gamma$ dimer and for adenylyl cyclase (Linder *et al.,* 1991; Taussig *et al.,* 1993). An as yet unidentified covalent modification on α_S also appears to affect its affinity for adenylyl cyclase (Kleuss and Gilman, 1997). The α subunit of the $\beta\gamma$ dimer is also subject to lipid modification by C-terminal prenylation (farnesylation or geranylgeranylation) (Yamane *et al.,* 1990; Yamane and Fung, 1993; Mumby *et al.,* 1990; Mumby and Linder, 1994). This α subunit modification is of importance for the interaction of $\beta\gamma$ with the α subunit and with effectors such as adenylyl cyclase (Iniguez-Lluhi *et al.,* 1992). The essence of the previous discussion is that both the "cyclic" α subunit GTP/GDP exchange reaction (Figure 4) and the posttranslational modification (e.g., reversible palmitoylation or subunit phosphorylation) can potentially affect the signaling process mediated by G-proteins; in principle, alterations at both of these potential levels of control (e.g., GTPase activity or palmitoylation) could have an impact on the regulation of cellular function. Further, the reversible palmitoylation modification could target these signaling constituents to discrete cellular locations (e.g., caveolae; see Section IV).

3. G-Protein-Regulated Effectors

a. Adenylyl Cyclase. The effector first known to be regulated by G-proteins is adenylyl cyclase, which is associated tightly with the plasma membrane. Of the nine isoforms of this enzyme now known to exist, all are activated by the α_S subunit (Sunahara *et al.*, 1996). Other effectors that are targets for α_S regulation may also include voltage-gated Ca^{2+} channels (Yatani *et al.*, 1987) and cardiac Na^+ channels (Schubert *et al.*, 1989). Early on in the studies of adenylyl cyclase, it was recognized that a number of agonists could cause an inhibition of cyclase acting via a pertussis toxin-sensitive G-protein (the toxin ADP-ribosylates the α_i subunit, thereby abrogating the receptor-triggered α_i $\beta\gamma$ dissociation process). For instance, in the platelet, this G_i-mediated inhibitory regulation of cyclase can be seen upon activation of the thrombin (PAR_1) and α_2-adrenergic receptors by thrombin and noradrenline, respectively. The mechanism of negative regulation of adenylate cyclase has turned out to be complex and dependent on the isoform of the enzyme with which the α_i or $\beta\gamma$ (both of which can regulate the enzyme negatively) interacts. The inhibition can be either via a direct interaction between the α_i subunit and the cyclase (e.g., isoforms V and VI) or indirect, due to the sequestation of active α_S by an excess of $\beta\gamma$, made available, for example, by activation of the quantitatively predominant G_i oligomer. This latter "$\beta\gamma$-mediated α_S sequestation" mechanism most likely explains the cyclase inhibitory action of thrombin and noradrenaline in a variety of cells. To complicate matters, the $\beta\gamma$ subunit itself can stimulate rather than inhibit selected cyclase isoforms (types II and IV). Thus, the mechanisms for the positive and negative regulation of adenylyl cyclase can vary considerably, depending on the cyclase isoforms present in a given cell type.

b. Phospholipase C. One effector that has a dramatic impact on the function of platelets, smooth muscle, and other tissues is phosphoinositide-specific phospholipase C. The β and γ isoforms of this enzyme lead to agonist-stimulated hydrolysis of cell membrane phospholipids, so as to yield the intracellular mediators, diacylglycerol (DAG) and inositol trisphosphate (IP_3) (see Fig. 4). The β isoforms of the enzyme can be stimulated by members of the α_q subfamily to hydrolyze phosphatidylinositol bisphosphate, and the $\beta\gamma$ subunit can activate phospholipase C isoforms independently, with a preference for activating PLC-β_3 (Park *et al.*, 1993 and summary by Fields and Casey, 1997). Thus, in some tissues, one could envision the activation of phospholipase C-β both via activation of the G_q family members and via the activation of the G_i members present in the membrane. In the case of the thrombin receptor, which can potentially couple to members of both G_i (e.g., α_z) and G_q family members, there could thus be a concurrent activation of phopholipase C-β (via α_q/α_{11} as well as via released $\beta\gamma$) along with a parallel $\beta\gamma$-mediated inhibition of adenylyl cyclase. Further, the activation of phospholipase C-β in tissues such as the platelet and smooth muscle would be possible both via "pertussis toxin-sensitive" G-proteins of the G_i family and via "toxin-insensitive" G_q family members. The stimulation of phospholipase C-β isoforms via G-protein-coupled processes would complement the activation of the phospholipase C-γ isoforms via tyrosine kinase-mediated pathways (see later). The concurrent activation of both groups of phospholipase C isoforms would yield an abundance of the "dual" intracelluar mediators inositol IP_3 and DAG (Berridge, 1993; see Section IIIA).

c. Other $\beta\gamma$-Regulated Effectors: Src Family Kinases and PI3-Kinase. In addition to the ability of the $\beta\gamma$ subunits to modulate adenylyl cyclase and phospholipase C-β, as summarized earlier, this G-protein-derived heterodimer is now known to regulate other effectors, such as the N- and P/Q-type Ca^{2+} channels (Clapham and Neer, 1993, 1997; Clapham, 1996). Of relevance to the function of a variety of tissues are data indicating that the $\beta\gamma$ subunit also appears to be involved in the activation of tyrosine kinase pathways involving c-Src family members and possibly pleckstrin homology (PH) domain-containing tyrosine kinases (Luttrell *et al.*, 1995; van Biesen *et al.*, 1996). Further, G-protein $\beta\gamma$ subunits have been observed to stimulate an isoform (PI3K-γ) of phosphoinositide 3-kinase present in human platelets (Thomson *et al.*, 1994; Abrams *et al.*, 1996). The modulation of Src family kinases via a $\beta\gamma$-mediated process may be of particular relevance to tissue function, providing for cross talk between the serpentine receptor systems and the receptor systems mediated via tyrosine kinase pathways. Activation of PI3-kinase, as alluded to earlier, can also occur via the interaction of its P85 subunit SH2 domains with phosphotyrosyl protein motifs. Upon activation, P13-kinase is thought to play a role in recruiting PH domain-containing signal proteins such as the serine threonine kinase AKT/PKB and the phosphoinositide-binding protein, Grpl, to the plasma membrane (Lemmon and Ferguson, 1998). Such P13-kinase-promoted membrane interactions very likely play a role in modulating intracellular tissue function and in regulating integrin-dependent "inside-out" adhesion processes. Essentially all of the G-protein-coupled signaling mechanisms discussed in this and previous sections are initiated by membrane-delimited reactions, triggered by the ligand-occupied receptor/G-protein interaction. Because some of the components of the system (e.g., the $\beta\gamma$ heterodimer) are modified by the addition of lipid moieties, these signaling reactions may take place in discrete membrane domains such as the caveolae (see Section IV).

F. Receptors with Intrinsic Ligand-Regulated Enzyme Activity

1. Generating an Intracellular Signal via Tyrosine Kinase Receptors

As outlined in Section II,A, one of the major cell-signaling paradigms involves receptors that are transmembrane agonist-regulated allosteric enzymes. The binding of the receptor-specific ligand to the extracellular domain produces a conformational change that is transmitted via a transmembrane portion to an intracellular domain; this intracellular receptor domain then exhibits increased enzymatic activity. For so-called "growth factor" receptors, such as the ones for insulin, EGF-URO, or PDGF, intrinsic receptor tyrosine kinase is a key property for agonist action. Other intrinsic receptor-associated enzyme activities that have been documented to date include serine/threonine protein kinases (e.g., the transforming-growth factor β receptor superfamily) and guanylyl cyclase activity [atrial natriuretic factor (ANF) receptor family, see Table II].

To date, the most extensively studied class of receptor enzymes are represented by the tyrosine kinase receptors for insulin [and, by extension, insulin-like growth factor I (IGF-I)] (Cheatham and Kahn, 1995; Czech, 1995), EGF-URO, and PDGF. These three receptors can serve as a paradigm for receptors that respond to a number of other polypeptide growth factors (e.g., fibroblast growth factor, nerve growth factor, colony-stimulating factor-1) and that possess intrinsic tyrosine kinase activity (Table II). At least 50 members of this receptor superfamily are known to exist (Heldin, 1996). The binding of a ligand to the extracellular receptor domain results in two rapid events: (1) oligomerization (dimerization) and (2) activation of the cytosolic tyrosine kinase domain (Malarkey *et al.,* 1995; Fantl *et al.,* 1993; Schlessinger and Ullrich, 1992). The dimerization/oligomerization event appears to be equally important for the activation of receptors for cytokines (see later), wherein intracellular tyrosine kinases (IPYK, E_3: Fig.

TABLE II

Ligand-Activated Transmembrane Receptors with Intrinsic Enzyme Activity

Agonist (receptors)	Enzyme activity	Effect
Insulin Epidermal growth factor Platelet-derived growth factor Fibroblast growth factor Colony stimulating factor-1	Tyrosine kinase	Tyrosine phosphorylation of cellular proteins
Activin Inhibin Transforming growth factor β	Serine/threonine kinase	Serine/threonine phosphorylation of cellular proteins
Natriuretic peptides (GC-A/B) Bacterial enterotoxins (GC-C) Sperm-activating peptides	Guanylyl cyclase	↑ cyclic GMP

1) are recruited into the signaling process (see later). As a consequence of the ligand-driven dimerization/kinase activation process, there is a receptor autophosphorylation event that results in the production of a number of phosphotyrosyl-containing domains in the receptor cytoplasmic tail. Distinct phosphotyrosine-containing sequence motifs can be found in the ligand-activated receptors for insulin, EGF-URO, and PDGF. Autophosphorylation of the insulin receptor on tyrosine residues has the effect of increasing its tyrosine kinase activity toward other intracellular substrates, such as the insulin receptor substrate 1. This insulin receptor substrate acts to amplify the insulin signal (Myers and White, 1996; White and Yenush, 1998, see later). Once activated, the kinase activity of the insulin receptor persists as long as the receptor remains autophosphorylated, even in the absence of additional insulin and on cellular internalization (Bevan *et al.,* 1995; Bergeron *et al.,* 1995). This persistent activation provides for considerable signal amplification. It is believed that an insulin receptor-targeted tyrosine phosphatase plays a role in terminating the insulin signal (Drake *et al.,* 1996; Band *et al.,* 1997). A number of experimental approaches, including site-directed mutagenesis of specific insulin receptor autophosphorylation sites, have shown that the intrinsic tyrosine kinase activity of the insulin receptor is essential for the mitogenic and metabolic actions of insulin (Murakami and Rosen, 1991; Wilden *et al.,* 1992; Cheatham and Kahn, 1995; Czech, 1995). This conclusion is supported strongly by the finding of naturally occurring missense mutations in the insulin receptor tyrosine kinase domain in individuals with insulin resistance (Flier, 1992; Taylor, 1992). For that reason, considerable effort has been devoted to the identification of the intracellular target proteins for the insulin receptor kinase (e.g., Shc, IRS-1 and IRS-2: Myers and White, 1996; White and Yenush, 1998) and for other receptors with intrinsic tyrosine kinase activity (e.g., for EGF-URO and PDGF).

As for the insulin receptors, structure–activity studies with the EGF-URO and PDGF receptors (e.g., see summary by Hollenberg, 1991) have established the key importance of the intrinsic receptor tyrosine kinase activity and receptor autophosphorylation for ligand action, and target-specific phosphotyrosyl docking sites have been identified for a number of effectors, including PLC-γ and GRB-2 (Carpenter, 1992; Hernandéz-Sotomayor and Carpenter, 1992; Fantl *et al.,* 1993; Heldin, 1996, 1997a). The PDGF receptor represents those receptors of this superfamily for which the tyrosine kinase domain is split into two subdomains, separated by an intervening sequence. Although the intervening sequence does not play a catalytic role, it can provide a docking site for SH2 domain-containing proteins (e.g., PLC-γ for the PDGF receptor: Fantl *et al.,* 1993; Heldin, 1997a). The differential binding of different effector moieties to distinct phosphotyrosyl domains on the different receptors is thought to play a role in the distinctive actions of the different receptor systems (e.g., insulin versus EGF-URO versus PDGF; see discussion by Schlessinger and Ullrich, 1992).

The cell activation mechanism mediated by receptors such as the ones for insulin, EGF-URO, and PDGF can be generalized as follows: (1) the receptor kinase catalytically activates a feed-forward enzyme, exemplified by the increased catalytic activity of phospholipase C-γ caused by activation of the EGF-URO receptor (Nishibe *et al.,* 1990). (2) The autophosphorylated receptor recruits the SH2 domain-containing noncatalytic "growth factor receptor-binding" protein, GRB2, to its tyrosine-phosphorylated motifs, and GRB2, in turn, via its SH3 domain, links to and promotes the activity of the Ras-targeted GDP/GTP exchange factor, mSOS (see Fig. 3). This second general mechanism has received intense scrutiny (McCormick, 1993). (3) In a noncatalytic manner, the tyrosine-phosphorylated receptor can interact with and promote the activity of SH2 domain-containing enzymes, as is thought to be the case for PI3 kinase activation via its SH2 domain-containing p85 subunit. (4) As a variant of the second and third general mechanisms, the receptor can phosphorylate a catalytically inactive docking protein on tyrosine residues; in turn, this tyrosine-phosphorylated adapter protein (e.g., Shc, IRS-1, or IRS-2) can interact with a variety of SH2 domain-containing proteins to propagate a cellular signal. Of prime importance is the interaction of the tyrosine-phosphorylated docking proteins with the SH2 domain of GRB2 and with the p85 regulatory unit of P13-kinase, in keeping with the process described earlier for mechanisms 2 and 3. (5) Finally, the tyrosine-phosphorylated receptor domains can act as docking sites for SH2 domain-containing tyrosine kinases, such as Src family members. The recruitment of c-Src itself to the membrane-localized autophophorylated PDGF receptor is thought to play a key role in the mitogenic action of PDGF (Erpel and Courtneidge, 1995). Data suggest that the targeting and tyrosine phosphorylation of growth factor receptor domains by a process *other than* receptor autophosphorylation may also permit a receptor such as the one for EGF-URO to act as a "scaffold" for signaling. It is presumed that this "scaffolding" process could set into play signaling processes akin to mechanisms 3 and 4. Thus, the "transactivation" of the EGF-URO receptor by tyrosine phosphorylation caused by either G-protein-coupled receptors or cytokine-related receptors (e.g., growth hormone) may play a signaling role in some circumstances (Daub *et al.,* 1996; Yamauchi *et al.,* 1997; Luttrell *et al.,* 1997).

The four basic mechanisms just discussed, including the possible "scaffolding" function, are thought to rationalize the signaling processes for those receptors that possess intrinsic tyrosine kinase activity. As will be seen from the discussion of cytokine receptor mechanisms (see Section II,G), comparable processes and common signal sequence motifs that are linked to phosphotyrosine residues in a sequence-specific manner (e.g., SH2/SH3, PTB domains; Pawson, 1995; Cohen *et al.,* 1995; Margolis, 1996; Borg and Margolis, 1998) are also used for receptor systems that do not, of themselves, possess intrinsic tyrosine kinase activity.

2. Regulating Receptors Possessing Intrinsic Tyrosine Kinase Activity: Silencing and Terminating the Signal

As for any cellular enzyme process, receptors are subject to complex feedback regulatory reactions. In the case of tyrosine kinase receptors, the intracellular domains can be targets for phosphorylation/dephosphorylation reactions taking place in the plasma membrane that regulate either the ligand-binding function or the kinase activity of the receptor. For instance, a kinase C-mediated phosphorylation of T_{654} of the EGF-URO receptor silences its abilty to bind to ligand; conversely, dephosphorylation of the insulin receptor by intracellular protein tyrosine phosphatases, as alluded to earlier, is thought to terminate its signal (Drake *et al.,* 1996; Band *et al.,* 1997). As with the signal generation mechanisms, SH2 domains (on the protein tyrosine phosphatases) are thought to play an important role for signal termination, via interactions with certain receptor phosphotyrosyl sequences (Tonks, 1996; Liu, and Chernoff, 1997; Byon *et al.,* 1997). Receptor phosphorylation/dephosphorylation events catalyzed by a variety of enzymes (e.g., kinase A, kinase C, Src family kinases, intracellular tyrosine phosphatases) no doubt play key roles in the regulation of all members of the tyrosine kinase receptor superfamily. Comparable phosphorylation/dephosphorylation events have been observed to play roles in the control of G-protein receptor function. In the case of G-protein-coupled receptors, some of the serine/threonine kinases are thought to be "receptor-specific" [e.g., the so-called G-protein receptor kinases (GRKs)] so as to be involved in ligand-specific receptor desensitization (Premont *et al.,* 1995; Freedman and Lefkowitz, 1996). However, comparable kinase A, kinase C, and protein phosphatase (both phosphotyrosine specific and phosphoserine/threonine specific) reactions undoubtedly also control the function of *all* classes of membrane-localized receptors, and the intriguing issues to be resolved for these phosphorylation/ dephosphorylation reactions lie (1) in the mechanisms for the targeting of the intracellular kinases/phosphatases to the membrane-localized receptor domains and (2) in the factors that govern the specificity of targeting the regulatory enzymes involved only to a restricted class or subtype of receptor.

Superimposed on the enzyme feedback loops that govern receptor binding and catalytic activity are the receptor internalization and recycling processes mentioned in Section II,B. Upon activation by its ligand, a receptor can migrate to so-called clathrin-coated "pits" on the cell surface. These specialized invaginations in the membrane, distinct from the caveolae discussed in Section IV, are sites where receptor-mediated endocytosis occurs. Pinched-off vesicles, enriched in receptor–ligand complexes, fuse with intracellular structures in which the receptor and ligand become encoupled. The compartment for the uncoupling of the receptor and its ligand (so-called CURL) is maintained at a comparatively low pH (4 to 5) that promotes ligand dissociation. Subsequently, a sorting process can take place whereby the CURL-bound receptor is either relocated to the cell surface or directed to a lysosomal compartment for degradation. One working hypothesis that has received some support proposes that the internalized receptor or an "active" fragment thereof may be relocated to the nucleus to regulate delayed cellular processes (e.g., gene transcription). Although unequivocal evidence has been obtained for the localization of receptors in the nucleus (Shah *et al.,* 1995; Smith *et al.,* 1997), a role for the internalized receptor in this type of signaling process remains an open question. Apart from the recycling of internalized receptors to the cell surface for reuse (as appears to be the case for this "transferrin receptor"), the complementary process of receptor messenger RNA transcription, translation, processing/glycosylation, and insertion into the membrane provides a tertiary level whereby a cell can regulate its sensitivity to a variety of agonists. This biosynthesis/membrane insertion process, which is regulated by the receptor signaling event itself, is beyond the scope of this chapter.

In summary, there are three levels at which receptor feedback mechanisms can regulate signaling: (1) at the level of the cell membrane, where feedback phosphorylation/dephosphorylation can rapidly (seconds to minutes) alter either ligand binding or the intrinsic receptor signaling capability to "desensitize" the signal; (2) at the level of receptor clustering and internalization, whereby receptor numbers (and thereby cell sensitivity) can be "downregulated" over a somewhat longer time period (minutes to tens of minutes); this "coated pit"-mediated internalization process may be linked to delayed intracellular signaling events; and (3) at the level of receptor biosynthesis, processing trafficking and membrane insertion (tens of minutes to tens of hours). These signal-modulating processes are thought to occur in a "second-tier" time frame that is slower than the initial dimerization/microclustering event that initiates a transmembrane signal. It is in this rapid time frame process that the caveolar structure may be found to play an important role (see Section IV).

3. Receptors with Intrinsic Serine/Threonine Kinase Activity

Hard on the heels of understanding the mechanisms for signaling by receptors with intrinsic tyrosine kinase activity has come the discovery of the transforming growth factor-β (TGF-β) family of serine/threonine kinase receptors that includes those for TGF-β itself as well as for cell regulators such as activin, inhibin, and Müllerian inhibiting substance (Massague, 1996; Massague and Weis-Garcia, 1996; Heldin, 1997b). Although the nature of the receptor substrates and the detailed mechanisms governing the actions of the serine/threonine kinase receptor family differ from those of the tyrosine kinase family of receptors, the essential signaling paradigm outlined by the mobile receptor model (Section II,B), and a number of the mechanisms summarized in Sections II,F,1 and II,F,2 can be seen to apply, at least in principle, although not in the details. The TGF-β receptor family, as understood currently, uses prinicipally mechanism No. 1 discussed in Section II,F,1 (ten Dijke *et al.*, 1996; Massague, 1996). Using an oligomeric receptor dimerization mechanism, the ligand (e.g., TGF-β) promotes the interaction of two dissimilar receptor serine/threonine kinase subunits (so-called R_I and R_{II}). The binding of TGF-β to the constitutively active R_{II} receptor subunit promotes its interaction and activation of the R_I serine/threonine kinase subunit. The "activated" receptor heterodimer then goes on to phosphorylate and activate cytoplasmic proteins (so-called SMADs) that can heterodimerize (e.g., with SMAD-4) to permit nuclear translocation and transcriptional activation. This type of signal paradigm (i.e., the activation of latent cytoplasmic transcription factors that translocate to the nucleus) mirrors mechanisms that are triggered by the activation of both cytokine receptors (see Section II,G) and growth factor receptors. For cytokine and growth factor receptors, the latent cytoplasmic transcription factors that become activated have been termed signal transducers and activators of transcription (STATs; see later). Thus, with the serine/threonine kinase family of receptors, a conservation of "signaling themes" can be seen to be in common not only with all other membrane-localized receptor kinases, but also with receptor systems wherein the ligand-binding oligomer is devoid of enzyme activity.

4. Guanylyl Cyclase Receptors

A third type of receptor with intrinsic enzyme activity is the particulate (i.e., plasma membrane bound, as opposed to cytosolic) guanylyl cyclase, which is activated by a number of agonist peptides, including atrial natriuretic factor (ANF or ANP) and a brain-derived homologue, BNP (Table II; Wong and Garbers, 1992; Drewett and Garbers, 1994; Garbers and Lowe, 1994; Anand-Srivastava, 1997). The mammalian guanylyl cyclase receptors A and B that bind ANP and BNP consist of an extracellular ligand-binding domain, a single transmembrane domain, an intracellular protein kinase-like domain, and a guanylyl cyclase catalytic domain at the carboxyl terminus that exhibits some sequence identity with the soluble forms of guanylyl cyclase. Cytosolic guanylyl cyclase is the target for nitric oxide activation. Although the receptor protein kinase-like domain has no catalytic activity, the binding of ATP to this domain is required for the ligand-stimulated activation of the receptors guanylyl cyclase activity (Chinkers *et al.,* 1991; Goraczniak *et al.,* 1992). Although two of the ANF-binding receptors (ANP-A and ANP-B; also termed GC-A and GC-B) possess intrinsic guanylyl cyclase activity, a third distinct ANP receptor subtype (the so-called ANP-C receptor) lacks this activity and has only a short 37 amino acid residue cytoplasmic portion. The ANP-C receptor is thought to regulate adenylyl cyclase and phospholipase C via a G-protein-coupled mechanisms (G_i); but the mechanism of G_i regulation via the single ANP-C receptor transmembrane domain has yet to be clarified (Anand-Srivastava, 1997). A distinct particulate guanylyl cyclase-containing receptor GC-C appears to be responsible for the pathogenic actions of a heat-stable *Escherichia coli* enterotoxin (STa) that causes diarrhea in mammals. A distinctive property of the ANF-A receptor is that it appears to be preassociated in the membrane in the absence of ligand as a noncovalent oligomer (Rondeau *et al.,* 1995); ligand binding in a 1:2 stoichiometric relationship activates the cytoplasmic guanylyl cyclase activity. Once activated, the guanylyl cyclase-linked receptors yield the intracellular messenger, cyclic GMP from GTP. Cyclic GMP has long been known as a "second messenger," akin to cyclic AMP, for the regulation of a variety of intracellular proteins, including cyclic GMP-dependent protein kinases and cyclic nucleotide phosphodiesterases (Lincoln and Cornwell, 1993). The cloning of homologous receptors from olfactory and retinal tissue, for which activating ligands have yet to be found (GC-D, GC-E and GC-F: Yang *et al.,* 1995, 1996), suggests that this receptor superfamily may have many more members to be characterized.

G. Receptors without Intrinsic Enzyme Activity That Act via Signal Adapter Proteins

1. Cytokine Receptors

As mentioned briefly in Sections II,A and II,F, a diverse group of receptors such as those for growth hormone, prolactin, GM-CSF, and interleukins (e.g., IL-3, IL-6) are transmembrane proteins that possess comparatively short cytoplasmic domains and do not exhibit intrinsic enzymatic activity. However, a common biochemical event in the action of these key intercellular mediates is the phosphorylation of a variety of intracellular substrates on tyrosine. A hallmark of this so-called "cytokine receptor superfamily" (summarized by Miyajima *et al.,* 1992, 1997; Sato and Miyajima, 1994; Hara and Miyajima, 1996), is the association of a high-affinity ligand-binding subunit (often termed a receptor α subunit) with a second receptor subunit that by itself is not able to bind the activating ligand (so-called β subunit). The second β subunit contains the sequence motifs required for signaling in its cytoplasmic domain. Together, the α and β subunits form a high-affinity complex with the activating ligand, which via distinct ligand sequences forms a cross-bridge between the two subunits. Common amino acid sequence motifs can be found in both extracellular and intracellular portions of the signaling subunits of the "cytokine" receptor family. In the special case of receptors for growth hormone and prolactin that belong to this superfamily, a single subunit, as a ligand-induced dimer (stoichiometry of 1:2, ligand:receptor), subserves both ligand binding and signaling functions (Wells and deVos, 1996). In other instances, distinct ligand binding α subunits (e.g., for IL-3, IL-5, and GM-CSF) can share a common signaling β subunit (see reviews by Nicola and Metcalf, 1991; Nicola *et al.,* 1997; Miyajima *et al.,* 1992; Sato and Miyajima, 1994). Other members of the cytokine receptor family, such as IL-6, require a distinct β subunit (so-called, glycoprotein 130 gp 130; Hirano *et al.,* 1997) that can nonetheless be shared by yet another set of agonists [e.g., Leukemia Inhibitory Factor (LIF), CNTF) with distinct α-subunit-binding components. Of particular note is that the binding subunits for one of the members of the cytokine family, ciliary neurotrophic factor, has no transmembrane receptor sequences, but is rather tethered to the cell surface via a glycolipid moiety. The signaling properties of the GPI-anchored receptors will be discussed further (Section IV,B) with respect to the potential roles of caveolae in signal transduction (Lisanti *et al.,* 1995).

An essential element for signaling by the cytokine receptor systems is the recruitment to and activation by the signaling "β subunit" of a family of so-called Janus tyrosine kinases (Wilks and Oates, 1996). These β-subunit receptor-associated tyrosine kinases (JAKs 1, 2, 3, and Tyk-2) can lead either to the phosphorylation of β-subunit sequences on tyrosine (analogous to the "autophosphorylation" event for the EGF-URO and PDGF receptor) or to the tyrosine phosphorylation of cytoplasmic substrates. These tyrosine-phosphorylated cytoplasmic substrates subsequently form homo- and heterodimers, which are translocated to the nucleus where they activate transcriptional events. As indicated earlier, because of this transcriptional activity, these tyrosine kinase substrates that mediate cytokine actions have been termed STATs (1 to 5). Thus, in principle, the general mechanisms whereby cytokine receptors create intracellular signals can be seen to be similar to a number of the generalized mechanisms discussed earlier for EGF-URO, PDGF, and TGF-β receptors (Sections II,F,1 and II,F,3). In keeping with the "mobile receptor" model outlined earlier (Section II,B), the α subunits can be seen as "classical" receptor recognition moieties and the β subunits can be seen as "signal-adapter" proteins that, like G-proteins, are able to couple the ligand-binding constituent to an effector signaling system. It is also in keeping with the "mobile receptor" paradigm that the affinity of the ligand for its receptor would be increased on interacting with the signal-adaptor β subunit. Notwithstanding, the types of protein–protein interactions governed by SH2 domain/phosphotyrosine binding and the roles of the recruited JAK kinases provide for a high degree of complexity in the signaling process. Further, because of the common protein–protein interactions between the tyrosine-phosphorylated β subunit of the cytokine receptor systems and other signaling intermediates that possess SH2 domains (e.g., GRB2, PLC-γ), it is not surprising that there are common biochemical responses (e.g., elevated intracellular calcium, via phospholipase C-γ activation; MAP kinase activation, stemming from the GRB2 $\rightarrow$ SOS $\rightarrow$ Ras pathway) triggered both by cytokines and by growth factors such as EGF-URO and PDGF.

2. Multicomponent Immune Receptors

In keeping with the theme of heterooligomeric receptor complexes, the T- and B-cell antigen receptors (TCR and BCR, respectively) are multicomponent receptors. TCR and BCR represent one of the most complex cell surface receptor systems studied to date. These receptors are composed of multiple subunits, all of which are required for the assembly and cell surface expression of a functional recognition/signaling complex. However, these receptors do not contain intrinsic enzymatic activity in their cell surface components. The TCR is composed of six different polypeptides, which include a ligand recognition heterodimer [$\alpha\beta$ or $\gamma\delta$ as well as subunits that have been designated CD3ε, CD3γ, CD3δ, and TCRδ (Wange and Samelson, 1996)]. The core units of the BCR are transmembrane immunoglobulin and the associated heterodimer Igα and Igβ (Reth, 1992). Engagement of either the TCR or the BCR receptor complex on their corresponding cell type with an antigenic determinant leads to cellular activation. The activation signal results in cellular proliferation and the acquisition of complex cellular functions (Gold and Defranco, 1994; Qian and Weiss, 1997). Within the cytoplasmic domains of the TCRδ and CD3 polypeptides of the TCR and within the Igα and Igβ molecules of the BCR are critical regions necessary for receptor signaling to occur. These receptor domains have been termed immunoreceptor tyrosine-based activation motifs (ITAMs) (Cambier, 1995). Key tyrosine residues within this motif become phosphorylated on either TCR or BCR activation and couple to the intracellular signaling pathway by acting as a high-affinity binding site for SH2-containing signaling molecules, as outlined in Sections II,C,1 and III,B. Two families of tyrosine kinases have been shown to be involved in both of these receptor-initiated events. Receptor-associated Src family kinases phosphorylate the key tyrosine residues within the ITAM, which then acts as a site for recruitment and subsequent activation of the ZAP-70/Syk protein tyrosine kinases. This recruitment requires the tandem SH2 domains in ZAP-70/Syk, each binding to the critically spaced phosphotyrosine residues within the ITAM (van Oers and Weiss, 1995; Weiss, 1995). Upon activation of these two families of kinases, a number of downstream signaling molecules can be triggered, resulting in the overall amplification of the initial signal. Multiple ITAMs present within a single receptor complex may provide the capacity to amplify the signal; there are three motifs within TCRγ and one in each of the CD3 polypeptides within the TCR. Because each TCR complex can contain a TCRγ dimer and two CD3 dimers, each TCR can therefore contain 10 ITAMs. In addition to playing a role in signal amplification, it is also possible that the distinct ITAM sequences in the different receptor subunits can exhibit specific interactions with different signaling molecules so as to provide differential signaling specificity (Wange and Samelson, 1996).

A novel general concept in signal transduction has come to light in the study of immunoreceptor biology: the ability of ligand binding to recruit and activate a coreceptor that either amplifies the signaling process or acts to inhibit cellular signaling (Tridandapani *et al.,* 1997). CD19 is a coreceptor for the BCR that is required for the normal production of antibody responses to T-cell-dependent antigens. Upon the co-ligation of CD19 and BCR there is a dramatic enhancement of intracellular signaling. Coreceptor stimulation can also provide a negative signal to the cell. Co-crosslinking of the BCR with the Fc receptors for IgG results in the abrogation of the proliferative signal normally induced by BCR activation. This inhibitory mechanism is not well understood, but requires yet another specific tyrosine-phosphorylated motif within the cytoplasmic domain of the Fc receptor, called the immunoreceptor tyrosine-based inhibitory motif (ITIM) (Amigorena *et al.,* 1992; Muta *et al.,* 1994). When phosphorylated, ITIMs recruit either an SH2-containing tyrosine phosphatase (SHP-1) or the SH2-containing inositol polyphosphate 5′-phosphatase (SHIP); both enzymes appear to be able to downregulate signaling, albeit by distinct mechanisms (Scharenberg and Kinet, 1996; Ono *et al.,* 1997). There is a growing family of proteins that bear ITAMs or ITIMs, and their activation and coactivation will allow for the appropriate modulation of cellular responses. The complexity of these systems is beyond the scope of this chapter, and the reader is encouraged to explore the following review articles for specific details (Reth, 1992; Cambier, 1995; van Oers and Weiss, 1995; Weiss, 1995; Scharenberg and Kinet, 1996; Wange and Samelson, 1996).

3. Noncatalytic Receptors That Activate Intracellular Proteinases

A novel cell activation mechanism that has come to light since the early 1990s also involves receptors that do not exhibit intrinsic ezymatic activity, but which recruit proteinases in addition to protein kinases into the signaling process. Receptors of this class comprise the tumor necrosis factor-α receptor superfamily, which includes the 55-kDa TNFR1 (CD120a), TNFR2 (CD120b), CD40, and the apoptosis-associated receptor (FAS/CD95/APO1). The signaling mechanisms used by this receptor superfamily represent yet another variation of the "mobile receptor" model, wherein specific sequences on the cytoplasmic portion of the TNF-α receptor isoforms are stimulated by ligand binding to interact, via adapter proteins, with intracellular effectors that, in this case, turn out to be either proteinases (Caspases: Riches *et al.,* 1996; Shu *et al.,* 1997) or a serine/threonine kinase (RIP or NIK; Hsu *et al.,* 1996; Malinin *et al.,* 1997). Ligand binding is believed to cause the formation of a receptor trimer. The main cellular responses monitored on activation of the receptor family members include (1) the induction of cell death by an apoptotic mechanism, (2) the activation of the transcription factor, NFκB, and (3) the activation of mitogen-activated protein kinase (MAPK) cascades (Riches *et al.,* 1996). All of these responses appear to be triggered by the ligand-induced microaggregation (trimerization) of receptor subunits, accompanied by the docking/activation of specific signal-directing adapter proteins bound to specific receptor cytoplasmic domains. In the model proposed to date (Wallach *et al.,* 1997), a cytoplasmic TNFR1 receptor sequence termed a "death domain (because of its correlation with TNF-α-induced apoptosis) becomes associated with a multisite docking protein, termed TNF receptor-associated death domain (TRADD). This TNF receptor scaffolding protein can then interact with a variety of other effector/adaptor proteins that differentially regulate the downstream effects of the ligands (i.e., TNF-α, FAS-ligand). One set of such adapter proteins, termed TNF receptor-activated factors (TRAFs 1–6) can interact with and activate an NFκB-inducing kinase (so-called NIK, related to the MAPKK kinases; Malinin *et al.,* 1997). This serine/threonine kinase is probably comparable to the distant enzyme recruited to the FAS receptor (acronym RIP) described previously (Hsu *et al.,* 1996). This secondary recruitment of a serine/threonine kinase into the TNF receptor family signaling pathway can be seen in principle to have elements in common with the recruitment and activation of the JAKs in the course of cytokine receptor signaling. Quite novel, however, is the ability of a distinct TNFR1/FAS-associated docking protein [Fas-associated death domain (FADD), also termed MORT-1] to interact with and promote the activation of the proteinase, Caspase 8. This intracellular proteinase is upstream in a catalytic proteinase cascade involving other caspase family members that ultimately lead to cellular apoptosis (Wallach *et al.,* 1997). Thus, in summary, the TNF-α receptor family, via a series of cytoplasmic adapter docking proteins, recruits into the signaling process not only protein Ser/Thr kinases having homology with the MEKK kinase family, but also a caspase proteinase cascade. This kind of multiplex signaling by a single receptor was envisioned clearly by the mobile receptor model outlined in Figure 2. The distinctive end response-triggered activation of members of the TNF receptor family (e.g., NFκB activation versus apoptosis) depend entirely on the differential binding specificity of specific sequence motifs in the set of effector proteins that link to the TRADD adapter protein. Evidently, receptor systems other than the TNF family (e.g., IL-6 signaling) can also make use of comparable adapter proteins (e.g., TRAF-6) that are in common with the other TRADD-binding proteins termed TRAFs (Cao *et al.,* 1996).

III. SIGNAL AMPLIFICATION

A basic question to deal with is how, once activated by any of the mechanisms outlined in Figure 1, does the cell amplify the initial signal that results in a generalized tissue response? As an answer to this question, four basic themes can be proposed to account for most of the signal amplification processes used by a variety of receptor systems: (1) a change in membrane potential, caused either by a ligand-regulated ion channel (receptor R_I, Fig. 1) or by G-protein-regulated ion channels, would, as elaborated on by Zierler and colleagues (Zierler, 1960, 1985; Zierler and Rogus, 1981), have a profound and instant effect on the orientation of all membrane proteins. This effect, as well as the propagation of membrane potential changes between cells in a tissue, would generate an amplified cell and tissue signal rapidly. (2) The initial receptor signal may generate one or more soluble, diffusable "signal messengers," that in turn regulate "cascade" enzyme reactions. Such "second messengers," including cyclic AMP, the first to be described (Sutherland, 1972), are discussed briefly in Section III,A. (3) The protein kinase activity of the receptor itself, or the series of receptor-triggered protein–protein interactions described in Sections II,C and II,F, can also, like cyclic AMP, lead to a protein kinase/phosphatase cascade, where each catalytic step amplifies the previous one. This type of phosphorylation cascade is well recognized for amplifying cell signals. (4) As described for the TNF-α receptor system, a proteinase cascade may be initiated. The large amplification properties afforded by proteinase cascades are well recognized in the complement and clotting cascade system. The following sections provide brief details about the soluble "second-messenger" systems and about those amplification mechanisms involving modular protein–protein interactions.

A. Membrane-Derived Diffusable Second Messengers

A number of diffusable messengers that amplify receptor signals originate as a result of plasma membrane-localized reactions. Of prime importance are the "classical" mediators: cyclic AMP, cyclic GMP, DAG, calcium, and inositol IP_3. Ceramide has also emerged as a potential "messenger" molecule. Each of these second messengers can result from the activation of a variety of receptor systems, as already touched on in preceding sections.

The "classical" source of a soluble second messenger can be seen in the receptor-mediated activation of the G_s-protein, which in turn activates adenylyl cyclase with the production of cyclic AMP from ATP (Rodbell, 1995). This G-protein mechanism comprises two initial amplification steps, followed by a protein kinase cascade amplification process. First, the recycling agonist–receptor complex has a "catalytic" effect on activation of the heterotrimeric G_s-protein complex (Levitzki *et al.,* 1993; Levitzki and Bar-Sinai, 1991), whereby a single agonist receptor moiety can activate hundreds of G_s-protein effectors. Second, amplification is provided by the turnover number of the effector enzyme (e.g., adenylyl cyclase) that can catalytically generate thousands of "messenger molecules" that go on to activate yet another round of protein kinase-catalyzed phosphorylation-mediated cascades. Thus, the third stage of amplification for cyclic AMP involves the activation of cyclic AMP-dependent protein kinases that continue the cell activation process, with overall amplification factors of about 10^5 or more relative to the initial stimulus (e.g., for G-protein regulation of signal transduction in the visual system, see Stryer, 1991). Like cyclic AMP, cyclic GMP, generated by activation of the ANF receptor family, can also set in motion a signal amplification process mediated by cyclic GMP-dependent protein kinases (PKGs).

The G_q-protein-mediated activation of membrane-localized phospholipase-Cβ provides yet another set of soluble mediators via a process that yields amplification factors comparable to or greater than the cyclic AMP system. In this case, the enzyme reaction yields not one, but two soluble mediators, DAG and IP_3, from the target substrate, phosphatidylinositol (Berridge, 1993). In turn, via the IP_3-mediated elevation of intracellular calcium and the DAG-regulated activation of protein kinase C (PKC), the initial receptor signal is amplified greatly. Elevated intracellular calcium not only can coordinately activate PKC along with DAG, but can also stimulate calcium/calmodulin-regulated kinases that, like kinase C, trigger further downstream signal-amplifying protein kinase cascades. The same set of soluble mediators (i.e., DAG, IP_3, and Ca^{2+}) can be released via the activation of phospholipase C-γ. As outlined in Sections II,F and II,6, this enzyme can be activated via its SH2 domain interactions with phosphotyrosine-containing proteins (Kamat and Carpenter, 1997). Thus, the activation of a wide variety of receptor systems (all of types 2, 3, and 4 in Fig. 1) can lead to the generation of a common set of "second messengers" and therefore some common amplified cellular responses.

Glycerolipid and sphingolipid mediators are also proposed to play a role in receptor-mediated signaling (Brindley *et al.,* 1996; Hannun and Obeid, 1997; Smyth *et al.,* 1997). For instance, the receptor-triggered activation of phospholipase D, rather than phospholipase C, yields phosphatidate, which can be metabolized to lysophosphatidate and DAG (discussed earlier). Phosphatidate and lysphosphatidate can activate cytoplasmic tyrosine kinases, providing for signal amplification. Further, the receptor-mediated activation of sphingomyelinase (e.g., by TNF-α) yields ceramide, yet another lipid believed to play a role in cell signaling. The potential roles of ceramide, phosphatidate, and lysophosphatidate as "second messengers" of agonist action are being explored actively at this point in time (Smyth *et al.,* 1997; Hannun and Obeid, 1997; Pena *et al.,* 1997). In essence, all of the "second messengers" discussed in this section can be seen to originate at the plasma membrane as a product of a plasma membrane-localized enzyme (e.g., adenylyl cyclase); for certain messengers, both the enzyme and its substrate may be localized in the membrane itself.

B. Module-Targeted Signal Transduction Pathways

One problem that a diffusable second messenger must overcome relates to the cellular location at which the signal amplification will take place. To some extent, this problem for diffusable messengers (e.g., cyclic AMP) is solved by prelocating the target enzymes via site-specific scaffolding and anchoring proteins (Lester and Scott, 1997). For receptors like the ones typified by types 3 and 4 in Fig. 1, the "diffusable messengers" are large molecular weight proteins rather than the comparatively small highly diffusable messengers such as cylic AMP. Thus, a different amplification mechanism has evolved employing modular adapter proteins, such as GRB2, discussed in Sections II,C and II,F. Through these "hand-to-hand"-like interactions between protein modules, the catalytic moieties can be recruited to the cell membrane at appropriate sites where the substrates are located. The scheme shown in Fig. 3 illustrates this principle for the activation of Ras by a "growth factor" receptor. The recruitment of mSOS to activate Ras resulting from the activation of either "growth factor" or G-protein-coupled receptors results subsequently in the downstream activation of the MAP kinase cascade that amplifies considerably the initial receptor signal (Graves *et al.,* 1997; Sugden and Clerk, 1997; Cobb *et al.,* 1996). Here again, a protein kinase cascade serves as the "signal amplifier" mechanism. In general, this adapter protein mechanism, employing specific sequence modules [e.g., SH2, SH3, PTB, and pleckstrin-homology (PH) domains] provides both for site specificity and the amplification of receptor signals (Pawson, 1995; Pawson and Scott, 1997).

A final amplification process that employs "module-targeted" signal amplification results from the activation of nuclear transcriptional events (Su and Karin, 1996). For this process, receptor triggering, via one of the processes outlined in Figure 1, leads to the phosphorylation/activation of cytoplasmic proteins (e.g., STATs, SMADs, or NFκB) that then translocate to the nucleus to initiate transcriptional events. This type of signal amplification occurs within a much longer time frame than those receptor-triggered amplification mechanisms that regulate rapid tissue responses such as changes in smooth muscle contractility. In summary, both "diffusable" low molecular weight second messengers and adapter-coupled receptor-signaling effectors use modular protein–protein interactions for the site-specific amplification of ligand-induced cell signals. Amplication can take place very rapidly at the cell membrane or at specific cytoplasmic sites; alternatively, an amplification at the level of gene transcription can occur over a slower time frame via the translocation of activated transcription factors from the cytoplasm to the nucleus.

IV. THE PLASMA MEMBRANE AS AN INTEGRATION SITE FOR SIGNAL TRANSDUCTION

A. Membrane Translocation of Signaling Molecules

As has been emphasized throughout this chapter, the plasma membrane is the focus for a number of critical events in signal transduction pathways. As mentioned in Section II,A, upon binding of ligand to its appropriate cell surface receptor, a cascade of signaling events is initated. The plasma membrane is not a passive structure, but one which dynamically helps to integrate and localize downstream signaling molecules, presumably to increase the efficiency of the signal transduction process after receptor stimulation has taken place. The inner leaflet of the plasma membrane acts as an anatomical site of molecular recruitment and clustering. Many signaling molecules are localized to the inner surafce of the plasma by a number of different molecular mechanisms. One mechanism by which membrane localization can be achieved results from the prelocalization of molecules based on posttranslational modifications (Casey, 1995). This concept has been introduced earlier with respect to the fatty acid modifications that occur on the G-proteins (Section II,E,2). In addition to G-proteins, the Src family of protein tyrosine kinases are attached to the plasma membrane by virtue of fatty acid modifications. Src itself is myristoylated, whereas many of the other Src family kinases, including Lck, Fyn, and Hck, are both myristoylated and plamityolated (Resh, 1994). A second class of proteins regulated by fatty acid modifications is explemfied by the small molecular weight GTPases, including Ras. These "small" G-proteins are prenylated with either the 15 carbon farnesyl or the 20 carbon, geranylgeranyl (Glomset and Farnsworth, 1994). In addition, many of these G-proteins contain a second motif required for plasma membrane localization: either an additional fatty modification such as palmityolation or a stretch of polybasic amino acids (Glomset and Farnsworth, 1994). These posttranslational modifications are essential for the cellular activity of these molecules, indicating the importance of the membrane localization of these constituents for regulating cell function. Although it still remains to be clarified, the different covalent modifications on these signal adapter molecules appear to target them to specific subdomains within the plasma membrane. For example, the Src family kinases that are dually acylated are localized to caveolar-like domains, whereas Src itself, which is only myristoylated, does not localize to these microcompartments (Robbins *et al.*, 1995) (see later).

A second way in which proteins can localize to the plasma membrane is by a redistribution either by direct binding to a membrane-localized protein (described in detail in Section II) or by specific membrane-targeting motifs. For example, Raf, a serine threonine kinase, interacts with Ras only when Ras is in its active GTP-bound configuration, and this relocalization of Raf to the plasma membrane is essential for its enzymatic activation (Marshall, 1995). Protein kinase C is an excellent example of a family of enzymes of which at least some isoforms translocate to the plasma membrane after receptor-mediated cell activation (Jaken, 1996). Stimulation of cells with hormones (or phorbol esters) that increase intracellular levels of DAG induces the translocation of several PKC isoforms from the cytosol to the particulate fraction, including the plasma membrane (Jaken, 1996). The different PKC isoforms have distinct subcellular localizations based on specific PKC-anchoring proteins located at these cellular sites. These proteins responsible for the PKC-anchoring process are collectively called RACKs (receptors for activated C-kinase) (Mochly-Rosen, 1995). RACKs bind to PKC at a domain that is distinct from the PKC substrate binding site, suggesting that RACKs aid in the subcompartmentalization of PKC isofoms, such that they now can interact with specific cellular substrates.

The signal-mediated formation of specific phospholipids within the plasma membrane is also an important mechanism leading to the recruitment of signaling molecules to the cell membrane. Various inositol phospholipds, such as phospahatidylinositol 4,5-bisphosphate (PIP2), resulting from cell activation, can bind to specific proteins that contain in their sequence a modular domain of approximately 100 amino acids known as the pleckstrin homogy domain (introduced earlier in Section II,E,3,c) (Gibson *et al.,* 1994; Lemmon and Ferguson, 1998). The binding of PIP2 to the PH domains is important in the translocation of proteins to the membrane by an interaction with the negatively charged head group of the polar phospholipid (Harlan *et al.,* 1994). Several classes of molecules utilize the PH domain for the agonist-induced translocation to membranes, including the serine threonine kinase Akt (PKB), the Bruton's tyrosine kinase, the GTPase dynamin, and the SOS Ras exchange factor, to name a few (Gibson *et al.,* 1994). It still remains to be determined if different phosphatidylinositol products have different specificites for specifc PH domains allowing for a specific recruitment of certain molecules. Furthermore, the production of inositol phospholipids such as PIP2 is not distrubuted randomly in cells, but instead is concentrated in a subcompartment of the plasma membrane known as caveolae (see later) (Pike and Casey, 1996; Hope and Pike 1996). The specific location of unique phospholipids in this membrane domain would further allow for their specificity in recruitment.

B. Compartmentalized Signaling at the Plasma Membrane

It is now clear that the plasma membrane is not homogeneous, but contains specific subcompartments based on their lipid and protein diversity. Considerable attention has been drawn to glycolipid rafts, which represent an assembly of glycosphingolipids and cholesterol (Simons and Ikonen, 1997). Caveolae are one such membrane microdomain that appear as small (50–100 nm) flask-shaped invaginations on the surface of a wide diversity of cell types (Parton, 1996). Caveolae have a unique glycolipid composition, are rich in glycosphingolipids, and are dependent on the presence of cholesterol in the membrane (Simons and Ikonen, 1997). The protein caveolin (now known as caveolin 1) is a resident structural protein of caveolae and was identified originally as a 22-kDa tyrosine phosphoprotein in chicken embryo fibroblasts infected with the Rous sarcoma virus (Glenney and Soppet, 1992; Rothberg *et al.,* 1992). There are now two other members of this family, caveolin 2 and caveolin 3 (Scherer *et al.,* 1996; Tang *et al.,* 1996; Way and Parton, 1995). Originally caveolae were thought to function only in receptor-mediated potocytosis (Anderson *et al.,* 1992). However, their potential functions are expanding rapidly to include (1) an alternative route of receptor-mediated endocytosis (Schnitzer, 1997), (2) a specific vesicular transporter for various toxins and viruses (Montesano *et al.,* 1982; Stang *et al.,* 1997), (3) transcytosis (Anderson, 1993), (4) a site for the clustering of GPI-linked proteins (Anderson *et al.,* 1992), (5) calcium signaling (Fujimoto *et al.,* 1992; Fujimoto, 1993), and (6) a localized site for signal transduction (Anderson, 1993; Lisanti *et al.,* 1994, 1995).

For the purposes of the content of this chapter, we will focus on the current knowledge on the potential role of caveolae as sites of signal transduction. It has been hypothesized that caveolae can serve as sites of signal integration based largely on the caveolar localization of specific molecules involved directly in various aspects of signal transduction. Both G-proteins and Src family kinases are localized within these membrane microdomains; this localization requires that both classes of proteins be palmitoylated. Furthermore, it has now been shown that some receptors may also be compartmentalized within caveolae, including the G-protein-coupled receptor, endothelin (Chun *et al.,* 1994), and various signaling-competent proteins that are attached by a glycosylphosphatidylinositol anchor (Anderson *et al.,* 1992: Simons and Ikonen, 1977). Regulated signal transduction in these discrete domains is an attractive hypothesis for achieving spatial and temporal specificity in signal transduction.

At least three potential mechanisms (Figure 6) can be suggested by which the caveolae could partake as a site for signal integration: (1) a cell surface receptor may reside outside of the caveolae. but be translocated there on ligand binding. This mechanism represents a dynamic targeting model and may also involve receptors moving from within to outside the caveolar domains. (2) The receptor may be a constitutive resident protein of caveolae and may initate signaling within this site. (3) The receptor system may not

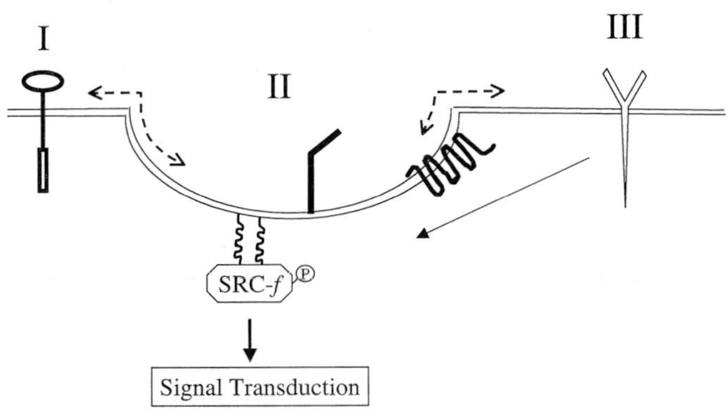

FIGURE 6 Caveolae as localized sites for signal transduction. Mechanisms by which caveolae serve as sites for signal transduction: (I) a cell surface receptor may reside oustside of the caveolae, but be translocated there on ligand binding, (II) the receptor may be a constitutive resident protein of caveolae and may initate signaling within this site, and (III) the receptor system may not itself be localized within caveolae, but on its activation may communicate a signal to the caveolae.

itself be localized within caveolae, but on its activation may communicate a signal to the caveolae by virtue of a second messenger, which could be either protein or lipid mediated. Evidence suggests that all three mechanisms or variations thereof are utilized for various receptor-initatiated signaling events. At least four receptor systems have been shown to translocate and become sequestered in a membrane compartment conisistent with caveolae or having analogous caveolar-like properties. Presently two G-protein-coupled receptors have been found to be localized spatially to caveolae on ligand binding. In cardiac myocytes, muscarinic acetlycholine receptors translocate to caveolae on binding agonist (Feron *et al.,* 1997). Second, the vasoactive peptide agonist bradykinin, when bound to the B2 bradykinin receptor, promotes the recruitment of the bound receptor into caveolae (de Weerd and Leeb-Lundberg, 1997). Concomitant with the sequestering of the B2 receptor, there was an increase in the enrichment of G-proteins localized within the caveolae, implying an active role for the caveolar compartment in signal transduction. In addition, the high-affinity immunoglobulin E receptor (FcεRI) is associated with caveolae-like domains after receptor aggregation (Field *et al.,* 1995, 1997). This aggregation-dependent translocation to these domains precedes its tyrosine phosphorylation and appears to be required for the initial events in signal transduction from this receptor (Field *et al.,* 1997). Furthermore, it has been shown that the B-cell peripheral membrane protein, CD20, a molecule involved in B-cell activation, is translocated rapidly to a caveolar-like fraction on cell surface cross-linking (Deans *et al.,* 1998). Several receptors have now been found to be localized within caveolae, including the G-protein-coupled receptor endothelin (Chun et al., 1994) and the growth factor receptors for PDGF and EGF (Liu *et al.,* 1996; Mineo, *et al.,* 1996; Liu *et al.,* 1997). Data suggest that signaling events initiated by these growth factors occur within these membrane microdomains (Liu *et al.,* 1996; Mineo, *et al.,* 1996; Liu *et al.,* 1997). In addition, several proteins that are attached to the outer leaflet of the plasma membrane by virtue of a GPI anchor have been shown to transmit an intracellular signal when bound to ligand or when cross-linked by specific antibodies (Stefanova *et al.,* 1991). These signaling-competent molecules include proteins such as Thy-1, CD14, CD59, and CD55. For many of these GPI-linked receptor systems, it still remains to be determined how a cell surface protein that does not span the plasma membrane is able to generate a transmembrane signal within the cell. It appears that the Src family of tyrosine kinases may play an early role in the signaling events by such GPI-linked receptors, as the Src family kinases are activated on receptor activation and are present in a complex that contains these GPI-linked receptors (Stefanova *et al.,* 1991; Brown, 1993). It has been speculated that transmembrane adaptors integrate the GPI-anchored receptor on the outside of the cell with the Src-family kinases that are localized on the inner leaflet of the plasma membrane. Finally, data suggest that insulin-mediated signaling results in the phosphorylation of several of the caveolar constituents, including the structural protein caveolin (Mastick *et al.,* 1995). In this scenario it is not clear whether the insulin receptor, the receptor-phosphorylated docking protein IRS1 (see Sections II,F), or both are localized anytime within the caveolar microdomain. It remains to be determined how signals are integrated within caveolae and how they would effect the specificity of signaling within a particular cell type. In addition to G-proteins and Src family protein tyrosine kinases, it is interesting to note that several other signaling molecules have been shown to localize to the cytoplasmic side of the caveolae, including PKCα (Mineo *et al.,* 1998), as well as Ras and Raf (Song *et al.,* 1996), suggesting a direct link to a number of downstream signaling pathways. We speculate that the Src family protein tyrosine kinases will provide a means by which caveolae can initiate signal transduction processes, probably in relationship to all three of the proposed mechanisms by which caveolae may serve as sites for localized signaling (Fig. 6).

V. SUMMARY

There has been an exponential growth in the identification of a large number of molecules that have been implicated in various aspects of signal transduction. These include a large family of specific cell surface receptors that come in a wide array of configurations, families of transducers including protein kinases and their antagonistic counterparts, and protein phosphatases, as well as a wide array of signal adaptor proteins. The challenge will be to understand how all of the different molecules integrate signals to a wide array of extracellular signals to elicit the appropriate phenotypic response, such as cellular proliferation and differentiation.

Acknowledgments

Work in the authors' laboratories relevant to this chapter is supported by grants from the National Cancer Institute of Canada (S.M.R.), the Medical Research Council of Canada (M.D.H.), and by a grant from the Alberta Cancer Board (S.M.R. and M.D.H.). S.M.R. is an Alberta Heritage Foundation for Medical Research Scholar.

References

Abrams, C. S., Zhang, J., Downes, C. P., Tang, X.-W., Zhao, E., and Rittenhouse, S. E. (1996). Phosphopleckstrin inhibits G$\beta\gamma$-activable platelet phosphatidylinositol-4, 5-bisphosphate 3-kinase. *J. Biol. Chem.* **271,** 25192–25197.

Amigorena, S., Salamero, J., Davoust, J. Fridman, W. H., and Bonnerot, C. (1992) Tyrosine-containing motif that transduces cell activation signals also determines internalization and antigen presentation via type III receptors for IgG. *Nature* (*London*) **358,** 337–341.

Anand-Srivastava, M. B. (1997). Atrial natriuretic peptide-C receptor and membrane signalling in hypertension. *J. Hypertens.* **15,** 815–826.

Anderson, R. G. W. (1993). Caveolae: Where incoming and outgoing messengers meet. *Proc. Natl. Acad. Sci. U.S.A.* **90,** 10909–10913.

Anderson, R. G. W., Kamen, B. A., Rothberg, K. G., and Lacey, S. W. (1992) Potocytosis: Sequestration and transport of small molecules by caveolae. *Science* **255,** 410–411.

Band, C. J., Posner, B. I., Duman, V., and Contreres, J. O. (1997). Early signaling events triggered by peroxovanadium [bpV(phen)] are insulin receptor kinase (IRK)-dependent: Specificity of inhibition of IRK-associated protein tyrosine phosphatase(s) by bpV(phen). *Mol. Endocrinol.* **11,** 1899–1910.

Bechade, C., Sur, C., and Triller, A. (1994). The inhibitory neuronal glycine receptor. *BioEssays* **16,** 735–744.

Bergeron, J. J., Di Guglielmo, G. M., Baass, P. C., Authier, F., and Posner, B. I. (1995). Endosomes, receptor tyrosine kinase internalization and signal transduction. *Biosci. Rep.* **15,** 411–418.

Berridge, M. J. (1993). Inositol trisphosphate and calcium signalling. *Nature* (*London*) **361,** 315–325.

Bevan, A. P., Burgess, J. W., Drake, P. G., Shaver, A., Bergeron, J. J., and Posner, B. I. (1995). Selective activation of the rat hepatic endosomal insulin receptor kinase. Role for the endosome in insulin signaling. *J. Biol. Chem.* **270,** 10784–10791.

Boeynaems, J. M., and Dumont, J. E. (1997). The two-step model of ligand–receptor interaction. *Mol. Cell. Endocrinol.* **7,** 33–47.

Borg, J. P., and Margolis, B. (1998). Function of PTB domains. *Curr. Top. Microbiol. Immunol.* **228,** 23–38.

Bourne, H. R. (1997). How receptors talk to trimeric G proteins. *Curr. Opin. Cell Biol.* **9,** 134–142.

Brindley, D. N., Abousalham, A., Kikuchi, Y., Wang, C. N., and Waggoner, D. W. (1996). "Cross talk" between the bioactive glycerolipids and sphingolipids in signal transduction. *Biochem. Cell Biol.* **74,** 469–476.

Brown, D. (1993). The tyrosine kinase connection: How GPI-anchored proteins activate T cells. *Curr. Opin. Immunol.* **5,** 349–354.

Byon, J. C., Kenner, K. A., Kusari, A. B., and Kusari, J. (1997). Regulation of growth factor-induced signaling by protein-tyrosine-phosphatases. *Proc. Soc. Exp. Biol. Med.* **216,** 1–20.

Cambier, J. C. (1995). Antigen and Fc receptor signalling: The awesome power of the immunoreceptor tyrosine-based activation motif. *J. Immunol.* **155,**3281–3285.

Cao, Z., Xiong, J., Takeuchi, M., Kurama, T., and Goeddel, D. V. (1996). TRAF6 is a signal transducer for interleukin-1. *Nature* (*London*) **383,** 443–446.

Carpenter, G. (1992). Receptor tyrosine kinase substrates: *src* homology domains and signal transduction. *FASEB J.* **6,** 3283–3289.

Casey, P. J. (1995) Protein lipidation in cell signaling. *Science* **268,** 221–225.

Cheatham, B., and Kahn, C. R. (1995). Insulin action and the insulin signaling network. *Endocr. Rev.* **16,** 117–142.

Chinkers, M., Singh, S., and Garbers, D. L. (1991). Adenine nucleotides are required for activation of rat atrial natriuretic peptide receptor/guanylyl cyclase expressed in a Baculovirus system. *J. Biol. Chem.* **266,** 4088–4093.

Chun, M. Liyanage, U. K., Lisanti, M. P., and Lodish, H. F. (1994). Signal transduction of a G protein-coupled receptor in caveolae: Colocalization of endothelin and its receptor with caveolin. *Proc. Natl. Acad. Sci. U.S.A.* **91,** 11728–11732.

Clapham, D. E. (1996). Intracellular signalling: More jobs for G beta gamma. *Curr. Biol.* **6,** 814–816.

Clapham, D. E., and Neer, E. J. (1993). New roles for G-protein $\beta\gamma$-dimers in transmembrane signalling. *Nature* (*London*) **365,** 403–406.

Clapham, D. E., and Neer, E. J. (1997). G protein beta gamma subunits. *Annu. Rev. Pharmacol. Toxicol.* **37,** 167–203.

Cobb, M. H., Xu, S., Cheng, M., Ebert, D., Robbins, D., Goldsmith, E., and Robinson, M. (1996). Structural analysis of the MAP kinase ERK2 and studies of MAP kinase regulatory pathways. *Adv. Pharmacol.* **36,** 49–65.

Cohen, G. B., Ren, R., and Baltimore, D. (1995). Modular binding domains in signal transduction proteins. *Cell* (*Cambridge, Mass.*) **80,** 237–248.

Conti-Tronconi, B. M., McLane, K, E., Raftery, M. A., Grando, S. A., and Protti, M. P. (1994). The nicotinic acetylcholine receptor: Structure and autoimmune pathology. *Crit. Rev. Biochem. Mol. Biol.* **29,** 69–123.

Cuatrecasas, P. (1969). Interaction of insulin with the cell membrane: The primary action of insulin. *Proc. Natl. Acad. Sci. U.S.A.* **63,** 450–457.

Cuatrecasas, P. (1972). Affinity chromatography and purification of the insulin receptor of liver cell membranes. *Proc. Natl. Acad. Sci. U.S.A.* **69,** 1277–1281.

Cuatrecasas, P., and Hollenberg, M. D. (1976). Membrane receptors and hormone action. *Adv. Protein Chem.* **30,** 251–451.

Czech, M. P. (1995). Molecular actions of insulin on glucose transport. *Annu. Rev. Nutr.* **15,** 441–471.

Daub, H., Weiss, F. U., Wallasch, C., and Ullrich, A. (1996). Role of transactivation of the EGF receptor in signalling by G-protein-coupled receptors. *Nature* (*London*) **379,** 557–560.

De Häen, C. (1976). The non-stoichiometric floating receptor model for hormone-sensitive adenylate cyclase. *J. Theor. Biol.* **58,** 383–400.

de Weerd, W. F. C., and Leeb-Lundberg, L. M. F. (1997). Bradykinin sequesters B2 bradykinin receptors and the receptor-coupled Gα subunits Gαq and Gαi in caveolae in DDT1 MF-2 smooth muscle cells. *J. Biol. Chem.* **272,** 17858–17866.

Deans, J. P., Robbins, S. M., Polyak, M. J., and Savage, J. A. (1998). Rapid redistribution of CD20 to a low-density detergent-insoluble membrane compartment. *J. Biol Chem.* **273,** 344–348.

Drake, P. G., Bevan, A. P., Burgess, J. W., Bergeron, J. J., and Posner, B. I. (1996). A role for tyrosine phosphorylation in both activation and inhibition of the insulin receptor tyrosine kinase *in vivo. Endocrinology* (*Baltimore*) **137,** 4960–4968.

Drewett, J. G., and Garbers, D. L. (1994). The family of guanylyl cyclase receptors and their lipids. *Endocr. Rev.* **15,** 135–162.

Erpel, T., and Courtneidge, S. A. (1995). Src family protein tyrosine kinases and cellular signal transduction pathways. *Curr. Opin. Cell Biol.* **7,** 176–182.

Fantl, W. J., Johnson, D. E., and Williams, L. T. (1993). Signalling by receptor tyrosine kinases. *Annu. Rev. Biochem.* **62,** 453–481.

Feron, O., Smith, T. W., Michel, T., and Kelly, R. A. (1997). Dynamic targeting of the agonist-stimulated m2 muscarinic acetlycholine receptor to caveolae in cardiac myocytes. *J. Biol. Chem.* **272,** 17744–17748.

Field, K. A., Holowka, D., and Baird, B. (1995). FcγRI-mediated recruitment of p53/56lyn to detergent-resistant membrane domains accompanies cellular signaling. *Proc. Natl. Acad. Sci. U.S.A.* **92,** 9201–9205.

Field, K. A., Holowka, D., and Baird, B. (1997). Compartmentalized activation of the high affinity immunoglobulin E receptor within membrane domains. *J. Biol. Chem.* **272,** 4276–4280.

Fields, T. A., and Casey, P. J. (1997). Signalling functions and biocemical properties of pertussis toxin-resistant G proteins. *Biochem. J.* **321,** 561–571.

Flier, J. S. (1992). Syndromes of insulin resistance. From patient to gene and back again. *Diabetes* **41,** 1207–1219.

Forman, B. M., and Evans, R. M. (1995). Nuclear homrone receptors activate direct, inverted, and everted repeats. *Ann. N. Y. Acad. Sci.* **761,** 29–37.

Freedman, N. J., and Lefkowitz, R. J. (1996). Desensitization of G protein-coupled receptors. *Recent Prog. Horm. Res.* **51,** 319–351.

Freychet, P., Roth, J., and Neville, D. M., Jr., (1971). Monoiodoinsulin: Demonstration of its biological activity and binding to fat cells and liver membranes. *Biochem. Biophys. Res. Commun.* **43,** 400–408.

Frye, L. D., and Edidin, M. (1970). The rapid intermixing of cell surface antigens after formation of mouse-human keterokaryons. *J. Cell Sci.* **7,** 319–335.

Fujimoto, T. (1993). Calcium pump of the plasma membrane is localized in caveolae. *J. Cell Biol.* **120,** 1147–1157.

Fujimoto, T., Nakade, S., Miyawaki, A., Mikoshiba, K., and Ogawa, K. (1992). Localization of inositol 1,4,5-trisphosphate receptor-like protein in plasmalemmal caveolae. *J. Cell Biol.* **119,** 1507–1513.

Garbers, D. L., and Lowe, D. G. (1994). Guanylyl cyclase receptors. *J. Biol. Chem.* **269,** 30741–30744.

Gibson, T. J., Hyvonen, M., Musacchio, A., Saraste, M., and Birney, E. (1994). *Trends Biochem. Sci.* **19,** 349–353.

Gilman, A. G. (1995). G proteins and regulation of adenylyl cyclase. *Biosci. Rep.* **15,** 65–97.

Glenny J. R., Jr,, and Soppet, D. (1992). Sequence and expression of caveolin, a protein component of caveolae plasma membrane domains phosphorylated on tyrosine in Rous sarcoma virus-transformed fibroblasts. *Proc. Natl. Acad. Sci. U.S.A.* **89,** 10517–10521.

Glomset, J. A., and Fransworth, C. C. (1994). Role of protein modification reactions in programming interactions bewteen ras-related GTPases and cell membranes. *Annu. Rev. Cell Biol.* **10,** 181–205.

Gold, M. R., and DeFranco, Al. (1994). Biochemistry of lymphocyte activation. *Adv. Immunol.* **55,** 221–295

Goraczniak, R. M., Duda, T., and Rameshwar, K. W. (1992). A structural motif that defines the ATP-regulatory module of guanylate cyclase in atrial natriuretic factor signalling. *Biochem. J.* **282,** 533–537.

Graves, L. M., Bornfeldt, K. E., and Krebs, E. G. (1997). Historical perspectives and new insights involving the MAP kinase cascades. *Adv. Second Messenger Phosphoprotein Res.* **31,** 49–62.

Gregory, H., Taylor, C. L., and Hopkins, C. R. (1982). Lutenizing hormone release from dissociated pituitary cells by dimerisation of occupied LHRH receptors. *Nature* (*London*) **300,** 269–271.

Guderman, T., and Nümberg, B. (1995). Receptors and G proteins as primary components of transmembrane signal transduction. Part 1 G-protein-coupled receptors: Structure and function. *J. Mol. Med.* **73,** 51–63.

Hannun, Y. A., and Obeid, L. M. (1997). Mechanisms of ceramide-mediated apoptosis. *Adv. Exp. Med. Biol.* **407,** 145–149.

Hara, T., and Miyajima, A. (1996). Function and signal transduction mediated by the interleukin 3 receptor system in hematopoiesis. *Stem Cells* **14,** 605–618.

Harlan, J. E., Hajduk, P. J., Yoon, H. S., and Fesik, S.W. (1994). Pleckstrin homology domains bind to phosphatidyl-inositol-4, 5,-bisphosphate. *Nature* (*London*) **371,** 168–170.

Hazum, E., and Keinan, D. (1985). Gonadotropin releasing hormone activation is mediated by dimerization of occupied receptors. *Biochem. Biophys. Res. Commun.* **133,** 449–456.

Heldin, C. H. (1996). Protein tyrosine kinase receptors. *Cancer Surveys* **27,** 7–24.

Heldin, C. H. (1997a). Simultaneous induction of stimulatory and inhibitory signals by PDGF. *FEBS Lett.* **410,** 17–21.

Heldin, C. H. (1997b). TGF-beta signalling from cell membrane to nucleus through SMAD proteins. *Nature* (*London*) **390,** 465–471.

Hernández-Sotomayor, S. M. T., and Carpenter, G. (1992). Epidermal growth factor receptor: Elements of intracellular communication. *J. Membr. Biol.* **128,** 81–89.

Hirano, T., Nakajima, K., and Hibi, M. (1997). Signaling mechanisms through gp130: A model of the cytokine system. *Cytokine Growth Factor Rev.* **8,** 241–252.

Hollenberg, M. D. (1991). Structure-activity relationships for transmembrane signalling: The receptors turn. *FASEB J.* **5,** 178–186.

Hope, H. R., and Pike, L. J. (1996). Phosphoinositide-utilizing enzymes in detergent-insoluble lipid domains. *Mol. Biol. Cell* **7,** 843–851.

Hsu, H., Huang, J., Shu, H. B., Baichwal, V., and Goeddel, D. V. (1996). TNF-dependent recruitment of the protein kinase RIP to the TNF receptor-1 signaling complex. *Immunity* **4,** 387–396.

Iniguez-Lluhi, J. A., Simon, M. I., Robishaw, J. D., and Gilman, A. G. (1992). G protein $\beta\gamma$ subunits synthesized in Sf9 cells. Functional charaterization and the significance of prenylation of γ. *J. Biol. Chem.* **267,** 23409–23417.

Jaken, S. (1996). Protein kinase C isozymes and substrates. *Curr. Opin. Cell Biol.* **8,** 168–173.

Kahn, C. R., Baird, K. L., Jarrett, D. B., and Flier, J. S. (1978). Direct demonstration that receptor crosslinking or aggregation is important in insulin action. *Proc. Natl. Acad. Sci. U.S.A.* **75,** 4209–4213.

Kahn, C. R., Baird, K. L., Flier, J. S., Grunfeld, C., Harmon, J. T., Harrison, L. C., Karlsson, F. A., Kasuga, M., King, G. L., Lang, U. C., Podskalny, J. M., and van Obberghen, E. (1981). Insulin receptor, receptor antibodies and the mechanism of insulin action. *Recent Prog. Horm. Res.* **37,** 477–538.

Kamat, A., and Carpenter, G. (1997). Phospholipase C-gamma1: Regulation of enzyme function and role in growth factor-dependent signal transduction. *Cytokine Growth Factor Rev.* **8,** 109–117.

Katzenellenbogen, J. A., O'Malley, B. W., and Katzenellenbogen, B. S. (1996). Tripartite steroid hormone receptor pharmacology; interaction with multiple effector sites as a basis for the cells and promoter-specific action of these hormones. *Mol. Endocrinol.* **10,** 119–131.

Kleuss, C., and Gilman, A. G. (1997). Gsalpha contains an unidentified covalent modification that increases its affinity for adenylyl cyclase. *Proc. Natl. Acad. Sci. U.S.A.* **94,** 6116–6120.

Langley, J. N. (1906). On nerve endings and on special excitable substances in cells. *Proc. Roy. Soc. Ser. B.* **78,** 170–194.

Lemmon, M. A., and Ferguson, K. M. (1998). Pleckstrin homology domains. *Curr. Top. Microbiol. Immunol.* **228,** 39–74

Lester, L. B., and Scott, J. D. (1997). Anchoring and scaffold proteins for kinases and phosphatases. *Recent Prog. Horm. Res.* **52,** 409–429.

Levitzki, A., and Bar-Sinai, A. (1991). The regulation of adenylyl cyclase by receptor-operated G proteins. *Pharmacol. Ther.* **50,** 271–283.

Levitzki, A., Marbach, I., and Bar-Sinai, A. (1993). The signal transduction between β-receptors and adenylyl cyclase. *Life Sci.* **52,** 2093–2100.

Lincoln, T. M., and Cornwell, T. L. (1993). Intracellular cyclic GMP receptor proteins. *FASEB J.* **7,** 328–338.

Linder, M. E., Pang, I. H., Duronio, R. J., Gordon, J. I., Sternweis, P. C., and Gilman, A. G. (1991). Lipid modifications of G protein subunits. Myristoylation of Go alpha increases its affinity for beta gamma. *J. Biol. Chem.* **266,** 4654–4659.

Linder, M. E., Middleton, P., Hepler, J. R., Taussig, R., Gilman, A. G., and Mumby, S. M. (1993). Lipid modifications of G proteins: Alpha subunits are palmitoylated. *Proc. Natl. Acad. Sci. U.S.A.* **90,** 3675–3679.

Lisanti, M. P., Scherer, P. E., Tang, Z.-L., and Sargiacomo, M. (1994). Caveolae, caveolin and caveolin-rich membrane domains: A signalling hypothesis. *Trends Cell Biol.* **4,** 231–235.

Lisanti, M. P., Tang, Z., Scherer, P. E., Kubler, E., Koleske, A. J., and Sargiacomo, M. (1995). Caveolae, transmembrane signalling and cellular transformation. *Mol. Membr. Biol.* **12,** 121–124.

Liu, F., and Chernoff, J. (1997). Protein tyrosine phosphatase 1B interacts with and is tyrosine phosphorylated by the epidermal growth factor receptor. *Biochem. J.* **327,** 139–145.

Liu, P., Ying, Y., Ko, Y.-G., and Anderson, R.G.W. (1996). Localization of platelet-derived growth factor-stimulated phosphorylation cascade to caveolae. *J. Biol. Chem.* **271,** 10299–10303.

Liu, J., Oh, P., Horner, T., Rogers, R. A., and Schnitzer, J. E. (1997). Organized endothelial cell surface signal transduction in caveolae distinct from glycosylphosphatidylinositol-anchored protein microdomains. *J. Biol. Chem.* **272,** 7211–7222.

Luttrell, L. M., Della Rocca, G. J., van Biesen, T., Luttrell, D. K., and Lefkowitz, R. J. (1997). Gbetagamma subunits mediate Src-dependent phosphorylation of the epidermal growth factor receptor. A scaffold for G protein-coupled receptor-mediated Ras activation. *J. Biol. Chem.* **272,** 4637–4644.

McCormick, F. (1993). How receptors turn Ras on. *Nature (London)* **363,** 15–16.

Malarkey, K., Belham, C. M., Paul, A., Graham, A., Mclees, A., Scott, P. H., and Plevin, R. (1995). The regulation of tyrosine kinase signaling pathways by growth factor and G-protein-coupled receptors. *Biochem. J.* **309,** 361–375.

Malinin, N. L., Boldin, M. P., Kovalenko, A. V., and Wallach, D. (1997). MAP3K-related kinase involved in NF-kappB induction by TNF, CD95 and IL-1. *Nature (London)* **385,** 540–544.

Margolis, B. (1996). The PI/PTB domain: A new protein interaction domain involved in growth factor receptor signaling. *J. Lab. Clin. Med.* **128,** 235–241.

Marshall, M. S. (1995). Ras target proteins in eukaryotic cells. *FASEB J.* **9,** 1311–1318.

Massague, J. (1996). TGFbeta signaling: Receptors, transducers, and Mad proteins. *Cell (Cambridge, Mass.)* **85,** 947–950.

Massague, J., and Weis-Garcia, F. (1996). Serine/threonine kinase receptors: Mediators of transforming growth factor beta family signals. *Cancer Surveys* **27,** 41–64.

Mastick, C. C., Brady, M. J., and Saltiel, A. R. (1995). Insulin stimulates the tyrosine phosphorylation of caveolin. *J. Cell Biol.* **129,** 1523–1531.

Mineo, C., James, G. L., Smart, E. J., and Anderson, R. G. W. (1996). Localization of epidermal growth factor-stimulated Ras/Raf-1 interaction to caveolae membrane. *J. Biol. Chem.* **271,** 11930–11935.

Mineo, C., Ying, Y.-S., Chapline, C., Jaken, S., and Anderson, R. G. W. (1998). Targeting of protein kinase Cα to caveolae. *J. Cell Biol.* **141,** 601–610.

Miyajima, A., Kitamura, T., Harada, N., Yokota, T., and Arai, K.-I. (1992). Cytokine receptors and signal transduction. *Annu. Rev. Immunol.* **10,** 295–331.

Miyajima, A., Kinoshita, T., Wakao, H., Hara, T., Yoshimura, A., Nishinakamura, R., Murray, R., and Mui, A. (1997). Signal transduction by the GM-CSF, IL-3 and IL-5 receptors. *Leukemia* **3**(Suppl), 418–422.

Mochly-Rosen, D. (1995). Localization of protein kinases by anchoring proteins: A theme in signal transduction. *Science* **268,** 247–251.

Montesano, R., Roth, J., Robert, A., and Orci, L. (1982). Non-coated membrane invaginations are involved in binding and internalization of cholera and tetanus toxins. *Nature (London)* **296,** 651–653.

Mumby, S. M., and Linder, M. E. (1994). Myristoylation of G-protein alpha subunits. *In* "Methods in Enzymology" (R. Iyengar, ed.), Vol. 237, pp. 254–268. Academic Press, San Diego.

Mumby, S. M., Casey, P. J., Gilman, A. G., Gutowski, S., and Sternweis, P. C. (1990). G protein gamma subunits contain a 20-carbon isoprenoid. *Proc. Natl. Acad. Sci. U.S.A.* **87,** 5873–5877.

Murakami, M. S., and Rosen, O. M. (1991). The role of insulin receptor autophosphorylation in signal transduction. *J. Biol. Chem.* **266,** 22653–22660.

Muta, T., Kurosaki, T., Misulovin, Z., Sanchez, M., Nussenzweig, M. C., and Ravetch, J. V. (1994). A 13-amino-acid motif in the cytoplasmic domain of Fc gamma RIIB modulates B-cell receptor signalling. *Nature (London)* **368,** 70.

Myers, M. G., Jr., and White, M. F. (1996). Insulin signal transduction and the IRS proteins. *Annu. Rev. Pharmacol. Toxicol.* **36,** 615–658.

Nicola, N. A., and Metcalf, D. (1991). Subunit promiscuity among hemopoietic growth factor receptors. *Cell (Cambridge, Mass.)* **67,** 1–4.

Nicola, N. A., Smith, A., Robb, L., Metcalf, D., and Begley, C. G. (1997). The structural basis of the biological actions of the GM-CSF receptor. *Ciba Found. Symp.* **204,** 19–27.

Nishibe, S., Wahl, M. I., Hernández-Sotomayor, S. M. T., Tonks, N. K., Rhee, S. G., and Carpenter, G. (1990). Increase of the catalytic activity of phospholipase C-γ1 by tyrosine phosphorylation. *Science* **250,** 1253–1256.

Ono, M., Okada, H., Bolland, S., Yanagi, S., Kurosaki, T., and Ravetch, J. V. (1997). Deletion of SHIP or SHP-1 reveals two distinct pathways for inhibitory signaling. *Cell (Cambridge, Mass.)* **90,** 293–301.

Park, D., Jhon, D. Y., Lee, C. W., Lee, K. H., and Rhee, S. G. (1993). Activation of phospholipase C isozymes by G protein $\beta\gamma$ subunits. *J. Biol. Chem.* **268,** 4573–4576.

Parton, R. G. (1996). Caveolae and caveolins. *Curr. Opin. Cell Biol.* **8,** 542–548.

Pawson, T. (1995). Protein modules and signalling networks. *Nature (London)* **373,** 573–580.

Pawson, T., and Scott, J. D. (1997). Signalling through scaffold, anchoring, and adaptor proteins. *Science* **278,** 2075–2080.

Pena, L. A., Fuks, Z., and Kolesnick, R. (1997). Stress-induced apoptosis and the sphingomyelin pathway. *Biochem. Pharmacol.* **53,** 615–621.

Pike, L. J., and Casey, L. (1996). Localization and turnover of phosphatdylinositol 4,5-bisphosphate in caveolin-enriched membrane domains. *J. Biol. Chem.* **271,** 26453–26456.

Premont, R. T., Inglese, J., and Lefkowitz, R. J. (1995). Protein kinases that phosphorylate activated G protein-coupled receptors. *FASEB J.* **9,** 175–182.

Qian and Weiss, A. (1997). T cell antigen receptor signal transduction. *Curr. Opin. in Cell Biol.* **9,** 205–212.

Resh, M. D. (1994). Myristylation and palmitylation of src family members: The fats of the matter. *Cell (Cambridge, Mass.)* **76,** 411–413.

Reth, M. (1992). Antigen receptors on B lymphocytes. *Annu. Rev. Immunol.* **10,** 97–121.

Riches, D. W., Chan, E. D. and Winston, B. W. (1996). TNF-alpha-induced regulation and signalling in macrophages. *Immunology* **195,** 477–490.

Robbins, S. M., Quintrell, N. A., and Bishop, J.M. (1995). Myristoylation and differential palmitoylation of the HCK protein-tyrosine kinases govern their attachment to membranes and association with caveolae. *Mol. Cell. Biol.* **15,** 3507–3515.

Rodbell, M. (1995). Signal transduction: Evolution of an idea. *Biosci. Rep.* **15,** 117–133.

Rondeau, J. J., McNicoll, N., Gagnon, J., Bouchard, N., Ong, H., and De Lean, A. (1995). Stoichiometry of the atrial natriuretic factor-R1 receptor complex in the bovine zona glomerulosa. *Biochemistry* **34,** 2130–2136.

Rothberg, K. G., Heuser, J. E., Donzell, W. C., Ying, Y.-S., Glenney, J. R., and Anderson, R. G. W. (1992). Caveolin, a protein component of caveolae membrane coats. *Cell* (*Cambridge, Mass.*) **68,** 673–682.

Sato, N., and Miyajima, A. (1994). Multimeric cytokine receptors: Common versus specific functions. *Curr. Opin. Cell Biol.* **6,** 174–179.

Savarese, T. M., and Fraser, C. M. (1992). *In vivo* mutagenesis and the search for structure–function relationships among G protein-coupled receptors. *Biochem. J.* **283,** 1–19.

Scharenberg, A. M., and Kinet, J. P. (1996). The emerging field of receptor-mediated inhibitory signaling: SHP or SHIP. *Cell* (*Cambridge, Mass.*) **87,** 961–964.

Scherer, P. E., Okamoto, T., Chun, M., Nishimoto, I., Lodish, H. F., and Lisanti, M. P. (1996). Identification, sequence and expression of caveolin-2 defines a caveolin gene family. *Proc. Natl. Acad. Sci. U.S.A.* **93,** 131–135.

Schlessinger, J., and Ullrich, A. (1992). Growth factor signalling by receptor tyrosine kinases. *Neuron* **9,** 383–391.

Schlessinger, J., Shechter, Y., Cuatrecasas, P., and Paston, I. (1978). Quantitative determination of the lateral diffusion coefficients of the hormone-receptor complexes of insulin and epidermal growth factor on the plasma membrane of cultured fibroblasts. *Proc. Natl. Acad. Sci. U.S.A.* **75,** 5353–5357.

Schnitzer, J. E. (1997). The endothelial cell surface and caveolae in health and disease. *In* "Vascular Endothelium: Physiology, Pathology and Therapeutic Opportunities" (G. V. R. Born and C. J. Schwartz, eds.), pp. 77–95.

Schubert, B., VanDongen, A. M., Kirsch, G. E., and Brown, A. M. (1989). Beta-adrenergic inhibition of cardiac sodium channels by dual G-protein pathways. *Science* **245,** 516–519.

Severson, D. L., and Hollenberg, M. D. (1997). The plasma membrane as a transducer and amplifier. *In* "Principles of Medical Biology: Membranes and Cell Signalling," Vol. 7B, pp. 387–419. JAI Press, Greenwich, Connecticut.

Shah, N., Zhang, S., Harada, S., Smith, R. M., and Jarett, L. (1995). Electron microscopic visualization of insulin translocation into the cytoplasm and nuclei of intact H35 hepatoma cells using covalently linked Nanogold-insulin. *Endocrinology* (*Baltimore*) **136,** 2825–2835.

Shu, H. B., Halpin, D. R., and Goeddel, D. V. (1997). Casper is a FADD- and caspase-related inducer of apoptosis. *Immunity* **6,** 751–763.

Simon, M. I., Strathmann, M. P., and Gautam, N. (1991). Diversity of G proteins in signal transduction. *Science* **252,** 802–808.

Simons, K., and Ikonen, E. (1997). Functional rafts in cell membranes. *Nature* (*London*) **387,** 569–572.

Smith, R. M., Harada, S., and Jarett, L. (1997). Insulin internalization and other signaling pathways in the pleiotropic effects of insulin. *Int. Rev. Cytol.* **173,** 243–280.

Smyth, M. J., Obeid, L. M., and Hannun, Y. A. (1997). Ceramide: A novel lipid mediator of apoptosis. *Adv. Pharmacol.* **41,** 133–154.

Song, K. S., Li, S., Okamoto, T., Quilliam, L. A., Sargiacomo, M., and Lisanti, M. P. (1996). Copurification and direct intercation of ras with caveolin, an integral membrane protein of caveolae microdomains. *J. Biol. Chem.* **271,** 9690–9697.

Stang, E., Kartenbeck, J., and Parton, R. G. (1997) Major histocompatibility complex class I molecules mediate association of SV40 with caveolae. *Mol. Biol. Cell* **8,** 47–57.

Stephenson, F. A. (1995). The GABA A receptors. *Biochem J.* **310,** 1–9.

Stefanova, I., Horejsi, V., Ansotegui, I. J., Knapp, W., and Stockinger, H. (1991). GPI-anchored cell-surface molecules complexed to protein tyrosine kinases. *Science* **254,** 1016–1019.

Strosberg, A. D. (1991). Structure/function relationship of proteins belonging to the family of receptors coupled to GTP-binding proteins. *Eur. J. Biochem.* **196,** 1–10.

Stroud, R. M., McCarthy, M. P., and Shuster, M. (1990). Nicotinic acetylcholine receptor superfamily of ligand-gated ion channels. *Biochemistry* **29,** 11009–11023.

Stryer, L. (1991). Visual excitation and recovery. *J. Biol. Chem.* **266,** 10711–10714.

Su, B., and Karin, M. (1996). Mitogen-activated protein kinase cascades and regulation of gene expression. *Curr. Opin. Immunol.* **8,** 402–411.

Sugden, P. H., and Clerk, A. (1997). Regulation of the ERK subgroup of MAP kinase cascades through G protein-coupled receptors. *Cell Signalling* **9,** 337–351.

Sunahara, R. K., Dessauer, C. W., and Gilman, A. G. (1996). Complexity and diversity of mammalian adenylyl cyclases. *Annu. Rev. Pharmacol. Toxicol.* **36,** 461–480.

Sutherland, E. W. (1972). Studies on the mechanism of hormone action. *Science* **177,** 401–408.

Tang, Z., Scherer, P. E., Okamoto, T., Song, K., Chu, C., Kohtz, D. S., Nishimoto, I., Lodish, H. F., and Lisanti, M. P. (1996). Molecular cloning of caveolin-3, a novel member of the caveolin gene family expressed predominantly in muscle. *J. Biol. Chem.* **271,** 2255–2261.

Taussig, R., Quarmby, L. M., and Gilman, A. G. (1993). Regulation of purified type I and type II adenylyl cyclases by G protein $\beta\gamma$ subunits. *J. Biol. Chem.* **268,** 9–12.

Taylor, S. I. (1992). Molecular mechanisms of insulin resistance. Lessons from patients with mutations in the insulin-receptor gene. *Diabetes* **41,** 1473–1490.

Taylor, J. M., Jacob-Mosier, G. G., Lawton, R. G., Remmers, A. E., and Neubig, R. R. (1994). Binding of an alpha 2 adrenergic receptor third intracellular loop peptide to G beta and the amino terminus of G alpha. *J. Biol. Chem.* **269,** 27618–27624.

ten Dijke, P., Miyazono, K., and Heldin, C. H. (1996). Signaling via hetero-oligomeric complexes of type I and type II serine/threonine kinase receptors. *Curr. Opin. Cell Biol.* **7,** 139–145.

Thomson, P. A., James, S. R., Casey, P. J., and Downes, C. P. (1994). A G-protein $\beta\gamma$-subunit-responsive phosphoinositide 3-kinase activity in human platelet cytosol. *J. Biol. Chem.* **269,** 16525–16528.

Tonks, N. K. (1996). Protein tyrosine phosphatases and the control of cellular signaling responses. *Adv. Pharmacol.* **36,** 91–119.

Tridandapani, S. Kelley, T., Cooney D., Pradhan, M. and Coggeshall, K. M. (1997). Negative signaling in B cells: SHIP Grbs Shc. *Immunol. Today* **18,** 424–427.

van Biesen, T., Luttrell, L. M., Hawes, B. E., and Lefkowitz, R. J. (1996). Mitogenic signalling via G-protein-coupled receptors. *Endocr. Rev.* **17,** 698–714.

van Oers, N. S., and Weiss, A. (1995). The Syk/ZAP-70 protein tyrosine kinase connection to antigen receptor signalling processes. *Semin. Immunol.* **7,** 227–236.

Wallach, D., Boldin, M., Varfolomeev, E., Beyaert, R., Vandenabeele, P., and Fiers, W. (1997). Cell death induction by receptors of the TNF family: Towards a molecular understanding. *FEBS Lett.* **410,** 96–106.

Wange, R. L., and Samelson, L. E. (1996). Complex complexes: Signaling at the TCR. *Immunity* **5,** 197–205.

Watson, S. P., and Arkinstall, S. (1994). "Factsbook on G Protein-linked Receptors and Their Effector Pathways." Academic Press, San Diego.

Way, M., and Parton, R. G. (1995). M-caveolin, a muscle-specific caveolin-related protein. *FEBS Lett.* **376,** 108–112.

Wedegaertner, P. B., Wilson, P. T., and Bourne, H. R. (1995). Lipid modifications of trimeric G proteins. *J. Biol. Chem.* **270,** 503–506.

Weiss, A. (1995). Signal transduction. Zapping tandem SH2 domains. *Nature (London)* **377,** 17–18.

Wells, J. A., and deVos, A. M. (1996). Hematopoietic receptor complexes. *Annu. Rev. Biochem.* **65,** 609–634.

White, M. F., and Yenush, L. (1998). The IRS-signaling system: A network of docking proteins that mediate insulin and cytokine action. *Curr. Top. Microbiol. Immunol.* **228,** 179–208.

Wilden, P. A., Kahn, C. R., Siddle, K., and White, M. F. (1992). Insulin receptor kinase domain autophosphorylation regulates receptor enzymatic function. *J. Biol. Chem.* **267,** 16660–16668.

Wilks, A. F., and Oates, A. C. (1996). The JAK/STAT pathway. *Cancer Surveys* **27,** 139–163.

Wong, S. K.-F., and Garbers, D. L. (1992). Receptor guanylyl cyclases. *J. Clin. Invest.* **90,** 299–305.

Yamane, H. K., and Fung, B. K. (1993). Covalent modifications of G-proteins. *Annu. Rev. Pharmacol. Toxicol.* **33,** 201–241.

Yamane, H. K., Farnsworth, C. C., Xie, H. Y., Howald, W., Fung, B. K., Clarke, S., Gelb, M. H., and Glomset, J. A. (1990). Brain G protein gamma subunits contain an all-trans-geranylgeranylcysteine methyl ester at their carboxyl termini. *Proc. Natl. Acad. Sci. U.S.A.* **87,** 5868–5872.

Yamauchi, T., Ueki, K., Tobe, K., Tamemoto, H., Sekine, N., Wada, M., Honjo, M., Takahashi, M., Takahashi, T., Hirai, H., Tushima, T., Akanuma, Y., Fujita, T., Komuro, I., Yazaki, Y., and Kadowaki, T. (1997). Tyrosine phosphorylation of the EGF receptor by the kinase Jak2 is induced by growth hormone. *Nature* (*London*) **390,** 91–96.

Yang, R. B., Foster, D. C., Garbers, D. L., and Fulle, H. J. (1995). Two membrane forms of guanylyl cyclase found in the eye. *Proc. Natl. Acad. Sci. U.S.A.* **92,** 602–606.

Yang, R. B., Fulle, H. J., and Garbers, D. L., (1996). Chromosomal localization and genomic organization of genes encoding guanylyl cyclase receptors expressed in olfactory sensory neurons and retina. *Genomics* **31,** 367–372.

Yatani, A., Codina, J., Imoto, Y., Reeves, J. P., Birnbaumer, L., and Brown, A. M. (1987). A G protein directly regulates mammalian cardiac calcium channels. *Science* **238,** 1288–1292.

Zierler, K. L. (1960). Effect of insulin on potassium efflux from rat muscle in the presence and absence of glucose. *Am. J. Physiol.* **190,** 1066–1070.

Zierler, K. (1985). Membrane polarization and insulin action. *In* "Insulin: Its Receptor and Diabetes" (M. D. Hollenberg, ed.), pp. 141–179. Dekker, New York.

Zierler, K., and Rogus, E. M. (1981). Effects of peptide hormones and adrenergic agents on membrane potentials of target cells. *Fed. Proc.* **40,** 121–124.

CHAPTER 12

Active Transport and Pumps

Robert L. Post
Department of Physiology, School of Medicine, University of Pennsylvania, Philadelphia, Pennsylvania 19104

I. INTRODUCTION

Membrane transport is movement of a substance across a cell membrane from an aqueous solution on one side of the membrane to an aqueous solution on the other side of the membrane. Cells are surrounded by a plasma membrane and also contain numerous intracellular organelles that are enclosed by their own membranes. Membrane transport includes solute movement across all of these membranes.

A. Forces Driving Membrane Transport

Three forces drive membrane transport. These are force of diffusion, electrical force, and force generated by a transport system intrinsic to the membrane.

1063-5823/99 $30.00

1. Diffusion

Diffusion is a randomization of the location of molecules in space that is driven by thermal agitation. The force is proportional to the absolute temperature (°K) and to the logarithm of the ratio of the concentration of the transported substance on one side of the membrane to that on the other side, i.e., to the difference in molar-free energy of a transported substance on the two sides of a membrane. The rate of diffusion also depends on the geometry of the space in which diffusion is taking place.

2. Electrical Force

The electrical force pertains to electrically charged solutes, namely ions. The force is the electrical potential across the membrane and is usually expressed in millivolts. If the cytoplasm is electrically negative with respect to the extracellular fluid, then there is an electrical force driving cations into the cell and anions out of the cell.

3. Equilibrium and Active Transport

When the force of diffusion is zero for an uncharged molecule or when the electrical force and the force of diffusion are equal and opposite for an ion, then that molecule or ion is at equilibrium and no net transport takes place in the absence of other forces. If net transport is observed under this condition, this observation is evidence for the third kind of force, a force generated by a transport system. Transport driven by this force is referred to as active transport.

4. Passive Transport

If a solute is not at equilibrium across a membrane and the only forces acting on it are the force of diffusion and the electrical force, and if there is a pathway by which transport can take place, then transport proceeds in a direction that brings the system closer to a state of equilibrium. This occurs either by net transport to make the concentrations on the two sides of the membrane more nearly equal or, in the case of ions, by the movement of electrical charge carried by the ions to change the membrane potential. In the case of ions, the change in membrane potential is usually produced by the movement of chemically insignificant amounts of ions. The direction of a movement that is driven only by the combined effects of diffusion and the elctrical force is "downhill." If net transport is observed to take place in a direction opposite to that in which the net effect of the forces of diffusion and the electrical force would drive it, i.e., "uphill," then this observation also is evidence for the third kind of force, which can generate uphill transport. This force is produced by transport systems intrinsic to the membrane.

B. Definitions of Active Transport

The expression "active transport" is used in two ways. One usage equates "active transport" with uphill transport. The opposite of this usage of "active transport" is "passive transport," which refers to downhill transport. This usage is in the context of energetics or thermodynamics of transport. The other usage of "active transport" is in the context of mechanism of transport and refers to a specific kind of stoichiometric transport system which is discussed later.

An interesting ambiguity arises in one special case. Suppose that an active transport system in a membrane is transporting a substance in a direction bringing the system closer to a state of equilibrium, i.e., "downhill." Is this process active transport? According to the energetics, the thermodynamic definition, it would be considered to be passive transport. Nevertheless, the transport is carried out by an active transport system. In a context of process, it is passive transport, whereas in a context of mechanism, it is active transport. This ambiguity has been a souce of confusion. In this chapter transport toward equilibrium will be called "downhill" and transport away from equilibrium will be called "uphill."

C. Types of Membrane Transport Systems

Membrane transport systems can be divided into two principal categories. These categories are catalytic systems on the one hand and stoichiometric systems on the other. Catalytic systems are pores and channels, whereas stoichiometric transport systems are carriers and pumps.

Catalytic systems carry out only downhill transport, movement toward equilibrium. The amount of substance transported is not related to the reaction sequence of the transport system, if there is one. The reaction sequence of a catalytic system consists of states in which it is open, permitting substances to pass through it, or closed, not permitting transport. There may be more than one of these states. A catalytic system may have only one state, open, and so not have a reaction sequence. When the system is open, the amount of transport varies directly with the degree of disequilibrium between the transported substances on the two sides of the membrane. If a transported substance is at equilibrium, there is no net transport even if the catalytic transport system is open. Away from equilibrium, under physiological conditions for instance, one sodium ion channel can transport thousands of sodium ions during a single open state. Thus the amount of substance transported can exceed the amount of the transport system greatly during a single cycle of its reaction sequence. An open catalytic transport system is accessible to transported substrate on both sides of the membrane simultaneously.

In contrast, stoichiometric transport systems can carry out transport uphill, away from equilibrium. The amount of substance transported is related numerically to the cycle of the reaction sequence; the number is characteristically an integer no larger than four solute ions or molecules per cycle. The reaction sequence has at least two states in which the transported substance has access to the transport system only from the solution on one side of the membrane. In one of these states the substrate has access from one side of the membrane and in the other state it has access from the other side of the membrane. In addition to these states there is an intermediate state in which the transported substrate does not have access to the solution on either side of the membrane. This state is the occluded state or occluded conformation.

Catalytic transport systems are pores and channels. Pores select their transported substrates according to the size of the transported molecule. Examples of pores include connexin (Goodenough *et al.,* 1996), porin (Schulz, 1996), and colicin (Gouaux, 1997). Channels are more selective. Characteristically they transport ions and select not only according to charge but also according to species. For example, the electrically excitable sodium channel of nerve membranes selects Na^+ in preference to K^+ by a factor of 10 to 1.

Stoichiometric transport systems are carriers and pumps. There are three kinds of carriers: facilitated diffusion, cotransporters (symporters), and countertransporters (antiporters). Cotransporters and countertransporters can perform "secondary active transport" by using energy from the downhill transport of one transported substrate to drive the uphill transport of another transported substrate. Pumps are distinguished from carriers by the linkage of transport to an external source of energy. Such sources are hydrolysis of a phosphate bond or capture of a photon. Pumps perform "primary active transport."

The physiological function of pumps is principally homeostasis, for instance, maintaining and restoring electrochemical (concentration) gradients across cell membranes. The physiological functions of channels are principally signaling and excretion or secretion of salts, which they do by dissipating electrochemical gradients.

D. Occluded Conformation of Stoichiometric Transport Systems

In the history of membrane transport, at an early stage the interesting question concerned permeability. What substances pass through what membranes and how easily? Rudolph Höber epitomizes this approach in Section 4 of his book (Höber, 1946). Later, interest focused on transport kinetics of physiologically important substances such as sugars (Widdas, 1952), and it became clear that specific transport systems were being characterized and that types of transport systems could be distinguished. For example, the rate of transport of glucose into erythrocytes became saturated as the concentration increased on either the extracellular side or the cytoplasmic side of the membrane. D-Glucose was transported much more rapidly than L-glucose and structurally related sugars were competitive inhibitors. Transport of glucose was always downhill, and a reaction sequence was devised. The reaction sequence implied that movement of a loaded carrier in one direction was accompanied by an equal and opposite movement of an empty carrier in the opposite direction. Widdas (1952) pointed out that if the empty carrier were initially loaded with a competing sugar at equilibrium across the membrane, then the addition of glucose would induce transport *uphill* of the competing sugar in a direction opposite to that of the transport of glucose. The prediction was fulfilled experimentally in two laboratories independently (Park *et al.,* 1956; Rosenberg and Wilbrandt, 1957). Uphill transport implies what is now called an "occluded conformation" in which the transported molecule lacks access to the solution on either side of the membrane (Post *et al.,* 1969). Indirect evidence for this reactive state was found in the behavior of the sodium and potassium pump and the state was named (Post *et al.,* 1972). The state was demonstrated by direct evidence (Glynn and Karlish, 1990) and now is an accepted feature of the reaction sequence of pumps.

II. PUMPS

A. Light-Driven Pumps

1. Bacteriorhodopsin

Bacteriorhodopsin is an intrinsic protein [a single chain of 248 amino acids (Khorana *et al.,* 1979)] in the cell membrane of bacteria that live in salt marshes, particularly *Halobacterium salinarum* (formerly *halobium*). After bacteriorhodopsin absorbs a photon, it transfers a proton from the cytoplasm to the extracellular medium (Lanyi, 1997). The proton passes from one side of the membrane to the other by way of a series of sites acting as relay stations. In sequence from cytoplasm to extracellular medium, these sites are at aspartate96, a Schiff base between the chromophore retinal and lysine216, aspartate85, glutamate204, and glutamate194. These residues are parts of two hydrogen bond networks, which include water molecules. One network is on the cytoplasmic side of the Schiff base and the other is on the extracellular side.

In a resting state, Asp96, the Schiff base, and Glu204 are protonated; Asp85 and Glu194 are not protonated and the retinal is in the all-*trans* conformation. After absorption of a photon at 568 nm, the retinal changes to a 13-*cis*,15-*anti* conformation. In this conformation the pK_a of the Schiff base decreases and transfer of a proton from the Schiff base to Asp85 is favored. Protonation of Asp85 reduces the pK_a of Glu204 and favors transfer of its proton to Glu194. Release of the proton from Glu194 to the extracellular medium in turn raises the pK_a of Asp85 and completes transfer of the proton from the Schiff base to Asp85. These changes complete the first phase of the transport reaction sequence. At this point a change in the conformation of the protein takes place that switches proton access to the Schiff base from the extracellular side of the membrane to the cytoplasmic side of the membrane. The unprotonated Schiff base now enters the cytoplasmic hydrogen bonding network containing Asp96. A further conformational change, a tilt of an α helix, increases hydration of Asp96 and reduces its pK_a sufficiently for it to donate its proton to the Schiff base. Transfer of the proton from Asp96 to the Schiff base allows the tilt to reverse and raise the pK_a of Asp86 so that it becomes protonated from the cytoplasmic medium. The last events in the cycle are a reisomerization of the retinal to a twisted all-*trans* conformation, transfer of a proton from Asp85 to Glu204, and relaxation of the twist in the retinal chain. Return of the pK_a of Asp85 to its initial value, about 2.5, makes the proton transfer effectively irreversible. Thus absorption of a photon by the chromophore decreases the pK_a of the Schiff base, enabling it to drive a proton through the extracellular hydrogen bond network into the extracellular medium. Discharge of this proton enables a conformational change that gives the deprotonated Schiff base access to a proton in the cytoplasmic hydrogen bond network. Further conformational changes deposit a proton from the network onto the Schiff base and deposit a proton from the cytoplasm onto the network. Finally the system relaxes back to the resting state.

2. Other Rhodopsins

H. salinarum contains other rhodopsins. Halorhodopsin uses the energy of light to transport chloride ions from the extracellular medium into the cytoplasm (Váro *et al.,* 1995). Sensory rhodopsins (SRI and SRII) use the energy of light of different wavelengths to direct migration of the bacteria away from injurious ultraviolet light (SRII) and toward orange light (SRI), which is the optimal wavelength to energize bacteriorhodopsin and halorhodopsin (Hoff *et al.,* 1997). Sensory rhodopsins communicate with intrinsic transducer proteins in the membrane (Spudich, 1994), which communicate with histidine kinases; the signal finally reaches a flagellar motor switch that alters the migration of the bacteria (Hoff *et al.,* 1997). If separated from their transducer proteins, the sensory proteins can pump protons (Spudich, 1994).

All of these rhodopsins are related to each other according to their primary sequence (Hoff *et al.,* 1997). They are also related to vertebrate rhodopsin, the visual pigment in the eye (Maeda *et al.,* 1997). Absorption of a photon by rhodopsin causes it to activate an intrinsic GTP-binding protein, transducin; activation of transducin is the next step in sending the signal eventually to the brain.

B. Pumps Driven by Electron Transport (Redox Pumps)

These pumps use energy from oxidation–reduction reactions to drive protons across a membrane. This category of pumps includes mitochondrial enzymes: NADH:ubiquinone oxidoreductase (complex I) (Walker, 1992; Brandt, 1997), ubiquinol–cytochrome c reductase (complex III) (Brandt, 1996), and cytochrome c oxidase (complex IV) (Tsukihara *et al.,* 1996; Yoshikawa, 1997). Ubiquinol–cytochrome c reductase and cytochrome c oxidase are members of a superfamily found also in prokaryotes (Calhoun *et al.,* 1994). Members contain two characteristic subunits named "I" and "II." Subunit II accepts electrons from a reducing substrate and donates them to subunit I. Subunit I pumps protons. The three-dimensional structure at 2.8-Å resolution has been solved for the enzyme from *Paracoccus denitrificans* (Iwata *et al.,* 1995) and for that from beef heart mitochondria (Tsukihara *et al.,* 1996).

The reaction mechanism of cytochrome c oxidase is becoming better understood (Wikström, 1989; Iwata *et al.,* 1995).[1] The overall reaction is the following:

$$\begin{aligned} &4\text{ ferrocytochrome c} + 8H^{+}\text{(inside)} \\ &\qquad + O_2 \rightarrow 4\text{ ferricytochrome c}^{+} + 4H^{+}\text{ (outside)} + H_2O. \end{aligned}$$

In the case of mitochondria, "inside" refers to the matrix and "outside" refers to the cytoplasm. In the case of prokaryotes, "inside" refers to the cytoplasm and "outside" refers to the periplasm. There are pathways within cytochrome c oxidase for the migration of electrons and pathways for the migration of protons. Electrons are donated from cytochrome c to a Cu_A center, which contains two copper atoms and is part of subunit II. From this center they pass to a heme-a center, which contains an iron atom and is part of subunit I. From this center they pass to a binuclear center, where

[1] See also ⟨http://arc-gen1.life.uiuc.edu/Bioph354/Cyt_ox.html⟩ by Antony Crofts on the Internet.

reactions with O_2 and H^+ take place. The binuclear center contains an iron atom in another heme, heme-a_3, and a copper atom, Cu_B.

The reaction sequence is approximately as follows. The enzyme passes through a series of states named O, P, and F. State O is the oxidized state of the binuclear center, Fe^{3+} Cu^{2+}. Two electrons from heme-a successively reduce the center to Fe^{2+} Cu^+. Then a dioxygen molecule is added to the center. The electrons migrate from the metals to the oxygen to form a Fe^{3+} Cu^{2+} peroxy intermediate, state P. Now a third electron and two protons from the matrix phase are added and the oxygen atoms dismutate. One atom combines with the protons and the electron to form a water molecule coordinated with Cu_B. The other atom forms a ferryl intermediate with the iron atom, $Fe^{4+}{=}O^{2-}$, state F. Finally a fourth electron and two more protons from the matrix phase are added. The iron returns to the Fe^{3+} oxidation level and a second water molecule is formed. The binuclear reaction center returns to state O.

Protons participate in the reaction in two ways: they combine with oxygen to form water (chemical protons) (see earlier) and they are transported outward across the membrane (pump protons). Two protons are pumped outward simultaneously with the consumption of two protons during each of two transitions. These transitions are from state P to state F and from state F to state O. Thus transport of pump protons is linked to the consumption of chemical protons (Verkhovsky *et al.,* 1997). Possible pathways for protons through the protein are through cavities between transmembrane α helices. Transmembrane α helices of subunit I are arranged in three groups (Iwata *et al.,* 1995; Tsukihara *eta l.,* 1996). Each group contains four helices arranged approximately in a semicircle as viewed from outside the membrane. The centers of the groups are arranged in an equilateral triangle. The groups contain the cavities. One group encloses heme-a and another encloses the binuclear center. Heme-a and heme-a_3 are located in the cavities halfway between the faces of the membrane. Hydrogen bond networks are possible pathways for protons to pass into and through the membrane.

C. ATP-Driven Pumps and ATP Synthases

1. Classification

Pedersen and Amzel (1993) classify these transport enzymes as P, F, V, or M types, omitting arsenite pumps. P-type pumps accept a phosphate group from ATP at a specific aspartyl residue as a step in their reaction sequence (Möller *et al.,* 1996). F-type ATP synthases and V-type ATPases are related to each other in structure, function, and homology of subunits

(Kibak *et al.,* 1992; Fillingame, 1996). They transport protons and synthesize or hydrolyze ATP, respectively. M-type ATPases are multidrug transporters (Sarkadi *eta l.,* 1996).

2. P-Type ATPases[2]

a. A Superfamily. These pumps accept a phosphate (P) group from ATP in their reaction sequence. The phosphate group is bound to a specific aspartyl residue. A consenus sequence of amino acids adjacent to this residue identifies this type of pump. This sequence is -Asp-Lys-Thr-Gly-Thr-Leu-Thr-. These pumps form a superfamily (Green, 1992; Horisberger, 1994; Möller *et al.,* 1996). Members of the superfamily are found not only in mammals, but also in amphibians, fishes, insects, crustaceans, protozoa, plants, fungi, and bacteria. These pumps characteristically pump inorganic cations but in some cases the function of the pump is not known because it was identified only by its primary structure. A chloride ion pump has been characterized in *Aplysia* (Gerencser and Purushotham, 1996). A flippase, an enzyme that moves amino phospholipids from one leaflet of the bilayer of a plasma membrane to the other leaflet, has been found to be a member of the superfamily (Tang *et al.,* 1996).

The superfamily can be divided into two principal types, I and II, on the basis of structure and kind of ion transported (Lutsenko and Kaplan, 1995; Möller *et al.,* 1996). Common to both types are six transmembrane and intervening domains. Type I P-type ATPases have two additional transmembrane domains and metal-binding sites at the N-terminal end of the common domains, whereas type II ATPases have four additional transmembrane domains at the C-terminal end of the common domains. Type I ATPases pump heavy metals whereas type II ATPases pump protons, alkali metal cations, or alkaline earth cations. The common domains include two cytoplasmic loops of about 140 and 330 amino acids each. The larger of these contains the active sites of phosphorylation and ATP binding. Four type II ATPases have been crystallized two dimensionally in the plane of the membrane. The crystal habits are diverse. Electron microscopy of calcium ATPase crystals shows a large cytoplasmic head bearing a projecting beak and joined to the membranes by a stalk (MacLennan *et al.,* 1997; Mintz and Guillain, 1997; Stokes, 1997). ATP binds to the membrane aspect of the beak whereas calcium ion transport sites are some distance away among transmembrane helices.

[2] To learn about developments in the field of P-type transport ATPases, consult the VIIIth International Conference on the "Na/K-ATPase and Related Transport ATPases" (1997). Ann. N.Y. Acad. Sci. Vol. 834.

b. Subunits. In addition to a principal subunit as described earlier, many P-type ATPases have additional subunits. The sites for K^+ transport in Kdp ATPase of *Escherichia coli* are in one of these additional subunits (Altendorf and Epstein, 1996). If a protomer is defined as the minimum combination of subunits that can function, there is evidence for structural and functional interactions between protomers in the reaction sequence and life cycle of P-type ATPases (Koster *et al.,* 1995; Mintz and Guillain, 1997).

c. Stoichiometry and Reaction Sequence. These features are rather well understood for the calcium ATPase of sarcoplasmic reticulum (SERCA) (MacLennan *et al.,* 1997; Mintz and Guillain, 1997) and for the sodium and potassium ATPase of plasma membranes (Na,K-ATPase) (Skou and Esmann, 1992; Cornelius, 1996). Stoichiometry and the reaction sequence of Na,K-ATPase will be outlined here, and SERCA follows a similar pattern. Gastric H,K-ATPase is also closely related to Na,K-ATPase; the two enzymes have homologous α and β subunits. Gastric H,K-ATPase acidifies gastric juice by pumping H^+ out of oxyntic cells and equimolar K^+ into the cells using a similar reaction sequence, but details are not as clear as in the other two enzymes (Shin *et al.,* 1996).

Na,K-ATPase transports three Na^+ outward and two K^+ inward across the plasma membrane per one molecule of cytoplasmic ATP hydrolyzed to cytoplasmic ADP and inorganic phosphate. Transport requires cytoplasmic Mg^{2+} as a catalyst. There are slow partial reactions and each of the ligands can be replaced by congeners to some extent. Na^+ and K^+ are transported in alternation and are bound preferentially by different conformations of the transport domain embedded in the membrane. These conformations of the enzyme are designated E1 for conformations with bound Na^+ and E2 for conformations with bound K^+. Transport steps include the acceptance of ions from one side of the membrane, an intermediate occluded state in which the transported ions do not have access to either side of the membrane, and delivery of the ions to the other side of the membrane. The state with occluded Na^+ is unstable in the native enzyme but can be stabilized by various treatments. The state with occluded K^+ is remarkably stable. Thus the reaction sequence for transport can begin with acceptance of 3 cytoplasmic Na^+ and their occlusion followed by their release to the extracellular medium. This release is followed by acceptance of 2 extracellular K^+ and their occlusion followed by release to the cytoplasm. Steps in the transport domain are linked closely to steps in the phosphorylation domain, which is located in a head piece attached to the membrane by a stalk. E1 and E2 in the transport domain are linked to corresponding E1 and E2 in the phosphorylation domain. E1 in the phosphorylation domain catalyzes phosphorylation of the active site aspartate residue by ATP reversibly. E2 in the phosphorylation domain catalyzes phosphorylation of the active site aspartate residue by inorganic phosphate reversibly. The principal features of the reaction sequence can be outlined by four partial reactions as follows:

$$3\ Na^{+}_{cyt} + Mg^{2+} + E1 \cdot ATP \rightleftarrows (Na_3)E1 \sim P(Mg) + ADP \qquad (1)$$

$$2\ K^{+}_{ext} + (Na_3)E1 \sim P(Mg) \rightleftarrows K_2 \cdot E2 - P(Mg) + 3\ Na^{+}_{ext} \qquad (2)$$

$$K_2 \cdot E2 - P(Mg) \rightleftarrows (K_2)E2 + Mg^{2+} + P_{inorganic} \qquad (3)$$

$$ATP + E2(K_2) \rightleftarrows E1 \cdot ATP + 2\ K^{+}_{cyt} \qquad (4)$$

In this notation, subscript "cyt" indicates ions in the cytoplasm and subscript "ext" indicates ions in the extracellular medium. Parentheses "( )" enclose ions occluded or tightly bound to the enzyme. "~P" indicates a high energy phosphate group and "−P" indicates a low energy phosphate group. As mentioned earlier, the conformation with occluded Na is unstable in the native enzyme but is included in the reaction sequence, as it is stable when the enzyme has been modified in various ways. The reaction transfers electric charge across the plasma membrane because 3 Na^+ carry more positive charge than 2 K^+. Furthermore, this charge is delivered from the cytoplasm to the bottom of an ion well embedded part way through the membrane (Rakowski *et al.,* 1997).

In SERCA the medium in the lumen of sarcoplasmic reticulum corresponds to the extracellular medium for Na,K-ATPase. The stoichiometry is 2 Ca^{2+} transported from the cytoplasm into the lumen per 3 H^+ transported from the lumen to the cytoplasm per 1 molecule of cytoplasmic ATP hydrolyzed (Mintz and Guillain, 1997). Mg^{2+} is required for high-affinity binding of ATP whereas in Na,K-ATPase Mg^{2+} is needed only for phosphorylation. Ca^{2+} occluded on SERCA is relatively stable in contrast to Na^+ occluded on Na,K-ATPase. Millimolar concentrations of lumenal Ca^{2+} stabilize $E1 \sim P$ in SERCA, whereas 0.1 molar or higher concentrations of extracellular Na^+ stabilize $E1 \sim P$ poorly in Na,K-ATPase because Na^+ also binds to $E2 - P$ to some extent (Suzuki and Post, 1997).

d. Regulation. The activity of pumps is regulated. Regulation is both by a balance between incorporation into and removal from the membrane and by direct modification of pump molecules. In the case of Na,K-ATPase in skeletal muscle, there are rapid and slow responses, which are regulated differently. Rapid responses appear within minutes whereas slow responses appear within hours to days. Rapid responses are induced by exercise, catecholamines, and insulin, for instance, while slow responses are induced by thyroid hormones and adrenal steroids (Clausen, 1996a,b). The mechanisms of these actions are being studied (Ewart and Klip, 1995). Two

P-type calcium pumps have well-defined regulatory mechanisms. These pumps are the calcium ATPase of plasma membranes, PCMA (Penniston *et al.,* 1997) and SERCA 2a of cardiac sarcoplasmic reticulum (Lee, 1996), which are regulated by small proteins. PCMA is regulated by calmodulin, a soluble protein. PCMA has an extended C-terminal tail that interacts with the rest of the enzyme to inhibit it. When calmodulin binds four calcium ions, it combines with this tail and relieves the inhibition. In this case the substrate for transport across the membrane, calcium ion, also catalytically stimulates activity of the pump. SERCA 2a is regulated by phospholamban, a membrane-bound protein. Phospholamban can be phosphorylated by various protein kinases. In its dephospho state it binds to SERCA 2b and inhibits it. In its phospho state it does not bind and does not inhibit. Protein kinases are activated by β-adrenergic agonists.

3. ATP Synthases or F-Type ATPases and V-Type ATPases

a. Introduction. V-type ATPases will be discussed later. They hydrolyze ATP and pump protons outward. F-type ATPases are similar to V-type ATPases in structure and share a common ancestor but their physiological function is the reverse of that of V-type ATPases, in that they do not to hydrolyze ATP but instead synthesize it with energy from downhill inward transport of protons.

b. F-Type ATP Synthases. These enzymes are also known as F_1F_o ATPases because their ATPase activity was studied first. They are found in the inner membranes of mitochondria, chloroplasts, and bacteria (Junge *et al.,* 1997). The stoichiometry of synthesis is one molecule of ATP formed per four protons transported inward in chloroplasts (Berry and Rumberg, 1996). "F_1" and Fo" are separable components of the synthase. Fo is a base piece embedded in the membrane, whereas F_1 is a head piece protruding from the membrane and attached to Fo by a stalk. F_1 hydrolyzes ATP when separated from Fo; Fo transports protons when separated from F_1. Energy is transmitted between F_1 and Fo components by a rotating shaft.

c. Structure of F_1. F_1 (Abrahams *et al.,* 1994; Junge *et al.,* 1997) is approximately spherical. It consists of five subunits in the following proportions: 3 α, 3 β, 1 γ, 1 δ, and 1 ϵ. In bovine mitochondria the subunits contain 510, 482, 272, 146, and 50 amino acid residues, respectively. The sphere of F_1 consists of the six α and β subunits, each shaped like the segments of an orange and arranged in a circular alternating pattern, $\alpha\beta\alpha\beta\alpha\beta$. These subunits surround a central cavity in which is located the γ subunit, the rotating shaft. The γ subunit extends out of the cavity into the stalk, which probably also contains the δ and ϵ subunits; these subunits are not well resolved by X-ray crystallography. δ is probably associated with the sphere and ϵ is probably associated with γ. Three different nucleotide-binding sites are located between the α and β subunits. One of these sites binds ADP and inorganic phosphate, another binds ATP, and the third site does not bind nucleotide. This arrangement is clockwise as seen from the matrix aspect of the synthase. The γ subunit is located asymmetrically in the cavity.

d. Reaction Mechanism of F_1. The reaction sequence follows a binding change mechanism (Boyer, 1997; Junge *et al.,* 1997). The γ subunit rotates clockwise as seen from the matrix aspect of the synthase. As the γ subunit rotates, the conformations of the nucleotide-binding sites change. The site-binding ADP and inorganic phosphate accepts a water molecule from these ligands and synthesizes bound ATP. The ATP-binding site then relaxes and releases ATP. The empty binding site accepts ADP and inorganic phosphate from the medium and binds them. Thus one molecule of ATP is synthesized per one-third of a rotation of the γ subunit and three molecules of ATP are synthesized per one complete rotation of the γ subunit.

e. Structure and Function of Fo. Fo (Senior, 1990; Fillingame, 1996) also must have a rotating part, a rotor, and a stationary part, a stator, but which subunits rotate and which are stationary is not known. Fo from *E. coli* is simpler than Fo from mitochondria. It comprises three subunits, a, b, and c, in molar ratios of 1:2:9–12, respectively. The subunits contain 271, 156, and 79 residues, respectively. The c subunit contains a unique residue that is an active site in proton transport, namely Asp61. This residue accepts a proton from the outside of the cell, delivers the proton to the inside, and returns deprotonated to its initial state. Very likely each subunit goes through this binding change sequence as the rotor turns. When the rotor has made one-third of a complete rotation, one molecule of ATP has been synthesized and 4 protons have been transported if the stoichiometry is the same as in the chloroplast synthase reaction. Presumably each of these protons is transported by a different c subunit. Thus for a complete rotation, three molecules of ATP are synthesized and 12 protons are transported through 12 different c subunits. These subunits are very likely arranged in a ring (Hatch *et al.,* 1995; Finbow and Harrison, 1997; Fillingame, 1996; Junge *et al.,* 1997).

f. V-Type ATPases. V-type ATPases (Stevens and Forgac, 1997) are found principally in membranes of vacuoles of eukaryotic cells, such as lysosomes, phagosomes, clathrin-coated vesicles, and tonoplasts. They are also found in the plasma membrane of certain kidney cells. They pump protons out of the cytoplasm using energy from the hydrolysis of ATP

(Nelson, 1996). V-type ATPases are related to F-type ATPases and are similar in structure. However, the subunit of V-type ATPases arose by gene duplication of an ancestor of the c subunit of F-type ATPases and so is twice as large (Kibak *et al.,* 1992). Nevertheless it has only one active site for the transport of protons, Glu139. There are only six copies of the c subunit per one copy of the other subunits. Thus the size of all the c subunits in V-type ATPases is about the same as the size of all the c subunits in F-type ATPases. Because there are one-half as many active sites for proton transport in V-type ATPases as in F-type ATPases, a putative rotor would rotate 60° per proton transported instead of 30°. Because the rotor turns 120° per ATP molecule hydrolyzed, the stoichiometry of the reaction is 2 protons transported per 1 ATP molecule hydrolyzed, as observed in beet tonoplast (Bennet and Spanswick, 1984). Thus the structure, the stoichiometry of the number of subunits interacting with both ends of a rotating shaft per complete rotation, determines the stoichiometry of the enzymatic reaction.

g. Relationship between Stoichiometry and Function. For a proton-pumping ATPase or a proton-driven ATP synthase at thermodynamic equilibrium, the free energy of hydrolysis of ATP equals the free energy of transport of one proton across the membrane multiplied by the number of protons transported per ATP molecule hydrolyzed or synthesized, i.e., by the stoichiometry of the enzymatic reaction (Nelson, 1996). Because a V-type ATPase transports one-half as many protons per ATP molecule hydrolyzed as an F-type ATPase, for a given free energy of hydrolysis of ATP at equilibrium the electrochemical gradient of protons across the membrane is twice as large for a V-type ATPase as for an F-type ATPase. A V-type ATPase can develop a larger electrochemical gradient of protons for a given free energy of hydrolysis of ATP than an F-type ATPase and so is better suited to pumping protons, whereas an F-type ATP synthase is better suited to the reverse reaction, synthesizing ATP (Finbow and Harrison, 1997; Junge *et al.,* 1997).

One other molecular machine employs a rotating shaft, which is the bacterial flagellum (Namba and Vonderviszt, 1997). However, this motor is 100-fold larger than the F-type and V-type ATPases and is analogous to their base pieces only structurally.

4. Multidrug Transporters

Multidrug transporters are members of a superfamily of proteins known as ABC transporters (Sarkadi *et al.,* 1996). "ABC" stands for "ATP-binding cassette." The ATP-binding cassette consists of four membrane-associated domains. Two domains consist of six hydrophobic, putatively membrane-spanning segments each with cytoplasmic amino-terminal and carboxy-terminal ends. These transmembrane domains are thought to be transport domains. The other two domains are cytoplasmic, bind ATP, and consist of highly conserved sequences of about 200 amino acids each. The domains may be linked covalently in various ways in the primary amino acid sequence (Higgins, 1992; Lomri *et al.,* 1996). ABC transporters include both pumps and channels.

Multidrug transporters are significant in treatment of cancer because they remove from cancer cells drugs designed to interfere with the growth of these cells. The best studied member of this family is multidrug resistance protein, MDR1, also known as P-glycoprotein. MDR2 and the multidrug resistance-associated protein, MRP, are less studied multidrug transporters (Stein, 1997). Substrates for transport are heterogeneous amphipathic molecules about 400–1100 Da in molecular mass. Different transporters transport different molecules but there is much overlap in their specificity. Substrates include xenobiotics, hormones, and glutathione conjugates. It is thought that substrates for transport are taken up from the lipid bilayer of the cell membrane rather than from the cytoplasm (Gottesman and Pastan, 1993) and are then extruded into the extracellular medium.

P-glycoprotein, MDR1, is a single glycosylated peptide chain (170 kDa, 1280 amino acids) comprising alternating transmembrane and ATP-binding domains. The amino-terminal domain is a transmembrane domain whose first cytoplasmic loop is glycosylated. The stoichiometry is one molecule transported per ATP molecule hydrolyzed (Eytan *et al.,* 1996). Both ATP-binding sites participate in hydrolysis, and a reaction sequence has been proposed (Senior *et al.,* 1995).

5. Arsenite Transport ATPase

a. Biology. Some bacteria protect themselves against arsenate poisoning by reducing cytoplasmic arsenate to arsenite and extruding the arsenite (Rosen *et al.,* 1992). Antimonite is also transported. Antimony compounds are used to treat visceral leishmaniasis (kala-azar), a significant disease in some tropical countries. Extrusion of antimonite by the arsenite-transport ATPase may be a factor in the resistance of leishmania parasites to antimony compounds.

b. Mechanism of Transport. The transport of arsenite can be driven by the hydrolysis of cytoplasmic ATP. The transport system comprises two proteins: ArsA (63 kDa) and ArsB (45 kDa). ArsB is an intrinsic protein of the plasma membrane and ArsA is bound to ArsB. The proteins can be studied separately. If ArsA is overexpressed relative to ArsB, the excess ArsA is soluble, can be purified, and shows arsenite-dependent ATPase activity. Antimonite is a better activator of ATPase activity than arsenite. When ArsA is not expressed, ArsB itself can extrude arsenite. Everted (inside out) vesicles that are prepared from strains of *E. coli* lacking ArsA take up arsenite when NADH is provided in the medium (Kuroda *et al.,* 1997). Presumably a membrane-bound electron-transport proton pump generates an electrochemical gradient of protons across the membrane. This gradient can drive the import of arsenite through ArsB either by facilitated diffusion, with the anion being attracted to a positive potential inside the vesicles, or by countertransport in exchange for protons. Thus this transport system is composed of separate proteins for transport, ArsB, and for coupling of transport to hydrolysis of ATP (ArsA).

c. Structure and Relatedness of Ars Proteins. The ArsA protein is composed of homologous amino-terminal and carboxy-terminal halves (Rosen *et al.,* 1992). The halves are homologous with nine other proteins found both in prokaryotes and in a eukaryote. None of these has a known transport function, although several have ATPase activity. Each of the halves of the ArsA protein has a consensus sequence for a nucleotide-binding site. Both sequences are functional and are required for ATPase activity, although in different ways. The ArsB protein has 12 membrane-spanning segments.

D. Carbohydrate Phosphotransferase Systems of Bacteria

These systems are found only in prokaryotes (Postma *et al.,* 1993; Saier and Reizer, 1994). Substrates of the overall reaction are an extracellular carbohydrate and cytoplasmic phosphoenolpyruvate; products of the overall reaction are cytoplasmic phosphorylated carbohydrate and cytoplasmic pyruvate. Thus transport is linked closely to phosphorylation of the transported substrate and only one high-energy phosphate group is consumed per phosphorylated sugar molecule accumulated. The systems consist of soluble cytoplasmic proteins and membrane-bound proteins. The systems are heterogeneous, differing principally in the carbohydrate that is transported. Selectivity for a specific carbohydrate depends on the membrane-bound protein. The membrane-bound proteins can be grouped into at least four classes on the basis of homology of the primary structure. Members of the glucose class transport glucose, maltose, *N*-acetylglucosamine, sucrose, and β-glucosides. Members of the mannitol class transport mannitol and fructose. Members of the lactose class transport lactose, galactose, and cellobiose. Members of the mannose class transport mannose, L-sorbose, and fructose. These classes also transport other sugars.

The reaction sequence consists of transfer of a phosphate group from phosphoenolpyruvate to a histidine residue on a soluble protein "EI" and from there to a histidine residue on another soluble protein "HPr." From HPr the phosphate group is transferred to domains specific for the transported carbohydrate, specifically to two domains in sequence, "EIIA" and "EIIB." A third domain, "EIIC," is intrinsic to the membrane and binds and presumably transports the carbohydrate. The extracellular (periplasmic) carbohydrate binds to a membrane-bound protein, EIIC, with high affinity and can be released to the cytoplasm very slowly without phosphorylation of the second carbohydrate-specific domain, EIIB. However, phosphorylation of EIIB and transfer of the phosphate group to the bound carbohydrate greatly accelerate release of the phosphorylated carbohydrate and further transport in subsequent cycles of the reaction sequence.

These domains are structurally diverse according to the species of bacteria. "EIIA" and "EIIB" may be single or separate soluble proteins or may be covalently part of a complex with a membrane-bound protein or proteins. The function and structures of many of the proteins of this transport system have been studied in detail (Postma *et al.,* 1993).

E. Oxaloacetate Decarboxylase Sodium Pump

Some bacteria, in a medium containing sodium, pump sodium out of the cytoplasm into the medium and use the electrochemical gradient of sodium across the membrane to energize other transport functions (Dimroth, 1987, 1997). In particular, *Klebsiella pneumoniae* pumps sodium using energy from the decarboxylation of oxaloacetate. The pump enzyme consists of three subunits, α (64 kDa), β (45 kDa), and γ (9 kDa). The β and γ subunits are intrinsic and are membrane bound, whereas the α subunit is attached to the other two. The first step in the reaction is catalyzed by the α subunit. This subunit has biotin attached to Lys(561). The biotin accepts a carboxylate group from oxaloacetate, thereby releasing pyruvate. In subsequent steps the membrane-bound subunits donate a proton from the medium to the carboxylate group releasing CO_2; at the same time they also transport 2 Na^+ out of the cell.

Another bacterium, *Propiogenium modestum,* pumps sodium outward using a methylmalonyl-CoA decarboxylase. In addition, this organism uses the resulting electrochemical gradient of sodium ions to synthesize ATP. It has an F-type ATP synthase driven by the inflow of sodium ions.

III. SUMMARY

In 1898 Earnest Overton proposed that the exchange of intracellular K^+ and extracellular Na^+ is involved in the contraction of muscle and understood that a mechanism to restore concentrations to the resting state would be needed (Chapter 1 of this book). In 1952, Hodgkin and Huxley gave a detailed description of exchange of K^+ and Na^+ in the action potential of a nerve cell. Four years later Skou demonstrated an enzyme activity that restores the resting state (Skou and Esmann, 1992). Now excitation is understood as due to the action of membrane-bound channels whose primary amino acid sequence is known and whose molecular mechanism is under investigation. Similarly, restoration of the resting state is understood as due to the action of membrane-bound pumps whose primary amino acid sequence is known and whose molecular mechanism is under investigation.

References

Abrahams, J. P., Leslie, A. G. W., Lutter, R., and Walker, J. E. (1994). Structure at 2.8 Å resolution of F_1-ATPase from bovine heart mitochondria. *Nature* (*London*) **370,** 621–628.

Altendorf, K., and Epstein, W. (1996). The Kdp ATPase of *Escherichia coli. In* "Biomembranes. A Multivolume Treatise. ATPases" (A. G. Lee, ed.), Vol. 5, pp. 403–420. JAI Press, Greenwich, Connecticut.

Bennet, A. B., and Spanswick, R. M. (1984). H^+-ATPase activity from storage tissue of *Beta vulgaris.* II. H^+/ATP stoichiometry of an anion-sensitive ATPase. *Plant Physiol.* **74,** 545–548.

Berry, S., and Rumberg, B. (1996). H^+/ATP coupling ratio at the unmodulated CF_oCF_1-ATP synthase determined by proton flux measurements. *Biochim. Biophys. Acta* **1276,** 51–56.

Boyer, P. D. (1997). The ATP synthase—a splendid molecular machine. *Annu. Rev. Biochem.* **66,** 717–749.

Brandt, U. (1996). Energy conservation by bifurcated electron-transfer in the cytochrome-bcl complex. *Biochim. Biophys. Acta* **1275,** 41–46.

Brandt, U. (1997). Proton-translocation by membrane-bound NADH: Ubiquinone-oxidoreductase (complex I) through redox-gated ligand conduction. *Biochim. Biophys. Acta* **1318,** 79–91.

Calhoun, M. W., Thomas, J. W., and Gennis, R. B. (1994). The cytochrome oxidase superfamily of redox-driven proton pumps. *Trends Biochem. Sci.* **19,** 325–330.

Clausen, T. (1996a). Long- and short-term regulation of the Na^+-K^+ pump in skeletal muscle. *News Physiol. Sci.* **11,** 24–30.

Clausen, T. (1996b). The Na^+, K^+ pump in skeletal muscle: Quantification, regulation and functional significance. *Acta Physiol. Scand.* **156,** 227–235.

Cornelius, F. (1996). The sodium pump. *In* "Biomembranes. A Multivolume Treatise. ATPases" (A. G. Lee, ed.), Vol. 5, pp. 133–184. JAI Press, Greenwich, Connecticut.

Dimroth, P. (1987). Sodium ion transport decarboxylases and other aspects of sodium cycling in bacteria. *Microbiol. Rev.* **51,** 320–340.

Dimroth, P. (1997). Primary sodium translocating enzymes. *Biochim. Biophys. Acta* **1318,** 11–51.

Ewart, H. S., and Klip, A. (1995). Hormonal regulation of the Na^+-K^+-ATPase: Mechanisms underlying rapid and sustained changes in pump activity. *Am. J. Physiol.* **269,** C295–C311.

Eytan, G. D., Regev, R., and Assaraf, Y. G. (1996). Functional reconstitution of P-glycoprotein reveals an apparent near stoichiometric drug transport to ATP hydrolysis. *J. Biol. Chem.* **271,** 3172–3178.

Fillingame, R. H. (1996). Membrane sectors of F- and V-type H^+-transporting ATPases. *Curr. Opin. Struct. Biol.* **6,** 491–498.

Finbow, M. E., and Harrison, M. A. (1997). The vacuolar H^+-ATPase: A universal proton pump of eukaryotes. *Biochem. J.* **324,** 697–712.

Gerencser, G. A., and Purushotham, K. R. (1996). Reconstituted Cl^- pump protein: A novel ion(Cl^-)-motive ATPase. *J. Bioenerg. Biomembr.* **28,** 459–469.

Glynn, I. M., and Karlish, S. J. D. (1990). Occluded ions in active transport. *Annu. Rev. Biochem.* **59,** 170–205.

Goodenough, D. A., Goliger, J. A., and Paul, D. L. (1996). Connexins, connexons, and intercellular communication. *Annu. Rev. Biochem.* **65,** 475–502.

Gottesman, M. M., and Pastan, I. (1993). Biochemistry of multidrug resistance mediated by the multidrug transporter. *Annu. Rev. Biochem.* **62,** 385–427.

Gouaux, E. (1997). Channel-forming toxins: Tales of transformation. *Curr. Opin. Struct. Biol.* **7,** 566–573.

Green, N. M. (1992). Evolutionary relationships within the family of P-type cation pumps. *Ann. N.Y. Acad. Sci.* **671,** 104–112.

Hatch, L. P., Cox, G. B., and Howitt, S. H. (1995). The essential arginine residue at position 210 in the α subunit of *Escherichia coli* ATP synthase can be transferred to position 252 with partial retention of activity. *J. Biol. Chem.* **270,** 29407–29412.

Higgins, C. F. (1992). ABC transporters: From microorganisms to man. *Annu. Rev. Cell Biol.* **8,** 67–113.

Hodgkin, A. L., and Huxley, A. F. (1952). A quantitative description of membrane current and its application to conduction and excitation in nerve. *J. Physiol. (London)* **117,** 500–544.

Höber, R. (1946). "Physical Chemistry of Cells and Tissues." The Blakiston Company, Philadelphia.

Hoff, W. D., Jung, K.-H., and Spudich, J. L. (1997). Molecular mechanism of photosignaling by archeal sensory rhodopsins. *Annu. Rev. Biophys. Biomol. Struct.* **28,** 223–258.

Horisberger, J.-D. (1994). "The Na,K-ATPase: Structure Function Relationship." R. G. Landes, Austin, Texas.

Iwata, S., Ostermeier, C., Ludwig, B., and Michel, H. (1995). Structure at 2.8 Å resolution of cytochrome c oxidase from *Paracoccus denitrificans. Nature (London)* **376,** 660–669.

Junge, W., Lill, H., and Engelbrecht, S. (1997). ATP synthase: An electrochemical transducer with rotary mechanics. *Trends Biochem. Sci.* **22,** 420–423.

Khorana, H. G., Gerber, G. E., Herlihy, W. C., Gray, C. P., Anderegg, R. J., Nihel, K., and Biemann, K. (1979). Amino acid sequence of bacteriorhodopsin. *Proc. Natl. Acad. Sci. U.S.A.* **76,** 5046–5050.

Kibak, H., Taiz, L., Starke, T., Bernasconi, P., and Gogarten, J. P. (1992). Evolution of structure and function of V-type ATPases. *J. Bioenerg. Biomembr.* **24,** 415–424.

Koster, J. C., Blanco, G., and Mercer, R. W. (1995). A cytoplasmic region of the Na,K-ATPase α-subunit is necessary for specific α/α association. *J. Biol. Chem.* **270,** 14332–14339.

Kuroda, M., Dey, S., Sanders, O. I., and Rosen, B. P. (1997). Alternate energy coupling of ArsB, the membrane subunit of the Ars anion-translocating ATPase. *J. Biol. Chem.* **272,** 326–331.

Lanyi, J. K. (1997). Mechanism of ion transport across membranes. Bacteriorhodopsin as a prototype for proton pumps. *J. Biol. Chem.* **272,** 31209–31212.

Lee, A. G. (1996). Cardiac Ca^{2+}-ATPase and phospholamban. *In* "Biomembranes. A Multivolume Treatise. ATPases" (A. G. Lee, ed.), Vol. 5, pp. 77–100. JAI Press, Greenwich, Connecticut.

Lomri, N., Fitz, J. G., and Scharschimdt, B. F. (1996). Hepatocellular transport: Role of ATP-binding cassette proteins. *Semin. Liver Dis.* **16,** 201–210.

Lutsenko, S., and Kaplan, J. H. (1995). Organization of P-type ATPases: Significance of structural diversity. *Biochemistry* **34,** 15607–15613.

MacLennan, D. H., Rice, W. J., and Green, N. M. (1997). The mechanism of Ca^{2+} transport by sarco(endo)plasmic reticulum Ca^{2+}-ATPases. *J. Biol. Chem.* **272,** 28815–28818.

Maeda, A., Kandori, H., Yamazaki, Y., Nishimura, S., Hatanaka, M., Chon, Y.-S., Sasaki, J., Needleman, R., and Lanyi, J. K. (1997). Intramembrane signaling mediated by hydrogen-bonding of water and carboxyl groups in bacteriorhodopsin and rhodopsin. *J. Biochem. (Tokyo)* **121,** 399–406.

Mintz, E., and Guillain, F. (1997). Ca^{2+} transport by the sarcoplasmic reticulum ATPase. *Biochim. Biophys. Acta* **1318,** 52–70.

Möller, J. V., Juul, B., and le Maire, M. (1996). Structural organization, ion transport, and energy transduction of P-type ATPases. *Biochim. Biophys. Acta* **1286,** 1–51.

Namba, K., and Vonderviszt, F. (1997). Molecular architecture of bacterial flagellum. *Q. Rev. Biophys.* **30,** 1–65.

Nelson, N. (1996). Vacuolar ATPase. *In* "Biomembranes. A Multivolume Treatise. ATPases" (A. G. Lee, ed.), Vol. 5, pp. 317–341. JAI Press, Greenwich, Connecticut.

Park, C. R., Post, R. L., Kalman, C. F., Wright, J. H., Jr., Johnson, L. H., and Morgan, H. E. (1956). The transport of glucose and other sugars across cell membranes and the effect of insulin. *Ciba Found. Colloq. Endocrinol. Proc.* **9,** 240–260.

Pedersen, P. L., and Amzel, L. M. (1993). ATP synthases. Structure, reaction center, mechanism, and regulation of one of Nature's most unique machines. *J. Biol. Chem.* **268,** 9937–9940.

Penniston, J. T., Enyedu, A., Verma, A. K., Adamo, H. P., and Filoteo, A. G. (1997). Plasma membrane Ca^{2+} pumps. *Ann. N.Y. Acad. Sci.* **834,** 56–64.

Post, R. L., Kume, S., Tobin, T., Orcutt, B., and Sen, A. K. (1969). Flexibility of an active center in sodium-plus-potassium adenosine triphosphatase. *J. Gen. Physiol.* **54,** 306s–326s.

Post, R. L., Hegyvary, C., and Kume, S. (1972). Activation by adenosine triphosphate in the phosphorylation kinetics of sodium and potassium ion transport adenosine triphosphatase. *J. Biol. Chem.* **247,** 6530–6540.

Postma, P. W., Lengeler, J. W., and Jacobson, G. R. (1993). Phosphoenolpyruvate: Carbohydrate phosphotransferase systems of bacteria. *Microbiol. Rev.* **57,** 543–594.

Rakowski, R. F., Gadsby, D. C., and De Weer, P. (1997). Voltage dependence of the Na/K pump. *J. Membr. Biol.* **155,** 105–112.

Rosen, B. P., Dey, S., Dou, D., Ji, G., Kaur, P., Ksenzenko, M. Y., Silver, S., and Wu, J. (1992). Evolution of an ion-translocating ATPase. *Ann. N.Y. Acad. Sci.* **671,** 257–272.

Rosenberg, T., and Wilbrandt, W. (1957). Uphill transport induced by counterflow. *J. Gen. Physiol.* **41,** 289–296.

Saier, M. H., Jr., and Reizer, J. (1994). The bacterial phosphotransferase system: New frontiers 30 years later. *Mol. Microbiol.* **13,** 755–764.

Sarkadi, B., Müller, M., and Holló, Z. (1996). The multidrug transporters—proteins of an ancient immune system. *Immunol. Lett.* **54,** 215–219.

Schulz, G. E. (1996). Porins: General to specific, native to engineered passive pores. *Curr. Opin. Struct. Biol.* **6,** 485–490.

Senior, A. E. (1990). The proton-translocating ATPase of *Escherichia coli. Annu. Rev. Biophys. Biophys. Chem.* **19,** 7–41.

Senior, A. E., Al-Shawi, M. K., and Urbatsch, I. L. (1995). The catalytic cycle of P-glycoprotein. *FEBS Lett.* **377,** 285–289.

Shin, J. M., Bayle, D., Bamberg, K., and Sachs, G. (1996). The gastric H^+/K^+-ATPase. *In* "Biomembranes. A Multivolume Treatise. ATPases" (A. G. Lee, ed.), Vol. 5, pp. 185–224. JAI Press, Greenwich, Connecticut.

Skou, J. C., and Esmann, M. (1992). The Na,K-ATPase. *J. Bioenerg. Biomembr.* **24,** 249–261.

Spudich, J. L. (1994). Protein–protein interaction converts a proton pump into a sensory receptor. *Cell (Cambridge, Mass.)* **79,** 747–750.

Stein, W. D. (1997). Kinetics of the multidrug transporter (P-glycoprotein) and its reversal. *Physiol. Rev.* **77,** 545–590.

Stevens, T. H., and Forgac, M. (1997). Structure, function and regulation of the vacuolar (H^+)-ATPase. *Annu. Rev. Cell Dev. Biol.* **13,** 779–808.

Stokes, D. L. (1997). Keeping calcium in its place: Ca^{2+}-ATPase and phospholamban. *Curr. Opin. Struct. Biol.* **7,** 550–556.

Suzuki, K., and Post, R. L. (1997). Equilibrium of phosphointermediates of sodium and potassium ion transport adenosine triphosphatase. Action of sodium ion and Hofmeister effect. *J. Gen. Physiol.* **109,** 537–554.

Tang, X., Halleck, M. S., Schlegel, R. A., and Williamson, P. (1996). A subfamily of P-type ATPases with aminophospholipid transporting activity. *Science* **272,** 1495–1502 (see erratum page 1597).

Tsukihara, T., Aoyama, H., Yamashita, E., Tomizaki, T., Yamaguchi, H., Shinzawa-Itoh, K., Nakashima, R., Yaono, R., and Yoshikawa, S. (1996). The whole structure of the 13-subunit oxidized cytochrome c oxidase at 2.8 angstrom. *Science* **272,** 1136–1144.

Váro, G., Zimányi, L., Fan, X., Sun, L., Needleman, R., and Lanyi, J. K. (1995). Photocycle of halorhodopsin from *Halobacterium salinarium. Biophys. J.* **68,** 2062–2072.

Verkhovsky, M. I., Morgan, J. E., Verkhovskaya, M. L., and Wikström, M. (1997). Translocation of electrical charge during a single turnover of cytochrome-c oxidase. *Biochim. Biophys. Acta* **1318,** 6–10.

Walker, J. E. (1992). The NADH: ubiquinone oxidoreductase (complex I) of respiratory chains. *Q. Rev. Biophys.* **25,** 253–324.

Widdas, W. F. (1952). Inability of diffusion to account for placental glucose transfer in the sheep and consideration of the kinetics of a possible carrier transfer. *J. Physiol.* (*London*) **118,** 23–39.

Wikström, M. (1989). Identification of the electron transfers in cytochrome oxidase that are coupled to proton pumping. *Nature* (*London*) **338,** 776–778.

Yoshikawa, S. (1997). Beef heart cytochrome c oxidase. *Curr. Opin. Struct. Biol.* **7,** 574–579.

Index

L

M

N

O

P

W

ISBN 0-12-153348-4

90018

9 780121 533489